Human Origins

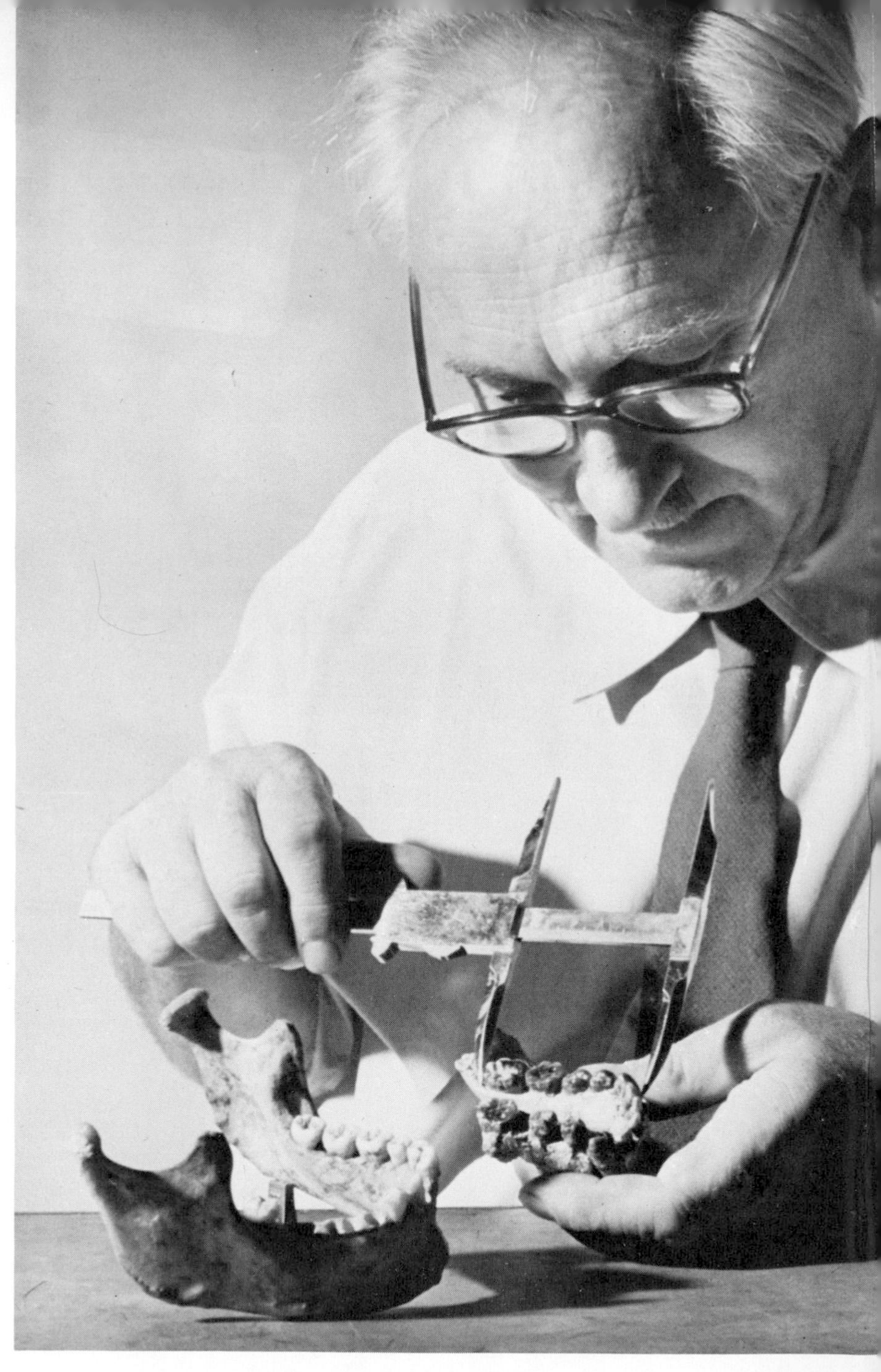

The late Dr. Louis Leakey, in whose honor this volume has been compiled. As an anthropologist and paleontologist he was an explorer of man's past, but as a naturalist he cared passionately about the future of our own species and the world that we inhabit. Here he is seen measuring th fossil mandible of *Homohabilis* from Olduvai Gorge. © 1975 National Geographic Society, photograph by Robert F. Sisson.

Human Origins

Louis Leakey and the East African Evidence

Edited by Glynn Ll. Isaac and Elizabeth R. McCown

Perspectives on Human Evolution, Volume III
A Publication of the Society for the Study of Human Evolution, Inc., Berkeley, California

A Staples Press Book
W. A. Benjamin, Inc., *Menlo Park, California* · *Reading, Massachusetts* · *London* · *Amsterdam* · *Don Mills, Ontario* · *Sydney*

This book is in the
W. A. Benjamin Series in Anthropology

Consulting Editor
Brian Fagan

The photograph on the front of the softcover edition of this book is of an oil painting by Jay H. Matternes, depicting his conception of a group of Australopithecus (Zinjanthropus) bosei. *It is reproduced by his kind permission.*

 Printed in the United States of America.
Published simultaneously in Canada.
Library of Congress Catalog Number 75–27746.

ISBN–0–8053–9941–0 Clothbound edition
ISBN–0–8053–9942–9 Paperbound edition
ABCDEFGHIJKL–HA–79876

W. A. Benjamin, Inc.
2727 Sand Hill Road
Menlo Park, California 94025

Human Origins:
Louis Leakey and the East African Evidence

is a book sponsored by:

Society for the Study of Human Evolution, Incorporated
Berkeley, California

Organized for the purpose of increasing and disseminating information in the field of human evolution through a series of books entitled Perspectives on Human Evolution.

Officers of the Society for 1975:
Vincent M. Sarich, President; Phyllis Dolhinow, S. L. Washburn, Alice Davis, Elizabeth McCown

Previous publications of the Society:

Perspectives on Human Evolution, Volume I.
S. L. Washburn and Phyllis C. Jay, Editors.
New York: Holt, Rinehart and Winston, 1968.

Perspectives on Human Evolution, Volume II.
S. L. Washburn and Phyllis Dolhinow, Editors.
New York: Holt, Rinehart and Winston, 1972.

Contents

Olduvai Gorge, Tanzania—an air view looking eastward from the junction of the main and side gorges to the Bal Bal depression. Louis and Mary Leakey have been working here for m than 40 years to recover a record of human evolution that spans almost two million years. © 1975 National Geographic Society, photograph by Emory Kristoff.

PREFACE

Why Is East Africa Important?

In recent years evidence from East Africa has transformed our knowledge about human evolution. The founding father of this movement in East Africa was unquestionably Louis Seymour Bazett Leakey, and for that reason this book has been prepared in his honor. In the volume we have tried to represent something of the ferment of activity that has taken place in East Africa, but clearly a book of this size can provide only a representative sample of the whole picture. However, we hope that the contributions will convey something of the vigor and imagination which are among the legacies that Louis has left in this field of inquiries.

Preparations for the volume were started with the intention that it be presented to Louis on his seventieth birthday in 1973 as a token of scientific esteem. His untimely death on October 1, 1972, prevented the realization of those plans, but we have continued with the volume so that it may serve as a posthumous expression of his colleagues' appreciation of his clear-sighted pioneering in the study of human evolution. He was a friend and mentor to most of the contributors, who all gained immensely from that association.

It is salutory to point out that two foundations have been formed to perpetuate the kind of work begun by Louis Leakey. Based in Los Angeles, the L. S. B. Leakey Foundation covers the full spectrum of

Louis' interests in man and nature. The more recently formed Foundation for Research into the Origins of Man, based in Delaware and New York, is concerned with the support of an international fellowship and research program, which will operate largely through the Louis Leakey Memorial Institute for Prehistory and Paleontology in Nairobi.

The editors wish to express their appreciation to all the contributors, not only for their valuable papers, but also for their agreeable and cooperative responses to our requests for revisions or additional photographic material.

Our special thanks go to Sherwood L. Washburn, the leading force behind the Perspectives series, for his continuing advice, encouragement, and expertise; and to Dr. Melvin Payne, President of the National Geographic Society, for his generous help in providing illustrations for this book. We also wish to thank Alice Davis, who has eased the task of getting this volume to press.

There are many factors that have contributed to the prominence of eastern Africa in studies of human evolution. First, the subcontinent seems to have been continuously occupied by men, early men, and their nonhuman ancestors since at least the Miocene. Second, the extraordinary geological circumstances of the region have favored the preservation of fossils, so that a record of evolution and its paleoecological setting can be found and studied. Third, a great diversity of primates can still be observed there in the wild to gain insight into adaptive mechanisms.

But the current importance of eastern Africa cannot be understood from the region's natural endowments alone. It is the life work of Louis Leakey that has focused the attention of science on this unique source of evidence. Leakey had the unusual qualities necessary to fire public imagination with a sense of the excitement of exploring man's origins. It is hard to estimate our debt to this extraordinary man.

It is only in recent years, however, that East Africa has received general recognition as a crucial source of evidence for human origins. When Mary Leakey discovered "*Zinjanthropus*" at Olduvai Gorge, in 1959, significant fossils relating to early evolutionary stages of men and near men were known from only a very few places, notably South Africa and Java. Raymond Dart and Robert Broom had found in five filled crevices and caverns in South Africa the bones of creatures which, like us, ran around on their hind legs, leaving their hands free to make and carry things and to get involved in other novel mischiefs. But these creatures, the Australopithecines, were small in stature and, one might suspect, hairy. They had brains only slightly larger than those of apes. In some ways the South African australopithecines seemed to constitute suitable prototypes for later, more evolved forms of near men, but there were serious problems. At least two varieties of australopithecine were represented in deposits of different ages. Was it possible that one kind had evolved into the other, or did they represent separate divergent lineages? How old were these fossils? Opinions ranged from Pliocene to Middle Pleistocene. That is, in the chronology in vogue at the time, it was believed that these fossils were from about one million years to three hundred thousand years old. Had any of the populations represented by these fossils actually been part of human ancestry? And if so, by what stages had the further transformations come about?

The next chapter in the human story was based on materials from halfway across the world, in Java. There, in 1891, Eugene DuBois found the series of fossils which he called *Pithecanthropus*. These fossils represented beings of much more human aspect, larger in stature than the australopithecines and with limbs and bodies almost indistinguishable from those of humans today, but with heavier, more muscle-bound heads incorporating larger faces, thicker bones, and a braincase about halfway between the size of that of *Australopithecus* and *Homo sapiens*. But what, if anything, was the connection between the Java fossils and the ones from South Africa? What were their ages?

This, then, was the situation—East Africa had yielded a few fossil hominid fragments to which scientists paid very little attention. The only thing that made East Africa look at all important for the study of human evolution at that time was the fact that Louis Leakey had been able to describe from Olduvai Gorge what appeared to be the world's longest known stratified sequence of cultural development. Archaeologists, at least, paid attention to this.

An avalanche of discoveries began in July 1959 when Mary Leakey found the fossil palate of a big-toothed australopithecine that had been exposed by the erosion of the lower layers of sediment at Olduvai Gorge. When she and Louis had finished excavating, they had an almost complete skull which they called *"Zinjanthropus" boisei*, though Louis later changed the name to *Australopithecus boisei*. Popular support and scientific attention became focused on East Africa, and the hunt was on. With its long-accumulated experience of the country and its genius for exploration, the Leakey family remained at the forefront during the new phase of discovery. Literally hundreds of hominid fossils have been recovered from more than a dozen localities. The number of new sites grows each year, and scientists have become almost blasé about individual fossil finds.

This spate of discoveries has completely transformed the entire fossil record of human evolution. The fossil series now spans a time range from 5 or 6 million years ago to the present, and as such constitutes the only long-term record that we have from a single region. In addition to the time depth and the wealth of well-preserved specimens, this series of East African fossils can be dated, their ecological setting reconstructed, and aspects of the behavior of the early men determined.

Because the fossil-bearing sediments are commonly interstratified with volcanic rocks, in 1960 Louis Leakey recruited Jack Evernden and Garniss Curtis from the University of California, Berkeley, to make geophysical age determinations by means of the potassium-argon method. Since then many other laboratories have been active, and East Africa now has the best-dated fossiliferous Pliocene and Pleistocene sequence known anywhere in the world.

The fossils come mainly from sediments deposited in basins of internal drainage that are strung out along the Eastern or Gregory Rift Valley. The basins preserve a wealth of geological, geochemical, and paleontological clues to the ecological circumstances under which particular hominids lived and died. A growing group of scientists is successfully teasing out this evidence, which promises to provide a paleoenvironmental sequence as detailed as any in the world.

The fossils which have been found are fragments of bone derived from a once living system; they are mere symptoms of the process of evolutionary change. In fact, what we want to discover are the dynamics of the process, and fortunately East Africa has provided another crucial line of evidence not readily available in other areas, namely, the archaeological traces of early behavior. Louis and Mary Leakey's excavations at Olorgesailie and in Bed I at Olduvai have demonstrated that, under circumstances such as prevail in East Africa, archaeologists can detect early stages not only of toolmaking, but also of hunting and food sharing, and from this data scientists can work out changing patterns of habitat and land use. This work is being extended in many localities in eastern Africa.

The foregoing is an impressive catalog of reasons for the growing importance of East African evidence to our understanding of human evolution, but it is not exhaustive. Stratigraphically, underneath the rich paleontological record are earlier fossils of great importance. Early and Middle Miocene faunas are well known from a time span of more than 20 million to 13 or 14 million years ago. They include a wealth of hominoid fossils that certainly have a strong bearing on understanding the origins of men and apes. Louis Leakey and others have argued that this series includes forms recognizable as specific human ancestors (hominids), although this is hotly debated. Nevertheless, as exploration proceeds, the Miocene record and the later Plio-Pleistocene records are being extended towards each other so that eventually we can expect these disparate fossil sequences to be linked.

East Africa also contains an intricate network of equatorial habitats that preserve many of the features of the environmental tapestry through which the strands of early human evolution were woven. In these habitats scientists are studying ecological dynamics and behavior that will illuminate prehistoric conditions. In particular, a wealth of primate species indicates alternative responses to different ecological settings. Notably, as Sherwood Washburn and Irven DeVore showed in the 1950s, baboons provide clear indications of some of the problems and the opportunities that confront primates living in the savannah. Then, too, the biogeographic range of man's closest living relatives, the chimpanzee and the gorilla, spans from West Africa into East Africa. The most detailed field studies of these relatives have been undertaken, in the last fifteen years, by George Schaller, Jane Goodall, Dian Fossey, Junichiro Itani, and Toshisada Nishida. Their work has broadened our perspective about what may have been involved in the early stages of human divergence.

In summary, by cutting through problems that were insoluble in South Africa and in Java, the East African evidence has transformed our knowledge of what happened in the course of human evolution. It provides a continuous dated record that serves as a standard of reference to which floating sets of evidences can be attached. What is equally important is that in East Africa scientists have learned how to do this kind of research. It is no longer necessary to wait for fossils to be found by accident. Suitable areas are being identified and searched by highly skilled personnel. This procedure gives indications that in East Africa at least the early hominids were not a particularly rare part of the fauna. Richard Leakey has shown that they were about as common as the large carnivores. Similarly, a rich record of paleoenvironments and of ecological relationships is being recovered. The rapid development of paleoanthropology over the past fifteen years shows clearly that, with perseverance, research in the next decade can reveal the story of human beginnings in detail that was undreamed of a few years ago.

East African flora and fauna, epitomizing the setting of human evolution. Courtesy, S. L. Washburn.

J. Desmond Clark:

African Origins of Man the Toolmaker

We do not yet know for certain whether mankind had an exclusively African ancestry, but the continent certainly has the longest and most complete record of its development. J. D. Clark's article is a general commentary on all the aspects of the paleontological and archaeological evidence that make Africa so important. This review is a highly intelligible, comprehensive introduction to topics pursued in more technical detail in subsequent sections of the book.

INTRODUCTION

Twenty years ago no one could have predicted the wealth or completeness of hominid and hominoid fossils now coming from eastern Africa. This evidence, often recovered in spectacular circumstances and as a result of ever more meticulous and thorough investigation of the fossil-bearing localities, amply confirms the belief of Darwin that it would be in the tropics, perhaps in Africa, that man would prove to have evolved from a simian ancestor. Indeed, greater precision in isotopic dating methods is now showing that toolmaking hominids were present in tropical Africa about two million years before the first toolmakers appear in Eurasia.

Man's closest living relatives among the great apes—the chimpanzee and gorilla—are both found in the lowland and montane rainforests of Africa, and an increasing body of evidence is showing how close man is biologically to these African apes. Primate studies, therefore, especially those concerning the molecular, anatomical, and behavioral characteristics of the gorilla and chimpanzee, are directly relevant to an understanding of the biological nature and behavior of the early hominids. Relevant also to a better understanding of the significance of this abundance of fossil and cultural evidence is knowledge of the habitat in which these ancestral forms lived and evolved.

Environment—the soils, geology, climate, and biome of a region—must have been of even greater importance in influencing the behavioral adaptations of early man than it has been since the appearance of modern man. The ways of life of the human populations occupying the five main biogeographical regions of Africa today are, as in the past, generally very different, and only one of these regions favored the evolution of the hominid line and led to efficient toolmaking. For climatic reasons the five biogeographical regions are complementary north and south of an equatorial, tropical zone, but they undergo much modification due to topography, winds, and ocean currents (Figure 1).

THE AFRICAN ENVIRONMENT

In equatoria is the humid, evergreen rainforest—"that towering multitude of trees . . . all perfectly still," as Joseph Conrad wrote in *The Heart of Darkness.* This equatorial forest zone is undergoing steady destruction at the hands of slash-and-burn cultivators and has done so for the past three thousand years. To the north and south of the forest and throughout much of the eastern area are the savanna lands—open forest, woodland, and grass savanna—which support an ungulate fauna that is fantastically rich in the number of species and in the large size of the populations. Here also on the higher plateaus, ridges, and mountains, are found the high-altitude evergreen forest and the montane grasslands that replace it when man interferes. The African savanna is one of the richest natural habitats in the world and stretches from the Atlantic to the Indian Ocean and from the southern border of the Sahara to the south coast of the Republic of South Africa.

The third of these biogeographical regions is the dry steppe country—the Sahel areas north and south of the Sahara, and the Karroo and Kalahari in the south of the continent. The vegetation of the steppe country consists of short, dry grass, succulents, low bushes, and thorn

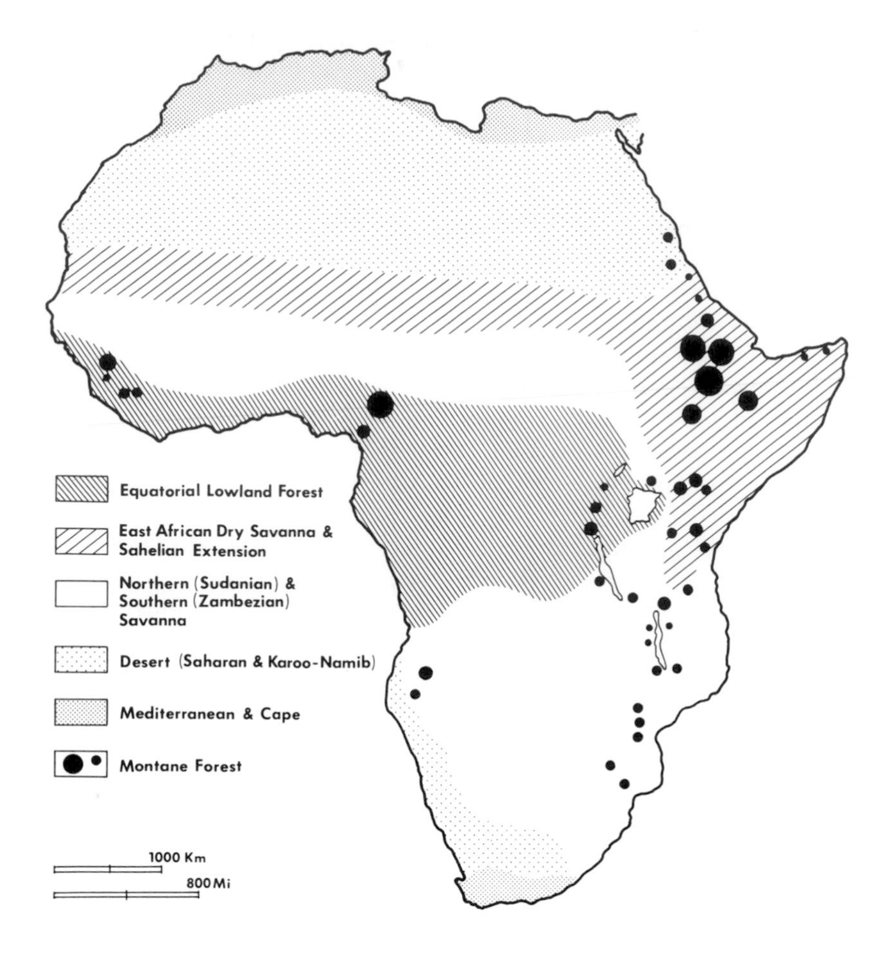

FIGURE 1
Main biogeographical regions of Africa. (After White 1965)

trees. It is often rich in food reserves and supports a mammalian fauna of medium- to small-sized animals.

The fourth zone is the desert of which there are several kinds—salt steppes, semidesert, and the dune fields and stony pavements of the desert proper. Although these constitute some of the most unfavorable parts of the continent for human settlement, they are nevertheless capable of supporting populations of economically self-sufficient hunters and herdsmen.

In the extreme north on the Mediterranean coast and in the south-western parts of the Cape at the other end of the continent is the fifth zone—country enjoying a winter rainfall and supporting a Mediterranean vegetation of evergreen forest and macchia. Generally, in north Africa this Mediterranean zone has had a profound effect upon the course of cultural development though lessened and dissipated in the tropics by the restricting influence of the Sahara desert.

All these zones, except perhaps the primary rainforests, are characterized by a great deal of variability incorporating different kinds of microenvironments, particularly in the savanna. Not only does composition of the northern savanna zone differ markedly from that of its southern counterpart, but in each the monotony of one predominant vegetation community—deciduous woodland or grass, for example—is relieved by the presence of relict or edaphic vegetation communities. These different communities, when taken together, provide a seasonal range of plant and animal sustenance more varied and substantial than is available in any of the other zones. The opportunities afforded by this kaleidoscopic savanna ecosystem encouraged experimentation and were behind the selective processes that determined the basis of early hominid behavior.

The effects of changes in world climate that characterized the Quaternary period can be seen clearly in the sedimentary record. It is apparent that all of the zones underwent modification, sometimes considerable, on several occasions in the past, the boundaries of each zone expanding or contracting in response to fluctuations in rainfall, humidity, temperature, wind velocity or direction, and other climatic factors. The readjustments in the vegetation communities that resulted from each such change can best be seen from micro-stratigraphical studies, from chemical alteration of sediments, and from fossil pollens. Paleoecological studies show that the climate of the Miocene was in general more uniform than that of the Quarternary. These more humid equable conditions permitted the spread of extensive forests in Africa and Eurasia and this particularly favored primate evolution, including that of certain semiarboreal forms of ape.

THE EARLIEST FOSSILS

Since evolutionary theory shows how the hominid line derives from an arboreal ancestor, and the chimpanzee and the gorilla are both forest dwellers, it might be in the forested, or formerly forested, regions of Africa that the fossil forms intermediate between ape and man will be mostly found. This is in fact what has now transpired in those parts of east Africa that are, or were once, covered by forest or a mosaic of forest and savanna vegetation. Very important fossil material is also known from the Fayum Depression in Egypt, thus showing the extent to which the forests had been able to spread during the mid-Tertiary in a region that is now desert (Simons 1965).

Unfortunately it is only in a comparatively few parts of the continent that fossils of the past 25 million years are preserved, and there is, for

FIGURE 2
The main faunal and fossil man localities in Africa (Miocene to earlier Pleistocene).

example, hardly a bone from the whole of the west African forest zone and from only very rare localities in the Congo basin. Similarly vast areas of savanna on the central plateau and in the east and south are also devoid of any fossil material because of the acid nature of the soil and groundwater. The main fossil-bearing regions of the continent are the Western and Gregory Rift areas of east Africa, the Maghreb of northwest Africa, certain localities in South Africa, and a few isolated localities in Egypt, the Sahara, Malawi, and Rhodesia (Figure 2). Far the richest of these is the east African region. Here the formation of deep troughs and basins, the fast accumulation in these of sediments, and their rapid burial with the fossils they contain, have combined to preserve a unique record extending back to the earlier Miocene some 22 million years ago.

Man is distinguished from other primates by his upright posture, bipedal locomotion, peculiarly prehensile forelimbs, his large brain, and the ability to make and use many kinds of tools. Each of these parts of the body appears to have evolved separately and at different times—thorax and arms first, then pelvis, legs and feet, and lastly, the head and brain. Man's humanity shows itself in the many complicated social and cultural patterns that are unique to his kind. Some of the stages whereby this transformation from a quadrupedal ancestor was effected can be adduced from a study of the fossil record as it is known from the discoveries in east Africa and Egypt.

Some 15–25 million years ago in the early to mid-Tertiary, there existed in Asia, Europe, and Africa a number of *Dryopithecus* ape forms—six species have now been recorded from east Africa (Andrews 1974). These fossils show modification of the limb bones and face indicating that they were adapted to living primarily in the trees, although *also* on the ground (Simons 1963). These creatures have been called "pre-brachiators" (Washburn 1968:48), and judging from their wide distribution, must have been living in a very favorable environment—the forest which in Africa constituted a mosaic with the savanna. This is well established by the finds of uniquely preserved fruits, seeds, and insects of this period in the Miocene deposits on Rusinga Island in Lake Victoria and those on the slopes of the volcanoes (Chesters 1957; Bishop 1968).

By some 14 million years ago, however, a more evolved form was present in India and east Africa, known as *Ramapithecus* (Simons 1961 and 1969; Tattersall 1969). Whereas the dryopithecines were unspecialized apes, morphological features indicate that *Ramapithecus* may have been a hominid. It is unfortunate that the remains consist mostly of fragments of the face and teeth and that nothing is known of the rest of the skeleton. However, the face has undergone considerable modification and has been reduced in length, the teeth having an arrangement and pattern that are essentially human. It has been inferred from this that *Ramapithecus* must have walked upright and must have had forelimbs adapted to using simple tools.

The remains of the later Miocene/Pliocene hominids are known only from rather small fragments of the face and from individual teeth, and it has been pointed out that the face may have become modified at a different time from the limbs. When, therefore, remains of the postcranial skeleton are found they may not be as evolved as has been suggested (Washburn 1968:14–15). Such observations are also very relevant for the interpretation of fossils of this kind *(Kenyapithecus africanus)* recovered by L. S. B. Leakey in a Lower Miocene context at Rusinga Island and Songhor, which suggest the possibility that hominids were present some 8–10 million years earlier still (Leakey, L. S. B., 1967). Recently good reason has been shown for thinking that man and apes may have shared a ground-dwelling, knuckle-walking existence up to the time that the human line developed bipedalism (Washburn 1967). Therefore, whether the hominid line had become separated from the common ancestor with the pongids as early as the mid-Tertiary must remain in doubt until more complete fossil material becomes available. Rather the recent molecular biological work on chromosomes, serum proteins, and hemoglobin (Klinger *et al.* 1963; Goodman 1963) and on the calibration of the immunological distance between man and other primates may indicate that the separation of the hominid and pongid lines more probably took place as recently as 4–5 million, certainly no earlier than 10 million years ago—that is, in the Pliocene (Sarich 1968; Wilson and Sarich 1969) (Figure 3).

The suggestion has been made that the late Miocene/Pliocene ancestor *(Ramapithecus)* may have been a ground-dwelling knuckle walker, like the

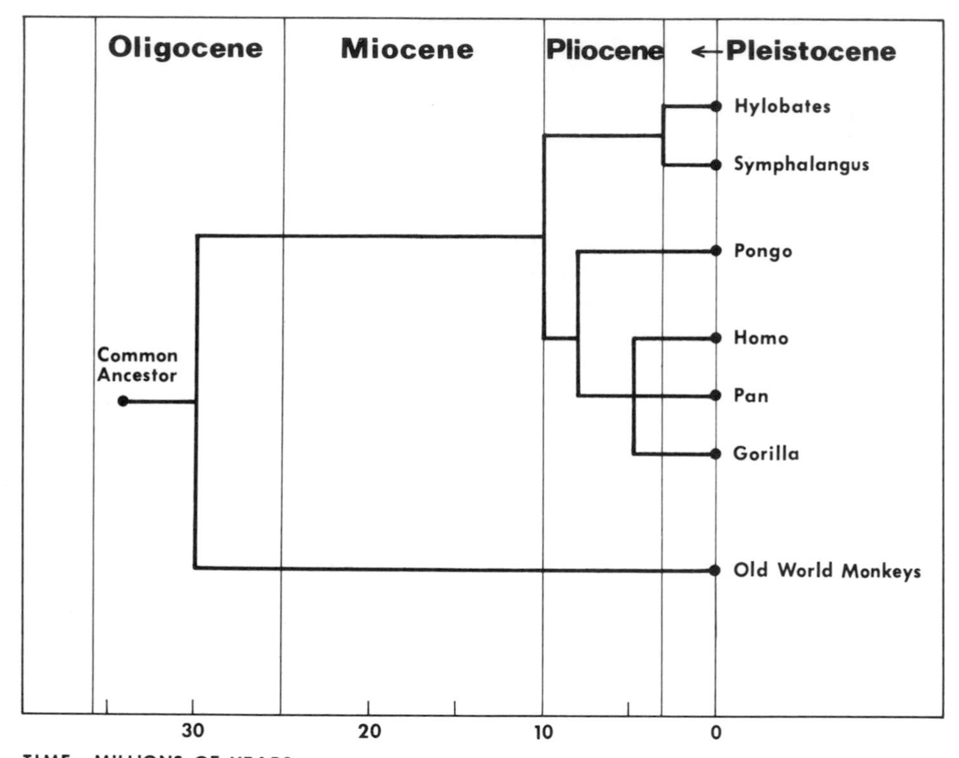

FIGURE 3
Times of divergence between the various hominids as estimated from immunological data. The time of divergence of hominids and Old World monkeys is assumed to be 30 million years. (Courtesy of V. M. Sarich and A. C. Wilson and *Science,* 1967) Copyright by the American Association for the Advancement of Science.

modern African great apes and that, like them, he escaped the extinction that overtook the smaller Miocene apes through competition with the forest-dwelling monkeys, because of his greater weight, adaptation to ground-dwelling and tool manipulation (Washburn 1967:23). Since the gorilla and chimpanzee are forest dwellers, it is most likely this was also the habitat of *Ramapithecus*—a habitat that then stretched from east Africa across to northern India where *Ramapithecus* is associated with plant and animal fossils, indicating a habitat of broad watercourses bordered by forest, giving way to savanna further from the rivers. Evidence at Fort Ternan also indicates there was considerable migration of land mammals between Africa and Eurasia in later Tertiary times, so it is likely that the environment was generally similar in both continents. If this was indeed so, it demonstrates to some extent the fundamental changes in climate, vegetation and fauna that have taken place since the later Tertiary, particularly in the Arabian peninsula and southwest Asia.

The east African form of *Ramapithecus* [*Ramapithecus (Kenyapithecus) wickeri*] from the Fort Ternan site east of Lake Victoria (Leakey, L. S. B., 1961) may have made and used simple tools, but adequate data are not yet avilable to judge this matter. It is claimed that this site shows evidence of what are suggested to be bone-bashing activities—a lump of lava with a battered and bruised edge was found associated with a long bone with a depressed fracture (Leakey, L. S. B., 1968). Only more and less ambiguous evidence can confirm the presence of this type of activity, but there is reason to think that *Ramapithecus (Kenyapithecus) wickeri,* though primarily vegetarian, was also a meat eater.

It is interesting that in the parts that are preserved (face and jaw) there is no very significant difference between *Ramapithecus* and the australopithecines that characterizes the next hominid stage. Continued selection for increased use of hind limbs for bipedal locomotion and of forelimbs for tool manipulation could have been responsible for the developments to be seen in the anatomy of the australopithecines in the earlier Pleistocene some 2 million or more years ago.

If *Ramapithecus wickeri* is a hominid, these morphological developments took place over a period of about 10–12 million years. On the other hand, using the chronology based on the immunological scale (Sarich 1968; Wilson and Sarich 1969), there are between 3 million and 1.5 million years during which the Hominidae, now separated from the pongid line, were evolving the lower limbs toward efficient bipedalism and the upper limbs toward more complex tool use and manufacture. If the latter estimate is correct then *Ramapithecus wickeri* would have been an aberrant ape with fortuitous dental resemblances to the hominids. Unfortunately comparatively few deposits of Pliocene age are known from Africa, so that we possess as yet almost no fossils of this time other than the

Miocene/Pliocene ones from the Fort Ternan site. However, since 1968, several extensive areas in northern Kenya notably in the Lake Baringo basin have been located. From one of these (Ngorora Formation) dating to < 12 ± million years B.P., a single hominid tooth was recovered, and in 1973, another locality (Lukeino Beds) c. 6.5 million years old, produced a further tooth believed to be hominid. Systematic investigation of these exposures is currently under way, and it can be expected that some significant new discoveries may not be far off. At the southwest end of the Rudolf basin at Lothagam, a jaw fragment that has been described as australopithecine, most probably the gracile form (Patterson *et al.* 1970) can be dated on the evidence of the associated elephants and pigs to c. 5 million years ago. Thus, this is the oldest known australopithecine fossil and shows the great antiquity of this hominid lineage.

It is not going to be easy to recover evidence for tool use since intentional and consistent fracture of stone to make a more efficient tool is likely to have been minimal or nonexistent at this time. In any case, only stone and bone, of all the materials that *may* have been used, can be expected to have survived. The evidence for tool use will be more convincing when naturally fractured stone, sometimes altered by use, is found in close association with bones (hominid or other) showing artificial modification, on surfaces where these materials could not have been deposited by geological or nonhominid agencies. There is every reason to suppose that such a mode of tool-using behavior and omnivorous feeding habits made available a much wider range of resources and thereby broadened and enriched the hominid environment. This encouraged renewed experimentation and so, in turn, brought about more rapid evolutionary change than would have been possible without the use of tools.

During the later Miocene and Pliocene, continent-wide crustal movement began to disrupt the earlier Tertiary pattern of internal drainage basins, replacing it with the present hydrographic system. At the same time there was faulting, deep rifting, orographic and volcanic activity along the line of the Great Rift Valley and certain other unstable parts of the continent. The onset of the Pleistocene period coincided with the later stages of these processes and, possibly, with the fragmentation of the evergreen forest vegetation and a significant lowering of temperature in Europe. The later Tertiary also saw the accumulation of the wind-blown Kalahari Sands over much of the western half of the subcontinent during a time of more arid climate, and drier conditions were generally widespread during the Pliocene. It may have been at this time also that the xerophytic Karroo vegetation expanded and spread over a large part of southern Africa at the expense of the lowland humid tropical forest. It appears likely that these events played an important part in accelerating hominid evolution some 9–6 million years ago.

FOSSILS FROM THE LOWER PLEISTOCENE

The fossil record from the Lower Pleistocene is very much better known than is that of the Pliocene in Africa (Figure 4). The australopithecines or man-apes, dating from the early Lower Pleistocene are known chiefly from limestone caves in the Transvaal in South Africa and from a number of localities in east Africa, in particular the Olduvai Gorge in Tanzania, east of Lake Rudolf in Kenya, and the lower Omo valley in southwestern Ethiopia. An advanced australopithecine has come to light in Chad, and there is an enigmatic fossil, *Meganthropus,* from Java that may belong to this group also, though in some respects it is further advanced (Tobias and von Koenigswald 1964). The discovery of the australopithecines in South Africa we owe primarily to two men—Raymond Dart and the late Robert

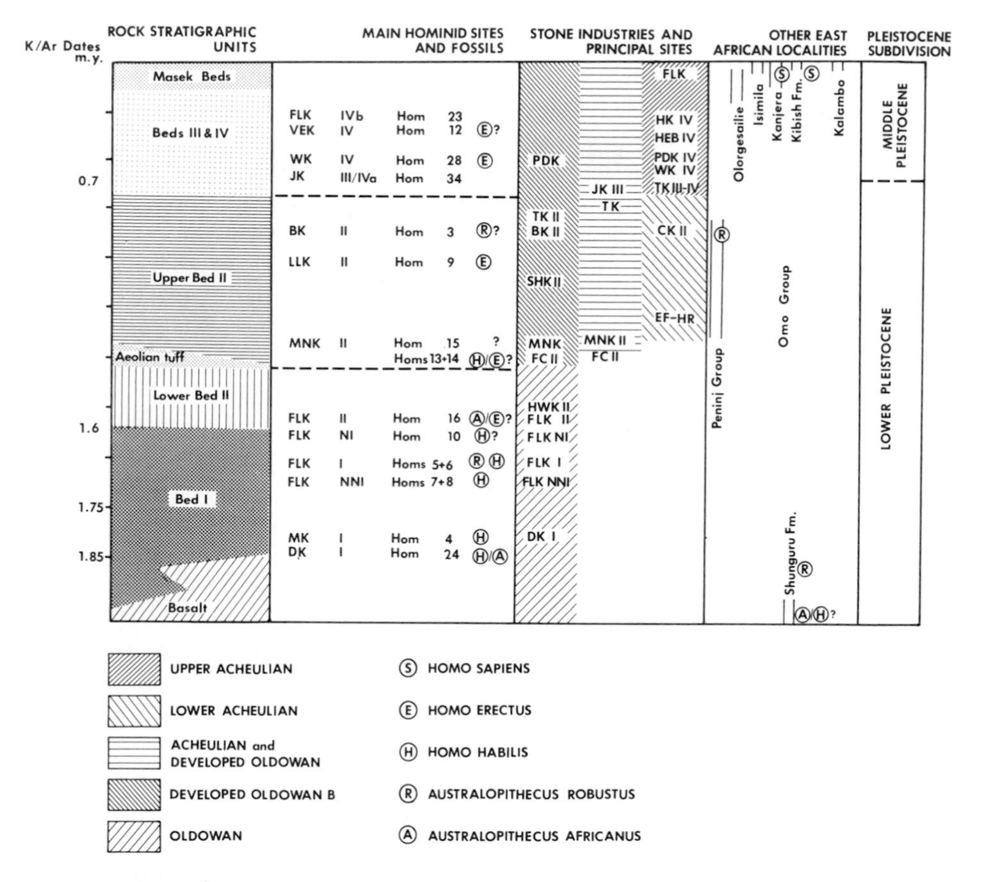

FIGURE 4

Stratigraphy, hominids, and stone industries at Olduvai Gorge and other East African localities

Broom. Later (from 1959) in east Africa, the tireless and persistent researches of the late L. S. B. Leakey, of Mary Leakey, their son Richard, F. Clark Howell and other investigators have resulted in the mass of material that is available today—in fact, no other fossil hominid genus is so well known as *Australopithecus.*

Australopithecines are early tool-making hominids combining a small brain (435–562 cc.) with a large and massive jaw which was the feature that made scientists first consider them to be fossil apes, related more closely to the modern apes than to man. The discovery of a nearly complete foot at Olduvai and of a number of the bones of a hand, together with a reasonably complete vertebral column and pelvis from Sterkfontein in South Africa and of complete and fragmentary limb bones from both regions, show that the early Pleistocene hominids walked erect and used their hands for manipulating tools. More recently, localities in East Rudolf have yielded several more complete femora and tibia, and from here and other sites in the Lake Rudolf basin have come further fragmentary remains of bones of the upper limb, including a complete ulna from the Shungura Formation in the Omo dating to 2.04 million years (Howell and Wood 1974).

Two races of australopithecines are represented in the South African caves—a slenderly built and smaller form, *Australopithecus africanus,* and a robust, larger form, *A. robustus,* originally named *Paranthropus* (Clark, W. E. Le Gros, 1967). The gracile form occurs at three sites—Taung, Makapan, and Sterkfontein—and the robust form at two—Swartkrans and Kromdraai. They are found cemented in breccia filling old caves in the limestone and are associated with many animal bones (Brain 1958). Paleontological evidence (Ewer 1967; Cooke 1963) suggests that Taung, Sterkfontein, and Makapan are the oldest,[1] followed by Swartkrans and then Kromdraai. The rhythm of climatic fluctuations obtained from sedimentation studies done on the breccias (Brain 1958) shows the consistency to be expected when based upon this sequence. *Australopithecus africanus,* the gracile form, appears therefore to precede *A. robustus.* Recently, however, the presence of at least one fossil showing characteristics of the robust form has been established at Makapan (Aguirre 1970), so that it now seems more probable that both forms are present in the older breccia localities as is also the case in east Africa in the Plio-Pleistocene time range. At Swartkrans a representative of the *Homo* lineage (cf. *Homo*

[1]It has been suggested that the Taung fossil is as recent as 0.87 m.y. B.P. and is a young *A. robustus* not *A. africanus* (Tobias 1973). This revised dating (Partridge 1973) is in large part based on geomorphological evidence of the estimated time when the cave from which the fossil came could have been formed. The validity of this method as a means of dating has, however, been called in question so that the age of this fossil must still remain uncertain.

erectus) occurs together with the robust australopithecine. Unfortunately it is not possible to date the South African fossils by isotopic methods, but the latest assessment of the age of the faunal assemblages from the earlier sites places them between 2.5 and 3.0 million years old and the younger Swartkrans site at 1.8 million years B.P.

These southern African discoveries were made mostly in the 1930s to 1950s, but in 1959 Louis Leakey made the first discovery of a hominid fossil from Bed I in the lower part of the long, lacustrine and terrestrial sequence of beds at the Olduvai Gorge. This discovery was the famous "Nutcracker Man"—*Zinjanthropus boisei,* now called *Australopithecus boisei* (Tobias 1967). The fossil was found in close association with a buried occupation site with broken animal bones and stone tools of which "Nutcracker Man" was, at first, hailed as the maker. In 1964 an almost perfect jaw, with teeth, of the robust australopithecine was found eroding from beds of Middle Pleistocene age at Peninj, west of Lake Natron on the Kenya/Tanzania border (Isaac 1967a). The age of this fossil is now shown to be most probably 1.35 million years.

Remains of another hominid had been found at Olduvai in 1960 from a level slightly below that from which *Zinjanthropus* came and near the same locality. They comprised parts of the side and back bones of the skull of a juvenile, that had been broken prior to fossilization. With them were a collar bone and fifteen hand bones from two individuals, an adult and a juvenile, and most of the bones of the foot of an adult individual. With these remains were ten worked stone tools and evidence of carnivore activity. The cranial fragments have been shown to belong to a somewhat larger-brained hominid (680 cc.) with a dental pattern different from that of *A. boisei* (or *Zinjanthropus*) but not very different from that of *A. africanus* (Leakey, L. S. B., 1960).

The associated hand and foot bones show close comparability to those of man, though with certain primitive features, suggesting so far as the foot is concerned that its owner was adapted to running but not, perhaps, to striding. Unfortunately the ends of the toes were missing, but a toe bone from higher up and near the top of Bed II shows that by this time striding was possible (Day and Napier 1964). The hand falls into a position immediately between that of *Homo sapiens* and the apes, and the great flexure and muscularity of the finger bones confirm the near ancestry with knuckle walkers. The Olduvai hand was small and has been described as having an opposable thumb (Napier 1962) that was not only capable of the power grip but also, probably, of the "precision" grip that made possible finer manipulation of objects in addition to the manufacture of simple tools. L. S. B. Leakey and his associates who studied these remains have described them as belonging to a new species of the genus *Homo* which they named *Homo habilis*—man with the ability to manipulate tools—and stone artifacts were found associated with these remains from the bottom of Bed I into the lower part of Bed II (Leakey, L. S. B., *et al.* 1964).

In addition to the finds from the type locality (FLK NNI) there are a number of others—individual teeth from Bed I and more complete cranial and jaw material from the lower part of Bed II (from sites MNK II and FLK II)—that have been ascribed to this form. The only nearly complete limb bones are a tibia and fibula from the same site that produced *Australopithecus boisei* (Davis *et al.* 1964). They are well adapted to bipedal walking, though with differences that suggest the manner in which these early men did so may not have been very like our own striding gait. Because teeth similar to those of *Homo habilis* were also present on this horizon, these leg bones have been attributed to that form rather than to the robust australopithecine. The more nearly complete thigh bones from East Rudolf confirm the bipedal, though clumsy, gait of these early hominids (Clark, W. E. Le Gros, 1967) but emphasize the differences in weight and height between the gracile and robust forms.

The other finds attributed to *Homo habilis* at the Olduvai Gorge include a crushed but reconstructed skull from the site of DKI in Bed I at the base of the sequence (Leakey, M. D., *et al.* 1971); a large part of the vault of a skull, and the greater part of a lower jaw and fragments of the upper jaw from one site (MNK II); and at another locality (FLK II: Maiko Gully) a skull, rendered fragmentary by the passing of a herd of Masai cattle when it lay exposed by erosion. Both these last finds came from the lower part of Bed II and probably belong in the earlier part of the Middle Pleistocene (Leakey, L. S. B., *et al.* 1964). They are considered to be morphologically more advanced than the finds from Bed I but no detailed descriptions have yet been published. Thus, the *Homo habilis* fossils, presently defined, span a long range of time, at least a million years.

What is the relationship of the *H. habilis* form to the gracile and robust australopithecines? Are they in part or entirely contemporaneous or are they chronologically distinct? The work that has been carried out in the Lake Rudolf basin, in East Rudolf, and the Omo since 1968 is providing substantially more evidence that, in part, helps to answer these queries but, at the same time, raises fresh ones. At Omo K/Ar dating of the volcanic tuffs and lavas interbedded with the thick sedimentary sequence has produced a consistent and unique sequence of dates. Isolated teeth, mostly from gracile but also from robust hominids, are present from c. 3.0 million years in the lower part of the Shungura Formation (c. 3.5–1.0 m.y.), and the parietals and occipital of a gracile juvenile are dated to slightly more than 2.0 million years. The robust australopithecine is also represented by incomplete cranial fragments and mandibles dating between 2.35 and 1.8 million years. Up to the fall of 1973 the East Rudolf localities had already yielded 107 specimens of fossil hominids (Leakey, R. E. F., 1974). These have been grouped into two main lineages—robust australopithecine and *Homo.* The former is represented by crania and mandibles, several almost complete and one entirely so, and some are more massive than any other known fossil of this group. In addition, postcranial bones of this form, in particular a femur, tibia, and humerus, are distinguished by their general robustness. Robust australopithecine fossils occur first in the lower member of the Koobi Fora Formation dating between 3.0 and 2.6 m.y. B.P. and, in particular, in the Upper and Ileret Members between 2.6 and 1.3 million years old. Throughout this time the australopithecine fossils show little morphological change.

Specimens representing the *Homo* lineage come from a similar time span and show greater morphological variability in particular as between the earlier and the later fossils. The earliest *Homo* fossil is represented by the large-brained individual known as KNM-ER 1470. This comprises an almost complete cranium with an estimated cranial capacity of c. 775 cc. (Leakey, R. E. F., 1973), and a large part of the face with the palate but with no teeth, which are missing. The tooth sockets show that these were fairly massive, and in the region of the face, this fossil shows robust, australopithecine-like features. KNM-ER 1470 comes from the lower member of the Koobi Fora Formation, as do a mandible and two nearly complete and one fragmentary slender femora that show resemblances to this bone in modern man. Other representatives of the *Homo* lineage come from the Upper Member of Koobi Fora and from Ileret, notably a crushed but largely complete skull and mandible (KNM-ER 1805), other adult and juvenile mandibles, and bones of the lower limbs. In 1973 the fragmentary skull of a smaller-brained (c. 500 cc.) adult gracile individual (KNM-ER 1813) was also found. Richard Leakey (1973) considers this to show resemblances to the gracile *Australopithecus africanus.* It is believed to come from the lower member of the Koobi Fora Formation and so is contemporary with both the robust form and with *Homo,* but as yet, it is not known whether this possible third lineage continues into the upper part of the sequence.

Crucial to an understanding of hominid phylogeny in the Plio-Pleistocene is the dating of the horizons yielding the fossils by means of isotopic and paleo-magnetic techniques. However, in some instances faunal correlations may provide a cross-check. Comparison of the well dated faunas from the Omo (see Howell, this volume) with the Lower Member fauna from East Rudolf (see Leakey and Isaac, this volume) has led some authorities to question the isotopic date of 2½ million years for the KBS Tuff at East Rudolf. Future work should resolve this uncertainty, but for the present, it appears that gracile and robust forms existed contemporaneously in the Rudolf basin from about 3.0 million years up to 1.3–1.2 million years ago. If the East Rudolf radiometric dating is not in error, the larger-brained *Homo* representatives first made their appearance by c. 3.0 million years ago. The great variability in brain size and other characteristics among the East African fossils of this period may be an indication that gracile australopithecine forms also existed for a time contemporaneously with the larger-brained *Homo.*

Comparative faunal studies also show that by the close of the Lower Pleistocene fossils from two other regions confirm the presence of a more advanced hominid lineage contemporary with the robust australopithecine. These are the front part of a skull and face from the Chad basin (Coppens 1967) and the face, palate, and front part of the skull and a crushed jaw from the Cave of Swartkrans in the Transvaal; at Swartkrans this fossil occurs together with stone implements and *Australopithecus robustus* (Robinson 1953). The Swartkrans remains have been attributed to a small species of *Homo erectus,* though Le Gros Clark and others consider that they might fall within the possible range of *A. africanus.* The remains from Chad are said to be transitional between *Australopithecus* and *Homo erectus.* Controversy has ranged widely around whether the robust and gracile forms represent the extremities of sexual dimorphism within a single lineage and whether the *Homo habilis* remains and those just referred to from Swartkrans can be included within or outside the range of *A. africanus.* Some, like Le Gros Clark, maintain that they can be included within the possible range of the man-apes. Others, including Louis Leakey, consider they are too advanced and lie outside this range and represent the oldest true *Homo.*

Although no detailed description has yet appeared, it has been suggested that the *Homo habilis* fossils from Bed II at Olduvai are more advanced than those from Bed I and nearer to *Homo erectus,* of which a large-brained form occurs at a site (LLK II) in the upper part of Bed II and also in Bed IV. On the other hand, the robust australopithecine, *Paranthropus,* continues apparently unchanged into the upper part of Bed II and thereafter disappears from the record.

There are few who do not believe that *Australopithecus robustus* represents a distinctive lineage and there is general agreement that this fossil form shows remarkably little morphological change from the beginning to the end of the record, eventually disappearing short of a million years ago, no doubt through unsuccessful competition with *Homo erectus.* Today there is an overall tendency for greater caution in attaching taxonomic labels to fossils, and indeed, such labels have doubtful value where such a rapidly evolving genus is concerned. Because of the number of "contemporary" fossil specimens that are available, it is now possible to appreciate that the variability within a single population at that time was much greater than had been expected. The record suggests, therefore, a model of small populations living in relative isolation, undergoing comparatively rapid diversification within the compass of three main lineages—gracile and robust australopithecines and *Homo.*

If the South African fossils are not younger than has been indicated, then the fossil record can best be explained by the view that the gracile form evolved into *Homo habilis* and so into *Homo erectus,* since it early developed the ability to make more efficient and complicated tools; whereas the robust form remained a tool user only and so stayed biologically unchanged eventually succumbing to competition with the toolmaker. If, on the other hand, the older sites in South Africa (Taung, Sterkfontein, and Makapan) are contemporary and later than those in Bed I at the Olduvai Gorge, then the gracile australopithecine fossils they contain cannot lie in the direct line of descent of man and would represent late survivals of the likely ancestor of *Homo habilis.* Such gracile forms were contemporary with both the robust and *Homo* lineages in east Africa between 3.0 and 2.6 million years ago if the taxonomic status of the KNM-ER 1813 fossil is confirmed. Possibly Olduvai hominid 24 from DKI at the bottom of the Olduvai sequence should also be included in the same lineage, along with some articular fragments of the knee area of another individual from a site dated by fauna to c. 3.0 million years at Hadar in the Afar Rift in Ethiopia. If the dating is correct, this evidence indicates that the *Homo* line must have already diverged from the gracile one more than 3.0 million years ago.*

Another way of looking at all this evidence is to consider the three lineages as mutually exclusive and contemporary populations throughout the Plio-Pleistocene time period. This seems less likely, except for the robust australopithecine, since all three appear to have shared similar habitats, though the evidence is not complete enough to show whether they exploited different ecological niches in the same way as do the chimpanzee and gorilla in their relationship to man today. These two possible interpretations are set out in diagrammatic form in Figure 5, the available evidence at present being more in favor of alternative B.

Every year, however, new discoveries make the evidence more complete and interpretation more reliable. If the intermediate position of the *Homo habilis* fossils between *Australopithecus africanus* and *Homo erectus* is confirmed, it is not of too great significance whether they are classified as an advanced australopithecine or a lowly form of *Homo,* although the cultural evidence seems to favor inclusion with *Homo.* The small and gracile australopithecine was already present in east Africa c. 5.0 million years ago, while the larger-brained *Homo* is not known in the record before 3.0 million years, perhaps not before 2.0 million years if the age of the East Rudolf hominid KNM-ER 1470 has to be revised. This suggests that it may have taken between 2 and 3 million years for the larger-brained hominids to evolve from the smaller-brained gracile form and a further one million years or less for the change to *Homo erectus* to come about.

*Further hominid remains from Hadar found in 1974 may belong to three distinct taxa.

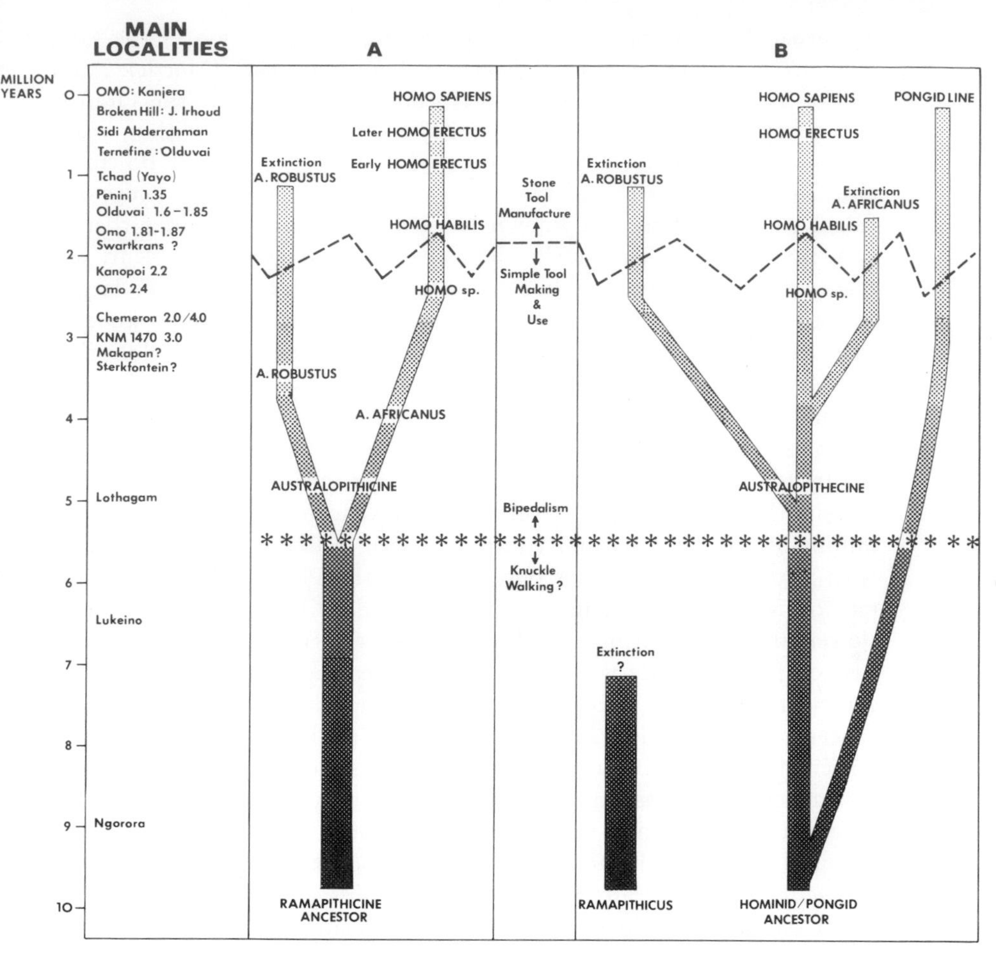

FIGURE 5

Two possible interpretations of the African fossil evidence for hominid evolution.

Such rapid evolutionary developments could have been possible only through the medium of culture and the feedback mechanism brought into play between physiological and cultural development. If this time range is substantiated, the changes that have come about in the bony structure show how truly significant was the acquisition of culture.

THE EARLIEST STONE TOOL ASSEMBLAGES

The earliest dated assemblages of stone tools come from two individual contexts in the KBS Tuff which is dated to 2.6 million years ago and separates the lower from the upper member of the Koobi Fora Formation (Leakey, M. D., 1970; Isaac *et al.* 1975). Two other primary context sites are situated in the upper part of the Shungura Formation of the Omo sequence and date between 1.9 and 2.0 million years (Merrick *et al.* 1973). The long, and until recently, unique sequence at the Olduvai Gorge begins at the bottom of Bed I and dates to c. 1.9 million years. Occasional artifacts have been found eroding from below the KBS tuff in East Rudolf, but with this and two other more doubtful exceptions, flaked stone tools are not known from deposits that are older than c. 2.0–2.6 m.y. B.P. They make their appearance relatively suddenly and there is probably a reasonably close correlation with fossils of the larger-brained hominid. If further work confirms this absence of flaked stone artifacts and larger-brained hominid fossils from deposits much older than 2.0–2.6 m.y., there is justification for associating the tools more specifically with *Homo,* thereby accounting in large part for the accelerated rate of genetic change this lineage underwent in the earlier Pleistocene.

In itself, stone tool manufacture is only the most easily recognized expression of a technological level well in advance of the simple range of minimally modified natural equipment that belongs more particularly with a tool-using level as is postulated for the Pliocene hominids. The rate of evolutionary change can be expected to have been appreciably slower, therefore, during the Pliocene when the transformation occurred that changed knuckle-walking to a fully erect posture and effected the coincidental increase in brain size and the changes in the hand that made conceptual and skilled stone working and utilization possible in the Lower Pleistocene. The immunological time scale suggests that it could have taken as little as from one to three million years to evolve from the ancestral form common to man and the African apes to a fully bipedal hominid form (Sarich 1968). A longer period of time is implied if the *Ramapithecus* fossils really are ancestral hominids.

It should not be imagined that this was a simple process of unilinear development, since the behavioral adaptations that would have been called for in the transition from a Pliocene forest dweller to a savanna dweller in the Lower Pleistocene, must surely have resulted in no small amount of diversification, though only the most successful survived. Similarly, in the initial stages of tool manufacture more than one form may have made simple tools, though only one form in which developing brain size permitted increasing complexity of artifact manufacture made the transition to *Homo erectus.*[2]

[2]The recent experiment by Richard Wright (1972), whereby a young orangutan through stages of demonstration, observation, and experiment was persuaded to strike off a flake and use it to cut the string that secured the lid of a food box, indicates that if a robust australopithecine with its superior brain capacity had had the opportunity to observe other hominids manufacturing and using stone tools, it would not have been beyond the ability of this form to do likewise.

THE WAY OF LIFE OF THE EARLY HOMINIDS

The chief sources of evidence on which to attempt to understand and reconstruct the behavior of the earliest toolmakers are the fossils themselves, the associated artifacts, animal remains, and features of the sites on which these occur and last, but not least, chimpanzee and gorilla behavior. To these should also be added any paleo-climatic evidence that can be deduced from associated fossils and a study of the lithology of the beds in which the remains are found.

On the paleo-climatic evidence from the Transvaal cave breccias there can be little doubt that the australopithecines were living in a habitat somewhat drier than that of the region today but becoming wetter at the upper end of the sequence (Brain 1958:119–22). At Olduvai the habitat has been shown to be not unlike the Serengeti Plains today—open and adjacent to a brackish or saline shallow water lake fed by freshwater streams from higher ground—while that of Chad was savanna associated with forest galleries and occasional patches of evergreen forest (Hay 1967a; Coppens and Koeniguer 1967). This was very different from the forest or forest fringe environment occupied by the Later Tertiary ancestral forms, so that considerable adaptability can be postulated for successful existence in dry savanna by this time. Since the social patterns of the great apes—in particular of the African ones—do not show a great range of variation, it is possible that these did not differ very much from those of the common ancestor they shared with man. Moreover, if the separation took place as recently as the immunological approach suggests, then chimpanzee and gorilla behavior is especially relevant to understanding that of the earliest hominids.

We know from the work of Jane Goodall (1964; 1968) among the free-ranging chimpanzees in the Gombe Stream Reserve in Tanzania, that chimps are highly social animals and also make and use a number of simple tools. They select twigs for extracting termites, sticks for breaking open tree ant nests to get out honey, sponges to mop up water from holes in hollow trees, and are capable of using stones to crack nuts. They have been observed throwing sticks and stones both over- and under-arm, in several different kinds of display against predators or at times of excitement (Kortlandt 1968). In addition they are reasonably adept at carrying objects and they construct sleeping platforms. To interpret man-ape behavior, therefore, this and other aspects of chimpanzee behavior must be taken into account as well as the closer anatomical relationship to ourselves shown by the australopithecine fossils. In other words, the greater brain size, presumed greater cortical complexity, and the character of the postcranial skeleton show that what the chimpanzee does the australopithecine must have done a good deal better—as well as much more that the chimpanzee cannot do.

At the Gombe Stream Reserve the chimpanzees live in open groups that vary in size between 33 individuals and 9 or less and are constantly changing in size and composition. Sometimes they comprise males, females and juveniles; sometimes only males or only females with young; sometimes adult males and females. The groups are largest when food is plentiful, and in particular, the males are long-ranging for purposes of foraging. When food is obtained the group is called together by vocalization and drumming and there is some evidence that the food is shared. The most permanent association is between mother, infant and juveniles, and adult males are attracted, not so much for sexual reasons, as by the mother-infant relationship. When individuals or small groups come together the greeting ceremonies (both among chimpanzees and gorillas) indicate the open nature of these societies, which is greater in the case of the chimpanzees than of the gorillas. In other words, a returning member of the group is welcomed and a new arrival must be, as it were, "vetted" before the group accepts him.

A change from life in the forest to one in the savanna would necessitate an alteration in the dietary habits from a primarily vegetarian to an omnivorous diet in which meat eating acquired increasing importance. As yet the chimpanzees of the Gombe Stream are the only ones that have been observed to hunt and kill fairly large animals for food. However, though this practice is now known to be commoner than previously thought, whether it is a general characteristic of chimpanzee behavior in the forests (Reynolds 1965) or only in savanna/forest mosaic, as at Gombe, is not yet known due to lack of adequate observation. It is to be expected, however, that the greatly increased amount of animal protein available and the less continuous supply of fruits in the savanna would have served to encourage the use of hard seeds and any hunting propensities possessed by the early hominids.

Another possible cause of the greater emphasis on animal foods now discernible may have been the need to supplement the dwindling vegetable resources in an environment that was becoming increasingly drier. Whether or not continent-wide climatic deterioration was a factor in bringing this about cannot, however, be shown as yet since the evidence for these changes is too poorly documented. Pollens from Omo indicate that the habitat was drier c. 2.0 million years ago (Bonnefille 1972). Again, there is no doubt that the microenvironment at the Olduvai Gorge in which the early hominids were living was essentially an arid one, though forest on the volcano slopes and along stream courses cannot have been far away. Whatever the process, therefore, it is certain that the possession of culture by the hominids ensured their success as predators and had by this time permitted groups of them to occupy a range of habitats outside the forest.

As our knowledge of the free-ranging behavior of the African apes grows, it becomes increasingly apparent that the australopithecines were not the aggressive "armed killers" that one school of thought has so vividly represented them to be. It is not yet possible to show that the Transvaal caves were the places where the man-apes lived, and the assemblages of broken animal bone with which they are associated cannot yet be shown to have resulted from the predatory and aggressive habits of the man-apes themselves (Dart 1957). In addition, it still remains to be proved that the differential preservation of the various skeletal parts in these caves was not due to the natural selective agencies that have recently been so effectively distinguished (Brain 1967a, b) or that the so-called wear on certain bones was not, similarly, a naturally controlled phenomenon. Some very persuasive voices have been raised in support of the idea that "man is a predator whose natural instinct is to kill with a weapon," and the associations of bones and hominids in the tropical caves and lake-side camps is used by these prophets of primeval violence as mute evidence of the "courage and cunning" that was said to be already part and parcel of our heritage two million years ago (Ardrey 1961).

Defense of territory lies behind the aggressiveness of many animals and of modern man, but the African apes do not appear to engage in vigorous defense of their territory, and as we have seen, their groupings are essentially open ones, the composition of which is frequently changing. Territoriality in human societies is, therefore, very likely to have come about appreciably later in the Pleistocene. The stone implements at the occupation sites can better be seen as evidence of man's reliance on artifacts of his own selection and manufacture primarily for obtaining and preparing food and for defense. Thus, leaving aside for the moment the controversial, osteodontokeratic so-called "implements" of bone from the South African caves, we find indisputable manufactured tools of stone that relate to life at the home base and represent, not weapons, but domestic equipment (Figure 6). The industry to which they belong has been termed *Oldowan* from the type site at the Olduvai Gorge. If there is thus little evidence to substantiate Hobbes' thesis that the natural state of man was one of ". . . warre, as of everyman against every man," there is just as little reason for crediting the earliest toolmakers with the compassion and humanity of Rousseau's "gentle savage."

The occupation sites of this time from East Rudolf, Omo, and in Bed I and at the base of Bed II at the Olduvai Gorge are unique in providing information on the nature of the early hominid living places (Leakey, M. D., 1967). These sites are small concentrations of occupation debris. One such oval area (FLK I with which *Zinjanthropus* was associated) measured twenty-one by fifteen feet and contained a quantity of broken bone fragments, stone implements, and chipping waste. At these sites are collected natural stones (manuports) and others that have been used for hammering and bashing. There are a number of flaked stone tools: chop-

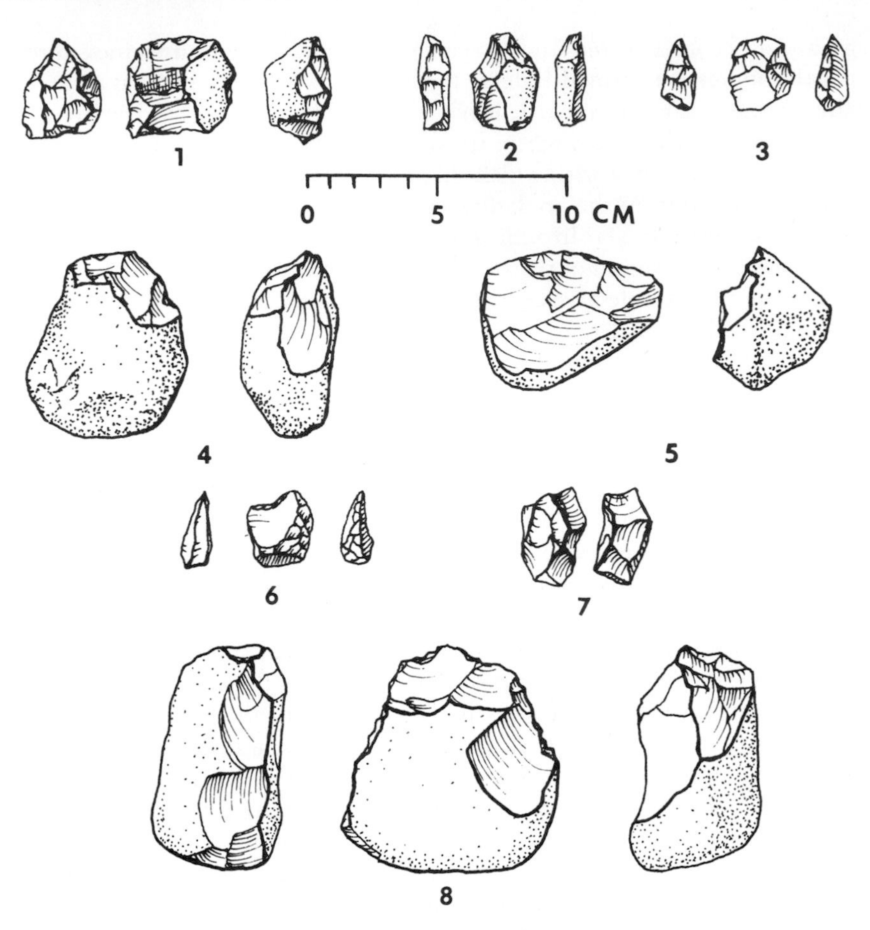

FIGURE 6
Quartz and lava tools of the Olduwan Industry from Bed I, Olduvai Gorge. 1. Bifacially worked chopper. 2. Flake scraper. 3. Minimally retouched flake. 4. Bifacially worked chopper. 5. Unifacial chopper. 6. Minimally retouched flake. 7. Modified chunk with utilized notch. 8. Proto-handaxe.

pers worked from one or both faces giving the tool an irregular chopping edge; polyhedrals, discoids, scrapers, burins, and spheroids as well as a number of utilized flakes. Many unmodified flakes also occur. Three basic forms of worked stone predominate—polyhedral bashing stones, choppers, and flake knives. There is nothing esoteric about their manufacture and they are all small implements with no "formality" about them. They show, however, clear evidence of a rudimentary knowledge of working stone for the production of flakes and chopping edges. Assemblages such as these must lie very close to the beginnings of stoneworking.

The East Rudolf and Omo sites are all situated close to the banks of streams draining to the lake plain. The range of artifacts is much the same as at Olduvai except that very few of the flakes show any evidence of retouch. As yet no choppers have been found *in situ* at the Omo sites although this may be due, in part, to the distance of suitable sized raw material from the delta habitats; at Koobi Fora there is little evidence of the bashed/utilized edges seen on the Oldowan choppers from Olduvai so that a designation of "core" or "core/chopper" rather than "chopper" is more appropriate here (Isaac *et al.* 1971). At one of the East Rudolf sites (HAS) the artifacts cluster round the dismembered remains of a hippo that had spilled down the bank while at all of the East Rudolf sites the artifacts are found mixed with the bone fragments from several different animal species and individuals.

It is a common misconception that the Oldowan tools were made from pebbles, and the term "Pebble Culture" has been used as synonymous with Oldowan. Where pebbles and cobbles abounded, these were frequently used, but the Olduvai evidence shows that angular lumps and flakes of lava and quartz were equally selected for making into tools. There are few other sites in addition to the Olduvai Gorge where well-dated Oldowan artifacts have been found. They occur in the oldest sediments of the coastal sequence of Atlantic Morocco and in stream sediments at Ain Hanech in Tunisia where a number of polyhedral spheroids and a few choppers are associated with a late Lower Pleistocene (Upper Villafranchian) fauna. A very few artifacts have been found also in lake beds in east Africa—at Kanam in the Lake Victoria basin and at Kanyatsi in the Lake Albert Rift. The associated fauna dates these to the earlier part of the Lower Pleistocene—earlier, therefore, than the beginning of the Olduvai sequence. "Pebble tools" have been reported from a number of other sites, usually from river gravels, but the evidence is insufficient or too incomplete to confirm their Lower Pleistocene age. Of course pebbles continued to be used for making tools in some regions until modern times.

Vegetable foods form about sixty to eighty percent of the food of hunters and gatherers today in warm and temperate climates (Woodburn 1968; Lee 1968). Such foods, however, leave very little evidence in the archaeological record. It is, nevertheless, possible to regard the numerous polyhedral and other heavier stones, not only as tools for breaking animal bones and dealing with tough parts of the skin (also eaten by present-day hunters), but also for breaking open nuts and preparing the otherwise unpalatable parts of plants by the breaking down of the fibrous portions. Experiment shows the so-called choppers to be effective for pointing a stick to use for digging out buried plant foods or small burrowing animals as well as for cutting by means of a sawing action. Unmodified flakes form very efficient knives for skinning and cutting. Occasionally wear on a bone also suggests use, though not intentional shaping by hominids. Two broken bones with Oldowan aggregates from Bed I (FLK NNI and FLK north) show striations and polishing similar to the kind of wear resulting from rubbing a hide or digging in the ground. A few of the "osteodontokeratic" bones show this kind of wear and there are others that cannot be readily explained as due to natural causes.

The Oldowan artifacts thus comprise the basic equipment necessary for obtaining a varied and unselected supply of plant and animal foods which were carried back to the home base. Ample evidence of this collecting behavior is provided by the animal remains on the sites, but as yet the full details have not been published. At one Olduvai site (FLK I) the remains have been described as coming mostly from small to medium-sized animals, though bones from larger animals are also present (Leakey, L. S. B., 1963). Usually individuals are represented by only very incomplete remains and this is particularly so in the case of the few large animals that occur. This may, perhaps, indicate the importance of scavenging as a means of supplementing the meat supply, though it could also reflect the habits of the hunters in that the animals may have been butchered and partly eaten or distributed away from the home base. There are also sites (DK I and FLK NI) where the full range of large to small animals is present. Thus, while it may be postulated that the australopithecines and early toolmakers were hunting mostly small game and scavenging larger animals, the evidence is as yet insufficient to provide the necessary confirmation.

Two butchery sites of large animals are known at the Olduvai Gorge from the top of Bed I (FLK NI, Level 6) and the base of Bed II at the same site. At the first, the nearly complete skeleton of an extinct elephant *(Elephas reckii)* was associated with one hundred and twenty-three artifacts, and at the second, the barely disarticulated remains of a Deinotherium had thirty-nine associated artifacts. However these creatures and the Koobi Fora hippo met their death, and they may equally have died from natural causes as have been killed, there is no doubt that by this time the makers of the Oldowan Industry had learned how to deal with the meat from large animals.

The degree to which hunting formed an essential part of australopithecine behavior is open to debate, but it need not have amounted to much more than the individual capturing of small mammals—young antelopes, pigs, or lizards, for example—by running them down and killing them with the bare hands or the aid of a stick or stone, as is still done today by some African peoples. However, some animals can hardly have been taken except by organized group hunting skills, and the recent evidence on chimpanzee hunting methods shows that group hunting must by now have formed an integral part of the behavior pattern. Successful hunting organization need not necessarily imply any complex communication system or the existence of language. The group hunting practiced by wild dog packs, though it makes little use of vocalization, is, nonetheless, highly successful. Chimpanzees are known to make use of some twenty-three distinct calls or vocalizations, besides a number of gestures, facial expressions and bodily postures and it is a legitimate assumption that toolmaking hominids would have had an even larger range of sounds and gestures at their command.

The *home base* is one of the most significant features of the early hominid sites since it represents a place of continuous, though temporary, occupation. The existence of a home base can probably be explained by the prolongation of preadult life and the greater dependence of the young upon the adults which emphasizes the mother/infant relationship. On the evidence of the eruption cycle of the teeth in australopithecines, this dependence continued for much the same length of time as in modern man (Mann 1968:64). Primate studies show that this stretching-out of the period before full maturity is reached is closely related to the learning of skills and behavior necessary for adult life. The young chimpanzee becomes independent at between seven and eight years of age and the transmission of the more complex learned skills of the man-apes must have required an even longer time. The home base can, therefore, be seen as the answer to the constraint imposed upon the mobility of the group by this longer period of learning. One particularly interesting base of this kind (DKI) from the very bottom of the Olduvai sequence contains a sharply defined area of stones with the usual stone and bone waste and Oldowan tools. This suggests an intentional piling, perhaps as some form of shelter or hide (Leakey, M. D., 1967:426).

There are no good means of estimating the size of a hominid grouping at this time, but the area covered by the home base suggests that the bands were not large, though it may be expected that the actual numbers and individuals underwent fairly continuous change. It has been calculated that the australopithecine fossils from Swartkrans represent between fifty and seventy individuals but, of course, this has to be spaced over the time taken for the breccia to accumulate (Mann 1968:69). Bands probably consisted of members of two or three compatible families of mothers and juveniles and perhaps three or four adult males. All these individuals were mutually dependent and shared the results of their hunting and foraging activities; indeed, food sharing is the basis of human society and one of the most fundamental differences between human and mammalian behavior. That it is practiced by chimpanzees (and, to a lesser extent, by wild dogs) is of great interest, but this is only a minor part of a very varied pattern of behavior; their society does not depend upon it as does human society. This is in marked contrast to the picture of the club-wielding aggressor that is usually painted.

If most of the females were concerned with looking after infants and juveniles and foraging for plant foods in the vicinity of the base, the males and females without young would have been directly engaged in hunting and the carrying back of the proceeds of the chase. The variety and number of individual animals represented on the living sites shows that these were not places of purely ephemeral occupation but must have served as a base for at least several days. The carrying of small or medium-sized animals presents little problem and larger ones can be cut up and the pieces divided among the hunters. But the gathering of vegetable products—fruits, nuts, roots and seeds, for example—requires the use of some kind of receptacle if any quantity is to be gathered. It seems one of the things that most infuriates a chimpanzee is his inability to carry away more bananas than he can hold in his arms, so it is likely that, even at this early time, the man-apes will have evolved some simple form of carrying device—a piece of bark or skin, a hollow tree burl or a tortoise shell—more specifically for the collection of vegetable foods.

These early sites are all close to water and could reflect man's inability to live far away from a permanent water supply. Certainly such localities were the most favored by man as well as by animals. Probably there existed no adequate way of storing and carrying water and waterside sites provided not only the best opportunities for hunting but (since the meat was of course not cooked) a regular supply of water to slake the thirst that invariably follows the eating of raw meat.

Such sites may have been sought after for these reasons, but their popularity also may have been connected with the thicker vegetation and tree growth usual at waterside localities in the tropics. This not only would have provided vegetable foods in greater abundance but would also have allowed tree climbing as one means of protection from the larger predators. Quite clearly, also, the use of sticks and stones must have been one of the ways in which the man-apes sought to protect themselves. The reduction in the size of the canine teeth that is a feature of australopithecine dentition was coincident with and a direct outcome of the increasing and more efficient use of tools. As we have seen, chimpanzees have been observed to throw sticks and use them as clubs against leopards as well as to throw stones. Many natural stones (manuports) occur on the living sites, and the circular concentration of natural stones at the DKI site at the base of Bed I at the Olduvai Gorge, besides any other purpose they may have served, could have provided a reserve supply of ammunition with which to keep scavengers away from food stocks, especially during the hours of darkness.

One of the dangers in interpreting the stones and bones that constitute the cultural remains associated with the earliest hominids on the living sites lies in the difficulty we experience in escaping from the preconceived notions that derive from membership in the evolved but rigid social system in which we live today. The opportunity of understanding the psychology and motive agencies that lie behind the behavior of present-day hunting-gathering peoples like the Bushmen and the Hadza is undoubtedly of the greatest relevance for understanding the behavior of prehistoric groups at a similar economic level, though few such studies have yet been made. But even the most lowly of present-day hunter-gatherers are far in advance of *Homo habilis* physiologically, intellectually, and culturally.

On the other hand, if we approach prehistory from the other direction as it were, through studies of primates, in particular of man's closest living relative, the chimpanzee, we have a means of gauging the minimal intellectual and cultural achievements possessed by the early Pleistocene hominids. The truth lies between these two, though probably closer to the level of the chimpanzee.

The close association between the development of the brain and the evolution of technical skills has been stressed (Washburn 1967), and as these skills advanced more efficient adaptive behavior resulted. In turn, this was made possible by the development of those parts of the brain connected with motor skills and the ability to communicate, so that culture and brain growth stand in a feedback relationship to one another. The comparative speed with which (it is inferred) the *Homo* line evolved from a gracile ancestor similar to *Australopithecus africanus,* while *A. robustus* remained biologically unchanged until overcome by competition, can only be due to skillful toolmaking and use originated by the gracile form.

The picture that now emerges, therefore, is one of small-brained, bipedal toolmakers, spread widely across the continent and living in small but variable, highly social, open groups existing by collecting vegetable foods and by organized hunting, the proceeds of which were shared with the other members of the group. The success of this behavioral adaptation lay in the way the learned skills of toolmaking were transmitted from parent to offspring and in the increasing experimental use to which the tools were put that led to new parameters, new adaptive behavior and parallel biological evolution.

To speak of hunting societies immediately suggests that hunting was not only the main activity of the group but also the chief way in which these populations obtained their food. Therefore, it should be emphasized from the beginning that no such implication is intended here. Ethnography shows that at a hunting and gathering stage of culture the total bulk of vegetable foods consumed by tropical peoples greatly exceeds that of meat, and evidence from the Central Kalahari Bushmen and the Hadza, for example, shows, as mentioned previously, that sixty to eighty percent of the food supply of these hunting peoples is vegetable.

Over many millennia, however, hunting and meat eating have had a very great attraction for man, and even in our modern advanced societies the hunter is still held in high repute. There is much evidence of the age-long significance attached to the hunting way of life, to the consumption of meat and the utilization of the other by-products of the chase. Among early hominid populations it would seem quite likely that a minimal consumption of raw meat was essential to ensure a balanced diet. If, therefore, hunting is stressed at this stage in the evolution of culture, it is because it appears as a comparatively new and very significant element in the pattern of behavior of early man as seen in the more numerous remains of food bones scattered around on his camp sites.

HOMO ERECTUS AND THE ACHEULIAN INDUSTRIAL COMPLEX

The same evolutionary processes that gave rise to the larger-brained *Homo* lineage in the Lower Pleistocene some time prior to two million years ago were also responsible for the genetic changes that gave rise to *Homo erectus* in the succeeding one million years and for the technological advances shown by his stone tools.

The result of these evolutionary processes can again best be seen in east Africa. Here the accumulation of sediments in the troughs of the Rift Valley lake basins and their rapid burial and preservation under volcanic dejecta document a unique record of this biological and cultural change (McCall *et al.* 1967). This is best seen at the Olduvai Gorge where in the long time period represented by Bed II the simple Oldowan tool kit is supplemented by a new Acheulian-type of assemblage. The first Acheulian assemblages here occur abruptly above a layer of aeolian tuff that separates Bed II into an upper and a lower half and which may also coincide with the extinction of certain archaic species of animals (Hay 1976b; Leakey, M. D., 1967:417–18, 432). Below this break the physical type is that of *Homo habilis* (from site FLK II) and the industry is Oldowan, but above the break one site (MNK II) low down in the stratigraphic sequence has yielded a hominid form provisionally assigned to *Homo habilis* but having, it has been suggested, some features that link it with the small-brained *Homo erectus* from the Djetis beds in Java (Tobias and von Koenigswald 1964). A third site (LLK II), higher in the sequence and *above* the aeolian tuff, has produced the skullcap of a large-brained *Homo erectus* form (cranial capacity about 1000 cc.). This fossil is contemporary with both Acheulian and Developed Oldowan assemblages. The *Homo* fossil from Swartkrans has also been ascribed to *H. erectus* (Robinson 1953) and is associated with a Developed Oldowan occurrence. Higher again in the sequence at Olduvai in Bed IV(a) (Site WK) the Acheulian is found in direct association with a femur and pelvic bone belonging to *Homo erectus* (Leakey, M. D., 1971a; Day 1971).

Homo erectus can also be directly connected with the Acheulian cultural tradition at three sites in northwest Africa—at Casablanca and Rabat on the Moroccan coast and at Ternifine on the Algerian plateau. The oldest are three fossil jaws and one parietal bone of a skull from Ternifine and, somewhat later in age, are jaw and skull fragments from Casablanca and Rabat. The variability to be seen in the different tool kits being used at this time during the Middle Pleistocene shows that *Homo erectus* was intellectually and technologically capable of a number of new cultural adaptations. He was a man with an expanded brain (between 775 and 1225 cc.) and the skull to contain it. His face was much more like our own than it was like that of the man-apes, though it was still massive. He had a postcranial skeleton that shows few significant differences from our own, and he possessed a much more extended range of abilities and indulged in a greater variety of activities than the evidence suggests to have been the case with *Homo habilis*.

The Acheulian Industrial Complex is the most widespread and, apart from the Oldowan, the longest-lived cultural tradition that we know. It first appears at the Olduvai Gorge above the volcanic tuff (Lemuta Member) separating the upper from the lower members of Bed II and is believed to date to c. 1.5 m.y. ago. The industry of picks and steep scraping tools from the Karari Plateau in East Rudolf (Isaac *et al.* 1975) is also probably related to the Acheulian and dates between 1.5 and 1.3 million years ago. In the Lake Natron basin, a little to the north, two other early Acheulian occupation floors have been found that date to shortly after 1.35 million years. The WK site in Bed IV at Olduvai dates shortly after the end of the last paleomagnetic reversal period c. 0.7 m.y. B.P. The Olorgesailie lake beds with an Upper Acheulian are dated to c. 0.45 million years. Near the top of the Olduvai sequence in the Mesak Beds a developed Acheulian is accorded an age of 400,000 years on amino-acid dating of bone. An upper Acheulian at Isimila in central Tanzania has an isotopic date of 0.26 m.y. while a similar industry at the Kalambo Falls in Zambia is > 0.19 m.y. old, using the amino-acid method on wood. Elsewhere there is evidence that an evolved form of Acheulian continued until between 125,000 and 115,000 years ago although by then it had acquired a more specialized and regional character. The Acheulian Industrial Complex probably lasted, therefore, for as long as 1.5 million years.

The man-ape and *Homo habilis* fossils found so far are confined to the tropical parts of the continent—east and South Africa and the Lake Chad basin. Whether these hominids were also located farther afield, occupying several different ecological niches, we as yet have no means of knowing. Most probably they were widely distributed in Africa outside the forest zone if we can use the evidence of the Oldowan tools, and it is not improbable that they will also be found in the tropical regions of Asia[3] and perhaps in southern Europe, too. It is apparent, however, that around one million years ago stone toolmaking hominids had spread widely, not only in the African continent but also into southern Europe and Asia, where the stone industries are likewise associated with *Homo erectus* and related fossils.

As yet the oldest dated cultural assemblages outside Africa are those from 'Ubeidiya in the Jordan Rift with a stratified sequence of Developed Oldowan and Acheulian assemblages and from Torre in Pietra in Italy dating to > 0.68 m.y. and 0.43 m.y., respectively. *Homo erectus* fossils are now known from southern Africa (Swartkrans) to central and western Europe, Indonesia, and China. The very diversity of these habitats suggests that this form was capable of a considerable degree of ecological adaptation ranging from the cold temperate conditions of Choukoutien in northern China, and Verteszöllös in the central Hungarian plain to the tropical forest/savannas of Java, and the grassland savannas of Africa. This degree of adaptability not only resulted in biological variability but also can be seen reflected in the composition of the tool kits. It is in large part from a study of the stone tools and their relationship to associated remains on the camp sites, as well as from studies of the nature of the camps themselves, that we can attempt to reconstruct the behavior patterns of Lower Palaeolithic times.

The questions to be answered are, for instance, how intensively and over how long a period were the camp sites used? How many people were involved? What were the patterns of settlement? or of movement? of territorial range? economic activity? communication and intergroup relationships? These and others can only be resolved after much more detailed excavation of complete camping places, recording of the relationships of the associated finds and analysis of the respective tool kits. However, a start at least can be made now and will serve to throw into relief those lines of research which can best help towards a more complete understanding of behavior.

It might be expected that in so long a time as that covered by the Acheulian tool tradition, some modification of the composition and complexity of the tool kits would have taken place. This has, indeed, happened but the degree of change is not of the magnitude that might be thought likely over more than a million years and all the evidence points to an extremely slow and gradual rate of cultural evolution. Attempts have been made to establish a series of cultural stages with regional significance, such as that in use in Morocco or those formerly employed at Olduvai or in the Vaal river basin, based on stratigraphic or geomor-

---

[3] *Meganthropus* from the Djetis beds in Java has been ascribed to *Australopithecus:* these beds have a possible age of 1.9 m.y. (Isaac and Curtis 1974).

phological data. Today, however, it is more generally recognized that the Lower Palaeolithic is divided into only three main industrial stages: an Oldowan, followed by a Lower and an Upper Acheulian. More sites of the Upper Acheulian are known than belong with the Lower stage, the components of which are chiefly characterized by the bold nature and small number of the flake scars, as well as by the more general lack of refinement which comes from using a hard hammer or anvil of stone for working the tools. Otherwise Lower Acheulian aggregates fairly closely resemble those of the Upper and later stage.

In the upper part of Bed II in the Olduvai Gorge, above the aeolian tuff, three kinds of assemblages have been shown to occur (Leakey, M. D., 1967). Firstly, there is a continuation of the Oldowan tradition of a flake and chopper-type of industry that was the only sort of tool kit present in the lower part of the sequence. In the upper part of Bed I and in Bed II there is some, though very little, development as compared with the assemblages from the lower part of the sequence at Olduvai. To the range of tools already noted, there is added a kind of simple pointed chopper, known as a proto-handaxe, rough bifacial forms, and rare handaxes together with some more evolved scraper forms and other small tools. When made of fine-grained stone, these take on a surprisingly sophisticated appearance. Spheroids are also present, and if the battering they have suffered comes from pounding foodstuffs to make them edible, these tools now imply a prolongation or regular resumption of such activities at the sites where they are found as well as a longer period of occupation.

As well as these Developed Oldowan assemblages others are now also found in a contemporary tradition—the Acheulian (Figure 7). Analysis details from one Lower Acheulian site at Olduvai (Site EF-HR) from the middle of Bed II show that the large cutting tools (handaxes and cleavers) which characterize the Acheulian Industrial Complex comprise nearly thirty-nine percent of the retouched tool classes, and is increased to over forty-seven percent if the large scrapers are included. The other classes comprise choppers (10.5%), spheroids (11.5%), utilized heavy duty equipment (21%), and the remainder are hammerstones and some miscellaneous specimens. This tool kit was associated with a large percentage (78%) of unmodified flakes in a total of 434 artifacts. The most significant feature about this Acheulian tool kit is that its appearance in the record is quite sudden and it is made from large flakes struck from cobbles or boulders, in this case of lava (Isaac 1969). This is in marked distinction to the Oldowan and Developed Oldowan, all of which are made from appreciably smaller fragments or cobbles.

The third type of assemblage that is present in Bed II combines both Acheulian and Developed Oldowan forms, though the handaxes are generally present only in small numbers and, on the whole, are less well

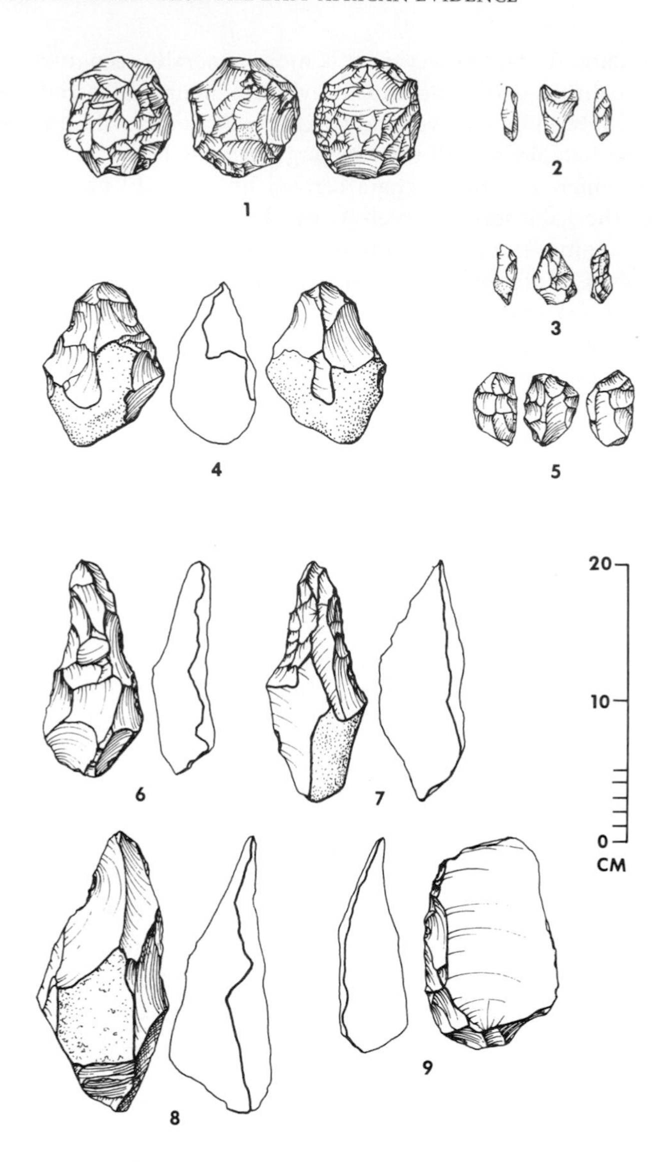

FIGURE 7

Developed Oldowan and Lower Acheulian implements. *Developed Oldowan:*
1. Multi-faceted polyhedral, limestone. (Pré-Acheuléen) (Ain Hanech, Algeria)
2. Double notched scraper, chert. 3. Flake with utilised notch and burin scar, chert. (Locality HWK II, Olduvai Gorge) 4. Proto-handaxe, quartzite. (Sterkfontein Extension breccia, Transvaal) 5. Denticulate core scraper, quartz. (Locality BK II, Olduvai Gorge) *Lower Acheulian:* 6. Handaxe—ficron type with chisel end. (Ternifine, Algeria)
7. Trihedral—pick-like handaxe. (STIC Quarry, Casablanca, Morocco) 8. Trihedral. (STIC Quarry, Casablanca, Morocco) 9. Cleaver on a Kombewa flake. (Ternifine, Algeria) Nos. 6–9 are all quartzite.

made. Characteristic are the numbers and variety of the small scraping tools made on flakes and fragments.

A fourth variant, not yet observed at Olduvai, comes from occupation sites in sediments below the Karari tuff in East Rudolf dating between 1.5 and 1.3 m.y. Handaxes and cleavers are very rare and, instead, the artifacts comprise a number of heavy duty tools—choppers, steeply trimmed core scrapers, and similar forms made on flakes and chunks: some of these forms show an interesting tendency towards bifacial flaking (G. Ll. Isaac and J. W. K. Harris, pers. comm.; Isaac 1975).

Acheulian tool kits did not evolve through transitional stages from the Oldowan in east Africa. The only other region with a comparable, long, cultural and stratigraphic sequence is the Moroccan coast and here also the earliest Acheulian assemblages make an equally sudden appearance. Judging from the wide distribution of the handaxe tradition, the most likely explanation would seem to be that this large tool element represents the addition of a new and important set of activities to the previously established way of life. The very nature of such an invention necessitates a sudden manifestation of its presence—large flakes result from breaking large blocks of stone. Whether this technique originated in east Africa from assemblages such as those from the Karari Plateau is unknown. At the same time, the ability to successfully strike off large flakes also represents an important technological development, a major step forward in man's mastery of his raw material. Therefore, it can be expected that, once invented, knowledge of this technique would spread rapidly throughout the Acheulian world.

At first the cores were randomly struck but it was not long before special techniques were developed for the removal of large flakes for making handaxes and other tools from boulders too large to be broken up in the normal manner. We find, therefore, what are called the proto-Levallois and Levallois methods of core preparation and the Kombewa method, giving the curious flake with double bulb of percussion. These occur with the Acheulian, both inside and outside Africa, wherever the size and nature of the raw material make it necessary; but which of these or other basic methods would, in fact, be used appears to have depended on traditional choice and the flaking properties of the material itself.

The three cultural phases or facies that we have just noted from Olduvai are found widely in the continent and persist from the beginning to the end of the Acheulian Industrial Complex; Karari-type industries are less well known but appear to have a similar range in space and time. It has been suggested that the Bed II evidence from Olduvai can be interpreted as showing the existence of two or more contemporary hominid forms representing separate phylogenetic streams (Leakey, M. D., 1967). While the evidence is too incomplete to rule this out, it would be necessary to postulate that where these forms did not come into competition they were each occupying separate ecological niches. The persistence of the robust form of australopithecine during the earlier half of the Middle Pleistocene is likely to have been possible only because there was no competition, and as soon as *Homo erectus* started to experiment and so begin to exploit the habitat of *Australopithecus robustus,* the latter's extinction was inevitable.

A more likely explanation, therefore, for these cultural facies is that they are related to different sets of activities—the large cutting tools being used for a purpose other than were the small flake tools or the choppers and spheroids. This explanation is strengthened by the clear evidence of selection in the use of raw materials. The large cutting tools of the Acheulian are more often than not made from tough hard rocks such as quartzite or various kinds of lavas. The small flake tools, on the other hand, are generally made from fine-grained rocks—quartz or chert, for example—that produce a sharp but brittle cutting edge and are more easily retouched to form a scraping edge. This is a pattern that is also reflected throughout the Oldowan sequence and can be seen at a large number of Upper Acheulian sites north and south of the Sahara. There can be little doubt, therefore, that we have here an intentional selection of raw materials best suited to the work for which the tools would be used. Both the tool kits in which the large cutting tools predominate and those in which they are mixed with Developed Oldowan forms are here designated as Acheulian: most assemblages are of the second type.

There are, as the map shows (Figure 8), many more designated Upper than Lower Acheulian sites known. However, their general geographical situation is the same, though the degree of technical skill evidenced in the tool assemblages is generally greater with the later ones. Archaeological occurrences of the earlier part of the Middle Pleistocene which include those with both Lower Acheulian and the Developed Oldowan tool kits, are more widely distributed in Africa than are occurrences with adequately attested Oldowan assemblages. They are present in northwest Africa, in east and South Africa, but there are few living places like those at Olduvai that remain undisturbed and in most cases the assemblages are contained in lakeside, river or marine sediments. The handaxes are usually either lanceolate or taper to a chisel-like end. Pointed, pick-like forms with a trihedral cross section, and cleavers with parallel sides and square butts are characteristic of the north African Acheulian sites of this time. One of the undisturbed sites is referred to as the STIC quarry at Casablanca, where a fine series of tools and food waste was found along the edge of a watercourse stratified above the seashore deposits of the Maarifian regression and sealed by the great consolidated Amirian dune rock deposits coeval with an early glaciation in Europe (Biberson 1961).

Another important site is Ternifine on the grass-covered Algerian plateau. Here the artifacts, both large and small, came from horizons now

FIGURE 8
Distribution of Oldowan and Acheulian sites. (After Clark, 1967 a)

below water level and lay on the bank and down the slope of a small lake fed by artesian water. Over fifty percent of the tools were worked pebbles and there were 107 cleavers and 110 handaxes out of a total of 652 artifacts; small tools were also present (Balout *et al.* 1967). This assemblage was associated with much bone food waste and three mandibles and a parietal bone of a north African race of *Homo erectus.*

At Peninj, in the Lake Natron basin, two further early Acheulian sites are situated on what was open and grassy floodplain on sandy ground adjacent to a seasonal stream course. Here fresh water would have been available and perhaps some tree growth would have provided shade along the stream as along the Peninj river today; the two aggregates are similar to that described above from the Olduvai Gorge (Isaac 1967a: 250:1).

At the Olduvai Gorge the Acheulian sites (those, as we have seen, where there is a preponderance of handaxes and cleavers) are situated near stream courses and gravel beds but away from the margin of the shifting *playas* or pans which had replaced the formerly stable salt lake (Hay 1967b:37). By contrast, the sites where the Oldowan assemblages were found lie on the lake-shore flats, close to streams of fresh water draining into the lake from the slopes of the adjacent volcanoes. This may show some major change in the choice of habitation sites, but it could be related to Acheulian man's need for suitable boulders from which large flakes could be struck (Isaac 1969).

In South Africa, a mixed assemblage of nearly three hundred Developed Oldowan (Acheulian) tools was found in breccia at the Extension Site at Sterkfontein, immediately adjacent to the older Type Site breccia which yielded the gracile australopithecine fossils (Robinson and Mason 1962). There were 287 artifacts and manuports; the artifacts being mostly bifacially worked choppers and polyhedral forms together with two possible handaxes. These artifacts cannot be regarded as coming from a living site, since, apart from seven flakes, there is no flaking waste present. Nevertheless, they were brought into the limestone cave for use, the quartzite cobbles being derived from the nearby Blaaubank stream. The few hominid teeth associated have been variously considered as representative of the gracile australopithecine and of *Homo habilis* (Tobias 1965). Similar tools come from the nearby cave breccia at Swartkrans associated with the robust *Australopithecus* and *Homo* (Brain 1958:88, 1967c; Robinson 1961).

Other, though disturbed, Lower Acheulian assemblages (Three Rivers and Klipplaatdrif) are found at the junction of the Klip river with the Vaal in South Africa, a little to the south of those just described (Mason 1962:119–43). Here the artifacts are found within or resting on river gravels and many of them are fairly heavily abraded, but the same tool forms are manifest as are present at the other sites that have been mentioned. There are handaxes, and more rarely, cleavers, crude, pointed, pick-like tools and often trihedral forms, spheroids, choppers, and a varying number of small, scraper-like artifacts or flakes. The retouch is by means of a stone hammer or anvil, the flakes are thick and broad while the negative scars on the large cutting tools are similarly bold and deep. As with other Lower Acheulian aggregates, these handaxes mostly show a minimal number of flake scars—often not more than eight or ten—while the edges are irregular, not straight.

Some bone and ivory fragments, split and minimally trimmed by percussion, are also found with Developed Oldowan aggregates at the Olduvai Gorge (Site MNK II) (Leakey, M. D., 1967:440) and Acheulian (Site SHK II) (Leakey, L. S. B., 1958). An especially convincing small, pol-

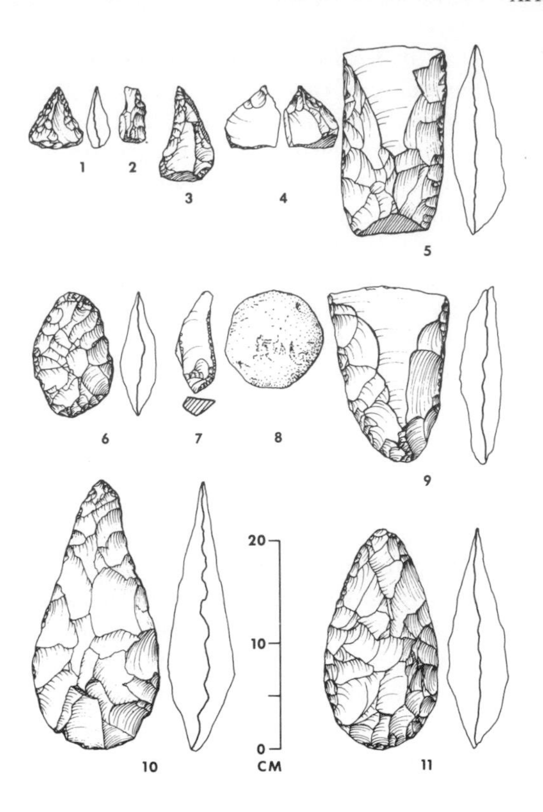

FIGURE 9
Upper Acheulian tools from the Kalambo Falls; large tools in quartzite, small tools in chert. 1. Convergent scraper. 2. Denticulated side scraper. 3. Concave side scraper. 4. Borer. 5. Parallel edged cleaver. 6. Ovate handaxe. 7. Flake knife with marginal retouch. 8. Spheroid. 9. Divergent edged cleaver. 10. Lanceolate handaxe. Dated to greater than 190,000 B.P. 11. Oviate handaxe.

ished bone fragment from Sterkfontein Extension Site shows fine striations and polishing that can only result from use (Robinson 1959).

The Acheulian sites that belong in the later part of the Middle Pleistocene show in general a greater diversity of retouched tools and an appreciable refinement in the technique employed for fashioning them.[4] Handaxes and cleavers are now made by what is called a "soft" hammer technique (that is, by using a hammer of hard wood, bone or antler instead of stone). This results in the removal of thinner and longer flakes and a considerably more refined end product with straight, regular sides on which much more labor and skill have been lavished than was strictly necessary to make a usable tool (Figure 9). These finely finished large cutting tools (handaxes and cleavers) may be some of the first evidence of an aesthetic sense in man and, although the general shape of the tool varies,

[4]Whereas the later Acheulian *in general* is more refined, the Industrial Complex as a whole shows no consistent trend from crude in the early to refined in the later stages. Some "refined" assemblages are known from the Lower Acheulian while "crude" assemblages are also found with the Upper Acheulian.

they are the first "formal" implements made to a regular pattern to appear in the cultural record. Additional evidence of the greater "awareness" of *Homo erectus,* perhaps, may also be seen in the piece of red coloring matter—hematite—carried onto one of the Developed Oldowan sites (BK II) at the top of Bed II at Olduvai (Leakey, L. S. B., 1958:1099).

Now also, considerably more refinement in the retouch of the working edge can be seen in the various scraper forms with the later Acheulian. The trimming is probably the result of more continuous use of the edge, which thus needs to be resharpened by flaking when it becomes blunt, together with a more skilful use of the tool on the material being worked. Although there are still many of the rough polyhedral spheroids on the camping sites, there are now others that are more perfectly shaped. Whether these formed part of some missile weapon, as has been suggested, is not known, but if their primary use was not for battering and bashing, this was, nevertheless, the method by which they were shaped. This carries with it implications for the pounding of vegetable foods and also for bone breaking and splitting, since these spheroids and broken bones occur together. What is again significant is that continued treatment by the "bouncing" action of stone on stone, which is the technique by which these spheroids were rounded (whether by intention or not), needs quite some time to bring about the reduction and rounding shown by some of them. Therefore, either the camping places were being occupied more intensively and for longer stretches, or they were being reoccupied seasonally over a longer period of time.

THE BEHAVIOR OF *HOMO ERECTUS*

The camping places themselves hold the clue to interpreting the behavior and composition of the prehistoric group that occupied them. Firstly, there are now many more sites of this time (Upper Acheulian) than of any previous period, and they occur mostly along stream and river courses, around lakes, and on the seashore. They are found in country that is today grassland, woodland savanna, semi- or complete desert; but they are generally absent from the forest zones. Most of the sites belong to what is now drier savanna and grassland where water and grazing are plentiful—in other words, where the greatest concentrations of the large herbivorous and gregarious game herds are found. Those camp sites situated in the Sahara and other desert regions are invariably associated with sediments that show their contemporaneity with wadi courses or springs, or with shallow pans, lakes or marshes between the dunes. Caves were sometimes occupied also—over long periods of time as is shown by the thirty feet of deposit at the Cave of Hearths in the northern Transvaal. The later Acheulian appears to have been associated with a long, fairly stable, period of somewhat wetter and more humid climatic conditions than those of the present time, and it is probable, though as yet unproven, that the earlier stages were contemporaneous with a series of wet and dry fluctuations.

Within the savanna ecosystem, however, as has already been said, several microenvironments existed—just as today—and the challenge of pressures from climatic fluctuation or other causes would have resulted in the increasing use of these resources. At the Kalambo Falls, for example, the Acheulian population could draw upon the *Brachystegia* woodlands, montane, and other evergreen dry forests, the *dambo* grasslands, the swamp land, and gallery forest along the river and the rocky slopes of the gorge, each with a different set of plant and animal communities (Van Zinderen Bakker 1969). At Olorgesailie, the biogeographic variability in Acheulian times can have been no less than it is today (Isaac 1969).

Not too much is known about the overall size of occupation sites since problems of removing large thicknesses of overburden up to now have generally precluded the excavation of a complete concentration of occupation waste. We have for comparison at least one Oldowan "floor" at the Olduvai Gorge (that which yielded *Zinjanthropus*) where the occupation debris—worked stone and broken bones—was concentrated within an area of roughly 3000 square feet. If we use as a criterion a recent estimate of the requirement of the central Kalahari !Kung Bushmen of about 140 square feet per person, we arrive at about twenty, or perhaps fewer, as the number of individuals using this FLK I *(Zinjanthropus)* site. If, however, we take the evidence for the median requirement of the Californian Indians, which is 25 square feet per person per house, and use *this* to calculate how many people may have been using the area of higher bone and stone density in the center part of the FLK I concentration (an area about 21 by 15 feet), we can estimate a group of some ten to twelve individuals at this site (Isaac 1969). If we use the same two criteria for the complete later Acheulian concentration at Olorgesailie in the Kenya Rift Valley (DE/89 Horizon B), using the Bushman figure, we get a total of not quite nine individuals and, using the much smaller requirements of the California Indians, we arrive at a total of about fifty people for an area of roughly 1250 square feet. Obviously, estimates of this kind must remain highly speculative in our present state of knowledge, but they may, at least, suggest some possible increase in overall group size.[5]

There is also now indication that structures of some kind may have been built at times. Some of the sites, as at Olorgesailie, show such a

[5]Acheulian sites with a high density of artifacts spread over a wide area do not necessarily, of course, indicate occupation by a large group. The numbers and spread of the artifacts may represent the accumulation of many years of continued re-occupation. Low density sites, therefore, are likely to give more accurate indications of the size of single period camps and so of group size, at least until some method is developed for estimating the length of time a particular site remained in occupation.

sharp delineation of the outer margin of the occupation waste as to suggest its having been contained within some fence or windbreak, though, admittedly, the mechanisms for natural concentrations are as yet only imperfectly understood (Isaac 1968, 1967b). A semicircle of stones on one of the occupation floors at the Kalambo Falls may have formed the base of a windbreak (Howell and Clark 1963), and looking outside the continent for the moment, the limestone blocks on the Latamne site in Syria may be indicative of some more substantial structure (Clark 1967b). At the more recently excavated site at Nice (Terra Amata), Lower Acheulian implements of stone and bone are associated with habitation places within an oval floor area of 15 by 36 feet marked by post holes and stone blocks; there are hearths associated also (de Lumley 1969).

An interesting and puzzling feature at some of the African sites is the accumulation of stony rubble that is associated with the artifacts. Mostly this rubble is concentrated in the same way as are the artifacts. In these instances there can be no doubt that the rubble was carried to the site by man and did not accumulate by natural means and it occurs on only some of the occupation floors. It has been suggested that this rubble may have provided the foundation for a home base situated in swampy ground as protection from predators; that it may have been used to hold down the base of a windbreak, that it was ammunition for keeping away scavengers and predators, or used for temporary storage of meat against scavenging animals by piling stones over it.

The *Zinjanthropus* "floor" (FLK I) at Olduvai and the Olorgesailie concentration DE/89 Horizon B are what have been termed "living sites." There are, however, other kinds of sites known from this time which show that *Homo erectus* engaged in several kinds of activities that left a record of several different sorts of tool kits. Firstly, there are the sites just mentioned where several different kinds of activities were being carried on and the tool kits are of the mixed Oldowan and Acheulian type, together with unmodified waste from the making of the tools.

Next, there are sites where the purpose of the tool kits was for butchering one or more large animals. Such, for example, are those found at Isimila in Tanzania (Sands 4 Horizon) (Howell *et al.* 1962) and one of the lower surfaces at Olorgesailie (Basal Bed A, Locality I:3) (Isaac 1968:259) where disarticulated and incomplete hippopotamus carcasses are associated with rare heavy-duty or occasional large cutting tools and small collections of flakes, only some of which show minimal modification and retouch. Similar assemblages with butchered large animals are better known from Europe, as at Torralba and Ambrona in central Spain (Howell 1966), where the remains of over twenty straight-tusked elephants and other large game are associated with the latter site. There is little doubt that in all these cases the equipment represents the tools made for use in butchering the animals.

Again, other sites have concentrations of large cutting tools with varying proportions of light-duty equipment but little or no bone. Such is the case on a partly excavated surface at Melka Kontouré in Ethiopia (Chavaillon 1967), or at the waterside site of Latamne in northern Syria where large cutting and light-duty tools were associated with large limestone blocks which, as we have said, suggested some kind of shelter construction: at each of these sites there was very little bone.

Other archaeological occurrences closely resemble the Developed Oldowan from Bed II at the Olduvai Gorge. A site in Bed III at Olduvai (JK III) (Kleindienst 1967), others from Olorgesailie (Basal Bed B[LSI], L.H.S. or "Hog" [LSXI]), and another from Broken Hill in Zambia (Clark 1959) have numbers of light-duty, small tools but few and poorly made large cutting tools.

Yet another facies of Upper Acheulian that makes its first appearance and has been noted in the Karari Industry can be distinguished by the number of rather crudely made heavy-duty tools—pick-like forms, some parallel-sided bifaces and large flakes—but the general lack of well-made handaxes and cleavers. Such assemblages are found in the peripheral parts of the Congo basin in northeast Angola and the Lower Congo and may be characteristic of localities of thicker vegetation and higher rainfall or, again, as places where wood and its by-products were worked (Clark 1968).

Other sites were specifically workshops, and examples of this kind are known from the Vaal river (Canteen Kopje) (Söhnge *et al.* 1937), from the Transvaal (Wonderboompoort) (Mason 1962), the Kalahari (Nakop) (Brain and Mason 1955), or the Sahara (Tazazmout) (Biberson 1965), where small concentrations or huge spreads of debris lie adjacent to the sources of the raw material—river boulders, an outcrop or a scree at the base of some scarp.

Each of these four main kinds of tool kit, when it is broken down into its component parts, shows a considerable amount of variation in the percentages of the individual tool forms. Sites such as Isimila and Olorgesailie or Kalambo Falls show this in the several successive Acheulian horizons they preserve. They show also that there is considerable variety in the attributes of the large cutting tools which may be pointed and long in one horizon while, in the succeeding level, they will be ovate in plan form. It has been suggested that such variability may reflect the idiosyncratic variation that was possible between individual groups or bands of toolmakers both contemporaneously and through time.

Estimation of the length of time a site was occupied is even more difficult than trying to determine the number of individuals who may have used it. Sometimes, as at the Kalambo Falls, the thin spread of sandy sediments that separates one concentration from another is likely to have been deposited in one season but, where such microstratigraphic evidence is lacking, the problem is more difficult. On the Olorgesailie floor (DE/89), referred to earlier, were found the remains of sixty-five baboons, the bones of which had been broken up, presumably for the marrow. It seems unlikely that these were all killed during one season's occupation of the site, and it is probable that such sites, favorably situated close to suitable sources of food, water, or other resources, were regularly visited over an extended period of time (Isaac 1968:259). The Hadza, incidentally, hunt baboons at night by surrounding their sleeping places and then letting fly with all available weapons (Woodburn 1968).

Regular visits over a period of years may account for the large number of elephant remains at the Lower Acheulian sites in central Spain or for the large concentrations of handaxes and cleavers on the gently sloping rocky surface leading down to the river at Power's site on the Vaal near Kimberley in South Africa (Söhnge *et al.* 1937). At the Kalambo Falls some seven or more occupation concentrations are found overlying each other separated by sandy and gravelly sediments; this is also the case at a number of other sites of this time, both in the open and in caves, showing that some favorable localities remained in use for quite long periods (Clark 1969:104–9, 172–4).

The quantities in which the large cutting tools are often found (for instance, they can be counted in hundreds at the famous "catwalk" site at Olorgesailie where there are more than five-hundred handaxes and cleavers) suggest, either a large population occupying the site on one occasion, which we have seen to be unlikely, or occupation and manufacture over a more extended period of time. Once the basic techniques of "hard" and "soft" hammer have been mastered, stone flaking is not laborious or difficult and it is likely that, when the band moved and since hunters must of necessity travel light, most if not all of the stone tool equipment was left at the camp where, at the favorable sites that were revisited, it accumulated over the years and was available for reuse.

Unfortunately we have as yet no direct means of knowing for what purpose many of these tools were used. We are now in a position to show a connection between light-duty tools (flake knives, small scrapers and chopping tools) and butchery practices, but the large cutting tools appear to be only incidentally associated with meat eating (Clark and Haynes 1970). Since they are often on waterside sites it seems more probable that they were general purpose tools connected with the collection of vegetable foods and animal food preparation, rather than primarily meat mattocks and flensers for cutting the flesh from the carcasses of large animals as was thought at one time.[6] Plant materials are rarely preserved in the early

[6]Relevant here, perhaps, is the importance in the diet of the gorilla of bark, sap and even wood which is chewed in large quantities (Diane Fossey: Personal communication).

sites, but gathering must always have been an important, indeed, the major source of food, and at the Kalambo Falls various edible fruits, seeds and nuts are found associated with the Acheulian living sites (Clark 1969, 1974).

Similarly, there is no precise way of estimating the area of country over which an Acheulian hunting band would range. A very approximate estimate can be obtained by establishing the location of the nearest source of a particular raw material found on an Acheulian occupation site. The distances established vary usually between three to five miles with a maximum of forty miles. By this means we arrive at an estimated range over an area of something less than twenty square miles at Olduvai for the periods represented by Beds I and II, and a range of between fifty and over twelve hundred square miles for Upper Acheulian sites in Africa and Europe. The maximum figure obtained derives from a Rhodesian watershed site at Lochard (Bond 1948).

Meat has now become a regular and important item of diet judging by the number of large and medium-sized animals represented by the bones on the camping places. Some of this meat was undoubtedly still obtained by scavenging, but most of it can more probably be seen as the direct outcome of the adoption of more efficient methods of hunting. Wooden spears with a simple point are known from two Middle Pleistocene sites in Europe, Clacton-on-Sea (England)[7] and Lehringen (North Germany), and it is a fair supposition that the spear was also used in Africa at this time. Stone throwing is likely to have been the most effective means of attack and defense and large quantities of collected natural stones are present, as we have seen, at the camping places.

Evidence of fire has not yet been found on Acheulian sites in Africa except in late contexts, though all probably > 0.2 million years old. Its use is well attested from several sites in Europe and Asia during the earlier Middle Pleistocene and in the colder, high-latitude regions—from Choukoutien, Torralba, Ambrona, Hoxne, and other places (Oakley 1955). It certainly seems probable that such a fundamentally important source of heat and power was similarly employed in the tropics, but on the older sites the resulting charcoals have probably been broken down and dispersed by soil fauna and so are not preserved.

The attraction that contained fire has for many animals and birds and the obvious potentiality for *Homo erectus* (if not for *Australopithecus*) must have led to early experimentation in the use of fire—for protection, for toolmaking and, perhaps, for hunting by setting fire to the grass and driving game as, for example, at Torralba. In this method of hunting it is not

[7]The robustness of this specimen suggests that it would have been equally, if not more, useful as a digging stick than as a spear.

only the fear of the fire but also the blinding effect of the smoke that accounts for the kills (Baker 1890). It is easier to accept the hypothesis that man may have been responsible for the extinction of a number of the earlier Pleistocene faunal species, if this was accomplished through the destruction of their habitat and food supply by the continued and uncontrolled use of fire, than by his killing them off by the more conventional use of hand weapons.

There may be a much wider range of animals represented at the Acheulian than at the Oldowan sites and a greater concentration of large and medium-sized animals. The large animals were butchered where they died, a method still often adopted by Bushmen. Analysis of the bone waste on these Acheulian sites suggests that the smaller animals were usually dismembered, in part, eaten where they were killed and, in part, carried back to camp in the same way as the Bushmen and Hadza do today, thus accounting for the partial remains of individual beasts among the food waste. Hunting activities on the scale suggested by the number of butchered *Pelorovis* and *Antidorcus* remains at the BK II and SHK sites at the top of Bed II at the Olduvai Gorge (Leakey, L. S. B., 1958), by the giant baboons at Olorgesailie and the elephants at the Spanish sites, indicate that some kind of better organized group hunting was probably practiced. Communal hunting on such a scale as well as the flaking skills and formal retouch exhibited by the stone implements imply some improved form of communication over that of the man-apes, and it has been suggested that these skills could only have been possible through the ability to speak, that is, the possession of language.

Speech implies certain musculatory modifications of the mouth which came about with the shortening of the jaw and the coordinated use of larynx, tongue, and lips to articulate sounds. But, as Bernard Campbell has said, "Speech. . . is more than complex sound. It is the act of codifying thought into sets of controlled and connected sounds, and such codification occurs in the cerebral cortex" (Campbell 1966). Since it is in the Middle Pleistocene that man's cranial capacity begins to increase more rapidly, it seems likely that it was during this time that those parts of the cerebral cortex that control speech ability—notably the inferior, posterior, parietal area—were undergoing genetic selection for communication by speech.

The cultural evidence suggests, however, that if Middle Pleistocene man were, indeed, possessed of the ability to speak, this was doubtless at a lower level of expression than language as we know it today. While hunting and the excitement that this engenders in the participants may have been one of the cultural stimulants to verbalization, efficient hunting, at least in its initial stages, can certainly be carried on with only minimal vocalization between the hunters. Wild dogs, hyenas and lions certainly use very little vocalization, and though hunting chimpanzees are undoubtedly more vocal, it is not necessary for quite efficient hunting that the communicative ability of Middle Pleistocene man should have been very much more complex than that of australopithecines or of chimps. That it *was* in advance can surely be postulated from the increasing brain size as well as the learning and mastery of the stone-flaking techniques and steadily expanding range of activities manifest in the different tool kits at this time.

One of the most striking things, however, about the broad cultural pattern of the Middle Pleistocene is its general "sameness" within the limits imposed by the stone industries. The overall similarity in pattern between the Acheulian and Developed Oldowan cultural assemblages wherever they are situated calls for a common level of behavior. Idiosyncratic differences and variation in the frequency with which the particular classes of artifact occur are very evident, as we have seen (Figure 10), but there is nowhere the fundamental, regional specialization in the tool kits that is present in the Upper Pleistocene. Handaxes from Europe, South Africa, or peninsular India are all basically similar tools, and this is also true for the rest of the heavy-duty and the light-duty elements. At the same time this "sameness" about the stone artifacts suggests that, although the period of learning necessary to become proficient in making them may have been drawn out, there is reason to suppose that some of this skill, at least, could have been learned by observation, experimentation, and imitation rather than by direct instruction. On the other hand, the degree of "standardization" shown by the handaxes and cleavers, for example, argues for at least limited instruction and so, perhaps, for some rudimentary form of speech.

Typologists have classified Lower Palaeolithic artifacts into many different types and subtypes, but the recent studies on toolmaking, recognition, and uses in ethnographic contexts throw serious doubt upon the validity of such classifications, if anything more than *difference* is implied by such groupings. It is the working edge and the angle of this edge and not the morphology of the artifact that are the most significant parts of the piece for the user among those ethnic groups who are still using stone tools. If this was also the case in the past, and there is every reason to believe that it was so, then it is possible to distinguish only some five or six sets of stone artifacts in the Acheulian toolkits that are likely to have been functionally distinct—choppers, spheroids, light- and heavy-duty tools, large cutting tools (handaxes and cleavers), and unretouched or minimally modified flake knives.

While, therefore, the wide distribution of occupation sites in the Old World shows the ability of man in the Middle Pleistocene to adapt to a wider range of ecological conditions, the absence of regional specialization and the overall standardization of the tool kits suggest a general pattern of behavior that was everywhere throughout the occupied world at much the

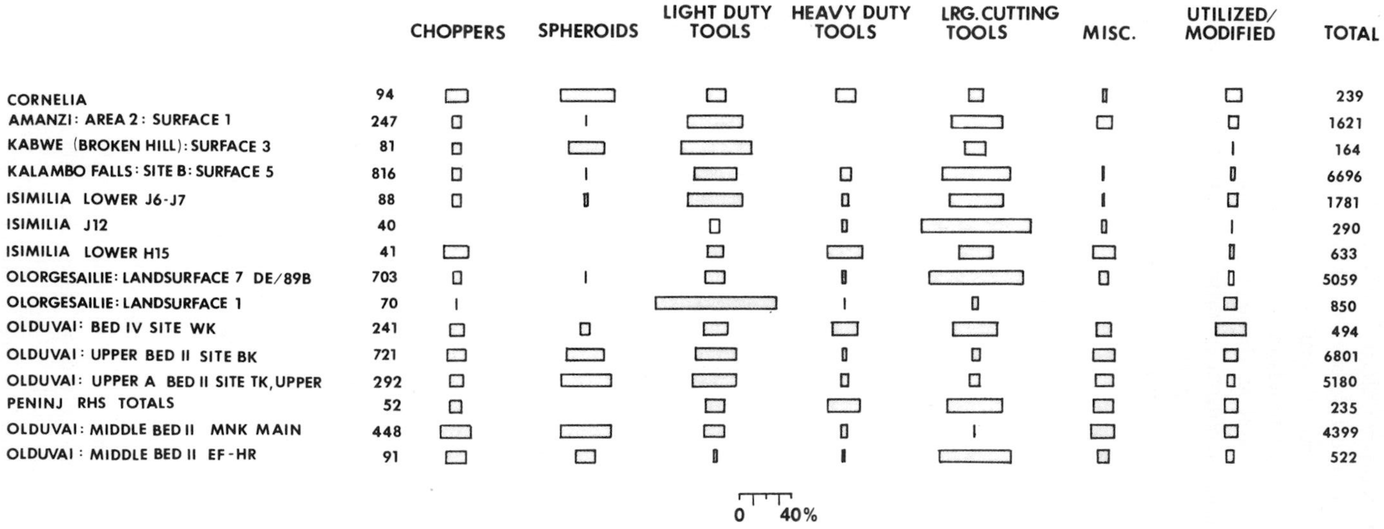

CHOPPERS SPHEROIDS LIGHT DUTY TOOLS HEAVY DUTY TOOLS LRG. CUTTING TOOLS MISC. UTILIZED/ MODIFIED TOTAL

Site		TOTAL
CORNELIA	94	239
AMANZI: AREA 2: SURFACE 1	247	1621
KABWE (BROKEN HILL): SURFACE 3	81	164
KALAMBO FALLS: SITE B: SURFACE 5	816	6696
ISIMILIA LOWER J6-J7	88	1781
ISIMILIA J12	40	290
ISIMILIA LOWER H15	41	633
OLORGESAILIE: LANDSURFACE 7 DE/89B	703	5059
OLORGESAILIE: LANDSURFACE 1	70	850
OLDUVAI: BED IV SITE WK	241	494
OLDUVAI: UPPER BED II SITE BK	721	6801
OLDUVAI: UPPER A BED II SITE TK, UPPER	292	5180
PENINJ RHS TOTALS	52	235
OLDUVAI: MIDDLE BED II MNK MAIN	448	4399
OLDUVAI: MIDDLE BED II EF-HR	91	522

0 40%

FIGURE 10
Comparison of the main components of Acheulian and other Middle Pleistocene industrial assemblages from various localities in sub-Saharan Africa, to show intersite variability.

same level of efficiency. The very slow rate of cultural change over more than a million years and the limited range of activities implied by the different tool kits and the contexts in which they are found suggest that *Homo erectus,* though he exploited a wide range of resources, did so only at a very low level of efficiency and with minimal ability to specialize.

However, although the rate of cultural evolution was extremely slow during the Middle Pleistocene, it was, nonetheless, one of continuing and quickening tempo and, by the time the Acheulian Industrial Complex came to an end, there is evidence that man had achieved social stability and intellectual abilities that led to more rapid diversification in many different directions and placed him firmly on the threshold of a new world.

References Cited

Aguirre, E.
1970 Identificacion de "Paranthropus" en Makapansgat. Cronica del XI Congreso Nacional de Arqueologia, Merida. 1969. pp. 98–124.

Andrews, P.
1974 New Species of *Dryopithecus* from Kenya. Nature 249:118–190.

Ardrey, R.
1961 African Genesis—a personal investigation into the animal origins and nature of man London: Collins.

Baker, (Sir) S. W.
1890 Wild Beasts and Their Ways: Reminiscences of Europe, Asia, Africa and America. London.

Balout, L., P. Biberson and J. Tixier
1967 L'Acheuléen de Ternifine (Algérie), gisement de l'Atlanthrope. L'Anthropologie 71:217–238.

Biberson, P.
1961 Le Paléolithique inférieur du Maroc atlantique. Rabat: Pub. Services des Antiquités du Maroc. Fasc. 17.

1965 Recherches sur le Paléolithique inférieur de l'Adrar de Maurétanie. *In* Actas del V Congreso Panafricano de Prehistoria y de Estudio del Cuaternario, Vol. I. L. D. Cuscoy, ed. Tenerife, 1963. pp. 173–189.

Bishop, W. W.
1968 The Evolution of Fossil Environments in East Africa. Trans. Leicester Literary and Philos. Soc. LXII:22–44.

Bond, G.
1948 Rhodesian Stone Age Man and His Raw Materials. S. Afr. Archaeol. Bull. III:55–60.

Bonnefille, R.
1972 Associations polliniques actuelles et Quaternaires en Ethiopie (Vallées de l'Awash et de l'Omo). Doctoral Thesis C.N.R.S. A07229. Paris. I:183–84.

Brain, C. K.

1958 The Transvaal Ape-Man-Bearing Cave Deposits. Transvaal Mus. Mem. 11:1–131.

1967a Bone Weathering and the Problem of Bone Pseudo-Tools. S. Afr. Journ. Sci. 63:97–99.

1967b Hottentot Food Remains and Their Bearing on the Interpretation of Fossil Bone Assemblages. Scientific Papers of Namib Desert Research Station, Pretoria, South Africa. No. 32:1–7.

1967c The Transvaal Museum's Fossil Project at Swartkrans. S. Afr. Journ. Sci. 63:378–384.

Brain, C. K. and R. J. Mason

1955 A Later African Chelles-Acheul Site Near Nakop, Southern Kalahari. S. Afr. Archaeol. Bull. 10:22–25.

Campbell, B. G.

1966 Human Evolution: An Introduction to Man's Adaptations. Chicago: Aldine.

Chavaillon, J.

1967 La Préhistoire Ethiopienne á Melka Kontouré. Archeologia 19:57–63.

Chesters, K. I. M.

1957 The Miocene Flora of Rusinga Island, Lake Victoria, Kenya. Paläontographica, Stuttgart, 101B:30–71.

Clark, J. D.

1959 Further Excavations at Broken Hill, Northern Rhodesia. Journal of the Royal Anthropological Institute, London 89:201–232.

1967a ed., Atlas of African Prehistory. Chicago: University of Chicago Press.

1967b The Middle Acheulian Occupation Site at Latamne, Northern Syria. Quaternaria 9:1–68.

1968 Further Palaeoanthropological Studies in Northern Lunda. Lisbon. Museu do Dundo Publicacoes Culturais No. 78.

1969 Kalambo Falls Prehistoric Site, Vol. I. Cambridge: The University Press.

1970 The Prehistory of Africa. New York. Praeger.

1974 Kalambo Falls Prehistoric Site, Vol II. Cambridge: The University Press.

Clark, J. D. and C. V. Haynes

1970 An Elephant Butchery Site at Mwanganda's Village, Karonga, Malawi and its Relevance for Palaeolithic Archaeology. World Archaeology I:390–411.

Clark, W. E. Le Gros,

1967 Man-Apes or Ape-Men? New York: Holt, Rinehart, & Winston, Inc.

Cooke, H. B. S.

1963 Pleistocene Mammal Faunas of Africa, with Particular Reference to Southern Africa. *In* African Ecology and Human Evolution. F. C. Howell and F. Bourlière, eds. Chicago: Aldine. pp. 65–116.

Coppens, Y.

1967 L'hominien du Tchad. Actas del V. Congreso Panafricano de Prehistoria y de Estudio del Cuaternario. L. D. Cuscoy, ed. Tenerife, 1963. pp. 329–30.

Coppens, Y. and J. C. Koeniguer
1967 Sur les Flores ligneuses disparues Plio-Quaternaires du Tchad et du Niger. C. R. Acad. Sc. Paris, Vol. 265, pp. 1282–85.

Dart, R. A.
1957 The Makapansgat Australopithecine Osteodontokeratic Culture. *In* Proceedings of the Third Pan-African Congress on Prehistory, Livingstone, 1955. J. D. Clark and S. Cole, eds., London: Chatto and Windus. pp. 161–171.

Davis, P. R. *et al.*
1964 Hominid Fossils from Bed I, Olduvai Gorge, Tanganyika: A Tibia and Fibula. Nature 201:967.

Day, M. H.
1971 Postcranial Remains of *Homo erectus* from Bed IV, Olduvai Gorge, Tanzania. Nature 232:383–87.

Day, M. H. and J. R. Napier
1964 Hominid Fossils from Bed I, Olduvai Gorge, Tanzania: Fossil Foot Bones. Nature 201:969.

Ewer, R. F.
1967 The Fossil Hyaenids of Africa: A Reappraisal. *In* Background to Evolution in Africa. W. W. Bishop and J. D. Clark, eds. Chicago: University of Chicago Press. pp. 109–122.

Goodall, J.
1964 Tool-Using and Aimed Throwing in a Community of Free-Living Chimpanzees. Nature 201:1264–66.

Goodall, J. van Lawick
1968 A Preliminary Report on Expressive Movements and Communication in the Gombe Stream Chimpanzees. *In* Primates. P. C. Jay, ed. New York: Holt. pp. 313–374.

Goodman, M.
1963 Man's Place in the Phylogeny of the Primates as Reflected in Serum Proteins. *In* Classification and Human Evolution. S. L. Washburn, ed. Chicago: Aldine. pp. 204–234.

Hay, R. L.
1967a Revised Stratigraphy of Olduvai Gorge. *In* Background to Evolution in Africa. W. W. Bishop and J. D. Clark, eds. Chicago. University of Chicago Press. pp. 221–225.

1967b Hominid-Bearing Deposits of Olduvai Gorge. Nat. Acad. Sci. Washington Publ., No. 1469:30–42.

Howell, F. C.
1966 Observations on the Earlier Phases of the European Lower Palaeolithic. *In* Special Number 68(2) of the American Anthropologist, J. D. Clark and F. C. Howell, eds. pp. 88–201.

Howell, F. C., G. H. Cole and M. R. Kleindienst
1962 Isimila: An Acheulian Occupation Site in the Iringa Highlands, Southern Highlands Province, Tanganyika. *In* Actes du IVe Congrès Panafricain de Préhistoire et de l'Etude du Quaternaire. G. Mortelmans and J. Nenquin, eds. Tervuren: Musée royal de l'Afrique centrale. pp. 43–80.

Howell, F. C. and J. D. Clark
1963 Acheulian Hunter-Gatherers of Sub-Saharan Africa. *In* African Ecology and Human Evolution. F. C. Howell and F. Bourlière, eds. Chicago: Aldine. pp. 458–533.

Howell, F. C. and B. A. Wood
1974 Early Hominid Ulna from the Omo Basin, Ethiopia. Nature 249:174–76.

Isaac, G. Ll.
1967a The Stratigraphy of the Peninj Group—Early Middle Pleistocene Formations west of Lake Natron, Tanzania. *In* Background to Evolution in Africa. W. W. Bishop and J. D. Clark, eds. Chicago: University of Chicago Press. pp. 229–257.

1967b Towards the Interpretation of Occupation Debris—Some Experiments and Observations. Kroeber Anthropological Society Papers 37. Berkeley. pp. 31–39.

1968 Traces of Pleistocene Hunters: An East African Example. *In* Man the Hunter. R. B. Lee and I. DeVore, eds. Chicago: Aldine. pp. 253–261.

1969 Studies of Early Culture in East Africa. World Archaeology I:1–28.

1975 Early Artefact Assemblages from the Koobi Fora Formation. *In* Earliest Man and Environments in the Lake Rudolf Basin: Stratigraphy, Paleoecology and Evolution. Y. Coppens, F. C. Howell, G. Ll. Isaac and R. E. F. Leakey, eds. Chicago: University of Chicago Press.

Isaac, G. Ll., R. E. F. Leakey and A. K. Behrensmeyer
1971 Archaeological Traces of Early Hominid Activities East of Lake Rudolf, Kenya. Science 173:1129–34.

Isaac, G. Ll., J. W. K. Harris and D. Crader
1975 Notes on the Archaeological Evidence from the Koobi Fora Formation. *In* Earliest Man and Environments in the Lake Rudolf Basin: Stratigraphy, Paleoecology and Evolution. Y. Coppens, F. C. Howell, G. Ll. Isaac and R. E. F. Leakey, eds. Chicago: University of Chicago Press.

Isaac, G. Ll. and G. H. Curtis
1974 Age of Early Acheulian Industries from the Peninj Group, Tanzania. Nature 249:624–27.

Kleindienst, M. R.
1967 Report on Excavations at Site JK2 Olduvai Gorge, 1961–2. Paper delivered at VIe Congrès panafricain du Préhistoire et de l'étude du Quaternaire, Dakar, 1967. Unpublished.

Klinger, H. P., J. L. Hamerton, D. Mutton and E. M, Lang
1963 The Chromosomes of the Hominoidea. *In* Classification and Human Evolution. S. L. Washburn, ed. Chicago: Aldine. pp. 235–242.

Kortlandt, A.
1968 Handgebrauch bei freilebenden Schimpansen. *In* Handgebrauch und Verständigung bei Affen und Frühmenschen. Stuttgart: B. Rensch. pp. 59–102.

Leakey, L. S. B.
1958 Recent Discoveries at Olduvai Gorge, Tanganyika. Nature 181:1099–1103.

1960 Recent Discoveries at Olduvai Gorge. Nature 188:1050–52.

1961 A New Lower Pliocene Fossil Primate from Kenya. American Magazine of Nat. Hist. Ser. 13:689–696.

1963 Very Early East African Hominidae and Their Ecological Setting. *In* African Ecology and Human Evolution. F. C. Howell and F. Bourlière, eds. Chicago: Aldine. pp. 451–52.

1967 An Early Miocene Member of the Hominidae. Nature 213:155–163.

1968 Bone Smashing by Late Miocene Hominidae. Nature 218:528–30.

Leakey, L. S. B., P. V. Tobias and J. R. Napier
1964 A new Species of the Genus *Homo* from Olduvai Gorge. Nature 202:7–9.

Leakey, M. D.
1967 Preliminary Survey of the Cultural Material from Beds I and II, Olduvai Gorge, Tanzania. *In* Background to Evolution in Africa. W. W. Bishop and J. D. Clark, eds. Chicago: University of Chicago Press. pp. 417–446.

1970 Early artefacts from the Koobi Fora area. Nature 226:228–230.

1971a Discovery of Postcranial Remains of *Homo erectus* and Associated Artefacts in Bed IV at Olduvai Gorge, Tanzania. Nature 232:380–83.

1971b Olduvai Gorge Volume III. Excavations in Beds I and II, 1960–1963. Cambridge; The University Press.

Leakey, M. D., R. J. Clarke and L. S. B. Leakey
1971 New Hominid Skull from Bed I, Olduvai Gorge, Tanzania. Nature 232:308–312.

Leakey, R. E. F.
1973 Evidence for an Advanced Plio-Pleistocene Hominid from East Rudolf, Kenya. Nature 242:447–450.

1974 Further Evidence of Lower Pleistocene Hominids from East Rudolf, North Kenya, 1973. Nature 248:653–56.

Lee, R. B.
1968 What Hunters do for a Living or How to make out on Scarce Resources. *In* Man the Hunter. R. B. Lee and I. DeVore, eds. Chicago: Aldine. pp. 30–48.

Lumley, H. de
1969 A Palaeolithic Camp at Nice. Scientific American 220:42–50.

Mann, A. E.
1968 The Palaeodemography of *Australopithecus*. Doctoral dissertation, Department of Anthropology, University of California, Berkeley.

Mason, R. J.
1962 Prehistory of the Transvaal. Johannesburg.

McCall, G. J. H., B. H. Baker and J. Walsh
1967 Late Tertiary and Quaternary Sediments of the Kenya Rift Valley. *In* Background to Evolution in Africa. W. W. Bishop and J. D. Clark, eds. Chicago: University of Chicago Press. pp. 191–220.

Merrick, H. V., J. de Heinzelin, P. Haesaerts and F. C. Howell
1973 Archaeological Occurrences of Early Pleistocene Age from the Shungura Formation, Lower Omo Valley, Ethiopia. Nature 242:272–75.

Napier, J. R.
1962 Fossil Hand Bones from Olduvai Gorge. Nature 196:409–411.

Oakley, K. P.
1955 Fire as Palaeolithic Tool and Weapon. Proc. Prehist. Soc. XXI, pp. 36–48.

Partridge, T. C.
1973 Geömorphological Dating of Cave Openings at Makapansgat, Sterkfontein, Swartkrans and Taung. Nature 246:75.

Patterson, B., A. K. Behrensmeyer and W. D. Sill
1970 Geology and Fauna of a new Pliocene Locality in North-Western Kenya. Nature 226:918.

Reynolds, V. and F.
1965 Chimpanzees of the Budongo Forest. *In* Primate Behavior, I. DeVore, ed. New York: Holt. pp. 368–424.

Robinson, J. T.
1953 *Telanthropus* and its Phylogenetic Significance. Amer. Journ. Physical Anthrop. 11:445–501.

1959 A Bone Implement from Sterkfontein. Nature 184:583–85.

1961 The Australopithecines and Their Bearing on the Origins of Man and of Stone Tool-Making. S. Afr. Journ. Sci. 57:3–13.

Robinson, J. T. and R. J. Mason
1962 Australopithecines and Artifacts at Sterkfontein. S. Afr. Archaeol. Bull. 17:87–125.

Sarich, V. M.
1968 The Origin of the Hominids: An Immunological Approach. *In* Perspectives on Human Evolution I. S. L. Washburn and P. C. Jay, eds. New York: Holt. pp. 94–121.

Sarich, V. M. and A. C. Wilson
1967 Immunological time scale for hominid evolution. Science 158:1200.

Simons, E. L.
1961 The Phyletic Position of *Ramapithecus.* New Haven: Postilla. 57:1–9.

1963 Some Fallacies in the Study of Hominid Phylogeny. Science 141:879–889.

1965 New Fossil Apes from Egypt and the Initial Differentiation of the Hominoidea. Nature 205:135–39.

1969 Late Miocene Hominid from Fort Ternan, Kenya. Nature 221:448–451.

Söhnge, P. G., D. J. L. Visser and C. van Riet Lowe
1937 The Geology and Archaeology of the Vaal River Basin. Geol. Survey. Mem. 35. Pretoria, South Africa.

Tattersall, I.
1969 Ecology of North Indian *Ramapithecus.* Nature 221:451–52.

Tobias, P. V.
1965 *Australopithecus, Homo habilis,* tool-using and tool-making. South Afr. Archaeol. Bull. XX(80) IV:167–192.

1967 The Cranium and Maxillary Dentition of *Australopithecus (Zinjanthropus) boisei.* Cambridge: The University Press.

1973 Implications of the New Age Estimates of the Early South African Hominids. Nature 246:79–83.

Tobias, P. V. and G. H. R. von Koenigswald
1964 A Comparison Between the Olduvai Hominines and Those of Java and Some Implications for Hominid Phylogeny. Nature 204:515–18.

Van Zinderen Bakker, E. M.
1969 The Pleistocene Vegetation and Climate of the Basin. *In* Kalambo Falls Prehistoric Site, Vol. I. J. D. Clark, ed. pp. 57–84.

Washburn, S. L.
1967 Behaviour and the Origin of Man. Huxley Mem. Lecture, Proc. R. A. I., 1967. pp. 21–27.

1968 The Study of Human Evolution. Condon Lectures, Eugene, Oregon.

White, F.
1965 The savanna woodlands of the Zambezian and Sudanian domains: An ecological and phytogeographical comparison. *Webbia,* XIX:651–681.

Wilson, A. C. and V. M. Sarich
1969 A Molecular Time Scale for Human Evolution. Proc. Nat. Acad. Sci. (U.S.A.) Vol. 63, pp. 1088–1093.

Woodburn, J.
1968 An Introduction to Hadza Ecology. *In* Man the Hunter. R. B. Lee and I. DeVore, eds. Chicago: Aldine. pp. 49–55.

Wright, R. V. S.
1972 Imitative Learning of a Flaked Stone Technology—The Case of an Orangutan. Mankind 8:296–306.

Louis Leakey, *"Zinjanthropus,"* and Phillip Tobias at the Centre for Prehistory, Nairobi, August 1966. (Photo: Shirley Coryndon.)

Phillip V. Tobias:

White African:
An Appreciation and Some Personal Memories of Louis Leakey*

Louis Leakey was a pioneer in the search for human origins. In addition to being a scientist he was an extraordinary person. In this biographic essay, Phillip Tobias provides an introduction to Leakey's contribution to research, as well as a glimpse of what it was like to be a close colleague and a friend.

Louis Seymour Bazett Leakey was born at Kabete near Nairobi in Kenya on August 7, 1903, and he died of a heart attack in London on October 1, 1972. His passing ends a quite remarkable career in which nearly half a century (of a lifetime of only 69 years) was devoted to contributions to paleontology, archaeology, and anthropology. It is undoubtedly a fact

*The original version of this biographical sketch and appreciation was presented as The L. S. B. Leakey Memorial Lecture in the Great Hall of the University of the Witwatersrand on August 3, 1973 under the combined auspices of the University, the L. S. B. Leakey Foundation (California), the Royal Society of South Africa, the Institute for the Study of Man in Africa, and the South African Archaeological Society.

that no single person has done more to unravel the story of man's past in Africa than the late Dr. Leakey. And no one man has done as much to spread about the world something of the glamour, the curiosity, the excitement of man's search for his own past—and the inevitability, the irresistibility of Africa's claim to have been the cradle of man. It is therefore wholly fitting that by this volume, we should do honor to this man, one of the greatest to have sprung from Africa's soil in recent times.

For the same reasons and, additionally, because of my deeply cherished fifteen years' association with him, I have agreed to present this appreciation of Dr. Leakey, but I do so with a heavy heart and a sense of desolation for the void he leaves. My personal sense of loss is great beyond words, for I have lost not only a fount of inspiration but, more importantly, a friend.

BIOGRAPHICAL SKETCH

Louis (pronounced Lewis) was a son of Kenya. His parents were missionaries who in 1901 left England to set up a station for the Church Missionary Society. Even earlier than that in 1891, Louis's mother, Miss Bazett, and two of her sisters had responded to a call from the Church Missionary Society to go out to East Africa to carry on evangelical tasks among the people there. After three years of work among the Moslem women at the coast, she became ill and was invalided back to England; she was told she must never again venture to tropical Africa. In 1899 she married Harry Leakey who was then a curate in a London parish, and in December 1901, after the birth of Louis's two sisters, the Reverend Harry Leakey sailed for Mombasa. From there he travelled to Kabete where he had been appointed to take over the newly established mission station. Mrs. Leakey, the two girls and a governess, Miss Oakes, joined him a few months later.

It was in the family's mud-and-thatch dwelling that their third child was born to the Leakeys on August 7, 1903—a premature boy who was christened Louis Seymour Bazett Leakey. So Louis was nurtured in an environment surrounded by Kikuyu neighbors with whom the family was building a close association and whose language they were all rapidly mastering. The fourth child, Douglas, was born in 1907. Early tutoring came from his father and Miss Oakes, but a great deal came, too, from his Kikuyu playmates and mission adherents—and equally from an Ndorobo hunter who taught him to track patiently and to stalk wild animals. He learned to make traps using the bark of shrubs for cord. Many other items of bushlore and veldcraft were acquired in this way.

The family spent a time on furlough back in England and Louis attended Gorse Cliff school for just over a year. In May 1913, the family returned to Kabete and Miss B. A. Bull, a London graduate, came back with them to continue their coaching.

The outbreak of World War I prevented eleven-year-old Louis from resuming his schooling in England. He later mused, in his early autobiography, *White African,* "If the Great War had not broken out, I should in the normal course of events have gone back to England to a public school when I was thirteen. I should presumably have passed through school in the usual way, and become a typical product of the public school system, and this book would never have been written."

The title of his autobiography comes from something said of Louis by Kikuyu Chief Koinange, "We call him the black man with a white face, because he is more of an African than a European, and we regard him as one of ourselves."

Said Leakey years later: "I was very flattered when I heard of [Chief Koinange's] remark for I have always considered myself more of a Kikuyu than an Englishman in many ways. I still often think in Kikuyu, dream in Kikuyu; and if my English is not all that it should be in the narrative which follows, my excuse must be that I would have preferred to write it in the Kikuyu language."

For six formative years, from 1913 to 1919, Louis's education was continued informally at Kabete. These years developed his passion for natural history and especially for ornithology—though he hoped eventually to become a missionary.

At the end of 1915, twelve-year-old Louis received as a Christmas present from a cousin in England, a book called *Days Before History* by H. N. Hall. This was a simple account of the later Stone Age people of Britain—and immediately it sparked a new enthusiasm in young Louis. Soon he had found that the area around his home was teeming with evidences of the African Stone Age.

In his early teens he amassed a large collection of artifacts, especially the shiny black ones made from natural volcanic glass or obsidian. Along with the stone implements, he collected animal bones and soon had the nucleus of a fine private museum. He was greatly encouraged and helped by Mr. Arthur Loveridge, who was the Curator of the first Kenyan Museum—he taught Louis and his sisters how to skin lizards and mammals and to "blow" eggs. Although it meant a two-hour bicycle ride to Nairobi, they would deliver many of their collected trophies to the Museum to which Louis—and his son Richard after him—would one day make such momentous contributions.

After the war, Leakey was able to return to more formal education in England, at Weymouth College. In 1922 he gained entry into St John's College, Cambridge. For Part I of his Tripos he decided to take Modern Languages. He was proficient in French but it came as something of a shock to that proudly conservative institution, when, for his second language, he chose Kikuyu! For this purpose, he was placed under the care of Mr. W. A. Crabtree of St Catherine's College. But Mr. Crabtree knew

Luganda, a somewhat similar East African Bantu language, so Leakey was instructed to instruct his Instructor in the Kikuyu language. And this he duly and solemnly did, teaching Kikuyu to Mr. Crabtree who was subsequently to set most of Louis's examinations.

The delights of rugby football dawned upon him about this time but they provided him with rather a rude awakening. During a match in the October term of 1923, in which Louis played for his College at wing, he was, as he put it, ". . . playing the game of my life, when somebody accidentally kicked me on the head, and I had to be carried off the field suffering from concussion. As soon as I felt better I foolishly insisted on returning to the game, and received a second blow which forced me to retire altogether." This twofold head injury finds a curious echo in the twofold intracranial trauma he sustained, over forty years after, in his late sixties.

As a result, he was advised to take a year's leave and to "convalesce in the fresh air." So it happened that in 1924, he assisted the Canadian paleontologist, W. E. Cutler, as a member of the British Museum East African Expedition to collect dinosaurs in Tanganyika Territory. The expedition gave Leakey the chance to learn Ki Swahili (the lingua franca of much of East Africa) and to gain valuable practical experience in handling fossils, not to speak of employing and managing porters, building grass houses, cutting trails and arranging for supplies. Leakey's account of the hardships endured on that expedition makes fascinating, even hair-raising, reading. The constant and lifelong battle against time is evident, even on this first expedition. It is strikingly illustrated in his understated account of his 269-mile forced march down the coast from Lindi to Dar-es-Salaam to make sure of not missing his boat back to England—he arrived with a few hours to spare. The pressing restlessness of his spirit strained ever with the great enemy, time. The glimmerings of what was to become a life pattern are there for all to read in the account of this first expedition to collect reptiles in Tanganyika.

Leakey's recovery from the hard knocks of the rugby field seemed complete when he returned to Cambridge to take a First in Languages and then, in May 1926, to gain a First in Archaeology and Anthropology in Part II of the Tripos.

Subsequently he led East African Archaeological Research Expeditions in 1926–27, 1928–29, 1931–32 and 1934–35. These ventures marked the beginning of a lifetime of work in a vast geographical zone which till then had been virtually unknown in the fields of prehistory and human evolution. They laid the foundations and established a framework for all that has later been learned of the archaeological and palaeontological sequence in East Africa.

In 1929 on his second expedition, he drove from Nairobi to Johannesburg, visiting numbers of prehistoric sites on the way. He attended the joint meeting of the British and South African Associations for the Advancement of Science. Thus it came about that his first "Outline of the Stone Age in Kenya" was published in the *South African Journal of Science* (1929)—this was only his fourth scientific publication.

On the road south, Leakey visited Broken Hill (now Kabwe), the mining town where, eight years earlier, a most important and very complete fossil human cranium had been found. This was the famous *Homo rhodesiensis* or, as we prefer to call this form of man today, *Homo sapiens rhodesiensis*. Leakey was fortunate enough to find that the Swiss miner, T. Zwigelaar, who had actually discovered it, was still employed at the mine—from him, Leakey reported obtaining some very interesting information. He also recovered two cases full of fossil bones from the old cave deposits which had been dumped in the vicinity by the present miners. On this same historical trip to the south, Leakey collected stone implements in the old river gravels below the Victoria Falls; in caves in the Matopo Hills south of Bulawayo where the late A. L. Armstrong was excavating; at Hope Fountain where the Rev. Neville Jones had done much archaeological work; and he visited Zimbabwe where Miss G. Caton-Thompson was excavating on behalf of the British Association. Afterwards he drove back to East Africa to show parties of the visiting scientists (including Professors Fleure, J. L. Myres, Balfour, Cox and P. G. H. Boswell) over his excavations in Gamble's Cave.

This cave is on a farm that formerly belonged to Mr. Gamble, and its name has been given to the Gamblian Pluvial or stratigraphic unit. It lies just south of the equator some nineteen kilometers from Lake Nakuru and a few kilometers from another important archaeological site, 'Nderit Drift.

In his excavations in the cave, Leakey had revealed a well-stratified deposit, eight meters thick, containing large numbers of stone implements and animal bones, human skeletal remains, pottery, charcoal and layers of wind-blown sand. The oldest remains in the cave are now known to be about 10,000 years old. There are parts of five human skeletons which had been deliberately buried some four meters from the surface—they were in a highly contracted position and the bodies had been treated with red ochre, thought to represent blood. They were tall people with long, narrow skulls, prominent chins and noses and they did not show negroid features. These Capsian people are considered to represent Mediterranean Caucasoids, and they have been called "proto-Hamites" as they resembled the present-day inhabitants of North and Northeast Africa.

From 1929 to 1934 Louis Leakey held a Fellowship at St John's College, Cambridge, and in 1934 he was Jane Ellen Harrison Memorial Lecturer there. He was a Leverhulme Research Fellow from 1933 to 1935. In

February 1936 he was Munro Lecturer at Edinburgh University—the 10 lectures he delivered formed the substance of his book, *Stone Age Africa* (1936).

Later, he was the Herbert Spencer Lecturer at the University of Oxford (February 10, 1961) and the Thomas Huxley Lecturer at the University of Birmingham (March 3, 1961). These two lectures were published in book form under the title, *The Progress and Evolution of Man in Africa* (1961). Subsequently, Leakey was the Regents' Lecturer at the University of California (1963), Silliman Lecturer at Yale University (1963–64), the George R. Miller Professor at the University of Illinois at Urbana (1965), Andrew R. White Professor-at-Large of Cornell University (1968), and Honorary Professor of Anatomy and Histology, Nairobi University College (1969).

His first marriage was to Miss Henrietta Wilfrida Avern (1928), known as Frida, by whom he had two children, Priscilla and Colin. Priscilla now lives in England. I met her in the sixties at a Royal Society Conversazione at Burlington House in London, where she helped Leakey, Shirley Coryndon and myself arrange an exhibit on the Olduvai fossils. Colin Leakey was until recently on the staff of Makerere University in Kampala—during 1972, Leakey managed to ensure the departure of Colin and his family from Uganda, only a short time before his own final, fateful journey to London. This first marriage was dissolved.

After the 1934–35 expedition, Louis married Mary Douglas Nicol, an archaeologist in her own right and his partner in life and work until his death. This remarkable lady was an indispensable member of a most unusual husband-and-wife team. Not the least of Mary's contributions was to give birth to three more sons—Jonathan, who was Curator of the Nairobi Snake Park while still in his teens, the discoverer at Olduvai Gorge of the type specimen of *Homo habilis,* and is today a snake farmer at Lake Baringa in Central Kenya; Richard, on whose shoulders the mantle of Louis has most securely descended and who, though barely into his thirties, has scientific control of one of Africa's richest fossil hominid areas, East Rudolf; and Philip who runs safaris and has a safari lodge in southern Kenya.

Throughout his life Leakey was wedded to his native country. Fluently conversant with Ki Swahili and Kikuyu, he identified himself with Africa and its peoples. Following the Independence of the Republic of Kenya in 1963 he became a Kenya citizen and was always a staunch supporter of the fledgling state.

Brought up in Kikuyu territory, he worked tirelessly for African welfare, organizing and leading fund-raising campaigns and striving for better conditions of life for underprivileged Kenyans. He served as a young man on a government committee which reported on Kikuyu Land Tenure (1929). His interest in this tribe was directed to a detailed investigation of Kikuyu customs which he undertook for the Rhodes Trustees from 1937 to 1939. Although the thousand-page report he wrote on this tribe has not been published, it is gratifying to know that it is at present being edited by Dr. Mary Leakey for posthumous publication. His intimate knowledge of the ways of the Kikuyu tribe led the Kenyan government to enlist the aid of Leakey during the years of the Emergency and the Mau Mau Movement. These experiences were reflected in his books, *Mau Mau and the Kikuyu* (1952) and *Defeating Mau Mau* (1954), as well as *First Lessons in Kikuyu* (1959) and *Kenya Contrasts and Problems* (1966).

During World War II, Leakey was Officer-in-charge of Special Branch 6 of the Nairobi Criminal Investigation Department. He took up calligraphy and soon became an expert on handwriting. Leakey was a Founder Trustee of the Kenya National Parks. He was a Council Member, Trustee and Executive Member of the East African Wildlife Society, and also served as a Council Member of the East African and Uganda Natural History Society. In 1931 he initiated the first of a series of annual scientific meetings of the Natural History Society at Nairobi. A lover of animals and pets, he was a dog judge and became President of the East African Kennel Club, while his wife Mary has become a Dalmation breeder of international repute.

From 1941 to 1961 Leakey served as Honorary Curator, and then as full-time Curator, of the Coryndon Memorial Museum, Nairobi (now the National Museums of Kenya). During this time, he initiated and acted as Secretary-General of the Pan-African Congresses on Prehistory and Pleistocene Studies. The first of these four yearly meetings was held at Nairobi in 1947 and was attended by delegates from all over the continent and abroad. The participants had the opportunity of seeing some of the important archaeological sites which the work of Leakey, his wife and colleagues had made famous.

There was Olorgesailie, an incredibly rich area of silts and old sand surfaces—no fewer than ten of them—and on each land surface the Leakeys located one or more camp sites of Stone Age man with hundreds of discarded stone implements and the fossilized bones of the animals which Stone Age man killed and ate.

There was Kariandusi River, an Acheulean factory site which had been found on Leakey's second expedition by J. Solomon and Elizabeth Kitson.

There was the Naivasha Railway Rock Shelter which was discovered to be a site of Stone Age habitation by A. J. Poppy and Mary Leakey in 1940—and which she had to excavate under pressure as the site was about to be destroyed by a realignment of the railway. This site contained implements up to 7500 years old and with them a tall, big-brained, narrow-headed man—like the proto-Hamitic skeletons from Gamble's Cave.

There was Lion Hill Cave near Lake Nakuru with its Elmenteitan Culture, like the Mesolithic industries from Gamble's Cave.

There was Little Gilgil River site with its beautiful obsidian implements of the same general upper Palaeolithic culture as so many other of the Rift Valley sites, like 'Nderit Drift and Naivasha Rock Shelter.

The participants in that 1947 Pan-African Congress had the chance to appreciate also the on-site field museums which the Leakeys had initiated at a number of East African localities.

CENTRE FOR PREHISTORY AND PALEONTOLOGY

In 1962 Leakey set up the Centre for Prehistory and Palaeontology under the Trustees of the Kenya National Museums and he was its Director until 1972. The Centre was formed "to meet a crisis—the rapid growth of collections of fossils and artifacts resulting from intensified research at the Olduvai Gorge in Tanzania and Fort Ternan in Kenya. The National Museum had neither the space nor the funds to accommodate the situation. . . ."[1] In the short space of ten years, the Centre, housed in a number of old buildings and temporary structures in the grounds of the National Museum, became a base for prehistorians and paleoanthropologists from many parts and one of the most important repositories of original hominid fossils in the world. Here have visited and studied in recent years Clark Howell, Glynn Isaac and Richard Hay of Berkeley, Vincent Maglio of Princeton, Adrienne Zihlman of Santa Cruz, David Pilbeam of Yale, Ralph Holloway of Columbia, Larry Robins and Milford Wolpoff of Michigan, Philip Rightmire of Binghampton, Bryan Patterson of Harvard, Michael Day of St Thomas's, London, Bill Bishop of Bedford College, London, Alan Gentry, Theya Molleson and Shirley Savage (formerly) of the British Museum of Natural History, Bernard Wood of Middlesex Hospital, London, Richard Hooijer of Leiden, Laszlo Vertes of Budapest, Yves Coppens and R. Lavocat of Paris, Alan Walker of Nairobi and now of Harvard, Eitan Tchernov of Jerusalem, Emiliano de Aguirre of Madrid, Holgar Preuschoft of Tübingen, Basil Cooke of Dalhousie and many others.

Here are housed hominid fossils from Olduvai (temporarily) and Peninj in Tanzania, from Baringo, Chemeron, Chesowanja, Kanapoi, Lothagam, East Rudolf, Rusinga, Songhor and Fort Ternan in Kenya. Thus, within his lifetime, Louis Leakey saw his Centre grow until it was bursting at its seams. It is a very suitable tribute that one of his sons, Richard, Administrative Director of the National Museum of Kenya, now proposes to establish in Nairobi, as successor to his father's Centre, a

[1]Excerpt from Brochure on the proposed "Louis Leakey Memorial Institute for African Prehistory" (Nairobi 1972).

Louis Leakey Memorial Institute for African Prehistory. This institution will depend upon an international financial and academic effort and its objects will be broadly:

> To foster human understanding through continued research into the origin and development of man. To contribute through its activities and relationships with other bodies to the growth of science in Africa.

GEO-ARCHAEOLOGICAL CONTRIBUTIONS

On his first East African archaeological expedition in 1926–27, Leakey excavated a number of Rift Valley sites, especially near Lakes Nakuru and Naivasha. These included the Nakuru and Makalia Burial sites and Gamble's Cave. He found old raised lake levels, well above the present-day levels of the neighboring lakes. He could date them by the stone implements contained in the gravels deposited on these high-level terraces.

His mind searched for causes—what could have made the East African lakes formerly so much more extensive and higher than they are today? This set him thinking of past climatic changes as the background to man's evolution in East Africa, in much the same way as the successive advances and retreats of glaciations had provided the Ice Age setting for man's development in Europe.

He was stimulated in such thinking by the independent observations and inferences of E. J. Wayland of Uganda and E. Nilsson of Sweden. Leakey subsequently developed the notion of a succession of East African wetter and drier periods (pluvials and interpluvials) in the Pleistocene. In the ultimate full flowering of this concept, he recognized four pluvial periods dubbed the Kageran, Kamasian, Kanjeran and Gamblian. This scheme of climatic alternations was severely criticized by Cooke, Flint, Pickering and others. Indeed, not all the details of Leakey's proposed climatic sequence, especially those affecting the earlier phases, have stood the test of time. Nonetheless, the notion of *some* past climatic changes in the African Pleistocene has survived. In any event, Leakey's hypothesis proved heuristically valuable, for it provoked intensive researches and the devising of newer techniques of geochronometric and paleoecological analysis. For Leakey was perspicacious and imaginative. He was the kind of scientist who erects hypotheses far ahead, as signposts for the common run of investigators, whose tedious lot it is to fill in the details, collect and interpret the facts, and eventually confirm or overthrow the hypotheses.

Just about the time that John Goodwin and C. ("Peter") van Riet Lowe were classifying the South African archaeological sequence (1929), Louis Leakey was doing the same for East Africa. Based on his excavations at Gibberish Cave, Makalia and Nakuru Burial Sites, Gamble's Cave, Lion Hill Cave and Kariandusi, between 1926 and 1929, Leakey presented "An Outline of the Stone Age in Kenya" at the 1929 meetings in Johannesburg. Subsequently it was expanded and amplified into his first book, *The Stone Age Culture of Kenya Colony* (1931), followed not long after by *The Stone Age Races of Kenya* (1935). He classified the many stone industries he had excavated into the handaxe culture (afterwards called the African Acheulean), a flake culture, a blade-and-burin culture, and mesolithic and neolithic industries. Later excavations at Kanjera, Olorgesailie, Olduvai and other sites were to add abundant new evidence on the African Acheulean and to uncover a yet earlier culture, the Oldowan.

OLDUVAI GORGE

This great treasury of fossils and implements had been discovered by the German entomologist, Kattwinkel, in 1911. Before World War I, Hans Reck of Berlin collected fossilized animal bones there; as these included extinct species and genera, considerable interest was aroused in Germany. The excitement was heightened by Reck's discovery of a human skeleton (Olduvai hominid I). At first this was thought to belong to the Middle Pleistocene, but later it was shown to be a relatively recent burial into older deposits. The outbreak of war and the subsequent transfer of Tanganyika to Britain as a mandated territory put to rest any further plans by Reck and his colleagues to explore and excavate the Gorge. After the war while still an undergraduate at Cambridge, Leakey visited Berlin in the course of a study he was making on African bows and arrows. There he met Reck (1925) and we can be sure that "Oldoway" was among the subjects discussed. In 1927, Leakey went to Mollison's laboratory in Munich to see the controversial Oldoway human skeleton and to compare it with some of the skeletons he had found on his first expedition. This resulted in a note to the weekly scientific journal, *Nature,* by Leakey (1928); it was the second of a lifetime series of some 35 notes, longer articles, and letters to this famous journal alone!

On his third East African Archaeological Expedition (1931–32), Leakey was accompanied by Professor Reck, as well as by A. T. Hopwood, D. MacInnes, and Bunny (later Sir Vivian) Fuchs. Reck had found only faunal remains previously; stone culture had eluded him. Leakey won a small wager with Reck when, within twenty-four hours of arriving at Olduvai, one of the Kikuyu aides found a fine handaxe. Soon after, one was found in situ—the forerunner of thousands which were to be excavated in the Gorge; so many, in fact, that in his 1951 book, *Olduvai Gorge,* Leakey was able to recognize no fewer than eleven stages in the evolution of the Great Handaxe Culture.

The story of the excavation of numerous sites in the Gorge by Louis Leakey, his wife, Dr. Mary Leakey, and their sons and other helpers, including their sharp-sighted African aides, is now so well known as not to bear repeating in detail. Indeed, the site has become so completely identified with Louis and his family that Peter Suffolk and Mr. Punch memorialized the link in a delightful sonnet:

When God at First—

When the first men were fashioned in the good Lord's forge,
He sent them, it seems, to the Olduvai Gorge,
There to be tested and kept an eye on
With the proto-lizard and proto-lion.
This-hyphen-pithecus and Homo That,
With the archaeo-elephant and palaeo-cat,
Lived there, and died, and were hidden away
Under layer on layer of African clay
Till countless millions of years should run
And Leakey discover them, one by one . . .
While, back in the heavenly forge, the Lord
Went back again to the drawing-board.
I sometimes wonder: suppose that I
Were digging out there, at Olduvai,
And I brought to light a significant bone
Of a kind I could positively call my own;
And under the bone, when I'd worked it free,
I found (let us say) an ignition key—
Should I declare it, as of course I ought,
Or should I just pocket it? Perish the thought!

(*Punch,* June 10, 1964, p. 866)

Suffice it to say that the beds exposed in the side walls of this great gorge span close on to two million years of human history. Within their layers are sandwiched implements of the earliest of human stone cultures—the Oldowan—and it increases through the layers in diversity and in the evident skill of its manufacturers. Then follow stages of the African Acheulean, the handaxe industries, and finally still later and more advanced cultures. Entombed within the same beds are the fossilized remains of animals, of which over 150 different species have been identified. Among them are at least three different species of hominids or members of the family of man. From the earlier part of the sequence, between 2.0 and about 1.5 million years B.P., have come two kinds of synchronic hominids. One is a very big-toothed, heavy-jawed and robustly built creature which, when Mary first discovered it in July 1959, was called by Louis, *Zinjanthropus* (man of East Africa, for which "Zinj" is an old name). It was an exciting discovery of an australopithecine, even bigger and heavier than those which R. Broom and J. T. Robinson had found at Kromdraai and Swartkrans in the Transvaal. Today, we call it *Australopithecus boisei.*

The second of the earlier kinds of hominid is the one we have called *Homo habilis*—a bigger-brained, almost perfect miniature of a man, who we now know lived in Africa from just less than three million years ago to something over 1 million years ago. He is represented by a number of specimens from Olduvai, some from East Rudolf in Northern Kenya and from Omo in Ethiopia, and probably, too, from Swartkrans and from the upper reaches of the Sterkfontein cave deposit.

The third species of hominid appears on the scene at Olduvai somewhat later—*Homo erectus.* The close study of these remains, with their associated cultural and faunal relics, in their geological and time setting, is throwing much light on our knowledge of human evolution between about 2 and 0.5 million years B.P. Louis Leakey's 1965 volume, *Olduvai Gorge (1951–1961),* inaugurated a new series of monographs to chronicle and interpret the spate of new discoveries.

The Leakeys' work at Olduvai has for a number of years been in the very competent hands of Mary Leakey, while Louis gave his attention to Fort Ternan in Kenya and to the arduous task of raising funds abroad to keep the Centre going. Thus, the exploration and systematic excavation of the Gorge is continuing. It remains a showpiece among East African sites. The yields since work was intensified in 1959–1960 have testified again to the foresight of Louis and to what has so often been called "Leakey's luck."

OTHER HOMINOID AND HOMINID DISCOVERIES

A much earlier chapter in the story of higher primates was written through the Miocene discoveries of Leakey. A site at Koru in western Kenya, not far from the Kavirondo Gulf of Lake Victoria, had yielded in 1923 the first known fossil apes from East Africa. On Rusinga Island in Lake Victoria Nyanza, Leakey first found a Miocene fossil ape on his 1931 expedition. It was about 20 million years old and the name originally given it, *Proconsul,* was a kind of scientific joke—for Consul was the name of a famous performing chimpanzee of those days, and Proconsul, then, would be his ancestor! From 1931 onwards, Leakey and his colleagues made a series of finds of Miocene pongids on Rusinga, at Koru, Donghor, Fort Ternan, and elsewhere. The specimens include members of several species of *Dryopithecus* (as Proconsul came to be called), an extinct genus of pongids or apes. Another early hominoid was found by Leakey at Fort Ternan on the farm of Mr. Fred Wicker, and Leakey called it *Kenyapithecus wickeri,* but it has been suggested that it is not different generically from *Ramapithecus,* known from India. The work of Leakey, amplified by that of E. Simons and D. Pilbeam, suggested that *Ramapithecus* should be regarded as an early genus of the hominid family—*Ramapithecus,* it was thought, might well be the Miocene ancestor of *Australopithecus!* Newer work has cast some doubt on the hominid status of *Ramapithecus,* but more fossils of this genus are needed before the issue can be solved one way or the other.

Among Leakey's earliest discoveries of fossil hominids were the crania of Kanjera (1932) and the mandible of Kanam (1932). These remains still excite controversy. This is partly because Leakey shocked the world by claiming that they were respectively Middle and Lower Pleistocene members of the species, *Homo sapiens,* or at least of a form of man very close to *H. sapiens.* They were found on the southern shore of the Gulf of Kavirondo, which points eastwards from the northeast corner of Lake Victoria. To get there overland was virtually impossible in those days so Leakey kept a boat moored at Kisumu on the southern side of the Gulf. The boat's name was *The Miocene Lady.* In 1957, after restudying the Kanam jaw in London, I wrote to Dr. Leakey to say I did not wish to publish anything on the Kanam mandible until I had seen the site itself. Would he take me there? To my joy, he replied that he had completed arrangements to take the Yale geologist, Professor Richard Flint, on an extended tour of all of the more important of his East African sites, including Kanam and Kanjera: I was more than welcome to join the party.

The story of that incredible excursion to over twenty-five sites in Kenya and Tanganyika remains to be written—it was a study in personalities as well as in prehistory and it was not without incident. When we first arrived at Olduvai, we nearly lost Professor Flint—as this big man walked, all unknowing, towards the only large tree in the vicinity, to answer a call of nature (we were a mixed party); but apart from its potential toilet usages, the large tree was the shady resting place of a lioness with two cubs. It was Leakey's shout that alerted Professor Flint when the lioness was already growling and breaking cover and saved him from adding his bones to the millions already in the Olduvai Gorge. Of that unique journey, more will sometime be said.

So it came about that, one day in June 1957, Professor and Mrs. Flint, Dr. and Mrs. Bill Bishop, Leakey, his African aide and I boarded *The Miocene Lady.* We set off for our first anchorage, across the Gulf, just off Kanam, Rawe and Kanjera, in the shade of the extinct volcano, Mt. Homa. It was a wild night—every few hours, our little boat was lashed by tropical storms which set us whirling crazily on our anchor chain—and we awoke to the sight of Leakey vigorously and actively trying to secure our craft, Leakey in his eternal off-white boiler suit, his silver mane of hair streaming out in the equatorial lightning, Leakey yelling, "Batten down the hatches" and similar, strange-sounding, nautical cries rising eerily above the growling thunder and hissing rain. The next time we awoke, it was the recipe as before, only this time the new storm was spinning us in the opposite—anti-clockwise—direction. There is another memory to add to that unforgettable snapshot—Leakey frying freshly caught lake fish in the kitchenette of the little craft, and dishing it up, sizzling and delicious, into our breakfast plates. We went ashore in a little rowing boat, and accompanied by some local Luo fishermen, we inspected the sites. Leakey showed us the fossil beds and the approximate positions where the skulls of Kanjera and the jaw of Kanam had been found. Later that day, Flint and I celebrated his release from the jaws of death at Olduvai, with a nude swim around *The Miocene Lady,* who discreetly kept her engine purring to shoo away any of the crocodiles with which the lake teems. Leakey preferred to wash himself on the shore, amid the bilharzia-infested reeds: he reasoned that he had had schistosomiasis so many times before, one more attack would do little extra harm.

The Kanam jaw turned out to be more primitive than had earlier been supposed, having features reminiscent of *H. erectus,* and it also appeared that it was not as old as the fauna among which it was found. The Kanjera crania, with their surprisingly "modern" brows, for long remained an enigma; only recently has K. P. Oakley (1974) disproved Leakey's claim that they were contemporary with the associated Middle Pleistocene fauna. Leakey accepted them as evidence for the early appearance of a sapient kind of man, a view he held until the end. Indeed, for Leakey, the brutish-looking *Homo erectus* was a sideline, a cul-de-sac of human evolution, off the beaten track leading to modern man. So too did he think of Neanderthal Man, with his grotesque, flat-topped, beetle-browed and enormous skull, as a caricature of a human being, not an ancestor of modern man. Leakey would have been the first to agree with Arthur Koestler when he wrote, in *Darkness at Noon:*

> There must have been laughter amidst the apes when the Neanderthaler came along. The highly civilized apes swung gracefully from bough to bough; the Neanderthaler was uncouth and bound to earth. The apes, saturated and peaceful, lived in sophisticated playfulness; or caught fleas in philosophic contemplation; the Neanderthaler trampled gloomily through the world, banging around with clubs. The apes looked down on him amusedly from their tree tops and threw nuts at him. Sometimes horror seized them; they ate fruits and tender plants with delicate refinement; the Neanderthaler devoured raw meat, he slaughtered animals and his fellows . . . . He transgressed against every law and tradition of the jungle. He was uncouth, cruel, without animal dignity—from the jungle point of view of the highly civilized apes, [he was] a barbaric relapse of history.

Yes, it sometimes seems that for Leakey, sapient man had always, or almost always, been and looked like modern man—that is why he welcomed the apparently modern-looking Kanam and Kanjera men as direct ancestors; and the miniaturized though modern-looking *Homo habilis* as a direct ancestor.

Indeed, very shortly before he left for London on his last journey, late in September 1972, he was shown the remarkable big-brained cranium (KNM-ER-1470) found by his son, Richard, east of Lake Rudolf. It was weathering out of a deposit which was dated to over 2.6 million years by the potassium-argon technique, and subsequently set at 2.9 million years by paleomagnetism determination! A cranial capacity estimate yielded a value of about 800 cc. Louis was delighted, for it seemed to him that Richard's discovery provided further evidence for the very early emergence of the genus, *Homo.* It was an extremely happy turn of the wheel that he saw this important new find shortly before he died. It was a smiling destiny that vouchsafed him the satisfaction of knowing that, not only through Mary's continuing work at Olduvai, but through Richard's activities east of Lake Rudolf, the unexampled family tradition of discovery was going forward.

LEAKEY THE MAN

Three abiding impressions of Dr. Leakey are of his singular energy, his immense enthusiasm, and his vision. The energy is reflected not only in the tally of his excavations and discoveries, or in the 20 books and over 150 articles he wrote. In everything he undertook that restless energy was evident. On safari, he walked hurriedly and worked indefatigably and single-mindedly. Many a younger man strove to keep up with him on an excursion, as he pressed on from site to site. He seemed to live in the intensive mood, a man of immense power and fantastic reserves.

He sparkled with an effervescent enthusiasm with which he was able to infect others. As a result, he inspired numbers of people to take an interest in man's past, to support his efforts and to join in the work. The list of projects he initiated, triggered-off or catalyzed is lengthy. It includes the National Primate Research Centre at Tigoni near Nairobi; the gorilla studies of Jill Donisthorpe in Uganda and of Dian Fossey in Rwanda; and the chimpanzee work of Jane Goodall at the Gombe Stream Reserve in Tanzania. In her book, *In the Shadow of Man,* Goodall fully acknowledges the enormous debt she owed to Leakey in her chimpanzee researches. For Leakey believed the understanding of extinct animals depended on the study of living ones, their structure, functioning and behavior. He believed too that while the study of the apes was fascinating in its own right, it had much to teach us about ourselves. It was a gentle philosophy learned from gentle creatures which permeates Goodall's writings on the chimpanzees and which Leakey defended powerfully in the last year of his life, in that famous Leakey Foundation dialogue between him and Robert Ardrey on the Killer Ape doctrine.

Leakey would have laughed in his breathless, panting, body-heaving way at Art Buchwald's report from Washington of a new book shortly to appear on the market and expected to cause a sensation in scientific circles: it is titled *The Naked Man* and it was written (says Buchwald) by Frederick III, a chimpanzee attached to the Rockefeller Institute. Frederick's thesis is that the apes have evolved from man and that modern *Homo sapiens* shows eating habits and a sex life similar to those practiced by modern apes. This view, writes Buchwald, has evoked a horrified response from many apes. A chimpanzee named Treetop, with the National Institute of Mental Health, has written a paper denouncing Frederick III's thesis—for Treetop the instinct in men to destroy is so strong that it is slander to class them in any way with apes.

Energy and enthusiasm are good qualities but they are not enough. It was Leakey's farsightedness that gave direction. Often he revealed an almost visionary quality—and it was this that gave purpose and goal to the efforts of the family team and their associates. He had a mind that comprehended detail, on the one hand, yet could grasp the grand scheme of things on the other. His syntheses were not always correct, but then, in this discipline as in others, it is abidingly true that he who never makes a mistake never makes a major contribution. Imbued with this visionary quality we see him as Leakey the romantic, Leakey the imaginative—the kind of scientist that science needs, along with the commoner variety.

Versatility is another trait we should add to the list: and to few men has it been given to help the advance of so many branches of science. Leakey was capable of operating simultaneously on many fronts and of infusing a zest for these campaigns into a large team of colleagues, assistants, students and supporters.

He has sometimes been criticized for publicity seeking. Yet the nature of Leakey's finds and of his concepts was always news. And if he was quick to publicize his discoveries, it might well be argued that this was a necessary means of arousing interest and so of eliciting financial support for his researches. Always, throughout his life, the funding of his research was a problem; always, from his Cambridge days onward, he knew the difficulty of gaining backing for researches the practical applications of which are often not readily apparent. This was the main motive behind those exacting lecture tours of America which undoubtedly shortened his life but which were necessary for the raising of funds to keep the Centre going. If publicity by means of press, radio and television helped this objective, then this consideration perforce outweighed all others. Moreover, as the late Dr. Broom used to say, it pays in another sense to work with the newspapers: every time another important discovery from Sterkfontein, Kromdraai or Swartkrans was announced in the press and over the radio, Broom would receive letters and phone calls about other potential sites and related helpful information. Doubtless this aspect, too, proved advantageous to Leakey. It was a process greatly enhanced by his consummate skill in communication. Both in speech and in writing he was fluent, impressive, persuasive, and possessed of a facile and clear style.

His strong will, positive personality, often seemingly dogmatic manner, all brought him into conflict with some of his fellow scientists. Often it was no more than light-hearted raillery as when *Punch* published an article on the discovery at Oboyoboi Gorge by Dr. C. J. M. Crikey of "what may well prove to be the oldest crown cork known to man" (March 8, 1961); or when, at the Pan-African Congress at Livingstone in 1955, M. D. W. Jeffreys overheard Leakey muttering to Sonia Cole, "Far too many damned amateurs here," whereupon "Doc" Jeffreys sat down and wrote a quatrain, after the style of the doggerel about the two famous families of Boston. Dr. Jeffreys' version ran:

> When archaeological tyros stroll
> On the plains of the wild Semliki,
> Then Tyros talk only to Cole
> And Cole talks only to Leakey.

This is noted here for the first time by kind and nonmalicious permission of "Doc" Jeffreys.

Sometimes the criticisms were vitriolic, as in a vituperative and ill-informed article which appeared in *The New Scientist,* not three months after Leakey's death.

Of those who judged him by the headlines and so missed the substance of the man, I can only say, "Theirs was the loss." But for those willing to look below the surface dispassionately, to explore the essence of the man's thoughts and claims, and not rest content with their first-flush expression, there was frequently great wisdom, sometimes sheer genius, and time and again an inspired and inspiring element to be dissected from the tissue of words, concepts and hypotheses. Anyone who did not have the patience so to dissect was the poorer for the omission—but many were enriched.

Louis Leakey's enormous contributions did not remain unrecognized. Among the awards and honors he received were the Cuthbert Peek Prize of the Royal Geographical Society (1933), the Andreé Medal of the Swedish Geographical Society (1933), the Henry Stopes Memorial Medal of the Geologists' Association, London (1962), the Hubbard Medal of the National Geographic Society, Washington, D.C., jointly with Dr. Mary D. Leakey (1962), the Richard Hopper Day Memorial Medal of the Academy of Natural Sciences of Philadelphia (1964), the Medal of the Svenska Sallskapet fur Antropologi och Geographi (1963), the Royal Medal of the Royal Geographical Society of London (1964), the Viking Fund Medal of the Wenner-Gren Foundation for Anthropological Research (1961–65), the Haile Selassie Award (1968), the Welcome Medal of the Royal African Society (1968), the Science Medal of Academy for Biological Sciences, Italy (1968) and the Prestwich Medal of the Geological Society of London, jointly with Dr. Mary D. Leakey (1969).

In 1958, an honorary Doctorate of Science was conferred on him by the University of Oxford. Further honorary degrees received by him were the Doctorate of Law of the University of California (1963), the Doctorate of Science of the University of East Africa (1965), and the Doctorate of Laws of Guelph University (1969).

The L. S. B. Leakey Foundation was established in the United States in 1968, to advance his work and similar researches by others. The Leakey Memorial Institute in Nairobi has already been mentioned.

A PERSONAL NOTE:

The Impact of an Older Scientist on a Younger One

Both Dr. and Mrs. Leakey, for the past fifteen years of his life, meant so much in my own life, that I make no apology for adding a personal and subjective note on the turning point in my career which was marked by my association with them. It all started in 1955. On my regular visits from Cambridge University to the British Museum (Natural History), I took the opportunity to study the superlative collection of original hominid fossils at South Kensington. Among them was the Kanam mandible: on my first study of it, I realized that its morphology had been misinterpreted. The "chin," long considered to betoken its modernity, seemed to me to be simply the consequence of a pathological lesion on the posterior surface of the symphysis and left body. Instead of its being "modern" in morphology, it turned out to have a number of features associated with jaws referred to *Homo erectus.* Dr. Leakey and Dr. K. P. Oakley kindly allowed me to undertake a new study of the Kanam jaw. This led to a short report in *Nature,* and to a longer report which I presented at the Fourth Pan-African Congress on Prehistory at Léopoldville (now Kinshasa).

The date was August 1959: only a few weeks after Mary Leakey had made the important discovery at Olduvai of the very robust cranium of *Zinjanthropus* (or *Australopithecus boisei).* The discovery was announced at the Opening Session of the Congress, in English by Leakey and in French by Mortelmans. A day or two later, I was presiding at the Human Palaeontology section, when Leakey presented his paper on what he then called *Zinjanthropuus.* After his paper I gave a few comments on the momentous discovery—it was on this occasion that I referred to the remarkable dentition as the "enormous set of 'nutcrackers'" in the upper jaw; thus was born the slightly unfortunate name, Nutcracker Man, used subsequently by the *Illustrated London News* and other periodicals.

Leakey was pleased with what had been made of the Kanam jaw, and with the straightening of the record both about Boswell's visit to Kanam in 1934–35, and about his report on it. After a showing of the new Olduvai cranium to a group of paleoanthropologists at the Congress, he and Mary Leakey approached me and said that they wished me to undertake the definitive study of the cranium. I was completely overwhelmed.

This began a fifteen-year working partnership with the Leakeys. During that time my career took a major turn in the direction of paleoanthropology. Previously, my interest had centered principally on genetics (including cytogenetics—chromosomes) and on the study of the Kalahari Bushmen and other living peoples of Africa. It is true that, in 1945, I had, on the advice of the late Professor Peter van Riet Lowe, organized and led the first expedition that brought back fossil baboons from Makapansgat, and several successive expeditions in 1946, which paved the way for the finding in 1947 of the first *Australopithecus* fossil there. My interest in human evolution had remained high, and I had carried out excavations at Makapansgat Limeworks, the Cave of Hearths and Rainbow Cave, Taung, Mwulu's Cave, and in the Sterkfontein area. Nevertheless, I did not carry out any laboratoy researches on the South African australopithecine remains as such. The reason was that I did not wish to encroach on a field which I considered to be the preserve of Professor Dart, my chief and predecessor, as well as of Dr. Robert Broom and, later, his assistant, Dr. J. T. Robinson. So I had eschewed laboratory work on these fossils and concentrated instead on the physical anthropology and genetics of the living and the recently dead. Thus, it was Dr. and Mrs. Leakey's generous offer which, properly speaking, brought me into the field of human evolution. Not only that, but they subsequently invited me to describe all the hominid fossils which, since 1959, have been emerging from Olduvai Gorge (as well as such other specimens as the Peninj mandible, the Chemeron temporal and the Kaphthurin jaw). This made an enormous difference to my whole career. It can be readily understood that my personal sense of loss is very deep.

This would be incomplete if I did not refer to Louis's illnesses. The strain of those hectic lecture tours told on this large, hyperactive and over-heavy man. He had to do it, for the funds he raised helped to keep his researches and his family going. And how popular his lectures were, as thousands flocked to hear him in the largest auditoriums in the United States: he was worshipped by countless students, his fanmail was astonishing, the adulation he received was such as few scientists have been accorded.

In March 1967, on one such lecture tour, he collapsed in Chicago—but it was exhaustion, they found in the Augustana Hospital, not his heart. Early in 1970, he insisted on embarking on another lecture tour of America, but he collapsed on the flight from Nairobi to London; this time it was a heart-attack. He recovered. The next year, 1971, he was at

Malindi on the Kenya coast, when he was attacked by a swarm of angry bees. Trying to ward them off, he received over 800 stings in his right arm. He collapsed, hitting his head and twisting his hip, and for days was at death's door. Again, he recovered, and the residual weakness in his right arm was attributed to the combined effect of hundreds of bee sting toxins. Later that year, 1971, once more he set off on yet another lecture tour—on his doctor's instructions, he agreed to be accompanied by one of his faithful secretaries, Cara Phelips. In California she was horrified to see Leakey push his chair back from the table over the edge of the relatively narrow speaking platform: before she could reach him, he had fallen heavily from the platform to the floor, knocking his head in the process. Within days he became incoherent, and it was evident that he had suffered an intracranial hemorrhage. At operation this was confirmed—but to the surgeon's amazement, signs of an earlier partly healed hemorrhage were found, an episode which had gone quite undiagnosed. Now the story of the bee stings and the weakness fell grimly into place. It was now evident that when he had collapsed at the time of the bees' attack, he had sustained an intracranial hemorrhage. It was this that had caused the residual weakness, which for almost a year had affected the mobility of leg, arm and finger, and prevented his writing at all. Yet his enormous strength and his dogged will had pulled him through that undiagnosed subdural hemorrhage . . . . These were the two episodes which awoke a faint echo of the double concussion sustained forty years earlier on the playing fields of Cambridge.

The four collapses had taken their toll. In September 1972 he had the satisfaction of seeing his son Colin safely out of Uganda and of chuckling happily over Richard's new 1470 Man from East Rudolf. Then, late in September, he took off for London, on his way once more to the United States for a further lecture tour. On Sunday morning, October 1, 1972, he had a heart attack in London. He died in St. Stephen's Hospital the same day.

In far-off Philadelphia, a wonderful little lady in her seventies called Anna Kennedy Winner, whom I am proud to call my friend, started a letter to me on October 6th:

> Last night I went over to the Academy of Music here, expecting to hear a lecture by Dr. Leakey, sponsored by the Academy of Natural Sciences of Philadelphia. It was not until after I had arrived that I learnt that Dr. Leakey had died Sunday morning, in London, on his way here. A lecture about him and his work was given by Dr. F. Clark Howell. I was more sorrowful than shocked, because when I saw him last January here, as I wrote to you, I had noticed that he seemed to be far from well. But I thought I ought to write to you, since I do not know any of the Leakey family, to express my profound regret at his passing, and the loss to the world which that will mean.

PART ONE

Connections between Prehistory and Natural History

Introduction to Part One

In the past the study of human evolution was a fairly dry museum pursuit involving for the most part such activities as the measurement of skulls and the comparative study of tooth form. However, in recent years there has been a mounting realization that a full understanding of human origins involves a broad appreciation of the biology and behavior of man's living relatives plus a deep understanding of the natural history and ecology of the realms in which the formative stages of protohuman differentiation took place. There has also been a growing awareness of the need to study the processes whereby parts of a complex living system are converted into fossils buried in sediments. Once these processes are understood, we can more realistically reconstruct the vanished systems of which the fossils are fragmentary relicts.

It is amazing to reflect that as little as ten or fifteen years ago virtually nothing was known of the details of the daily lives of man's closest living relatives, the great apes. It now seems self-evident that such knowledge plays a crucial part in any balanced comprehension of human origins. Recognizing this even before primate studies had become especially

A young chimpanzee uses a grass stem as a tool for catching edible termites.
© 1963 National Geographic Society, photograph by Hugo van Lawick.

fashionable, Louis Leakey sought out young, imaginative persons with interests in animals and natural history and he helped them get into the field and observe the African apes. In particular, two of the field projects inspired by Leakey have become longterm research endeavors of tremendous importance. One of these is represented in this volume by an essay contributed by Jane Goodall, which deals with some of the features of her chimpanzee studies—both in regard to their own intrinsic interest and also in relation to their implications for our thinking on the subject of human origins (see also Goodall 1968, 1971; Fossey 1972, 1974). A forthcoming volume in this series will deal specifically with current research on the great apes *(Perspectives on Human Evolution* Volume 5, edited by D. Hamburg and J. Goodall, in prep.).

In recent years East Africa has been the focus of intensive studies of wildlife and of the field ecology of game animals. Although outside the scope of this volume, the results constitute an important comparative context for the enquiry into human origins. Two exmples of recent contributions of this kind from East Africa are Kruuk's and Schaller's monographs on the spotted hyena and the lion (Kruuk 1972; Schaller 1972). These works show something of how carnivores make a living and thereby help to throw into relief the kinds of opportunities and problems faced by early hominids (see also Schaller and Lowther 1969). Another set of relevant data—information on the food potential of various kinds of wild animal populations—is contained in a paper by Talbot (1966). Large numbers of important primary studies have been published in recent years in such periodicals as *East African Wildlife Journal, Ecological Monographs, Journal of Mammology, Journal of Ecology, African Wildlife, Mammalia, Behaviour, Journal of Wildlife Management, Zoologica Africana, Journal of Animal Ecology.* Since few summaries or syntheses have as yet appeared, readers with broad interests in the setting of human evolution must browse through these journals to gain appreciation of current research.

As has been said, to understand the record of evolution we also need to understand how the fossil record was formed. Research into the processes intervening between once-living mammal communities and the samples of fossil bone that we find eroding out of ancient sediments constitutes a new science that rejoices in the name "taphonomy." Among the pioneers of these kinds of investigations has been C. K. Brain of the Transvaal Museum, Pretoria, who has prepared a summary of important results which had previously been published in rather inaccessible periodicals. Brain's findings should immediately make it clear that deductions from features of bone assemblages should only be made in the context of broadly based comparative data on the characteristics of bone assemblages accumulated with and without the intervention of man. The deductions cannot safely be made from an armchair.

Other important taphonomic studies are now underway in East Africa by workers such as Andrew Hill (unpublished) and Kay Behrensmeyer (1975). When their studies are fully published, they can be expected to have considerable impact on paleoanthropology.

That the present is the key to the past is a truism that applies no less to human evolution than to the study of geology.

References Cited

Behrensmeyer, A. K.
1975 The taphonomy and paleoecology of Plio-Pleistocene vertebrate assemblages east of Lake Rudolf, Kenya. Bulletin of the Museum of Comparative Zoology 146:473–578.

Fossey, D.
1972 Vocalizations of the Mountain Gorilla *(Gorilla gorilla beringei).* Animal Behaviour 20:36–53.
1974 Observations on the Home Range of One Group of Mountain Gorillas *(Gorilla gorilla beringei).* Animal Behaviour 22:568–581.

Kruuk, H.
1972 The Spotted Hyena: A Study of Predation and Social Behavior. Chicago: University of Chicago Press.

Lawick-Goodall, J. van
1968 The Behaviour of Free-living Chimpanzees in the Gombe Stream Area. Animal Behaviour Monographs 1(3)161–311.
1971 In the Shadow of Man. Glasgow: Wm. Collins Sons and Co., Ltd.

Schaller, G. B.
1972 The Serengeti Lion—A Study of Predator Prey Relations. Chicago: University of Chicago Press.

Schaller, G. B., and G. R. Lowther
1969 The Relevance of Carnivore Behaviour to the Study of Early Hominids. South Western Journal of Anthropology 25:307–341.

Talbot, L. M.
1966 Wild Animals as a Source of Food. Special Scientific Report, Wildlife No. 98. U.S. Department of the Interior. (Reprinted in Perspectives on Human Evolution 2, S. L. Washburn and P. Dolhinow, eds.).

A social grouping of chimpanzees—an old male places his hand on the head of a juvenile female. Both have their mouths full of food. © 1965 National Geographic Society, photograph by Hugo van Lawick.

Jane Goodall:

Continuities between Chimpanzee and Human Behavior

Humans are prone to stress the uniqueness of their species, but recent field studies have discovered unsuspected amounts of behavioral patterns held in common between mankind and its closest living relatives, the chimpanzees. In this essay, Jane Goodall sketches the way that Louis Leakey encouraged her in undertaking studies of chimps in the wilds of East Africa. Some aspects of the behavior that she and others have discovered seem to be of particular relevance for understanding human origins. These include tool use, hunting, prolonged infant dependency, and family ties. Goodall shows that the chimpanzee studies provide the basis for imaginative, but yet controlled reconstructions of early hominid behavioral patterns.

Louis Leakey, first and foremost, was a paleontologist indefatigible in his search for the remains of man's earliest ancestors and of the prehistoric creatures who shared with them that primeval world. This volume bears ample testimony to the magnitude of the contribution of this one man and his team to our understanding of human evolution in Africa. Louis, however, was not concerned solely with relicts from the past: his was a mind that constantly spanned the bridge of time between the world as we know it today and its ancient beginnings, a mind that continually strived to interpret the present more fully that we might have a more insightful view of prehistory.

I first met Louis when he was Curator of the National Museum (then Coryndon Museum) in Nairobi where he and his staff made major contributions to our understanding of the fauna of present day Africa. Concepts regarding animals and their behavior were constantly being examined and questioned, new theories brought forward and investigated. Much of Louis' own efforts at that time were directed to instigating research into the behavior of contemporary animals as it relates to their skeletal structure and habitat. Only by such understanding, Louis argued, could one hope to gain clues into the probable behavior of creatures known to us only through their bones.

It was this line of reasoning which first stimulated Louis' interest in the behavior of the anthropoid apes, closest living relatives of man. We should realize, at this point, that Louis was some way ahead of the majority of his contemporary scientists in realizing the potential value of primate field studies for the understanding of human behavior. Today we take it for granted that scientists should isolate themselves in a variety of jungle, mountain or desert habitats in order to study monkeys and apes in their natural surroundings. But when Louis first proposed that a young girl should set off on her own to observe the habits of mountain gorillas in the wild forests of Uganda, the world (or most of it) thought him irresponsible and slightly crazy. Even those scientists who realized that such studies would be crucial were overawed by the supposed difficulties and dangers inherent in such undertakings. For Louis, however, born and brought up in the wilds of Africa, these concerns simply did not enter his mind. Why should lions or elephants or snakes pose any greater danger than cars on busy city streets? And so in 1956 Rosalie Osborn undertook the first of the Leakey series of anthropoid field studies for which he will always be remembered.

That first study was very brief, but the others are longterm. Birutė Galdikas-Brindamour has been studying the orang-utan in Malasia for three years already; Dian Fossey has been with mountain gorillas for seven years and is still there; and the observations on chimpanzee behavior at the Gombe Stream Research Centre in Tanzania were into the fifteenth year in July 1974.

When Louis first suggested in 1957 that I should try to study chimpanzees at Gombe, little was known of their behavior in the natural habitat. The only major field study, the pioneering venture of Dr. Henry Nissen in 1930 (Nissen 1931), lasted for only two-and-a-half months. From the behavior of captive chimpanzees it was known that there were some striking similarities between their behavior and that of our own species, but the full range of those similarities had not yet begun to be realized.

Recent biochemical and anatomical research has revealed striking similarities in the biology of man and chimpanzee in relation to the number and form of the chromosomes, the blood proteins and immune responses, the structure of the DNA, and perhaps most importantly, the anatomy and circuitry of the brain. These similarities, taken in conjunction with the behavioral similarities which I shall describe, suggest that man and chimpanzee shared a common ancestor at some point in the remote past. We may argue that behavior which is common to modern man and modern chimpanzee probably occurred also, in similar form, in that common ancestor—and thus in Stone Age man as well.

This chapter will attempt to indicate some of the ways in which an understanding of chimpanzee behavior in the field may provide clues regarding the probable behavior of early man, particularly concerning those aspects about which we can deduce little from the fossil record. Since I have recently been involved in two major papers on this same theme (Lawick-Goodall and Hamburg, in press; Lawick-Goodall, in press) I shall here present a briefer resume of arguments discussed more fully in the other papers.

USE OF OBJECTS AS TOOLS

When I began my research in 1960 there were three incidental accounts of wild chimpanzees using tools (Beatty 1951; Merfield and Miller 1956; Savage and Wyman 1843–4), but I doubt if many people suspected the extent to which these apes do, in fact, make use of natural objects as tools in their daily lives. With the notable exception of man, chimpanzees use more objects for a wider variety of purposes than any other creature.[1] I can still remember how excited Louis was when he heard about my first tool-using observations.

At Gombe, many instances of tool use occur in a feeding context: grass stems or thin twigs are used to "fish" termites from their nests; sticks are pushed into underground or arboreal nests of various species of ants; sticks may also be used as levers to enlarge the openings of underground bees' nests. A leaf "sponge" (Louis' own terminology) is used for sopping up water in the hollow of a tree which the chimpanzee cannot reach with his lips, and a sponge of this sort was once used for cleaning out the inside of a baboon skull after a kill. Twigs and sticks are also used as investigation probes, again sometimes in a feeding context. Thus a twig may be inserted into a hole in dead wood: after sniffing the end the chimpanzee either discards the wood, or presumably in response to some olfactory cue, tears it apart and (usually) finds and consumes a beetle or

[1] I have discussed elsewhere tool using in mammals and birds (Lawick-Goodall 1970) and specifically in chimpanzees (Lawick-Goodall 1973).

Tool use was formerly considered the hallmark of mankind, but research at Gombe Stream Reserve and elsewhere has shown that it is also part of the repertoire of chimpanzees. Here two adults use stems and twigs to fish for termites. © 1963 National Geographic Society, photograph by Jane Goodall.

wasp larva. Frequently a number of openings in a termite nest are investigated thusly before the chimpanzee begins to work in ernest. Non-food objects may also be examined in this way: one individual slowly pushed the end of a dead palm frond towards the head of a dead python and then sniffed the point of contact. Leaves may be used to wipe dirt or blood from the body. And, finally, sticks and stones may be used as weapons in aggressive contexts.

In addition to using the objects around them in the above ways, the chimpanzees sometimes modify the material to make it more suitable for the purpose in hand. Thus, during termite fishing, leaves may be stripped from a twig, a strip of bark may be shredded, or the blades may be stripped from a wide length of grass. When drinking from a hollow, the chimpanzee almost always crumples the leaves by chewing them briefly before using them as a sponge, thus increasing their water-carrying capacity considerably. These simple modifications may be considered the primitive beginnings of toolmaking.

In other areas of Africa chimpanzees have been seen to use objects as tools which have not been observed in the repertoire at Gombe. Leafy twigs may be used as fly whisks in Uganda (Sugiyama 1969) and rocks may be used as hammer stones in Liberia (Beatty 1951) and Central Africa (Savage and Wyman 1843–4). It seems likely then, especially in view of the importance of social learning in the development of the young chimpanzee (which I shall discuss) that the tool-using performances of different chimpanzee populations may be regarded as simple cultures (Lawick-Goodall 1973).

HUNTING BEHAVIOR AND THE SHARING OF FOOD

Chimpanzees hunt, kill and eat medium-sized mammals quite often at Gombe and have also been observed to hunt in other areas (Lawick-Goodall 1968, 1970; Teleki 1973; Nishida 1968; Kano 1971; Kawabe 1966; Suzuki 1971). At Gombe, one community (with about eight to ten males of over eleven years of age) may catch between 30 and 40 prey animals a year when they hunt relatively more frequently. (Not all attempts are successful, of course). Hunting is usually a male occupation although, on two occasions, females have been observed to catch baby bushpigs themselves. Sometimes a kill is an opportunistic event: one chimpanzee comes across a suitable victim, seizes and kills it. At other times, however, the chimpanzees may set out to hunt with deliberation and may show quite sophisticated cooperation in their stalking technique. If, for instance, the proposed prey is a young juvenile baboon, separated from the rest of the troop and up in a tree, one chimpanzee may begin, very slowly to climb towards him. Other chimpanzees may stand below that tree and also below other trees which might act as escape routes for the quarry. Even when a chase became very tense, with the hunter following his prey back and forth between the crowns of two palm trees, the other chimpanzees did not leave their guard positions until the baboon made a wild leap for safety: at this point the hunters converged to try to seize him.

To date no weapons have been used during hunting, nor have any objects been used as cutting tools to assist in dismembering the carcass. In fact, the only time a tool was used in conjunction with meat eating was when a leaf sponge was used to wipe clean the inside of a skull.

Chimpanzees appear to be extremely fond of meat: after a successful kill there is usually much excited calling which serves to attract other individuals who happen to be in the area to the scene of the kill. All those present gather around the male (or males) in possession of the carcass or pieces of meat and beg for shares. Begging is frequently successful. Sometimes the possessor of the carcass tolerates another feeding from it at the same time as himself; sometimes he permits a small piece of meat to be pulled off by another; frequently he spits a chewed mouthful into the outstretched hand of a beggar (usually after he has finished with it himself).

Less often the possessor of meat himself detaches a fragment and places it in the outstretched hand of a supplicator. Once a male actually tore in two the body of an infant baboon he had caught and gave one half to an insistently begging male companion.

The success or failure of begging depends mainly on the individuals concerned and the relationship between them, and partially on the amount of food available. Some males are far more "generous" than others, and some individuals are much more persistent in their attempts to get a share for themselves. Sharing between chimpanzees who associate frequently is more common than sharing between individuals who seldom see each other: a female is more likely to obtain a share when she is in oestrus than when she is not sexually attractive. And, finally, some chimpanzees are much more persistent than others in their attempts to gain pieces of a kill. Presumably this reflects a greater liking—or perhaps a greater physiological need—for meat.

THE LONG PERIOD OF IMMATURITY

From the very first Louis impressed upon me the value of longterm studies: I believe that he knew how long I would be working at Gombe even before I knew it myself! It is, of course, the longterm nature of this study which has revealed so many similarities between chimpanzee and human behavior—similarities which might have passed unsuspected after a field study of only a couple of years, however intensive those years might have been.

The surprisingly long period of immaturity in the chimpanzee is a case in point. The infant does not start to walk until he is six months old, and he seldom ventures more than a few yards from his mother until he is over nine months old. He may ingest a few scraps of solid food when he is six months, but solids do not become a significant part of his diet until he is about two years of age and he continues to nurse until he is between four-and-a-half and six-years old. Moreover, while he may travel short distances under his own steam when he is about four-years old, he continues to make long journeys riding on his mother's back until he is five or six, and may jump onto her back if suddenly alarmed when he is even older. He continues to share his mother's nest at night until he is about six or until a new sibling is born, at which time he usually begins to sleep in his own nest near his mother.

The young male usually begins to make his first intentional independent journeys away from his mother when he is about eight- or nine-years old. Prior to this he is likely to become anxious if he accidentally becomes separated from her for a while. Even after he has made the preliminary break and has travelled and slept in a number of groups away from his mother, he continues to return and associate with her very frequently during the next couple of years and, indeed, is likely to spend a good deal of time with her throughout her life. The female continues to associate almost constantly with her mother for an even longer time period—usually she does not travel (intentionally) independently of her mother until she becomes sexually attractive to the adult males at about eleven years of age. In between periods of oestrus she continues to associate with the mother and family.

Chimpanzees go through a period of adolescence which commences around puberty (about nine years of age) and continues until the individual has attained social maturity (about thirteen in a female, when she is able to give birth successfully, and about fifteen in a male, when he is able to take his place in the social hierarchy of the mature males). Adolescence has often been regarded as a culturally determined period and, therefore, unique to man. However, it is also quite a clear-cut phase in the chimpanzee life cycle, marked by distinctive changes in behavior and physiology (Pusey, in prep.).

For the male chimpanzee, adolescence is stressful at times. Initially he becomes larger and stronger and, along with this growth spurt, he typically becomes more aggressive. This aggression is directed mostly towards females, particularly towards the lower-ranking adult females. After a couple of years most adolescent males are easily able to intimidate all but the strongest or most aggressive females.

At the same time, however, the young male must show increasing caution in his dealings with his superior males who become increasingly intolerent of him at the onset of adolescence. Yet despite his growing apprehension, evident in many of his interactions with the adult males, he does appear to be extremely fascinated by them. His early journeys away from his mother are usually with adult males and, although he occupies a peripheral position in a group of males, he spends a good deal of time intently watching their behavior. Gradually, from about eleven years of age onward, the adolescent male begins to work his way further and further into the adult male society. He begins to join them in their sessions of social grooming, to perform his charging displays along with the other males rather than at a distance, as he did when he was younger. Eventually he begins to challenge the lower-ranking mature males in aggressive contexts. Usually he can be classified behaviorally as a socially mature individual by the time he is about fourteen- or fifteen-years old.

The female, as mentioned, stays in close association with her mother for longer than the male. If there is a new infant in the family she has much opportunity for learning the role of the adult female without leaving her mother. Many adolescent females, particularly before they begin regular oestrus cycles, show a good deal of fascination for infants and spend much time playing with them, carrying them around and grooming them—provided this is tolerated by the mother.

In the chimpanzee female, as in the human female, there appears to be a period of "adolescent sterility" (Graham 1970; McGinnis 1973). At Gombe this period, from the time of menarche to the first successful impregnation, lasts for one to two years. Towards the end of this time the young female, during periods of oestrus, frequently wanders from her home community, seemingly of her own volition, to mix and mate with males of a neighboring community. Usually, after her oestrus period is over, she returns to her home community, but occasionally a female actually transfers, apparently permanently, to a community other than that in which she was raised. Japanese scientists working in an area south of Gombe have reported the same phenomena of female transfer between communities (Nishida and Kawanaka 1972). This is of particular interest because, in other nonhuman primates studied thus far, it appears to be the role of the young male to transfer to a new group and thus ensure exchange of genes between different groups of a population (e.g. Schaller 1963; Sade 1972; Boelkins and Wilson 1972; Ransom 1972).

SOCIAL LEARNING

In the human species the long period of immaturity is usually considered adaptive in relation to social learning, and this is undoubtedly of major significance for the young chimpanzee also. There is much that he needs to learn regarding the complex society into which he is born—appropriate responses to different communicative patterns in different individuals, tool-using and hunting techniques, and so on. During the period when he is able to rely on the presence of an adult experienced female to lead him to appropriate food sources and respond correctly to dangerous situations, he is free to devote time and energy to play, exploration and the observation of adult behavior.

I have discussed social and observational learning fully elsewhere (Lawick-Goodall 1973; Lawick-Goodall and Hamburg, in press). Here I should summarize by stating that in the laboratory it has been shown that primates are, indeed, capable of learning by direct observation of the behavior of another individual (e.g., Butler 1965) and, in the wild, young chimpanzees do spend much time watching others intently, particularly their mothers. Often, after watching some activity such as a male charging display or a complex tool-using performance, an infant may then try to perform the same actions. Subsequently he may practice the behavior time and again.

The period of adolescence, which may be regarded as a prolongation of social immaturity, is probably also helpful in perfecting some of the behaviors that will be crucial in adult performances such as child-raising techniques for the female, charging displays, subtleties of courtship behavior for the male, and so on.

THE FAMILY

The chimpanzee family comprises a mother and her offspring since the father, having sired an infant, plays no further specific role in its upbringing, although he is generally protective towards all small infants in his community. However, between mothers and their offspring, and between some siblings (usually those of like sex) strong affectionate bonds may develop just as they may in human families.

A mother chimpanzee is typically extremely tolerant of her infant and seldom administers physical punishment. She often shares food with her infant, and may continue to do so through his juvenile period. Not only does she protect her child during his infancy, but she may also hasten to the assistance of an older offspring. One old mother (Flo), when she heard her son of about fifteen-years old screaming, hurried over. He had been threatened by an older male and was retreating: when his mother arrived, however, he became more courageous and together mother and son chased the aggressor away. When Flo's other son was about twenty-years old, he hurt his wrist during a status conflict with another male. Flo was very old at the time, but when she heard his loud screams she ran for some quarter of a mile to the scene of action. She could do nothing positive to help, but her mere presence seemed to calm her son: he quieted down, limped away with her, and remained virtually out of contact with other adults until he was quite recovered two weeks later. Other mothers have been seen helping their sons and, similarly, sons may go to the assistance of their mothers.

The relationship between mothers and their growing daughters may be similar to that described for mothers and sons. On the other hand, some older juvenile and adolescent females seemed fearful of their mothers, particularly during feeding situations.

Close affectionate bonds often develop also between brothers as they mature; elder brothers, in particular, tend to support their younger brothers in a variety of social situations. The same kind of relationship may develop between sisters (it has in the only pair studied to maturity), but to date no such bonds have been observed between grown brothers and sisters.

When a mother dies leaving a dependent child, the nature of the bond between family members may be dramatically illustrated. The infants themselves may show behaviors associated with states of depression in humans—huddling, loss of appetite, rocking from side to side and withdrawal from active social behaviors, particularly play. Infants under three years of age obviously cannot survive without maternal milk, but in the case of three older individuals it seemed that the psychological shock caused by the abrupt severance of their relationships with their mothers was a crucial factor in their own subsequent deaths (Lawick-Goodall 1968, 1970; Thorndahl, in prep.).

Infants with older siblings were "adopted" by them after the death of their mothers. A juvenile female waited for her small brother when she travelled from place to place, and allowed him to share her nest at night. Despite this, the infant (between four and five years) showed increasingly abnormal behavior and died one-and-a-half years after losing his mother. A second orphan of similar age was adopted by an adolescent sister who actually permitted the infant to ride everywhere on her back. This youngster gradually got over her signs of depression and, one year after the mother's death, showed more or less normal behavior for her age. It is tempting to speculate that the added social security provided by close physical contact with an experienced female was a crucial factor aiding recovery.

It is worth commenting here that older non-related females played no role in caring for orphans, even in the case of a three-year-old who had no elder sibling and who spent much of the two-month period she survived her mother wandering about alone. On the other hand, a fourteen-month-old infant was cared for by her juvenile *brother,* who struggled to carry her everywhere he went, but since she was still totally dependent on milk she only lived two weeks. This is yet another pointer to the significance of family relationships.

An unusual case was that of Flint. He was eight-and-a-half-years old when his mother Flo died, but very dependent upon her. He showed immediate signs of depression (huddling, loss of appetite, social withdrawal) and survived Flo by only three weeks. During this period when he was physically weakened due to lack of food and general apathy, he developed a form of gastroenteritis which also contributed to his death (Thorndahl, in prep.)

NONVERBAL COMMUNICATION PATTERNS

For me, one of the most striking behavioral similarities between man and chimpanzee lies in the nonverbal communication system—the repertoire of postures and gestures by which one chimpanzee communicates with his fellows. In many cases, it is not only the gesture which is so similar to that of a human (such as kissing, patting on the back, holding hands) but also the contexts in which such patterns are likely to occur.

When a chimpanzee is frightened he may reach to touch or embrace a companion, and he seems to derive comfort from such contact. When two chimpanzees are suddenly excited—if, for instance, they come across an unexpected supply of food—they are likely to indulge in much contact-seeking behavior of this sort, touching, embracing, kissing, patting until it seems they are calmed by the physical contact with each other. This apparent need for physical contact with another in times of stress is often vividly illustrated by a young chimpanzee who has been threatened or attacked by another. The victim may approach the aggressor, screaming and tense, and take up a submissive, crouching posture in front of him. In response to such behavior the aggressor typically reaches out to touch, pat or even embrace the screaming or whimpering subordinate. The effect of such a reassurance gesture on the victim is usually immediately apparent: his screams diminish, his whimpers gradually cease, and he slowly relaxes his tense posture. Sometimes he seems quite calm by the time he moves away.

When chimpanzees have been separated for a while they may, if they are close associates, show friendly behavior which we may call greeting. One individual may pat, embrace or kiss another, or they may reach out and hold hands.

Some of the patterns and contexts of chimpanzee aggression are also similar to some of our own. A chimpanzee who threatens another may make vigorous movements of upraised arm, he may run towards the other in an upright posture, sometimes waving both arms, he may throw rocks or other objects, often with good aim, or he may brandish a stick. During an actual attack he may bite, pull hair, scratch, punch, hit or kick.

Aggression may occur when two chimpanzees are competing for social status, for favored food in short supply or, very occasionally, for a female. A chimpanzee may become aggressive if a member of his immediate family is threatened or attacked. He may become irritable, and thus aggressive, if he is in pain, or if subordinates make too much noise and commotion nearby. Very often, when one chimpanzee is threatened or attacked by a superior whom he dare not fight back, he redirects his aggressive feelings against a subordinate who happens to be nearby. In addition, the sight or sound of chimpanzees of a neighboring community may cause aggressive displays and, if a "stranger" is encountered, he or she may become the victim of a savage attack (e.g. Bygott 1972).

DISCUSSION

Even the foregoing brief and much over-simplified account of some of the behavior patterns of free-living chimpanzees reveals quite striking points of similarity between their behavior and our own. And, according to our original hypothesis, it is these behaviors—shared by modern man and modern chimpanzees—which may have been present in the common ancestor and in Stone Age man himself. Unfortunately it seems unlikely that we shall ever find a way to prove or disprove speculations of this sort concerning the behavior of our remote forebears.

Probably it is universally believed, nowadays, that early men used twigs, sticks, leaves, and stones as tools long before they ever shaped the first stone artifact. At the present time the Bushmen of the Kalahari and a tribe of South American Indians use leaves as sponges in a similar manner to the chimpanzees.

We have learned a vast amount about the manner of hunting and killing prey animals practiced by our stone age ancestors from the paleontological evidence. We can examine the fossilized remains of bones from their living sites to deduce which species they ate, and we can trace the gradual sophistication of their use of weapons in hunting and tools in dismembering animals. An understanding of hunting behavior in the modern chimpanzee can, however, throw some light on one topic that is frequently discussed—whether man began his carnivorous history as a hunter, or as a scavenger.

As has been dealt with fully elsewhere (Lawick-Goodall and Hamburg, in press; Lawick-Goodall, in press), all evidence from the comparative study of the living primates suggests that man was a hunter before he was a scavenger. Scavenging can be difficult and extremely dangerous (Lawick-Goodall and Lawick 1970). Large carnivores, while they might run from a human today (for they have come to fear us) were far less likely to run from a group of ape-like creatures armed only with sticks and stones. Many of the primates studied in the field have been observed to catch live mammals or birds for food, but there are no reliable reports of scavenging meat from the kill of other predators. Indeed, the chimpanzees at Gombe have, several times, been offered fresh meat which they have consistently refused. Without doubt early man once he had acquired a taste for meat, and particularly after his weapons had become more sophisticated, would have taken meat from other predators if he had had the chance, for a characteristic of the human species is, and undoubtedly always has been, opportunism. But familiarity with the savanna and its predators, and with primate behavior, strongly inclines me to the belief that hunting came first in our earliest ancestors.

It seems almost certain that Stone Age man had a long period of immaturity, a time when he could rely on the protection and guidance of his elders while he acquired the skills necessary for survival as an adult. We can fairly safely picture a Stone Age toddler crawling around his mother while she prepared vegetable or insect food for him, or moving close to a group of males as they distributed meat from a kill. We can picture him picking up a nut dropped by his mother and hitting it, as she had been doing, with a stone; or manipulating an abandoned scraper, perhaps scratching with it on the ground, or against some piece of animal skin lying around. From time immemorial human children must have been carefully watching their elders and imitating their behavior.

We can suppose, too, that our early ancestors went through a period of adolescence when the young males and females, having left childhood behind, prepared to take their places in adult society. Perhaps the adolescent boys accompanied older males on hunting forays, learning the secrets of their wild environment, watching the fascinating ways of grown men, longing (if their intellect was sufficiently developed) for the time when they, too, could be acclaimed successful hunters. The adolescent girls probably stayed with their mothers, helped to look after the younger babies, gradually acquired the competence necessary for mothering their own future infants. Were they, like adolescent female chimpanzees, restless after puberty? Did they feel any urge to wander away from the familiarity of the home range in search of new mates? Such a biological heritage would certainly be helpful to girls in many cultures who at marriage have to abruptly leave the familiarity of their homes where, until that point, they have led a very sheltered life.

It seems entirely reasonable to suppose that strong emotional bonds prevailed within Stone Age families; that grown sons maintained close contact with their mothers and young siblings, that brothers travelled together on hunting expeditions in search of vegetable foods or animal prey. Undoubtedly, too, family members aided one another during squabbles within the group.

Perhaps those gestures shared by modern man and modern chimpanzee occurred, in similar form, in our early human ancestors. It certainly seems likely, to me, that Stone Age men and women, when fearful, touched or embraced one another, that a submissive youngster was reassured by a pat on his head or shoulder after some misdeed, that members of a group, when they met, greeted one another with a kiss or an embrace, a hand on the shoulder, or a clasping of hands. For these patterns occur today in many human cultures throughout the world just as they do in chimpanzees in their native forests.

Without doubt many of the circumstances which trigger aggression in man today caused quarrels and conflicts in our ancestors just as they do in chimpanzee society—competition for leadership, for food, for females; irritability; defense of a family member or close friend. Perhaps, too, our Stone Age ancestors became hostile and angry when they encountered strangers from neighboring groups.

In conclusion I should like to add one further comment: even if one is not prepared to accept the usefulness of the study of chimpanzee behavior for the interpretation of the lifestyle of early man, Louis Leakey's contribution to the science of primatology itself must be considered of inestimable value. He invested much time and energy to the inspiration, encouragement, organization and funding of longterm studies of the three anthropoid apes. Dian Fossey has collected an unbelievable wealth of data regarding the social behavior of the great mountain gorilla—for the publication of which her colleagues are waiting eagerly. Biruté Galdikas-Brindamour, who began her study of the orang-utan in Malasia in 1971, has, like Dian Fossey, an incredible amount of as yet unanalyzed data of great significance. Then there is the Gombe Stream Research Centre where students from three continents are engaged in the longitudinal study of chimpanzee behavior. The growth of Gombe is, in itself, a monumental tribute to the man whose vision and foresight made it possible. In short, leaving aside the significance of anthropoid behavior to the better understanding of our own, Louis has helped the world to a better appreciation of some of its most amazing creatures—the closest living relatives of *Homo sapiens.*

References Cited

Beatty, H.
1951 A note on the behaviour of the chimpanzee. J. Mammal. 32:118.

Boelkins, R. C. and P. P. Wilson
1972 Intergroup social dynamics of the Cayo Santiago Rhesus *(Macaca mulatta)* with special reference to changes in group membership by males. Primates 13:125.

Butler, R. A.
1965 Investigative behaviour. *In* Behaviour of Non-Human Primates. A. M. Schrier, H. F. Harlow and F. Stollnitz, eds. New York, London: Academic Press.

Bygott, J. D.
1972 Cannibalism among wild chimpanzees. Nature 238:410–411.

Graham, C. E.
1970 Reproductive physiology of the chimpanzee. *In* The Chimpanzee, Vol. 3. G. Bourne, ed. Basel, Switzerland: Karger.

Kano, T.
1971 Distribution of the primates on the eastern shore of Lake Tanganyika. Primates 12:281–304.

Kawabe, M.
1966 One observed case of hunting behaviour among wild chimpanzees living in the savanna woodland of western Tanzania. Primates 7:393–396.

Lawick-Goodall, J. van
1968 The behaviour of free-living chimpanzees in the Gombe Stream area. Anim. Behav. Monog. 1:161–311.

1970 Tool-using in primates and other vertebrates. *In* Advances in the Study of Behaviour, Vol. 3. D. S. Lehrman, R. A. Hinde and E. Shaw, eds. New York, London: Academic Press.

1973 Cultural elements in a chimpanzee colony. *In* Precultural Primate Behavior. E. W. Menzel, ed. New York: Karger

In press *In* The Quest for Man. M. M. Goodall, ed. London: Phaiton Press.

Lawick-Goodall, J. van, and D. A. Hamburg
In press Chimpanzee behavior as a model for the behavior of early man. *In* American Handbook of Psychiatry, Vol. 6. New Psychiatric Frontiers. D. A. Hamburg and K. H. Brodie, eds. New York: Basic Books.

Lawick-Goodall, J. van, and H. van Lawick
1970 Innocent Killers. Boston: Houghton Mifflin. London: Collins.

McGinnis, P. R.
1973 Patterns of sexual behaviour in a community of free-living chimpanzees. Unpublished doctoral dissertation, University of Cambridge.

Merfield, F. G., and H. Miller
1956 Gorillas Were My Neighbours. London: Longmans.

Nishida, T.
1968 The social group of wild chimpanzees in the Mahali Mountains. Primates 9:167–224.

Nishida, T., and K. Kawanaka
1972 Inter-unit group: relationships among wild chimpanzees of the Mahalio Mountains. *In* Kyoto University African Studies. T. Umesao, ed. Kyoto: Kyoto University Press.

Nissen, H. W.
1931 A field study of the chimpanzee. Comp. Psychol. Monogr. 8:1–22.

Pusey, A.
In prep. Behavior changes in adolescent chimpanzees.

Ransom, T. W.
1972 Ecology and social behaviour of baboons in the Gombe National Park. Unpublished doctoral dissertation, University of California, Berkeley.

Sade, D. S.
1972 A longitudinal study of social behavior of rhesus monkeys. *In* The Functional and Evolutionary Biology of Primates. R. Tuttle, ed. Chicago: Aldine, Atherton.

Savage, T. S., and J. Wyman
1843–1844 Observations on the external characters and habits of the Troglodytes niger (geoff) and on its organization. Boston J. Nat. 5:417–443.

Schaller, G.
1963 The Mountain Gorilla. Chicago: University of Chicago Press.

Sugiyama, Y.
1969 Social behaviour of chimpanzees in the Budongo Forest, Uganda. Primates 10:197–225.

Suzuki, A.
1971 Carnivority and cannibalism observed among forest-living chimpanzees. J. Anthrop. Soc. Nippon 79:30–48.

Teleki, G.
1973 The omniverous chimpanzee. Sci. Amer. 228:1.

Thorndahl, M.
In prep. Family relations in wild chimpanzees.

A hottentot breaks a goat bone for marrow in the traditional manner, using a stone hammer and anvil.

C. K. Brain:

Some Principles in the Interpretation of Bone Accumulations Associated with Man

Can we recognize bone tools used by early men and distinguish hominid food refuse from other natural bone accumulations? C. K. Brain describes comparative studies that help to answer these questions.

The mammalian skeleton is constructed from individual bones varying considerably in shape and structure. In the living animal the skeleton functions as an integrated whole but, upon death of the animal, the individual bones frequently become disarticulated and may then contribute to a bone accumulation as individual parts. Bones with some relevance to the study of early man are generally those which have succeeded in surviving for a considerable time and which have somehow evaded the destructive processes constantly at work in the natural environment.

It is with the question of survival of bones that this paper is particularly concerned. Clearly the individual elements in a mammal skeleton are extremely variable as to their strength characteristics and durability. The rounded and compact form of an antelope's astragalus makes this bone far more resistant to destruction than does the thin and blade-like structure of the same animal's scapula. When subjected to certain influences, the astragalus may survive while the blade of the scapula disappears.

Although bones from a great variety of animals contribute to bone accumulations, those assemblages associated with early man in Africa are often dominated by remains of antelope. Particular attention will therefore be given to survival characteristics of skeletal parts from animals belonging to the family Bovidae.

THE MAKAPANSGAT STUDY

The study of bone accumulations in Africa has not had a long history. In fact, Professor R. A. Dart's analysis of bones associated with *Australopithecus* at Makapansgat Limeworks in 1957 was a pioneering investigation and revealed some interesting and unexpected facts (Dart 1957). The 7159 bone fragments analyzed were found to represent parts of at least 433 animals; 293 of these were antelope while the rest included a wide variety of animals among which were 45 baboons, 20 pigs, 17 hyaenas, 7 porcupines, and 5 australopithecines. Dart concluded that the bones had been collected originally by the hominids who used them both as food and as tools.

It was found that 91.7% of all the bone fragments came from antelope, the 293 individual animals being grouped as follows: 39 large (like eland), 126 medium (like wildebeest), 100 small (like gazelle) and 28 very small (like duiker). From the point of view of the overall bone accumulation, the antelope are by far the most important group and are the only animals which need concern us in the present discussion.

The analysis of the parts of the skeletons revealed some unexpected facts. Parts of the skulls were, for instance, exceptionally common, making up 34.5% of all the recognizable fragments. Vertebrae, on the other hand were unaccountably rare, a total of only 163 (or 1.4% of what there should have been) being found. Among these scarce vertebrae, the atlas and axis were abnormally abundant, while tail vertebrae were not represented at all.

Similar remarkable disproportions were found in the limb bones, parts of the forelimbs being much more abundant than those of the hind limbs. Turning to the individual bones in the legs, Dart found that some ends of such bones were more common than others. In the humerus, 336 distal ends were found but only 33 proximal ones, a ratio of 10:1. Similarly in the tibia, the ratio of distal to proximal ends was 119:64.

What did these disproportions mean? In an attempt to explain them Dart suggested that the australopithecines brought back only certain parts of the prey animals to the cave. They concentrated particularly on bones which made good tools—mandibles for saws and scrapers, distal humeri for clubs. Parts missing from the fossil collections were either not brought back at all or, as in the case of tails, were used for special purposes outside the cave. They served as "whips or signals in hunting."

The presence or absence of certain bones in an assemblage can certainly reflect deliberate hominid activity and may result from particular butchery techniques. But this may not necessarily be the case. A mammalian skeleton subjected to natural destructive processes will deteriorate in a characteristic manner. Certain parts will typically survive while others disappear. If the analysis of a bone accumulation shows a pattern of survival and disappearance similar to that which may be predicted to result from natural influences, it will not be necessary to invoke the agency of deliberate hominid selection. It may even be unwise to do so.

The natural destructive influences referred to here include the feeding activity of hominids and other carnivores, the action of porcupines and the normal processes of weathering and disintegration. They specifically exclude cultural activity such as the selection of bones as tools or butchery procedures which impose an artificial bias on the sample.

In searching for a contemporary situation in which the effects of "natural" destructive agencies on a bone accumulation might be studied, the writer was led to Hottentot villages in the Central Namib Desert where the remains of goats provide information relevant to the question.

THE KUISEB RIVER HOTTENTOTS AND THEIR FOOD REMAINS

Stretching across the Namib Plain from the escarpment in the East to Walvis Bay in the west is the Kuiseb river, dry throughout the year, except after sporadic rain. Scattered along its banks over a distance of about 100 miles are eight Hottentot villages, each consisting of a small group of beehive-shaped huts built from the bark of acacia trees which grow in the river bed. These villages house a Topnaar Hottentot population of about 130 whose way of life, pedigree and blood groups have been studied in some detail (Jenkins and Brain 1967).

The Namib Plain, where it is traversed by the Kuiseb River, is extremely inhospitable. To the south is an immense area of high dunes while northwards featureless gravel plains extend for many miles. The result is that the human population is closely tied to the riverbed itself, from which water is obtained in shallow wells. The economy of the Hottentots is built around their goat herds and these, in turn, are dependent on the Kuiseb River for survival. They subsist very largely on the dry seedpods of *Acacia albida* and villages are spaced in a linear fashion along the river, at intervals determined by the number of goats kept at each. In 1966 there was a total of approximately 1750 goats in the Kuiseb villages. These provide the Hottentots with milk, meat, and skins.

The aridity of the environment results in a general absence of vegetation around the villages. Occupational relics, such as bones, are easily seen, and it was appreciated in 1966 that an analytical study of these bones could be of great interest from the point of view of fossil interpretation (Brain 1967a). Although they have accumulated in natural circumstances, the situation is remarkably simple and controlled: perhaps 95% of all bones to be found around the villages came from goats, these being the only mammals normally used for meat. The bones are simply broken for extraction of marrow and are then discarded, no bone tool use being practiced. Once discarded, the bones are further gnawed by dogs, after which the remnants are left to bleach on the gravelly desert surface. Apart from Pied Crows, other scavengers are not involved.

Finally, the great merit of the Kuiseb River situation lies in the fact that the accumulation is constantly being added to and that the process can be observed directly. There can be no doubt as to what influences the bones have been subjected to—problematical aspects can be verified on the spot.

PROCESSES INVOLVED IN BUILDING UP THE BONE ACCUMULATION

Fairly detailed information on Hottentot butchering technique and eating habits is now available both from direct observation and from questioning of local people. An apparently typical goat-processing procedure will be described; these observations were made in the Zoutrivier village in February 1966. The goat was led to a particular tree where slaughtering is normally carried out. Several Hottentots held the goat down on its side while another cut its throat with a pocketknife. The blood was caught in an enamel basin and fed to two waiting dogs who lapped it avidly. Once dead, the goat was suspended by its hind feet from an overhanging branch and the skin removed completely, being split along the midventral line, along the insides of the limbs and around the neck just behind the horns. It was salted and pegged out in the shade. The abdominal cavity was opened next and the viscera removed; the stomach was slit open, its contents emptied out and its lining washed. This, together with the liver and kidneys, was said to be a delicacy. The intestine, once the contents had been squeezed out, was kept for the making of sausage. Other abdominal organs were fed to the dogs.

Turning again to the carcass, the front legs were removed complete with the scapulae; hind limbs were taken off with the innominate bones

FIGURE 1
An aged Topnaar woman from the Kuiseb River. She was part of the group who accumulated the goat bones described in this paper.

attached, by cutting through both the pubic symphysis and the sacroiliac joints. The feet were severed from the legs at their metapodial/phalangeal joints; these were taken by the children who cooked them themselves over a fire. Ribs on one side of the carcass were separated at their vertebral articulations. Finally the head was removed, a knife being used to sever the axis from the third cervical vertebra. The atlas and axis vertebrae remained attached to the occiput.

All meat is normally cooked before it is eaten, either by boiling in large metal pots or by direct roasting over the fire. The head was dealt with in a characteristic manner—the horns were broken off at their bases by a sharp blow from an axe and were discarded. The complete head was then boiled for several hours in a pot standing over the fire. All edible meat was picked from it and eaten, after which the braincase was smashed in the occipital region with a hammerstone for removal of the brain. The skull and mandibles were then passed on to the dogs.

As the eating progressed, all marrow-containing bones were broken. They were held on a rock anvil and hammered with another stone. Neither the anvil nor hammerstone is an artifact in the usual sense of the word—they are simply suitable pieces of rock which happen to be at hand. The Hottentots habitually eat in a squatting attitude on the ground. Typical utensils are small pocketknives, rock anvils, and hammerstones.

Once discarded by the Hottentots, the goat bones were gnawed sporadically for many days by the dogs, all of which were jackal-like in size. In 1966 it was found that a total of 40 dogs were kept at the eight Kuiseb River villages. Jackals themselves are now extremely rare in this part of the desert and do not seriously enter into the picture.

Pied Crows are fairly common along the Kuiseb River and, when they can, will certainly carry off scraps of meat, sometimes with bones adhering to them. On one occasion a crow was seen flying from the Zoutrivier village with most of a goat's tail in its bill.

When lying in fully exposed positions on the gravel surface, bone fragments become bleached and degreased within three months. Exposure to the sun results in weathering of the bone surface and a soft, chalky superficial layer develops. Gnawing of the bones by gerbils of the genus *Desmodillus* (whose burrows are often concentrated around old goat kraals) is not uncommon. Where bones lie on sand which is constantly disturbed by the feet of animals, a remarkable polish may develop on their surfaces (Brain 1967b).

The feeding behavior of Kuiseb River Hottentots is a mixture of long-standing tradition and European influence. The anvils and hammerstones are perhaps atypical of Stone Age counterparts in that they are not specifically fashioned for their purpose. Folding pocketknives, enamel basins, and iron cooking pots appear to be standard equipment in all the villages. The use of bones as tools was not practiced by any of the Hottentots living in the villages at the time of the study.

It seemed advisable to be able to separate the damage done to goat bones by Hottentots themselves from that caused by their dogs (Figure 3). A goat was consequently bought from one of the inhabitants of the Zoutrivier village and was then given back to the people of the community. Over two days they consumed, in their traditional manner, all that was edible of the goat and returned the bones without allowing their dogs access to them. The goat was a young male, estimated to be one-year old,

FIGURE 2
A hottentot dog, one of those which contributed to the damage observed on the goat bones.

FIGURE 3
Occupational debris, including goat bones, outside one of the Topnaar Hottentot villages on the Kuiseb River. Characteristic huts, made of the bark from *Acacia albida* trees, are visible in the background.

in which the second molar teeth were about to erupt. The following is a summary of the observed damage to the skeleton:

Skull: the 7-inch horns were broken off at their bases to allow cooking of the head; the occiput was smashed to allow removal of the brain; snout and palate were broken off as a unit; mandible undamaged.

Vertebrae: the head was removed by chopping through the axis; the atlas and part of the axis remained attached to the occiput. Very little damage was done to the other cervicals; thoracic vertebrae suffered fairly extensive damage to their dorsal spines and transverse processes.

Lumbars: slight damage to transverse processes.

Sacrum: undamaged.

Caudal: only the first survived, all the rest had been chewed and eaten.

Ribs: slight damage to their distal ends only.

Scapula: undamaged.

Pelvis: chopped through pubic symphysis and across actabula. No other damage.

Humerus: both shafts were broken transversely through their middles for extraction of marrow. One proximal end was completely chewed away, one left complete; both distal ends were undamaged.

Radius and ulna: both severely shattered by stone impact.

Femur: heads and trochanters removed and proximal shaft ends chewed; both shafts were broken through the middle; both distal epiphyses removed and distal shaft ends chewed.

Tibia: both shafts were broken through middle. Some damage to each end.

Metapodials: all four proximal ends complete; all distal epiphyses removed and distal shaft ends chewed back severely.

Carpal and tarsal bones: undamaged.

Phalanges: undamaged.

Apart from the results of stone impact, it was surprising to find that the Hottentots are capable of inflicting quite considerable damage to bones with their teeth (Figure 4). Fifteen tail vertebrae were chewed and swallowed, while limb bones, such as femora and metapodials, suffered severely at their ends. It is doubtful if the condition of Hottentot teeth would be as good as that of hunter-gatherer peoples. The staple Hottentot diet, apart from occasional meat, is mealie-meal porridge which very likely results in accelerated dental decay. It is to be expected that Stone Age people would have done even greater damage to bones with their teeth than is the case with Kuiseb River Hottentots.

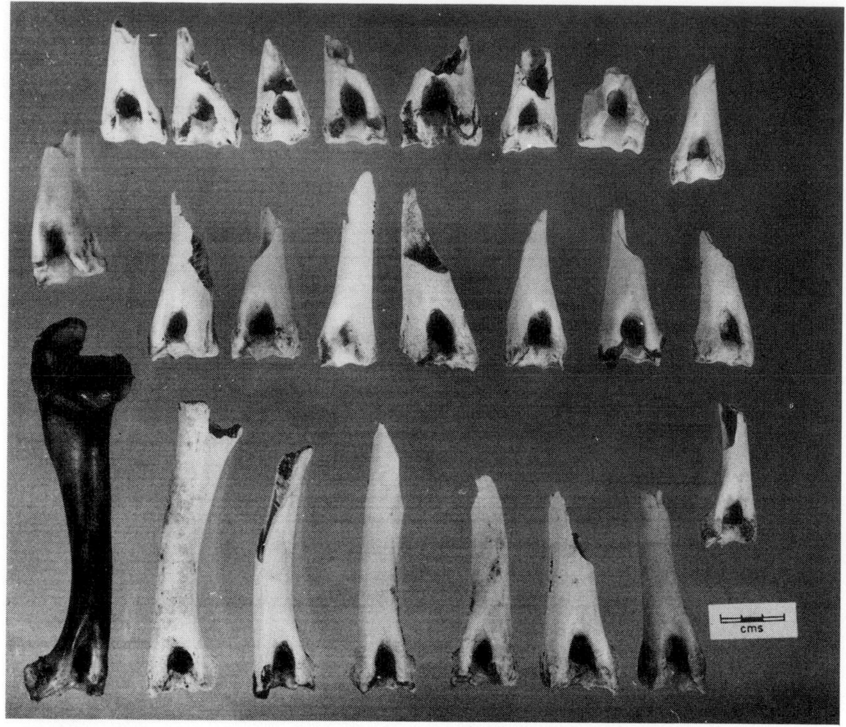

FIGURE 4
Distal humeri of goats from the Kuiseb River remains, showing characteristic spiral fractures of the shafts. A complete humerus is shown at bottom left.

THE COMPOSITION OF THE BONE ACCUMULATION

During 1966 and 1967 a collection of 2373 goat bone fragments was made in Hottentot villages along the Kuiseb River. This collection has been described elsewhere (Brain 1967a) and is made up of parts shown in Table 1.

The minimum number of individual goats which contributed to the sample is, when estimated on horns, 190. Since the bone accumulation was originally described, it has been found that the figure of 190 is deceptively high. The reason for this is that in the extreme aridity of the Kuiseb River environment, horn is almost indestructible and lasts for many years after the last trace of bone has disappeared. Part of the original sample came from two deserted village sites which had not been occupied for over ten years. These yielded horns to the almost complete exclusion of other skeletal parts.

The average rainfall on the Kuiseb study area is less than one inch per year. In more normal climatic areas, with rainfalls of over 10 inches per

TABLE 1

The composition by body part of a sample of goat bone fragments collected from Hottentot villages along the Kuiseb River, 1966–67.

	Skeletal part		Totals
Skull	Horns and cores	385	
	Cranial fragments	70	
	Maxillary fragments	57	512
Mandible	Complete half mandibles	38	
	Mandibular fragments	150	188
Loose teeth		15	15
Vertebrae	1st cervical (atlas)	12	
	2nd cervical (axis)	14	
	Other cervical	12	
	Thoracic	21	
	Lumbar	31	
	Sacral	1	
	Caudal	0	
	Fragments	24	115
Ribs		174	174
Scapula	Head portion	28	
	Other fragments	31	59
Pelvis	Acetabular portion	34	
	Other fragments	21	55
Humerus	Proximal ends	0	
	Distal ends	82	
	Shaft fragments	114	196
Radius + ulna	Complete bones	3	
	Proximal ends	62	
	Distal ends	19	
	Shaft fragments	123	207
Femur	Proximal ends	18	
	Distal ends	9	
	Shaft fragments	88	115
Tibia	Proximal ends	13	
	Distal ends	72	
	Shaft fragments	152	237
Metacarpal	Complete bones	8	
	Proximal ends	24	
	Distal ends	15	
	Shaft fragments	53	100
Metatarsal	Complete bones	9	
	Proximal ends	30	
	Distal ends	11	
	Shaft pieces	51	101
Astragalus	Complete	16	16
Calcaneum	Complete	14	14
Phalanges	Complete	21	21
Bone flakes		248	248
TOTAL			2373

annum, horn disappears rapidly, exposing the bony core which is composed of easily destroyed spongy bone. It is now obvious that, while the goat bone sample is being considered as an entity in itself, horns may reasonably be included; nevertheless, if it is to be used for comparison with bone accumulations from other areas, the incidence of horns will appear deceptively high.

The purpose of this article is to compare the goat bone sample with a fossil accumulation from Makapansgat. There is evidence that although the Makapansgat climate was drier than it is today (Brain 1958), it certainly did not approach the aridity of the Namib Plain. For purposes of discussion therefore, horns will be omitted from the bone accumulations under review.

Following horns, the most numerous single skeletal parts present are mandibles. It was found that the 188 fragments could be divided into 53 left-mandibles and 64 right-half ones. These indicate a minimum of 64 individual goats which have contributed to the sample. Following aging criteria quoted by Cornwall (1956) for sheep, the age structure of the sample has been worked out from the mandibles. The aging criteria used are as follows:

1st molar unerupted: under 6 months.
1st molar in use; 2nd unerupted: 6 to 12 months.
2nd molar in use; 3rd unerupted: 12 to 20 months.
3rd molar in use: over 20 months.

On this basis, indicated ages for the left and right mandibles are given in Table 2.

TABLE 2
Ages for right and left mandibles (Kuiseb River bone sample).

AGE CLASS	NUMBER OF GOATS	
	LEFT SIDE	RIGHT SIDE
Under 6 months	1	0
6 to 12 months	17	23
12 to 20 months	7	6
Over 20 months	28	35
TOTALS	53	64

The age groups are plotted in Figure 5. It is clear that goats are slaughtered predominantly either when young, less than one year of age, or when fully adult at two or more years old. Such indications are confirmed by verbal statements of the Hottentots. As a generalization one may say that almost half the goats represented in the bone sample are immature animals.

The combined feeding action of Hottentots and dogs on the bones has resulted in the disappearance of some parts of the skeletons and survival of others. It has also resulted in some very characteristic damage to certain parts. Such damage will now be considered briefly:

Skull: the braincase has been broken open by stone impact to allow removal of the brain. In most cases the occiput or floor of the skull has been broken out resulting in the production of a basin-like fragment. At Makapansgat, Dart has suggested that bowl-like skull fragments have been used as receptacles. Among the goat bones, their presence and form is clearly coincidental. In most cases the palates have been detached from the braincases completely; mandibles are generally little damaged except round their angles and lower margins.

Vertebrae: these show damage particularly on their spines and processes.

Ribs: these have generally been chewed at both ends.

Scapulae: extensive damage has normally been done to the flat blades.

Pelves: these have characteristically been gnawed down to little more than acetabular portions.

Damage to limb bones is reflected best by the presence or absence in the sample of their ends (to be discussed shortly). Shafts have typically been broken through by hammerstone impact and spiral fractures are common. Such fractures are a feature noted by Dart at Makapansgat. Carpal, tarsal and phalangeal bones, when they occur, are typically undamaged.

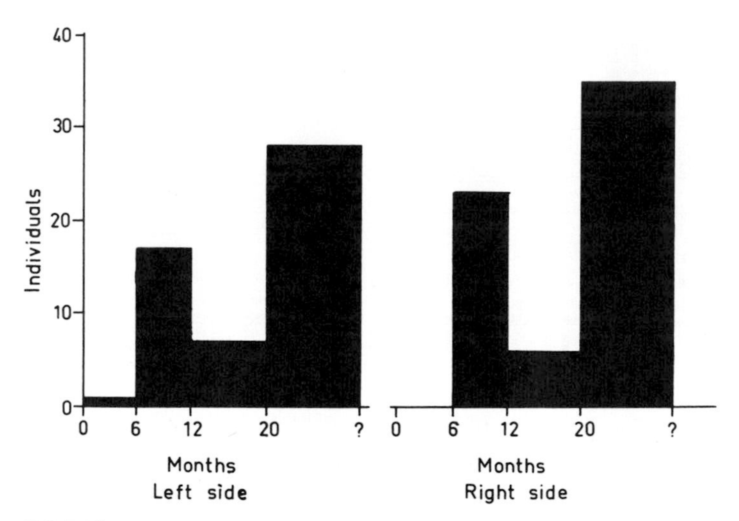

FIGURE 5
Histogram showing the numbers of individual goats in each age class as estimated by tooth eruption in left and right half-mandibles.

SURVIVAL AND DISAPPEARANCE OF SKELETAL PARTS

The survival of parts of the goat skeletons in the sample under review is clearly based on the durability of such parts. Certain elements in the skeletons disappear when subject to the combined chewing of Hottentots and their dogs, others do not. The percentage survival of different parts is therefore a measure of their resistance to this kind of destruction.

Working on a minimum number of 64 individual goats it is possible to calculate the original number of each skeletal part which must have existed, and from this one may estimate the percentage survival of the part in the sample. In the case of ribs, for instance, 26 of which are found in a single goat skeleton, the original number contributed by 64 goats must have been 1664. Only 179 have been found, indicating a 10.2% survival.

Table 3 shows different parts of the goat skeleton arranged in descending order of survival. These results are plotted in Figure 4a. It will be seen that the parts most resistant to destruction are mandibles and distal ends of humeri, which are the most numerous. Proximal ends of humeri and caudal vertebrae have proved so vulnerable to have disappeared entirely.

TABLE 3
Percentage survival of different parts of the goat skeleton (Kuiseb River bone sample).

PART	NUMBER FOUND	ORIGINAL NUMBER	PERCENTAGE SURVIVAL
Half mandibles	117	128	91.4
Humerus, distal	82	128	64.0
Tibia, distal	72	128	56.3
Radius and Ulna, proximal	65	128	50.8
Metatarsal, proximal	39	128	30.4
Scapula	35	128	27.4
Pelvis, half	34	128	26.6
Metacarpal, proximal	32	128	25.0
Axis	14	64	21.9
Atlas	12	64	18.8
Metacarpal, distal	23	128	18.0
Radius and Ulna, distal	22	128	17.2
Metatarsal, distal	20	128	15.6
Femur, proximal	18	128	14.1
Astragalus	16	128	12.5
Calcaneus	14	128	10.9
Ribs	170	1664	10.2
Tibia, proximal	13	128	10.1
Lumbar vertebrae	31	384	8.1
Femur, distal	9	128	7.0
Cervical 3–7 vertebrae	12	320	3.8
Phalanges	21	768	2.7
Thoracic vertebrae	21	832	2.5
Sacrum	1	64	1.6
Caudal vertebrae	0	1224	0
Humerus, proximal	0	128	0

THE PREDICTABLE PATTERN OF SURVIVAL IN LIMB BONES

It is clear that those parts of the goat skeletons which survive best are the unchewable ones. Nevertheless, in the case of limb bones, percentage survival can be related in quantitative terms to particular qualities. In the case of the humerus for instance, survival of the proximal end is nil, while that of the distal end amounts to 64.0%. As has previously been discussed (Brain 1967a), survival of part of a long bone can be related to the times at which each epiphysis fuses to the shaft. In the case of the goat, the distal epiphyses fuses when the animal is four-months old; fusion of the proximal end is not complete until 17 months. This means that when a year-old goat is eaten, the distal end of the humerus will be fully ossified and unchewable, while the proximal end remains cartilagenous.

In addition to fusion times, structural considerations are very important. The proximal end of the humerus is wide, thin-walled, and filled with spongy bone; the distal end is comparatively narrow and compact. Such qualities may be expressed quantitatively, in terms of specific gravity of each end of the bone. Experimental procedure is as follows. The shaft of a dry, defatted humerus is cut through at right angles to its axis, midway along the length of the bone. Each end is weighed individually, the cut ends of the hollow shaft are then filled with plasticine. Any other openings are similarly filled. The volume of each end is then measured by submersion in water and specific gravities are calculated. It is found that the proximal end of a goat humerus has a sp. gr. of approximately 0.6; that of the distal end is about 1.0. There is a clear and direct relationship between specific gravity of the end of a long bone and its percentage survival.

Table 4 gives figures for percentage survival, specific gravity and fusion time for each end of the goat limb bones listed. These figures are plotted in Figure 6. It will be seen that percentage survival is related directly to specific gravity of the part concerned, but inversely to the fusion

TABLE 4
Percentage survival, specific gravity and fusion time for each end of the goat limb bones (Kuiseb River bone sample).

PART	PERCENTAGE SURVIVAL	SPECIFIC GRAVITY	FUSION TIME (MONTHS)
Humerus: proximal	0	0.58	17
Humerus: distal	64.0	0.97	4
Radius and Ulna: proximal	50.8	1.10	4
Radius and Ulna: distal	17.2	0.97	21
Femur: proximal	14.1	0.75	18
Femur: distal	7.0	0.72	20
Tibia: proximal	10.1	0.82	25
Tibia: distal	56.3	1.17	15

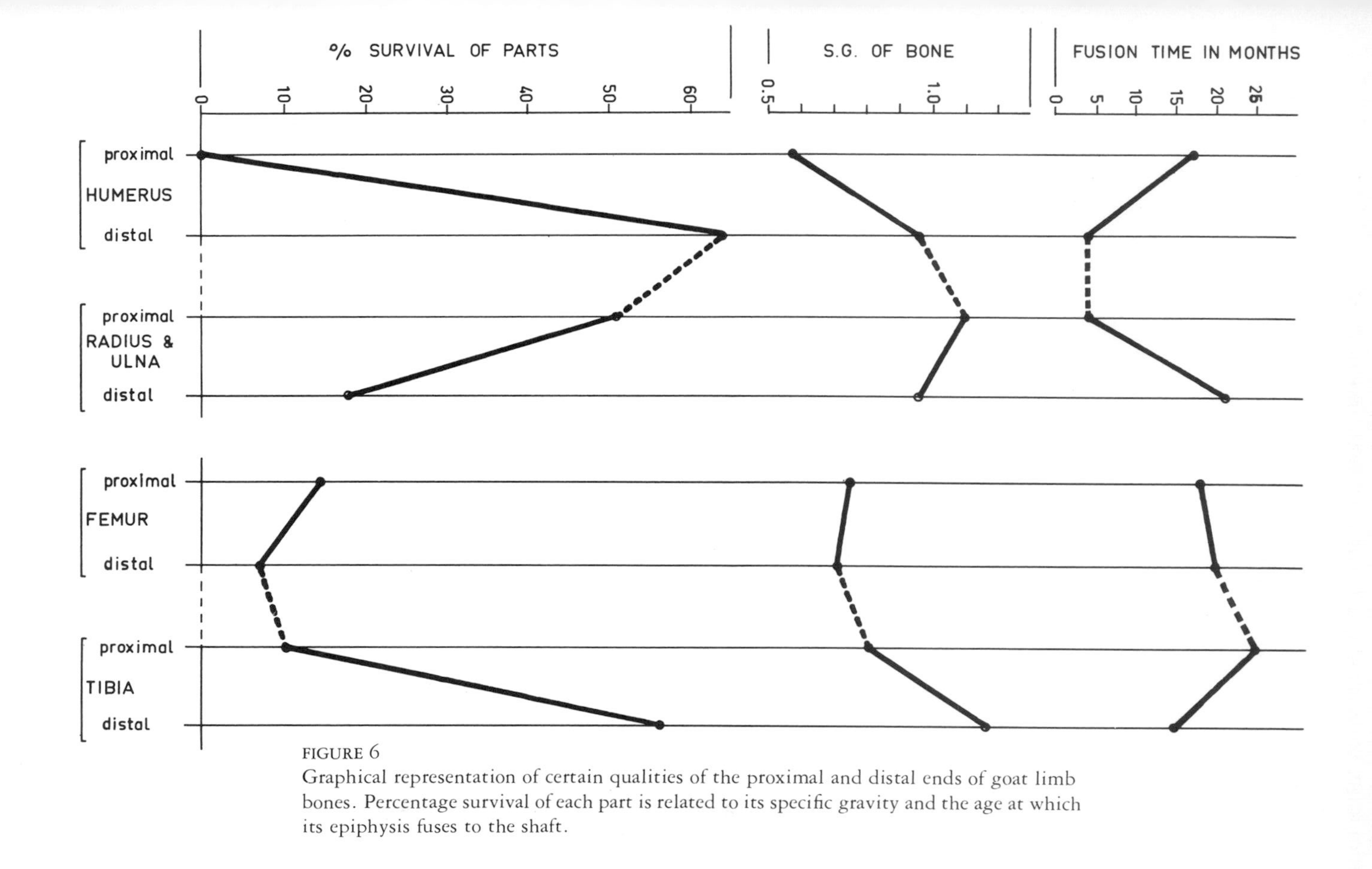

FIGURE 6
Graphical representation of certain qualities of the proximal and distal ends of goat limb bones. Percentage survival of each part is related to its specific gravity and the age at which its epiphysis fuses to the shaft.

time expressed in months. The conclusion to be drawn is simply that survival is not haphazard, but is determined by inherent qualities of the parts.

SURVIVAL OF PARTS IN THE MAKAPANSGAT BONE SAMPLE

Dart's 1957 analysis was based on remains from 293 antelope. His estimation of minimum numbers of individual animals of different sizes was as follows:

Large antelope, based on 74 radial fragments: 39 individuals.

Medium antelope, based on 238 humeral fragments: 126 individuals.

Small antelope, based on 191 mandible fragments: 100 individuals.

Very small antelope, based on 53 mandible fragments: 28 individuals.

TABLE 5
Percentage survival of antelope skeleton (Makapansgat bone sample).

PART	NUMBER FOUND	ORIGINAL NUMBER	PERCENTAGE SURVIVAL
Half mandibles	369	586	62.9
Humerus, distal	336	586	57.3
Radius and Ulna, proximal	279	586	47.6
Metacarpal, distal	161	586	27.4
Metacarpal, proximal	129	586	22.0
Scapula	126	586	21.5
Tibia, distal	119	586	20.3
Radius and Ulna, distal	114	586	19.5
Metatarsal, distal	110	586	18.8
Metatarsal, proximal	107	586	18.3
Pelvis, half	107	586	18.3
Calcaneus	75	586	12.8
Tibia, proximal	64	586	10.9
Astragalus	61	586	10.4
Femur, distal	56	586	9.6
Axis	25	293	8.5
Atlas	20	293	6.8
Humerus, proximal	33	586	5.6
Sacrum	16	293	5.5
Femur, proximal	28	586	4.8
Cervical 3–7 vertebrae	47	1465	3.2
Lumbar vertebrae	30	1758	1.7
Phalanges	47	3516	1.3
Ribs	66	7618	0.9
Thoracic vertebrae	24	3809	0.6
Caudal vertebrae	1	4688	0

Using the total number of 293 individuals, it has been possible to calculate the percentage survival of different parts of the skeleton, as has been done for the Kuiseb River goat bones. Skeletal parts, listed in descending order of survival are given in Table 5 and plotted in Figure 7.

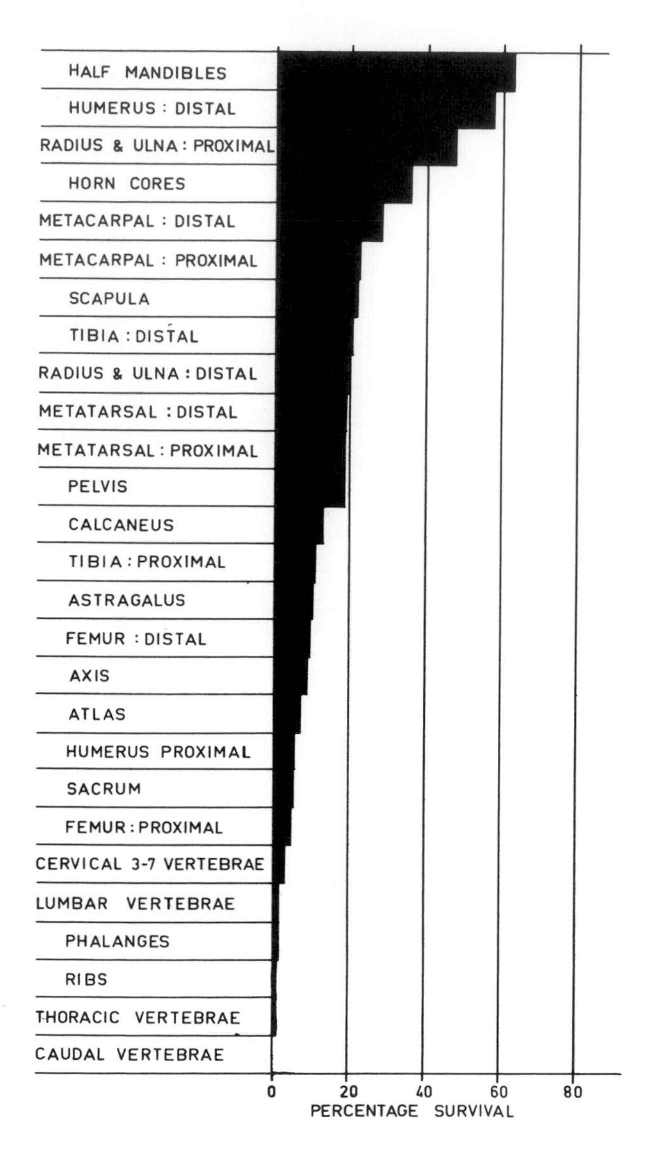

FIGURE 7
Histogram showing the percentage survival of parts of bovid skeletons from Makapansgat. The sample consists of bones from a minimum number of 293 individuals.

THE MAKAPANSGAT/GOAT COMPARISON

The order of survival of different skeletal parts in the goat-bone sample is plotted in Figure 4a; that of the Makapansgat antelope remains in Figure 8. It will be seen that the form of the two histograms is similar. In both,

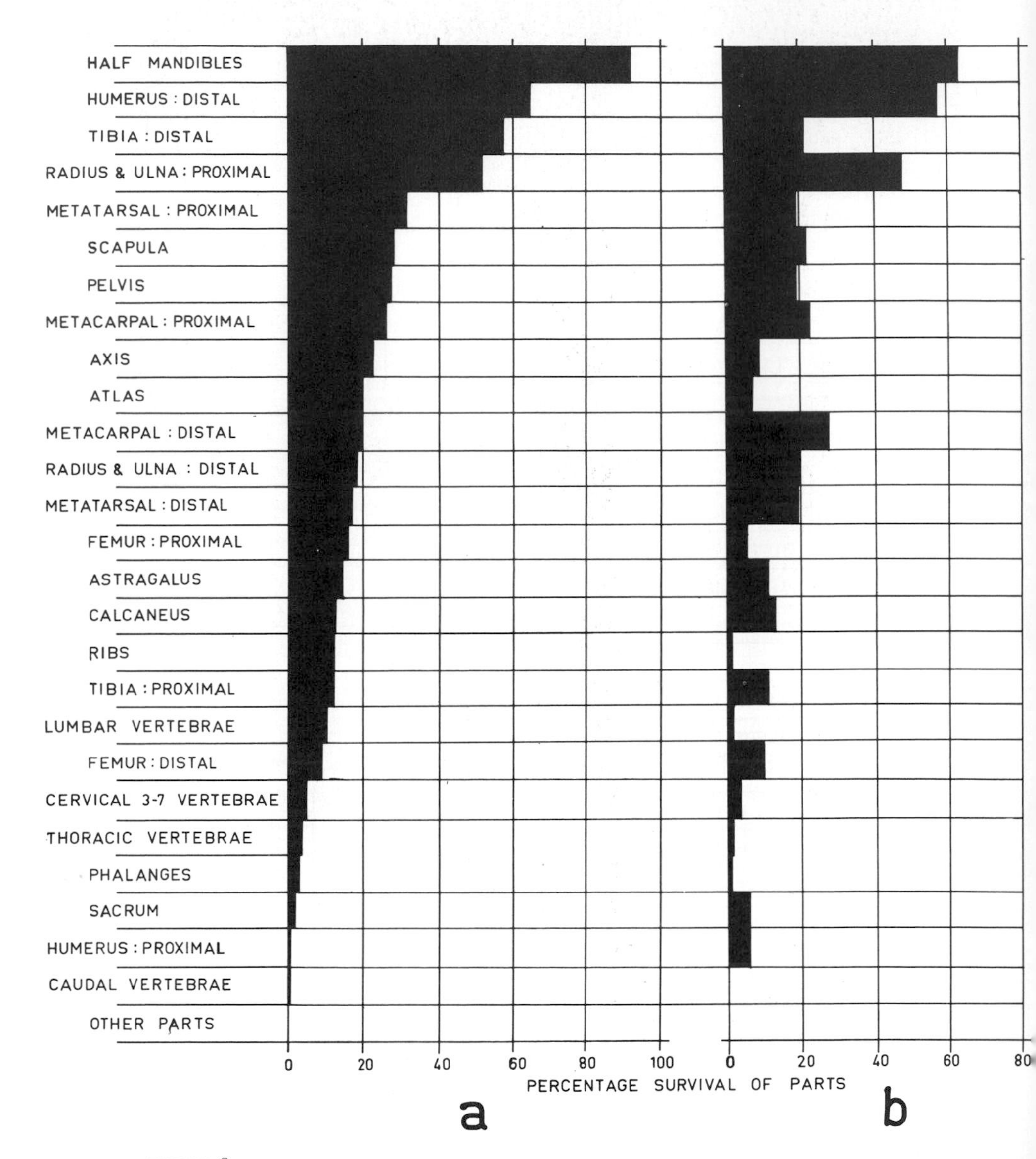

FIGURE 8
(a) Histogram showing percentage survival of parts of goat skeletons from the Kuiseb River. Calculations are based on a minimum of 64 individuals. (b) Percentage survival of parts of bovi skeletons from Makapansgat, arranged in the same order as for (a).

the parts with the highest percentage survival are mandibles, followed by distal humeri. At the lower end of the survival curve in both collections are such parts as thoracic and caudal vertebrae. In spite of the broad similarity between the form of the two histograms, the detailed order of survival of parts differs in the two cases.

For the purposes of direct comparison, the percentage survival figures for the Makapansgat are replotted in Figure 4b so that they follow the order laid down by the goat bones. It will be seen that although the two histograms are not identical, the trends in survival order are broadly similar.

When comparing these results it should be borne in mind that the Makapansgat sample is made up of bones from animals ranging in size from eland to steenbok. They have almost certainly been subjected to destructive treatment of a variety of kinds, including feeding and tool-using activities of australopithecines, as well as scavenging by carnivores. By contrast, the goat-bone sample is made up of bones from one species of small bovid, subject only to feeding activities of men and domestic dogs. In view of this, the overall similarity in composition of the bone collections is remarkable. It is a reflection of the predictable pattern of survival which manifests itself when whole bovid skeletons are subjected to destructive treatment.

CONCLUSIONS

In his pioneering study of the bone accumulation at Makapansgat, Dart found serious disproportions in the parts of skeletons preserved as fossils. Certain parts were common, others were hardly represented at all. In an attempt to explain these disproportions Dart postulated that the missing bones were simply not brought back to the cave at all by the resident australopithecines. He writes, "the disappearance of tails was probably due to their use as signals and whips in hunting outside the cavern. Caudal and other vertebrae may also have disappeared because of the potential value of their bodies as projectiles and of their processes (when present) as levers and points." Likewise, "the femora and tibiae would be the heaviest clubs to use outside the cavern; that is probably why these bones are the least common. Humeri are the commonest of the long bones; probably because they would be the most convenient clubs for the woman-folk and children to use at home" (Dart 1957).

The evidence of the Kuiseb River goat bones strongly suggests that the disproportions which Dart encountered in fact do not require any special explanation. Artificial selection of certain skeletal parts need not be postulated. If, for instance, antelope were hunted as they came to drink at a waterhole in the entrance of the cave and were then consumed by australopithecines and scavenging carnivores, a considerable bone accumulation could have been built up in the lower parts of the cavern. The bones preserved would have been those most able to survive the destructive treatment to which they had been subjected. The Makapansgat sample, like that from the Kuiseb River, does in fact consist of resistant skeletal elements, whose frequencies follow a predictable pattern.

Acknowledgments

Grateful acknowledgment is made to the Wenner-Gren Foundation for Anthropological Research and the C.S.I.R. for their support of this research.

References Cited

Brain, C. K.

1958 The Transvaal Ape-Man-Bearing Cave Deposits. Mem. Transv. Mus. No. 11:1–131.

1967a Hottentot food remains and their bearing on the interpretation of fossil bone assemblages. Scient. Pap. Namib Desert Res. Stn. No. 32:1–11.

1967b Bone weathering and the problem of bone pseudo-tools. S. Afr. J. Sci. 63:97–99.

1969 The contribution of Namib Desert Hottentots to an understanding of australopithecine bone accumulations. Scient. Pap. Namib Desert Res. Stn. No. 39:13–22.

Cornwall, I. W.

1956 Bones for the Archaeologist. London: Phoenix House. 1–255.

Dart, R. A.

1957 The Osteodontokeratic Culture of *Australopithecus prometheus*. Mem. Transv. Mus. No. 10:1–105.

Jenkins, T. and C. K. Brain

1967 The peoples of the lower Kuiseb Valley, South West Africa. Scient. Pap. Namib Desert Res. Stn. No. 35:1–24.

PART TWO

Geology and the Reconstruction of the World of Early Man

Introduction to Part Two

The fossil record of human evolution comes to us from ancient rocks thus making geological science both literally and metaphorically a foundation stone in the study of human origins. From geology we derive first and foremost information on the sequence of fossils, and wherever possible their ages in years. However, from careful imaginative work we can also establish the particulars of the geography and environments in which our ancestors and early hominid relatives lived, died, and evolved.

The five essays in this section exemplify the characteristics of current geological and paleoenvironmental studies. They show the extensive interrelationship of geology and paleontology that enlivens current research. The first essay, "East Africa as a source of fossil evidience for human evolution," was specifically written as an introduction to this geological section.

Sediments accumulate on the floor of the Rift Valley, layer upon layer, but repeated earth movements shift portions of the deposits to positions where they are subject to erosion. The formation of this small canyon near Lake Natron exposed the Peninj australopithecine mandible.

Lake Turkana, formerly known as Lake Rudolf. (A satellite image NASA ERTS E-1193-07221-702.)

Glynn Ll. Isaac:

East Africa as a Source of Fossil Evidence for Human Evolution

Why have so many early-man fossils been found in East Africa? How do we know how old they are? What influences have environmental changes had on the course of human development? These and related questions are dealt with in Glynn Isaac's essay which serves as a general introduction to this section.

The singular importance of East Africa as a source of evidence relating to human origins is due in several crucial ways to a series of extraordinary geological circumstances, which have served to preserve a sequence of fossil hominoids and hominids that spans more than 20 million years. The record is exceedingly complex, and one of Louis Leakey's many contributions to science was his success in attracting a number of younger geologists who have made detailed studies that begin to give us an understanding of the processes involved in the accumulation of the rich sedimentary and fossil record. The fruits of this kind of geological research are clearly exemplified in four essays in this volume: John and Judy Van Couvering provide an excellent review of the combined implications of geology and faunal lists for understanding Miocene conditions in East Africa. W. W. Bishop presents a critical examination of the "Pliocene," paying particular attention to tectonic controls on sedimentation and paleoenvironmental implications. F. C. Howell provides a comprehensive summary of the results of one of the largest interdisciplinary research projects yet undertaken in pursuit of human origins. His paper explains the intricate interrelations between geology, paleontology, and paleoanthropology. Finally, R. L. Hay's summary of the "Environmental setting of hominid activities in Bed I, Olduvai Gorge" is an example of the kind of detailed ecological reconstruction that is possible if prolonged periods of careful and resourceful geological studies are undertaken at one locality.

In this short essay I will try to provide material for non-geological readers that will help to furnish a background to the other four geological essays. The topics to be covered that most closely affect paleoanthropological understanding are tectonics and sedimentation; chronology and chronometry; paleogeography and environments.

The specific geological processes that have given rise to East Africa's extraordinarily extensive record are the earth movements and vulcanism associated with the Rift Valley system. The tectonic complex was first recognized as such last century (Suess 1891; Gregory 1896), but it was not until the last decade that studies of it got beyond a reconnaissance stage. A great intensification of research followed the discovery of the mid-oceanic ridges and rifts and the realization that the rifts were part of global tectonic systems (e.g., Takeuchi *et al.* 1970; Calder 1972; Wilson 1972; Cox 1973). The surge of interest has given rise to large quantities of new technical data, most of which has not been synthesized and interpreted. However, some summaries have begun to appear and are cited here, and these contain further technical references. Brief histories of Quaternary research, with bibliographies can be found in Cooke (1958) and in Bishop (1971).

TECTONICS, VULCANISM AND SEDIMENTATION

Baker, Mohr and Williams (1972) have provided an excellent and concise position paper on studies in the Eastern Rift, viewed from the stance of hard rock geologists and geophysicists. This provides stratigraphers and paleontologists alike with a useful body of background information and with references to detailed sources. This summary and others show that the Eastern Rift system transects areas where the earth's crust has been uplifted into gentle domes or "blisters." One such dome forms the highlands of Ethiopia and the Harrar plateau; another forms the highlands of Kenya and northern Tanzania (see Figure 1). The uplift of these regions

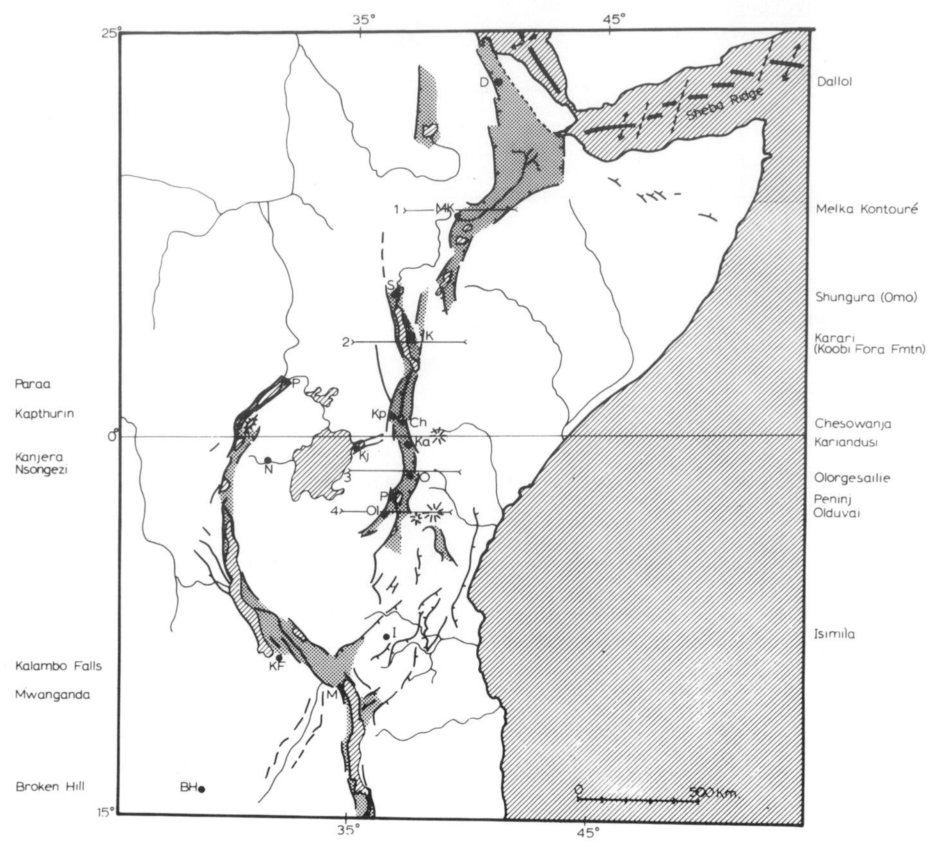

FIGURE 1
The East African Rift system (after Baker *et al.* 1972), with the location of some important paleoanthropological sites marked.

seems to have occurred during the last 10–15 million years and much of it during the last few million years. Through the domes, the Rift Valley slices as a complex series of "cracks" and "troughs" which look on the surface like a chain of closed-lake basins bounded by rugged escarpments that step up to high plateaus. The down-faulted troughs are far too large to be explained simply as the fallen keystones of the arches created by updoming, and it is widely believed that the earth's crust is actively being extended or stretched across the line of the Rift system. Baker *et al.* (1972:54) give estimates of 30 km. of extension in Ethiopia and 10 km. in Kenya over the past 10–20 million years. This implies a rate of stretching of 0.4–1.0 mm. per year, which is one or two orders of magnitude lower than the rate typical of the more active mid-oceanic ridges (see Figures 1 and 2).

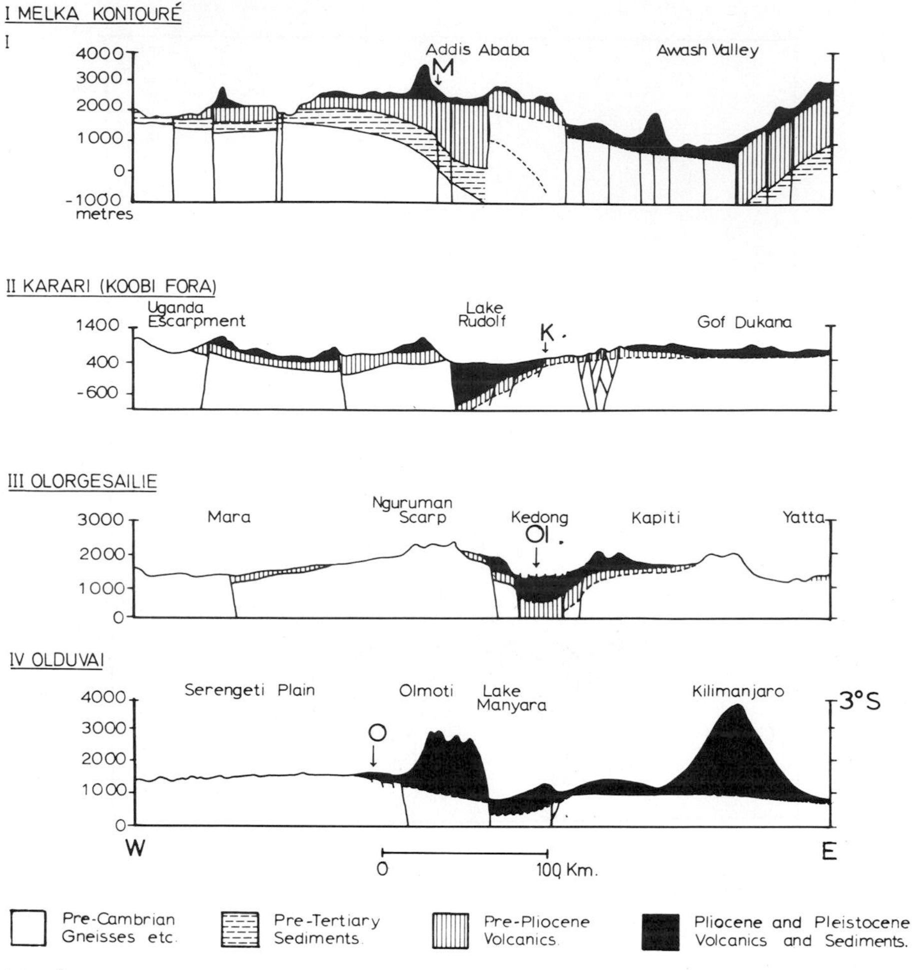

FIGURE 2
Diagrammatic sections across the Rift show the structural context of various paleoanthropological sites (from Isaac, in press).

Associated with the crustal extension and deformation, there has been widespread vulcanism so that the visible rocks of the margins of the eastern Rift and the floor of the trough are largely volcanic. There are vast flood sheets of lavas and huge central volcanoes, as well as gaping calderas and explosion craters. Inasmuch as volcanic eruptions can create highlands, block drainages, or provide sediment in the form of ash, vulcanism and tectonics interact to have major effects on the stratigraphic record.

The geotectonic situation has several consequences of the greatest importance for paleontologists and paleoanthropologists; these include, on the one hand, factors of preservation and accessibility, and on the other, factors connected with environments and ecology.

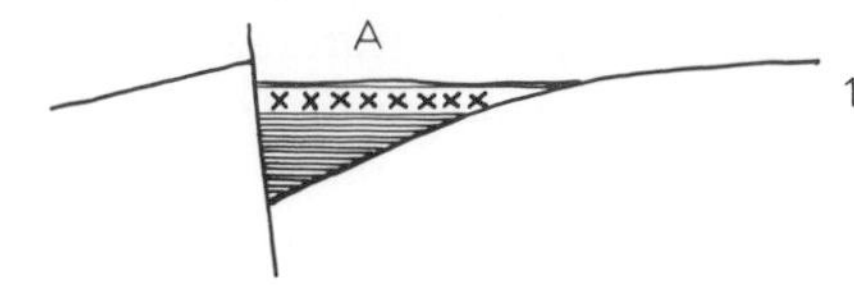

1 Formation A being deposited in a half graben with preservation of fossils.

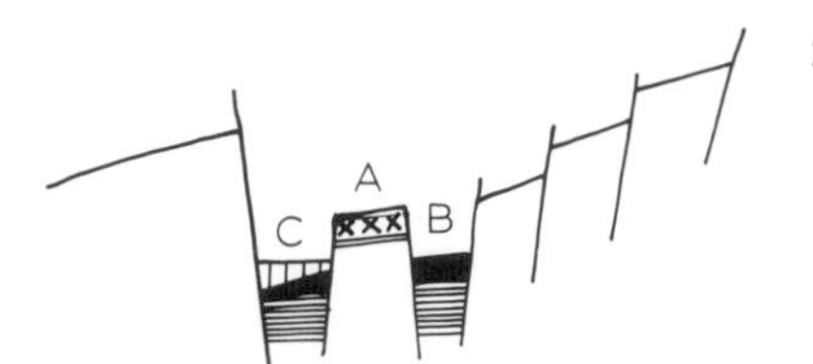

2 Subsidence and fault fracture continues resulting in the covering of Formation A by lavas, B, and younger sediments, C, but a fragment is uplifted and eroded on a horst.

FIGURE 3
Representation of processes in the Rift which first preserve a fossil record and then make it available by erosion.

PRESERVATION AND ACCESSIBILITY

Because the Rift troughs and closed basins are very effective traps, sediments accumulate in them and fossils are often deeply buried and so preserved. Moreover, the restless reshuffling of tectonic blocks brings a proportion of previous accumulations back up to relatively elevated positions and exposes them to erosion so that their contents become potentially visible (Figure 3). The alternating processes of deposition and fracture with partial uplift have varying rates and configurations, so that in different parts of the Rift suites of rocks of different ages are exposed. By studying these and fitting each locality into its correct chronological and paleoenvironmental position, a composite history of changing conditions and of evolution can be built up. The picture can never be complete, but one may acquire a good representative sample of the kinds of configurations which were involved at successive stages.

ENVIRONMENTAL AND ECOLOGICAL EFFECTS OF THE RIFT

The Rift system not only creates a restless set of agencies which alternately preserves geological records and then lifts parts of them up for erosion and eventual destruction; it also creates landscapes with an unusual environmental diversity. Sharp topographic relief coupled with hydrographic and rain-shadow effects mean that over comparatively short distances, conditions may range from grassland-steppe to dense rivurine forest, or from scorching desert to cold alpine meadows.

Careful studies make it clear that this diversity was a feature of the region through the Tertiary and Quaternary. With continuing crustal deformation, environmental conditions in any locality have often been subjected to extensive change, so that we have to think of biological evolution in the region as a varied series of genetic threads running through a complex mosaic of habitats that were continually undergoing kaleidoscopic change.

The other geological essays in this volume each provide specific information on the way in which tectonics and vulcanism have combined and interacted to create and modify the paleontological record that is available. More extensive general treatments of these questions are also to be found in other essays by W. W. Bishop (e.g., 1963, 1965, 1971) and Isaac (1969).

CHRONOLOGY AND CHRONOMETRY

In 1961 Leakey, Evernden and Curtis published a report on potassium-argon age determinations made of feldspar crystals collected from strata closely associated with the *Zinjanthropus* fossil horizon at Olduvai. The age estimate obtained was 1.7 million years, a value 2 or 3 times larger than most paleoanthropologists and stratigraphers would previously have guessed. There was a storm of protest, but since that pioneer dating enterprise, hundreds of other K-Ar age determinations have been made on volcanic rocks that are interbedded with the fossiliferous sequences of East Africa. These have confirmed the high antiquity of the fossils. Because of this geophysical work, the East African sequence is now better calibrated with respect to age than any other paleoanthropologic sequence yet studied. Papers in the volume edited by Bishop and Miller (1972) provide excellent summaries of the methods and of the results. They also make it very clear that this kind of dating, while possible, is not always straightforward. There are many sources of gross error and distortion and continual cross-checks are essential.

An extremely valuable way of checking dates and of adding precision to the dating of sequences has come to the fore in the last few years—namely paleomagnetic stratigraphy. The method hinges on the fact that during the late Cenozoic, the earth's magnetic polarity has changed back and forth at fairly frequent but irregular intervals (Cox, Dalrymple and Doell 1967; Cox 1968, 1969). Over the past four or so million years it has been possible to determine the history of polarity changes with some precision thereby establishing so-called "reversal chronology." Throughout that time, there has been no period longer than about 600,000 years without a polarity change, and the duration and frequency of events is unique for each major segment of late Cenozoic time.

At first, reversal chronology was applied largely to lavas in East Africa and elsewhere (Grommé and Hay 1963; Isaac 1967; Musset, Reilly and Raja 1965). However, it has subsequently been established that if great care is taken, polarity stratigraphy can sometimes be determined in sediments also. This has been done for a number of important East African sections such as Olduvai (see Leakey, M. D., this volume), East Rudolf (Brock and Isaac 1974), and the Omo (Shuey *et al.* 1974). It is also being applied to others.

In addition, small numbers of age estimates have been based on other methods such as Thorium/uranium ratios determined for bone or carbonates (e.g., Butzer, Brown and Thurber 1969; Roubet 1969; Howell *et al.* 1972) and by fission-track dating (Fleischer, Price, Walker and Leakey 1965). Both of these methods yield results consistent with the long chronology indicated by K/Ar. (See Bishop and Miller 1972 for critiques of the methods and the results).

In spite of its rich endowment with datable volcanic rocks, not all important fossil-bearing beds in East Africa have yet been geophysically dated. For these, relative age estimation by the established methods of bio-stratigraphy remain the only possible procedure. Louis Leakey himself drew up some of the first approximations of East African mammal stratigraphy (e.g., Leakey, L. S. B., 1951). Subsequently, others have either based classification and seriation on total faunal assemblages (Cooke 1958, 1963, 1967; Coppens 1972) or have considered groups such as pigs and/or elephants to be particularly sensitive time markers and have used these (Cooke and Maglio 1972; Cooke 1975). Correlations based on the characteristics of fossil assemblages can be used for dating sections the age of which is otherwise unknown, and they can also be used as a cross-check on geophysical age determinations. Clearly there is potential for conflict between paleontology and geophysics here, and there are already some debates in progress. In the Rudolf basin, some paleontologists prefer faunistic correlations that make the Lower Member at East Rudolf contemporary with Members E, F and G at the Omo (see Coppens *et al.* 1975). In spite of the difficulties with K/Ar at East Rudolf, Fitch and Miller (Coppens *et al.* 1975) are confident of an age for the Lower Member that would make it equivalent to Members B and C. For this position there is good support from paleomagnetism (Brock and Isaac 1974) and the interpretations of other paleontologists (Maglio 1972). Only further research can resolve this debate. (Figure 4, drawn by A. K. Behrensmeyer, illustrates this important dilemma.)

It is clear that as paleontological research becomes more detailed and as independent geochronometric dating becomes more refined, such apparent "conflicts" will arise again. For example, it is well-established that there was a comparatively abrupt change in fauna at Olduvai which took place at the time of deposition of the Lemuta member in lower Bed II

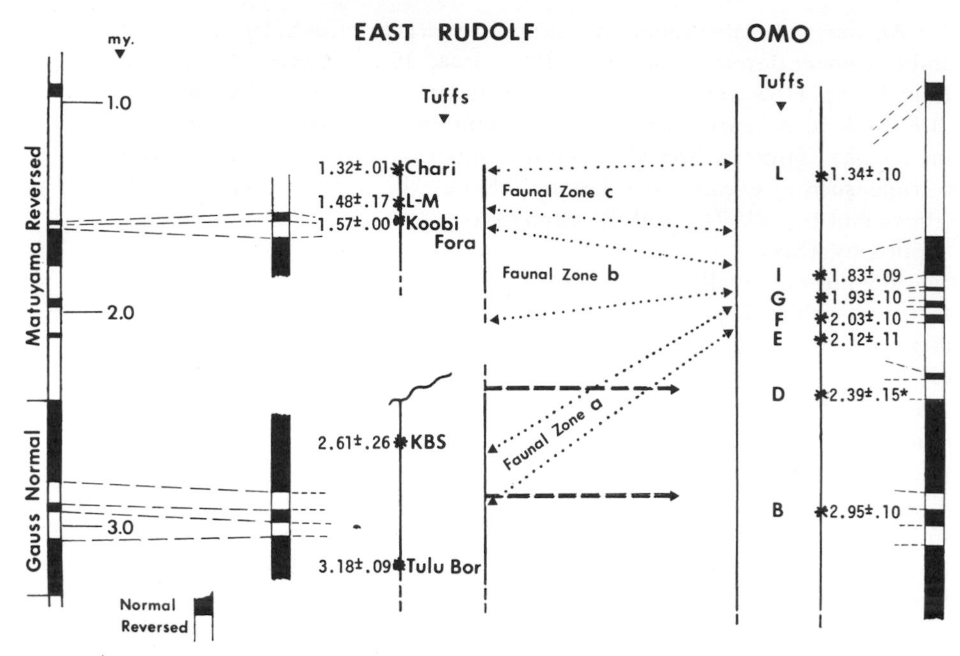

FIGURE 4
Representation of ideas about time relations between the Omo and East Rudolf. Horizontal lines represent time correlations proposed on the basis of current geophysical dates. Diagonal lines represent aspects of similarity of fauna. The question arises as to whether the discrepancies are due to errors in the dates, errors in assessment of faunal resemblance, or to the shift of an ecological/biogeographic zone boundary from one region (ER) to another (Omo).

(Leakey, M. D., 1971). The change is commonly attributed to the onset of more steppic conditions in post-Lemuta times. Now the question arises, where was the Bed II fauna during Bed I times? If a Bed II-type fauna is found in an adjoining sedimentary basin, is it to be equated in time with Bed I or with Bed II? This example serves to show that with a complex mosaic of habitats and with environmental oscillation through time, chronological and ecological correlations can become confused.

Figure 5 provides a summary of the chronological relations among formations in East Africa providing important paleoanthropological evidence from the last 2 million years. This chart is based, where possible, on direct K/Ar dates which are shown, or on K/Ar dates for related strata which are not shown. Where paleomagnetic stratigraphy has been determined, a simplified version of this is shown by stippling (normal) and hatching (reversed). For sources and critical discussion, reference should be made to Bishop (1972), Howell (1972), Isaac (1972) as well as to Isaac (in press) where a similar table is presented. The chart shows that in spite of uncertainty about some sites, the main time relations among fossil and artifact-bearing sediments are better established in East Africa than in any other region. In fact, the matching of other regions to the standard East

African sequence has become a major need in paleoanthropology. For example, where do the various Transvaal australopithecine caves fall on the East African time sequence? What age range do the Java finds span? These are questions to which we do not know the answers at present, though various authorities have favored guesses.

Paleomagnetism has provided geochronologists with a powerful new tool, and one of its first uses has been the recognition in all five continents and the ocean floor of the time plane furnished by the Brunhes-Matuyama polarity change which occurred about 700,000 years ago. It now appears that most of the classic Pleistocene succession of Europe is subsequent to this time plane. Certainly the entire artifact and human fossil record of the temperate zone appears to be of Brunhes age. Figure 6 presents a chart drawn up at a recent Burg Wartenstein symposium (Butzer and Isaac, in press). It shows age estimates based on dates offered by well-informed delegates for important hominid fossils in the entire Old World. The importance of East Africa is apparent—it is the only region where a long sequence has been dated at intervals through its entire span.

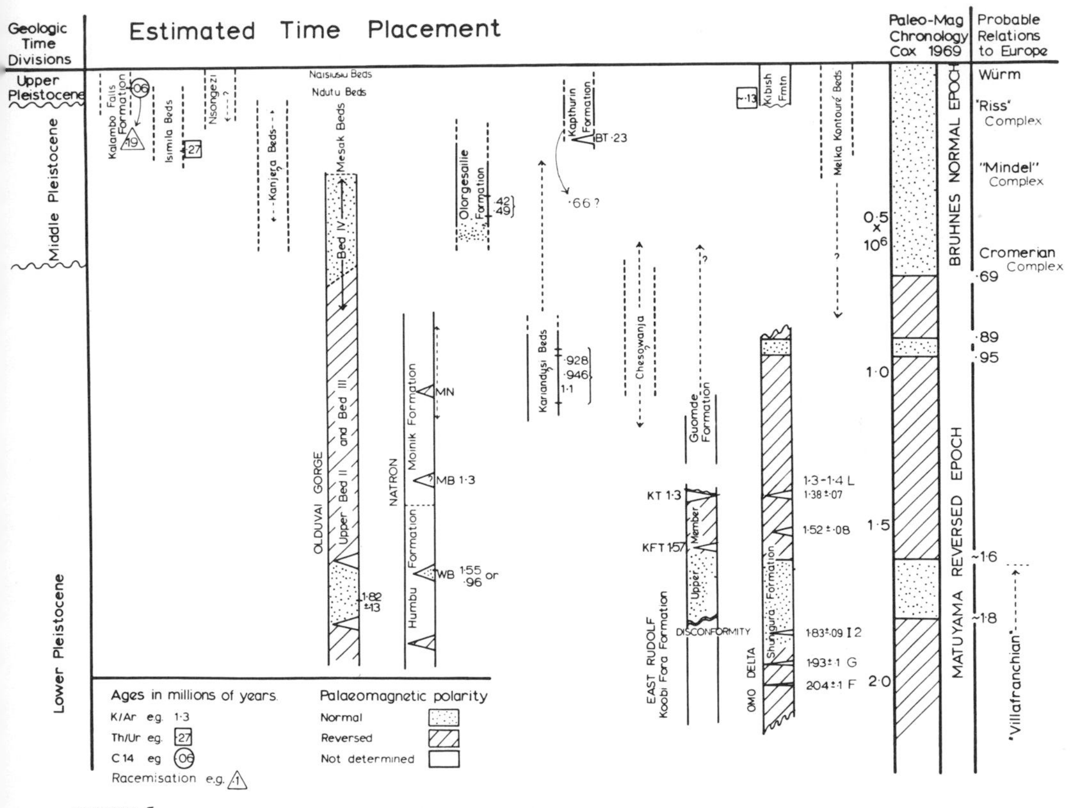

FIGURE 5
The estimated and geophysically measured age relations among Pleistocene formations in East Africa (2.0–0 m.y.) (modified from Isaac, in press).

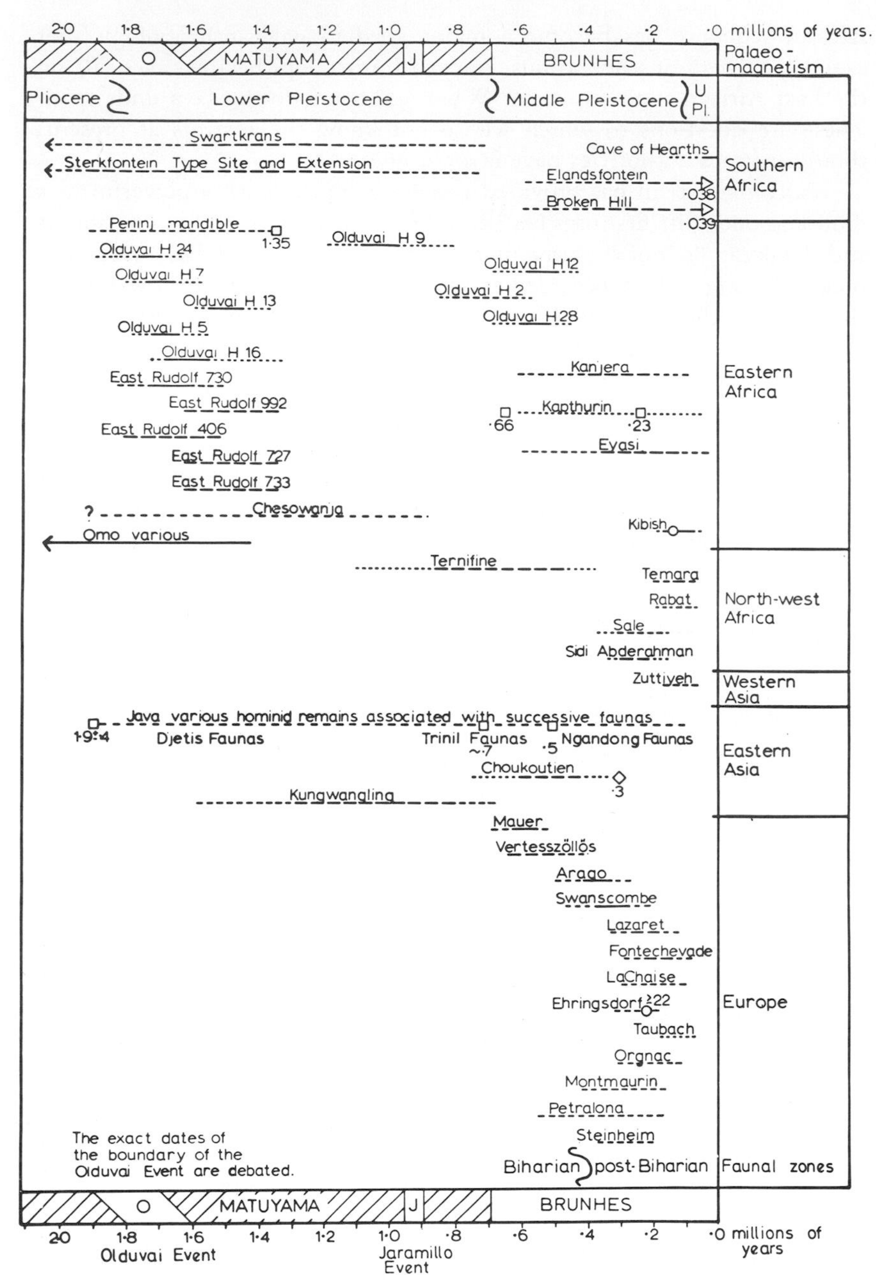

FIGURE 6
Measured or estimated age ranges for selected important hominid fossils of the earlier Pleistocene (2.0–0.1 m.y.). The length of the East African record and its relatively precise dating is apparent. Squares = K/Ar dates; circles = Th/Ur; triangles = C^{14} and diamonds = amino-acid age estimates (from Butzer and Isaac, in press).

PALEOENVIRONMENTS

During the second half of the nineteenth century and the first decades of the twentieth, evidence for what seemed to be a simple and dramatic succession of ice ages was discovered in Europe and America. Afterwards, as research in Africa and other regions got underway, this European image of the Pleistocene was adopted as a sort of norm. Evidence for extreme, widespread climatic fluctuations was expected as a matter of course. Just as the European Quaternary was believed to be marked by a succession of "glacials" and "interglacials," so the African sequence was imagined to consist of a series of "pluvials" and "interpluvials."

Neither the European model nor the African derivative is incorrect, but modern research has shown that both were gross oversimplifications of what was in fact a very complex history of climatic oscillations. We now know that instead of the neat classic system of four major Pleistocene glacials separated by three interglacials, there must have been at least eight interglacials during the last 700,000 years. (See Kukla, Brunnacker and Shackleton in Butzer and Isaac, in press.) The African sequence if anything was more complex.

There is abundant evidence of changing conditions from layer to layer in East African sedimentary sequences, but change in climate is only one of the causes; and many of the most dramatic shifts are now known to be due to the extensive earth movements which have already been discussed.

Even when climatic shifts can be detected and isolated, they turn out to be subtle ones. For instance, R. L. Hay has shown that throughout the Pleistocene, the Olduvai sedimentary basin was subject to semi-arid conditions (Hay 1967a,b, 1971). However, within this warm dry regime there were significant fluctuations such as the "drying up" that is documented in Upper Bed I and Lower Bed II (Hay, this volume). The change was real enough, but as yet there is no easy way to link it with any global climatic history. The former European practice of making climatic change the framework of chronology simply cannot be applied even in the best studied early Pleistocene localities.

There are other reasons why European perspectives have proved inappropriate and need to be replaced. The East African environment is today a very complex mosaic of interpenetrating habitat types (Figure 7). There is no clear-cut zonal patterning corresponding to the latitudinal zonation of Eurasia. Under climatic fluctuations, the mosaic and its component biota has been far more stable than were European floras and faunas. The mosaic appears to have acted as a kind of buffer against extreme effects of climatic oscillations on living forms. The relative proportions of forest, grassland and arid steppe may well have changed repeatedly, but all of these habitat types continued to be present in all regions.

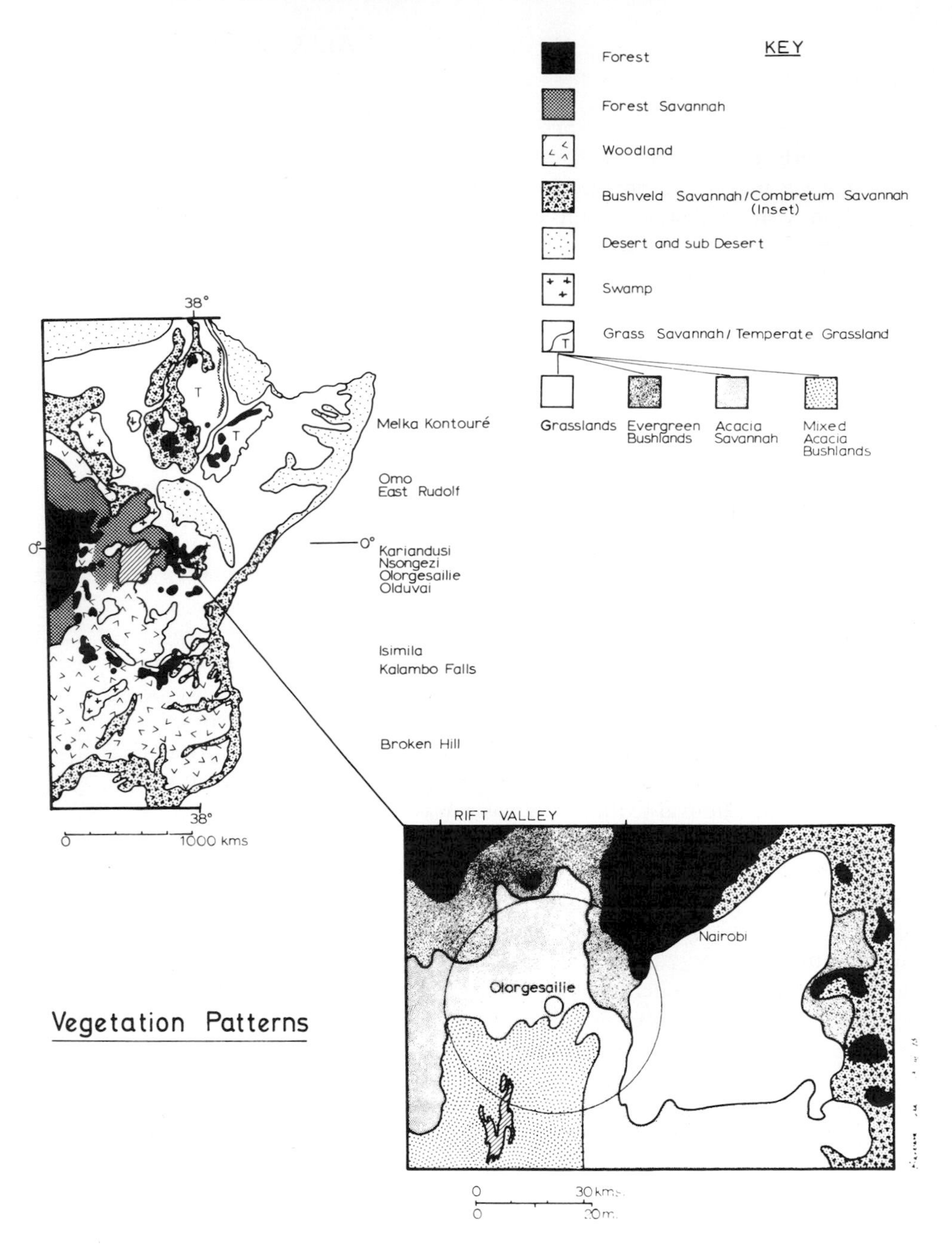

FIGURE 7
The environmental complexity of Eastern Africa as indicated by vegetation zones. This complex mosaic has probably acted as a buffer against over-drastic effects of climatic fluctuations. The inset shows the detail for the vicinity of one important site, as an example (from Isaac, in press).

From a European point of view, it is often said that African fauna is archaic—in the sense that it is more like a Tertiary fauna than is the modern European fauna. However, the matter can be put another way—Africa's biota is diverse and stable, while Europe's fauna consists of the impoverished survivors of a series of climatic traumas.

These are important points to bear in mind when thinking about the environmental context of the earlier stages of human evolution. Hominids were subject to restless fluctuations in their habitats, but not to complete ecological disruptions.

We are a very long way from being able to string together a complete correlated history of the numerous environmental oscillations which undoubtedly occurred in Africa. Indeed, it may never be possible, and it may not be essential for our understanding of evolution. This is part of the message of W. W. Bishop's essay in this volume.

Since the critical reviews of Cooke (1958) and Flint (1959a,b) research endeavor has turned from preoccupation with narrow historicism to attempts to reconstruct examples of specific paleoenvironmental settings in which hominid populations can be shown to have lived. From a series of such cases we may come to a far better understanding of ecological relationships and their dynamics than we ever could by retaining a preoccupation with global climatic history. R. L. Hay's essay in this volume is an excellent example of such a study. It is clear that his ability to offer a detailed reconstruction of the habitat far outweighs the disadvantage of not being able to assign Bed I to one or another pluvial. The Van Couverings' paper provides further examples of such reconstructions for an earlier time range.

In order to achieve the reconstructions, an increasingly wide range of techniques and disciplines is being used. These include detailed sedimentary studies, mineralogy, pollen analysis, and various physical and chemical techniques. As mentioned in the introduction to Part Two, a new kind of investigation termed *taphonomy* is now widely regarded as of great importance. This involves examination of the processes whereby a potential fossil record forms today, with consequent heightened ability to interpret the environmental meaning of fine details of ancient fossil beds (Behrensmeyer 1973, in press; Hill and Walker 1972). As an alert naturalist, Louis Leakey was always a vigorous and unorthodox advocate of using the present as a key to the past, and taphonomy is largely a systematization of that alertness.

What we now know about paleoenvironments and evolution in East Africa is like the exposed tip of an iceberg—it is abundantly clear that vast numbers of fossiliferous and artifact-bearing localities remain to be found. It is also clear that careful and imaginative investigations can answer more and more detailed questions regarding the circumstances surrounding evolution in East Africa.

References Cited

Baker, B. H., P. A. Mohr, and L. A. J. Williams

1972 Geology of the Eastern Rift System of Africa. Geological Society of America. Special Paper 136.

Behrensmeyer, A. K.

1973 The Taphonomy and Paleoecology of Plio-Pleistocene Vertebrate Assemblages east of Lake Rudolf, Kenya. Ph.D. thesis, Harvard, Cambridge.

In press The Habitat of Plio-Pleistocene Hominids in East Africa; Taphonomic and Micro-Stratigraphic Evidence. *In* African Hominidae of the Plio-Pleistocene, C. Jolly, ed.

Bishop, W. W.

1963 The Later Tertiary and Pleistocene in Eastern Equatorial Africa. *In* African Ecology and Human Evolution. F. C. Howell and F. Bourlière, eds. Chicago: Aldine, pp. 246–275.

1965 Quaternary Geology and Geomorphology in the Albertine Rift Valley, Uganda. *In* International Studies on the Quarternary. H. E. Wright Jr. and D. G. Frey, eds. Boulder: Geological Society of America, Special Papers 84. pp. 293–321.

1971 Late Cenozoic History of East Africa in Relation to Hominoid Evolution. *In* Late Cenozoic Glacial Ages. K. K. Turekian, ed. New Haven and London: Yale University Press. pp. 493–528.

1972 Stratigraphic Succession 'versus' Calibration in East Africa. *In* Calibration of Hominoid Evolution. W. W. Bishop and J. A. Miller, eds. Edinburgh: Scottish Academic Press. pp. 219–246.

Bishop, W. W., and J. A. Miller, eds.

1972 Calibration of Hominoid Evolution. Edinburgh: Scottish Academic Press.

Brock, A., and Glynn Ll. Isaac

1974 Paleomagnetic Stratigraphy and Chronology of Hominid-bearing Sediments east of Lake Rudolf, Kenya. Nature 247:344–348.

Butzer, K. W., F. H. Brown, and D. L. Thurber

1969 Horizontal sediments of the Lower Omo Valley: the Kibbish Formation. Quaternaria XI:15–29.

Butzer, K. W., and G. Ll. Isaac, eds.

In press After the Australopithecines; Stratigraphy, Ecology and Culture Change. The Hague: Mouton. (World Anthropology Series)

Calder, N.

1972 The Restless Earth. New York: Viking Press.

Cooke, H. B. S.

1958 Observations Relating to Quaternary Environments in East and Southern Africa. Alex du Toit Memorial Lecture No. 5. Geological Society of South Africa, Annexure to Vol. LX:1–73.

1963 Pleistocene Mammal Faunas of Africa with Particular Reference to Southern Africa. *In* African Ecology and Human Evolution. F. C. Howell and F. Bouliere, eds. Chicago: University of Chicago Press. pp. 65–116.

1967 The Pleistocene Sequence in South Africa and Problems of Correlation. *In* Background to Evolution in Africa. W. W. Bishop and J. D. Clark, eds. Chicago: University of Chicago Press. pp. 175–186.

1975 Suidae from Pliocene/Pleistocene Strata of the Rudolf Basin. *In* Earliest Man and Environments in the Lake Rudolf Basin: Stratigraphy Paleontology, and Evolution. Y. Coppens *et al.,* eds. Chicago: University of Chicago Press.

Cooke, H. B. S., and V. J. Maglio
1972 Plio-Pleistocene Stratigraphy in Relation to Proboscidean and Suid Evolution. *In* Calibration of Hominoid Evolution. W. W. Bishop and J. A. Miller, eds. Edinburgh: Scottish Academic Press. pp. 303–329.

Coppens, Y.
1972 Tentative de Zonation du Pliocène et du Pléistocène d'Afrique par les grands Mammifères. Comptes rendus des séances de l'Académie des Sciences 274:181–184.

Coppens, Y., F. C. Howell, G. Ll. Isaac, and R. E. F. Leakey, eds.
1975 Earliest Man and Environments in the Lake Rudolf Basin; Stratigraphy, Paleoecology and Evolution. Chicago: University of Chicago Press.

Cox, A.
1968 Lengths of Geomagnetic Polarity Intervals. Journal of Geophysical Research 73:3247–3260.

1969 Geomagnetic Reversals. Science 163:237–244.

Cox, A., Ed.
1973 Plate Tectonics and Geomagnetic Reversals. San Francisico: W. H. Freeman.

Cox, A., G. B. Dalrymple, and R. R. Doell
1967 Reversals of the Earth's Magnetic Field. Scientific American 216:44–54.

Fleischer, R. L., P. B. Price, R. M. Walker, and L. S. B. Leakey
1965 Fission track dating of Bed I, Olduvai Gorge. Science 148:72–74.

Flint, R. F.
1959a On the Basis of Pleistocene Correlation in East Africa. Geological Magazine 96:265–284.

1959b Pleistocene Climates in East and Southern Africa. Bulletin of the Geological Society of America 70:343–374.

Gregory, J. W.
1896 The Great Rift Valley. London: John Murray.

Grommé, C. S., and R. L. Hay
1963 Magnetization of Basalt, Bed I Olduvai Gorge, Tanganyika. Nature 200:560–561.

Hay, R. L.
1967a Hominid-bearing Deposits of Olduvai Gorge. *In* Time and Stratigraphy in the Evolution of Man. A symposium sponsored by the Division of Earth Sciences, National Academy of Sciences, National Research Council, Washington, D.C.

1967b Revised Stratigraphy of Olduvai Gorge. *In* Background to Evolution in Africa. W. W. Bishop and J. D. Clark, eds. Chicago: University of Chicago Press. pp. 221–228.

1971 Geologic Background of Beds I and II: Stratigraphic Summary. *In* Olduvai Gorge, Volume III, M. D. Leakey. Cambridge: The University Press.

Hill, A., and A. Walker
1972 Procedures in Vertebrate Taphonomy: Notes on a Uganda Miocene Fossil Locality. Journal of the Geological Society 128:399–406.

Howell, F. C.
1972 Pliocene/Pleistocene Hominidae in Eastern Africa—Absolute and Relative Ages. *In* Calibration of Hominoid Evolution. W. W. Bishop and J. A. Miller, eds. Edinburgh: Scottish University Press. pp. 331–368.

Howell, F. C., *et al.*
1972 Uranium Series Dating of Bone from the Isimila Prehistoric Site, Tanzania. Nature 237:51–52.

Isaac, Glynn Ll.
1967 The Stratigraphy of the Peninj Group—Early Middle Pleistocene Formations west of Lake Natron, Tanzania. Background to Evolution in Africa. W. W. Bishop and J. D. Clark, eds. Chicago: University of Chicago Press.

1969 Studies of Early Culture in East Africa. World Archaeology 1:1–28.

1972 Comparative Studies of Pleistocene Site Locations in East Africa. *In* Man, Settlement and Urbanism. P. J. Ucko, R. Tringham and G. W. Dimbleby, eds. London: Duckworth. pp. 165–176.

In press Stratigraphy and Cultural Patterns in the Middle Range of Pleistocene Time in East Africa. *In* After the Australopithecines. K. W. Butzer and G. Ll. Isaac, eds. The Hague: Mouton.

n.d. Plio-Pleistocene Artefact Assemblages from East Rudolf, Kenya. To be submitted to Nature. (In preparation.)

Leakey, L. S. B.
1951 Olduvai Gorge. A Report on the Evolution of the Handaxe Culture in Beds I–IV. Cambridge: The University Press.

Leakey, L. S. B., J. F. Evernden, and G. H. Curtis
1961 Age of Bed I, Olduvai Gorge, Tanganyika. Nature 191:478–479.

Leakey, M. D.
1971 Olduvai Gorge, Volume III. Excavations in Beds I and II, 1960–1963. Cambridge: The University Press.

Maglio, V. J.
1972 Vertebrate Faunas and Chronology of Hominid-bearing Sediments east of Lake Rudolf, Kenya. Nature 239:379–385.

Mussett, A. E., T. A. Reilly, and P. K. S. Raja
1965 Palaeomagnetism in East Africa. *In* East African Rift System; Report of the Upper Mantle Committee—UNESCO Seminar Nairobi 1965. Nairobi: University College. Part II, pp. 83–94.

Roubet, C.
1969 Essai de datation absolue d'un biface-hachereau paléolithique de l'Afar (Ethiopie). L'Anthropologie 73:503–524.

Shuey, R. T., F. H. Brown, and M. K. Croes
1974 Magneto-stratigraphy of the Shungura Formation, Southwestern

Ethiopia: Fine Structure of the Lower Matuyama Polarity Epoch. Earth and Planetary Science Letters 23:249–260.

Suess, E.
1891 Die Brüche des Östlichen Afrika. Denk. K. Akad. Wiss, Wien. Math. Nat. V.58. pp. 555–584.

Takeuchi, H. S., S. Uyeda, and H. Kanamori
1970 Debate about the Earth. San Francisco: Freeman, Cooper.

Wilson, J. T., ed.
1972 Continents Adrift. San Francisco: W. H. Freeman.

Typical flora and fauna of East Africa. Courtesy, S. L. Washburn.

W. W. Bishop:

Pliocene Problems Relating to Human Evolution

If man has evolved from arboreal primates of former times, why did his ancestors leave the trees? Some writers have sought to explain this by a theory that the Pliocene was a dry epoch in which forests were greatly reduced, thereby forcing their inhabitants to leave or perish. William Bishop examines the geological record for evidence of such a drastic change.

The Pliocene problems outlined in this essay may be summarized as:

(1) those associated with the difficulty of relating the Pliocene as defined in marine strata to continental deposits.

(2) those concerned with the establishment of firm ages for continental deposits in East Africa and the reassessment of the time boundaries of the Pliocene in that area.

(3) those relating to the slow process of substituting sound environmental interpretations, based upon detailed stratigraphical, sedimentological and paleontological investigations, for those based solely upon theoretical climatic or evolutionary considerations.

Only ten years ago the principal problem concerning the Pliocene in Africa south of the Sahara was that very few continental deposits were known that could be firmly assigned to that age. This was particularly true of mammal- and hominid-bearing strata, which although much sought after, remained in very short supply. At that time the Pliocene, following the various radiometric time scales of Holmes (1937, 1959) or Kulp (1961), was thought to range from a base established, respectively, at 16 m.y., 11 m.y. and 13 million years ago. The upper Plio-Pleistocene limit was equally variable and much has been written concerning it.

When Ardrey (1961) wrote, "Through all the *changeless millions of Pliocene years* Africa has stood still" (italics added), he was referring to this long time scale and also to the absence of fossil evidence from that period. In 1967 Gautier noted that the "Pliocene may have suffered amputation at both its time-boundaries in sub-Saharan Africa" (Gautier 1967:82). This trend in paring down the length of the Pliocene has continued. It now spans only about 3 million years, from 5.0 m.y. to 1.8 m.y. (Berggren 1973) or, in round figures, to about 2.0 m.y. The problem is to relate Berggren's sequence, as defined for marine strata and based upon planktonic foraminifera and calcareous nanno-plankton, to mammal-bearing continental strata.

Good evidence of continental "Pliocene" deposits (whether defined on the older time scale or using only the shortened span) has been discovered during the past five or six years in East Africa. This takes the form of well-exposed and isotopically dated sedimentary sequences yielding abundant mammalian faunas. However, these new discoveries have as yet made little impact upon the textbooks. In addition, the period is still frequently interpreted in archaeological and anthropological literature in the light of climatological mystique. This is similar to the climatic inference of pluvials that dominated African Quaternary stratigraphy until these ideas were swept away only a few years ago.

THE MYTH OF A PLIOCENE DROUGHT

It is difficult to say exactly where or when the notion of a Pliocene drought originated. Even more uncertain is when the notion of such a drought was first related to man's evolution and assigned a role in developing the human character.

In recent years, Wayland traced an alleged battle of "desert versus forest" in East Africa and concluded that "clearly the 'normal' climate of East Africa is arid and the wet phases are superimposed thereon" (Wayland 1940:332). Claiborne (1970:164), writing of the relations between "climate, man and history," concluded that the Pliocene was obviously cooler and drier.

It would be easy but rather pointless to continue to list the many references to this recurrent theme. I propose rather to quote several passages from Robert Ardrey's *African Genesis* (Ardrey 1961). In doing so I am not criticizing Ardrey, for he is here merely acting as a literary catalyst giving expression to the views that he found in the voluminous literature. These he distilled and recorded. I know that he will forgive this culling of seven quotations from *African Genesis,* a book which I found, and indeed still find, most stimulating.

> [1.] "The gentle Miocene vanished. Through million-year span after million-year span the rainy seasons shortened, arrived with less certainty, brought dwindling moisture. Rivers shrank. Lake levels fell. Winds ripped the savannah. And *about twelve million years ago came the Pliocene.*" (italics added)
>
> [2.] "What happened to the human stock in the Pliocene's dry inferno? The dusty bankrupt kept no records. For lack of lakes, we are left no lakebeds; for lack of streams, no valley terraces; for lack of ground-water, no lime-filled caverns. And for lack of all, *we are left no fossils from the desperate Pliocene days.*" (italics added)
>
> [3.] "The record now blank is one that time and research may some day fill in. *Rare Pliocene fossil beds will be found.*" (italics added)
>
> [4.] "We have the beginning of the story in the Miocene Eden, and the end in the changful Pleistocene. And the middle, we know *must be the parching African Pliocene.*" (My italics added)
>
> [5.] "The gentle Miocene vanished slowly, but it vanished. And the climatic deterioration of the encroaching Pliocene brought a crisis to the whole ape world."
>
> [6.] "Still the Pliocene drought deepened, and now even the bush dwindled with the forests."
>
> [7.] "Such was our life on the changeless highlands of Pliocene Africa. We were bad-weather animals, made by the natural disaster that had unmade the ape of the forest. It was a hard life, but for millions upon millions of years it was the only life we knew." (Ardrey 1961:287–301)

These quotations serve to outline various myths which have grown up concerning the Pliocene as relating to human evolution. It now remains to examine the evidence from those areas of East Africa that in the last few years have yielded the now by no means "rare Pliocene fossil beds."

EAST AFRICAN PLIOCENE EVIDENCE

Sedimentary sequences that span the Pliocene as now defined (from 5 to 2 m.y. or rather less), or as previously assigned to the Pliocene (from 12 or

10 to about 1 m.y.), are known from the following areas of East Africa:

1. The Lower Semliki Valley and Kaiso areas of the Lake Albert Rift.
2. The Omo Valley, East Rudolf and southwestern areas of the Lake Rudolf Basin.
3. The Baringo Basin of the Kenya Rift.

The Lower Semliki Valley and Kaiso Areas of the Lake Albert Rift

These areas have not yet yielded any hominid fossils or artifacts of Pliocene age, but they expose a long sedimentary sequence that has provided good mammalian assemblages (Bishop, 1965, 1969, 1971; Cooke and Coryndon 1970).

The Albertine Rift is a classic graben. The simplicity of the physical setting is reflected in the sedimentary record which, since mid-Miocene time, has been typically that of a fault-bounded lake basin. Evidence of volcanicity is absent, and tectonic activity appears to have been restricted to repeated movements of the boundary faults. Along much of its shoreline, Lake Albert laps directly against the escarpment walls that delimit the graben. The sediments accumulating on the present lake bed represent the uppermost stratum of a long record of lacustrine deposition. Geophysical investigations suggest a maximum thickness of over 8000 feet of deposits in the center of the graben, and a bore hole penetrated over 4000 feet of sediments before running into the rift wall.

The Albertine Rift has yielded no evidence of hominoid fossils or artifacts earlier than the mid- and later Pleistocene. Fossiliferous ironstones occur as lenses and probably represent former shallow basins or swampy lagoons joined by sluggish rivers with abundant vegetation to provide food for numerous gastropods. From time to time individual basins virtually dried out, killing the freshwater mollusca and fish. Later inundation and a fresh supply of detritus sealed the horizon, and comparatively rapid sedimentation provided conditions suitable for the burial and eventual preservation of bones of any mammals or crocodiles which may have died at or near the margins of the shallow basins or swamps.

The consistent chemical, lithological, faunal, and stratigraphical nature of the Kaiso Formation, wherever it occurs in Uganda and the Congo, raises the problem of the wider environmental setting in which the deposits were formed. The shallow-water nature was not questioned by earlier workers. Wayland (1926) envisaged sedimentation keeping pace with rift subsidence and sometimes exceeding it but considered the fossiliferous ironstones to be evidence of desiccation. Fuchs (1934) assigned the Kaiso sediments to a first pluvial and considered that toward the end of deposition the climate became more arid, while Wayland (1934) refers the "Kaiso bone beds" (which he thought of as a single unique horizon) to an interpluvial period.

In contrast, Solomon (1939) wrote, "The fossiliferous horizons do not necessarily indicate a marked climatic change." Heinzelin (1955) suggested that the deposits gave evidence of a humid climate, with a period of aridity marked by beds of gypsum in the Congo. Despite the diversity of the above views, the upper fossiliferous horizons of the Kaiso Formation were assigned to a first interpluvial by the Pan African Congress (Clark and Cole 1957:p.xxxi).

I agree with Solomon and believe that climatic interpretations are not warranted, in view of the rift valley location of the deposits and the fact that the formation is some 2700 feet thick with numerous fossiliferous ironstone horizons. The nature of the climate prevailing at the time of deposition of the formation is not proven, although intermittent emergence undoubtedly took place (Flint 1959).

The characteristic lithologies of the Kaiso deposits indicate lacustrine environments into which little coarse or unweathered detritus was brought. Repeated minor movements of the boundary faults account for the rhythmic sedimentation seen in the Kaiso Formation and also for the input of coarser detritus from the escarpment and the onset of more rapid sedimentation associated with the ironstones. A parallel might be sought in the rhythmic sedimentation seen in carboniferous strata. A sequence of clay, succeeded by sand, gravel and oolite grading up again into fine sands, silts and clays, constitutes a typical Kaiso cyclothem which is repeated many times. Identifications of mammalian fossils by Cooke and Coryndon (1970) allow one to suggest that the lower Kaiso assemblage, far from being Pleistocene in age, may be as old as about 5± m.y., as judged from comparisons with faunas for which isotopic ages have been established in the Eastern or the Gregory Rift. Similarly, the assemblages of mammals that occur toward the top of the formation may be as young as 3± or 2± m.y. Thus, the Kaiso Formation of the Albertine Rift provides a typical Pliocene sequence as now defined. It offers no evidence of aridity but contains a series of sediments which reflect the tectonic setting of a large rift valley lake.

The Omo Valley, East Rudolf and Southwestern Areas of the Lake Rudolf Basin

The Omo valley has been the subject of detailed international study since 1966. A team from the United States, under the direction of Professor F. Clark Howell, combined with a French team directed first by the late Professor C. Arambourg and then by Dr. Y. Coppens. There was some initial participation by a Kenyan team led by R. E. F. Leakey. The difficult terrain in Kenya to the east of Lake Rudolf has been investigated since 1968 by teams from various countries under the leadership of R. E. F. Leakey, and based upon the Kenya National Museum in Nairobi.

In addition to these two important regions which are adjacent to each other to the north and east of Lake Rudolf, the area immediately southwest of the Lake has been investigated by research workers from Harvard University (Patterson 1966; Patterson *et al.* 1970). Three important localities have been studied. They are, from north to south, Lothagam, Ekora, and Kanapoi.

Lothagam, Ekora and Kanapoi. In the vicinity of Lothagam Hill an isolated, westward-tilted fault block exposes a succession almost 1000 meters thick of which sediments, comprising three distinct lithological units, account for well over 600 meters. The fauna from the oldest unit, Lothagam 1, is suggested by Patterson *et al.* (1970) as being older than the early Kaiso mammal assemblages. A potassium-argon date of 3.71–0.23 m.y. for a sill intruded at the top of this oldest unit gives a minimum age for the fauna. Lothagam 2 is much thinner, consisting of silts, clays, and bedded tuffs, and is interpreted as being of lacustrine origin. Vertebrate fossils are rare but beds of gastropods are common. Lothagam 3 marks a return to fluviatile conditions. The scanty fauna shows remarkable similarity with that of Kanapoi where an overlying basalt has yielded three potassium-argon dates ranging between 2.9–0.3 and 2.5–0.2 m.y. This gives an upper age limit for this fauna and, on a basis of its similarity to other dated faunas, an age of about 4− m.y. is suggested for Kanapoi (Patterson *et al.* 1970) which also shows similarities to the Chemeron Formation fauna (see below). The Ekora fauna is shown to be younger than that of Kanapoi, and the elephants suggest that the assemblage may be rather earlier than the oldest part of the main Omo succession.

Thus, the three localities provide a Pliocene sequence which at the top is rather earlier than the base of the Omo fossiliferous sequence (about 4 m.y.). The oldest Lothagam strata are perhaps 5.5 m.y. although there is no isotopic date to confirm this.

A specimen of *Australopithecus* sp. cf. *A. africanus,* from near the top of Lothagam 1, and a specimen of cf. *Australopithecus* from Kanapoi extend the hominid record back in time from the lowest Omo levels. The sedimentary sequences suggest conditions of lacustrine and fluviatile deposition similar to the pattern seen in the rift valley at the present day.

The Omo Valley. The classical "Omo Beds" are now formally designated as the Shungura Formation. They consist of over 500 meters of clays, silts, sands, and tuffs, predominantly of deltaic origin, in a series of cyclic units. Tuff bands provide useful marker horizons and have been labeled in ascending order from A to L. Six potassium-argon determinations on feldspar-bearing pumices between horizons B and I_2 range in age from 3.75 m.y. to 1.84 m.y. (Brown and Nash 1975). All the dates have low probable errors and occur in correct order of stratification, thus providing a unique calibration for this fossiliferous sedimentary sequence.

At the White Sands and Brown Sands localities in the Omo Valley, the Usno Formation is exposed. The deposits are silts and clays representing deltaic or floodplain deposits and sands and gravels of fluvial origin. The Usno Formation is probably the time equivalent of part of the Shungura Formation. A potassium-argon age of 3.1± m.y. for a basalt underlying the sequence provides a maximum age for the commencement of deposition.

Two other lithological units older than the Shungura and Usno Formations have been recognized in the Omo Valley and termed the Nkalabong and Mursi Formations. The type area of the latter is at the Yellow Sands locality some 50 miles north of the present shore of Lake Rudolf. The unit, which constitutes the oldest sedimentary formation in the area, consists of over 150 meters of deltaic and prodeltaic sediments. These have been divided into three members and are overlain by a basalt member. The basalt has yielded potassium-argon ages of 4.05 m.y. and 4.25 m.y. The Nkalabong Formation rests unconformably upon the weathered and eroded surface of the basalt. Deltaic beds are virtually absent but fluviatile current-bedded silts and clays indicate former courses of the Omo River. A lapilli tuff in member II has yielded a potassium-argon date of 3.95± m.y. These two formations extend the Omo sequence a little further back in time.

I am indebted to Professor F. Clark Howell for the information that the Omo area has yielded a total of 182 hominid specimens of which 161 are from the Shungura Formation (from 84 different fossil localities) and 21 are from the Usno Formation. This sample is composed of 162 isolated teeth, 7 postcranial bones, and 13 specimens representing parts of crania, maxillae or mandibles.

The whole of the Usno Formation falls into the Pliocene as now defined together with most of the Shungura Formation, depending upon whether one accepts an age of 2.0 m.y. or 1.8 m.y. as the base of the Pleistocene. For example, only four hominid specimens from the Shungura Formation come from Member H or higher. Tuff H is dated at about 1.9 m.y. Some 50 specimens come from Member G. Stone artifacts occur in Member F and can be thought of as uppermost Pliocene on the new dating. Some 41 hominid specimens are from Members B and C, well into the Pliocene and in excess of 2.37–2.56 m.y.

The Omo Valley Formations collectively sample hominid and mammalian fossils from the upper half of the Pliocene. The paleoenvironments recorded by the sediments suggest dominance of the record throughout this period by input from an Omo river with dimensions possibly similar to those of the present-day river. Further detailed work may enable finer resolution to be established. At present, however, there is no evidence that local conditions fluctuated greatly from those represented by the "norm" in the lower Omo valley at the present day. The Kibish Formation records high still-stands of Lake Rudolf (Butzer and Thurber 1969), but there is as yet no evidence of prolonged drought from this area for the period between about 4.0 and 2.0 m.y.

East Rudolf. The East Rudolf area has yielded 125 hominid fossils since research commenced in 1969. Of these 50 are postcranial specimens, 16 are isolated teeth, and 59 are complete or partial cranial, maxillary or mandibular pieces. The differences in proportion of these elements when compared with Omo probably reflect contrasting environments of preservation in the two areas.

The paleoenvironments established for East Rudolf range from fluviatile to lake margin and deltaic. The potassium-argon age recorded for the KBS tuff, an important marker in the sedimentary record which has stone artifacts on the surface of its tuff-filled channels at several localities, is 2.61–0.26 m.y. (Miller and Fitch 1970). This date has recently received support from other dates for tuffs higher in the sequence (at about 1.5 m.y.) and for the underlying Tula Bor tuff just in excess of 3.0 m.y. These dates are published in a volume on the Rudolf Basin (Coppens, Howell, Isaac and Leakey 1975). With other dates from the lavas in the East Rudolf area, these new dates suggest that the age range for the main sequence of sediments is from about 4.0 m.y. to about 1.3 m.y. (see also Maglio 1972 for mammal age evidence). The sequence faunas, dates and paleomagnetic data emphasize that recognizable tools range back to at least 2.6 m.y. with the abundant hominid record extending well into the Pliocene to about 3.0 m.y.

The environments, although different from those at Omo where the presence of a proto-Omo river has been the dominant factor, suggest a pattern of interplay between the lake and small rivers from the east and northeast. This is again reminiscent of the present-day situation in the area.

The Baringo Basin

This basin of the northern Kenya rift valley provides a long sequence from later Tertiary through Quaternary strata with seven superimposed fossiliferous sedimentary units occurring through 3000± meters of succession in the main scarp and foothills of the Tugan Hills (Table 1). Many of the interbedded lavas are rich in potassium while sanidine feldspars occur in angular pumice clasts and cobbles at a number of horizons within the fossiliferous sedimentary units. The sequence is ideal for comparison of calibration based largely upon potassium-argon dating with that derived from mammalian paleontology.

Study of the sediments and mammalian faunas of the Baringo area and their arrangement in a calibrated sequence has been made possible by the following cooperative programs.

Geological Mapping. The mapping program of the East African Geological Research Unit (EAGRU), undertaken at the request of the Kenya Government, is based at Bedford College. This work under the direction of Professor B. C. King has ensured a systematic coverage of the whole Baringo area. The mapping and naming of sedimentary units, and the initial discovery of many fossiliferous localities, have resulted from field work carried out by graduate students since 1965. Second-phase studies by postgraduates, under my supervision and also based at Bedford College but whose more specific concern has been the study of sediments and faunas, could not have been contemplated without the groundwork undertaken through the EAGRU project (King and Chapman 1972; Bishop *et al.* 1971).

Potassium-argon Dating. Dr. J. A. Miller of the Department of Geodesy and Geophysics at Cambridge University has collaborated in a dating program directed deliberately to the calibration of fossiliferous sedimentary units. Additional dates relevant to the mammalian faunas have been established by Dr. G. H. Curtis, Department of Geology and Geophysics, University of California, Berkeley, and by Dr. N. J. Snelling, Institute of Geological Sciences, London (Bishop 1972); Bishop *et al.* 1971).

Paleomagnetic Studies. A Model-70 portable fluxgate magnetometer was used to establish preliminary paleomagnetic polarity in the field. In addition Dr. A. Khan of the University of Leicester undertook a program to check in the laboratory the polarity of selected samples from lava flow units for which fluxgate readings had been established. More recently a group from the sub-department of Geophysics at the University of Liverpool, under the leadership of Dr. P. Dagley, has undertaken in the area west of Lake Baringo drilling and sampling of 611 cores from 149 lava flow units. Detailed investigations of direction of magnetism will be carried out, and where possible, the strength of the magnetizing fields and the potassium-argon ages of the cores will be measured. The results of these geophysical investigations should yield valuable new data on paleomagnetic polarity and isotopic ages. Recently, preliminary samples have been collected from the Baringo area to determine the remanent polarity of sediments aged less than 5 m.y.

Mammalian Paleontology. Various specialists have kindly identified and described material from the fossiliferous units as follows: Primates: Professor M. H. Day, Mr. R. E. F. Leakey, Professor P. V. Tobias, Dr. A. C. Walker; Proboscides: Dr. J. Harris, Dr. V. J. Maglio; Rhinocerotidae: Dr. D. A. Hooijer; Suidae: Professor H. B. S. Cooke; Hippopotamidae: Mrs. S. Coryndon (Savage); Bovidae: Dr. A. W. Gentry, Dr. R. Hamilton.

THE SEQUENCE WEST OF LAKE BARINGO

The area is situated between the midline of the Gregory Rift, approximately through Lake Baringo, and its western boundary fault, marked by the fall of the Elgeyo escarpment from 2750 to 1200 meters (Figure 1). The principal relief feature in the area is the tilt block of the Kamasia or Tugen Hills. On the east, faults step down strata from the crest of the range at 2600 meters to below Lake Baringo (1050 meters). On the west, a dissected dip slope descends into the Kerio Valley (1200 meters).

Preliminary descriptions of sediments and faunas have been given by Bishop and Chapman (1970), Martyn (1967, 1969), Bishop *et al.* (1971). In many areas of the Gregory Rift earlier strata are obscured by mid- to later Pleistocene sediments and lavas. However, along the eastern face of the Tugen Hills, a long sequence is exposed from mid-Miocene through Pliocene and Pleistocene into Holocene strata. This succession is seen to rest upon Basement Complex rocks. Almost 3000 meters of later Cenozoic volcanics and sediments are present in this section of the rift. The older part (1800 meters) of this succession outcrops in the upper part of the main escarpment and is capped by lavas of the Kabarnet Trachyte Formation, aged about 7.0 m.y. The younger rocks are exposed in the grid-faulted region between the scarp foot and Lake Baringo.

The mammalian faunas as known at the time were recorded in Bishop *et al.* (1971) and Bishop (1972). In the latter paper the isotopic dating evidence was reviewed critically against the stratigraphic and mammalian faunal sequences. The faunal assemblages from several of the formations have been extended since 1972, and the work of collection, description, and identification is still proceeding. The faunas are not described further in this paper but the reader is referred to the lists in the two publications cited above.

The sequence of seven superimposed sedimentary units, of which five have yielded rich mammalian assemblages (Table 1), may be considered as a series of control units for which the correct time sequence is known from the long stratigraphic succession established by geological mapping.

TABLE 1
Mapped Units of West Baringo Succession

SUPERIMPOSED SEDIMENTARY UNITS YIELDING FOSSIL MAMMALS	$^{40}K-^{40}Ar$ AGES AND ERROR BARS		SUPERIMPOSED VOLCANIC UNITS	
Kapthurin formation* †	(0.23)			
(Baringo trachyte extruded		(0.5)		Loyamarok phonolite
early in K times)	(0.75)			Baringo trachyte
		(1.0)		Hannington phonolites etc.
		(2.0)		Ndau mugearite
Chemeron Formation* †	(2.1)			
(Ndau mugearite interbedded near top of formation; weathered and eroded KB underlies)				
Aterir Beds				
(Overlies latest flows on flank of Ribkwo volcano)		4.4		
	4.75		R	Ribkwo volcanics
		5.1	KB	Kapataina basalts
Lukeino Formation* †				(Some Ribkwo flows near top of unit)
(widespread mapped unit; overlies KT and is overlain by KB)		6.7		
Mpesida Beds	7.2		KT	Kabarnet trachyte
(local unit interbedded with KT)		7.7		(overlies Sumet phonolite)
	7.95		S	Sumet phonolite
		9.3		
	(9.75)			
Ngorora Formation* †				
(widespread unit; overlies T and is overlain by S)		12.0		
	(12.1)			
Muruyur Beds	13.15		T	Tiim phonolite
(interbedded between flows of Tiim phonolite)		14.3		

Code letters from Figs. 2 and 3 (Bishop 1972).

In Table 1 the sedimentary units yielding an abundant mammalian fauna are marked † and those which have yielded hominoid fossils are shown with an asterisk (*). Of the deposits assigned at present (or in the past) to the Pliocene, the Ngorora Formation and the Lukeino Formation have so far yielded only a single molar crown and the Chemeron Formation part of the temporal of a hominid. However, all have provided good assemblages of fossil mammals.

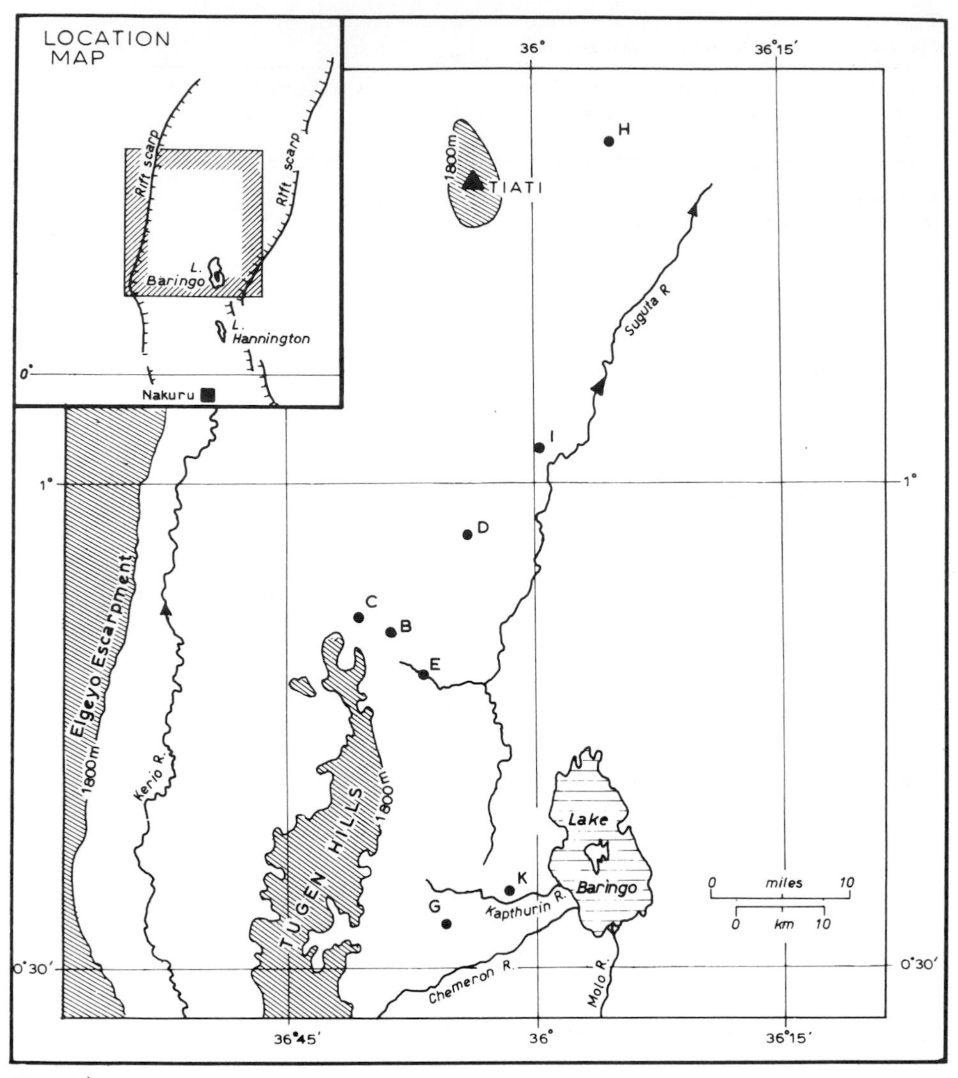

FIGURE 1
Inset: Location of Baringo area in relation to the rift valley and the equator. Main map: Location of type areas of mammal-bearing sedimentary units mentioned in text: B—Muruyur Beds, C—Ngorora Formation, D—Mpesida Beds, E—Lukeino Formation, G—Chemeron Formation, H—Aterir Beds, I—Karmosit Locality, K—Kapthurin Formation.

The Pliocene, as now defined, would range in the sequence from the basal lavas of the Ribkwo volcano, or the upper part of the Kaparaina Basalts, through to the base of the Ndau Mugearite flow (Table 1). The environments indicated by the sediments of the Aterir Basin and the Chemeron Formation show no indication of prolonged drought. Indeed, they show that lacustrine and fluviatile sedimentation continued throughout the history of the two basins. The deposits range from river-channel gravels to diatomaceous silts. The fluctuations observed do not exceed, at their extremes, the conditions that can be observed in modern sedimentary analogues in the rift valley.

If one extends the analysis back to 10 or 12 m.y. to include the Pliocene as dated previously there is still a remarkable similarity to the present-day pattern of variation in the modern environments. No pronounced or prolonged dry periods can be identified. The Ngorora, Mpesida, and Lukeino sedimentary basins all show a similar range from lacustrine to fluviatile environments and from fish-bearing diatomites to coarse fluviatile channel gravels. In the Ngorora Formation one facies of the deposits (the Kapkiamu Shales) records largely chemical precipitation in an enclosed basin. The fish fossils suggest that the lake may have been soda rich. However, stratigraphical mapping shows that this is merely an embayment or a separate arm of the whole Ngorora fluvio-lacustrine basin. The lake undoubtedly fluctuated in size from time to time, but a largely fresh-water basin with abundant stream channels yielding varied mammalian assemblages throughout its history seems to have been adjacent to and contemporary with the Kapkiamu lake.

Whether one considers a "greater" or a "lesser" Pliocene, no evidence from the sediments or the fossils supports aridity and certainly not persistent aridity for the Baringo area during the Pliocene.

CONCLUSION

After reviewing the observed sedimentary and faunal evidence in the three principal areas of East Africa where a good Pliocene record is present, one can only conclude that cold water must be poured on the notion of Pliocene aridity. Attractive though the idea may have seemed, human strength of character was neither tempered in the heat of a Pliocene drought nor quenched in the Pleistocene pluvials.

Acknowledgments

When I first came to East Africa in 1956 my wife and I were fortunate to receive kindly advice and helpful tutoring from Dr. Louis Leakey and his wife, Dr. Mary Leakey. From that time they remained friends whose guidance was sought frequently and never in vain. I offer these thoughts on the Pliocene and its problems as a tribute to the late Louis Leakey. It was he who introduced me to the wide spectrum of East African late Cenozoic studies—a spectrum on which he was knowledgeable on virtually all the component colors and an accepted authority on several of them.

References Cited

Ardrey, R.

1961 African Genesis—a personal investigation into the animal origins and nature of man. London: Collins.

Berggren, W. A.

1973 The Pliocene time scale: Calibration of planktonic foraminiferal and calcareous nannoplankton zones. Nature 243:391–397.

Bishop, W. W.

1965 Quaternary geology and geomorphology in the Albertine rift valley, Uganda. *In* International studies of the Quaternary. H. E. Wright, Jr. Geol. Am. Spec. Papers, 84.

1969 Pleistocene stratigraphy in Uganda. Geol. Survey Uganda, Memoir X. Entebbe.

1971 The late Cenozoic history of East Africa in relation to hominoid evolution. *In* The Late Cenozoic glacial ages. K. K. Turekian, ed. New Haven/London: Yale University Press. pp. 493–527.

1972 Stratigraphic succession 'versus' calibration in East Africa. *In* Calibration of Hominoid Evolution. W. W. Bishop and J. A. Miller, eds. Edinburgh: Scottish Academic Press.

Bishop, W. W., and G. R. Chapman

1970 Early Pliocene sediments and fossils from the northern Kenya rift valley. Nature 226:914.

Bishop, W. W., G. R. Chapman, A. Hill and J. A. Miller

1971 Succession of Cainozoic Vertebrate Assemblages from the Northern Kenya Rift Valley. Nature 233:389–394.

Brown, R. F., and W. Nash

1975 Radiometric dating and tuff mineralogy of Omo Group deposits. *In* Earliest Man and Environments in the Lake Rudolf Basin. Stratigraphy, Paleoecology and Evolution. Y. Coppens, F. C. Howell, G. Ll. Isaac and R. E. F. Leakey, eds. Chicago: University of Chicago Press.

Butzer, K. W., and D. L. Thurber

1969 Some late Cenozoic sedimentary formations of the lower Omo basin. Nature 222:1132.

Claiborne, R.

1970 Climate, Man and History. New York: Norton and Co.

Clark, J. D., and S. Cole, eds.

1957 Proceedings of the Third Pan-African Congress on Prehistory, Livingstone, 1955. London: Chatto and Windus.

Cooke, H. B. S., and S. Coryndon

1970 Pleistocene mammals from the Kaiso Formation and other related deposits. *In* Fossil Vertebrates of Africa, Vol. 2. L. S. B. Leakey and R. J. G. Savage, eds. London: Academic Press.

Coppens, Y., F. C. Howell, G. Ll. Isaac and R. E. F. Leakey, eds.

1975 Earliest Man and Environments in the Lake Rudolf Basin. Stratigraphy, Paleoecology and Evolution. Chicago: University of Chicago Press.

Flint, R. F.

1959 On the basis of Pleistocene correlation in East Africa. Geol. Mag. 96:265.

Fuchs, V. E.
1934 The geological work of the Cambridge expedition to the East African lakes 1930–1931. Geol. Mag. 87:149.

Gautier, A.
1967 New observations on the later Tertiary and early Quaternary in the western rift. *In* Background to Evolution in Africa. W. W. Bishop and J. D. Clark, eds. Chicago: University of Chicago Press. pp. 73–87.

Heinzelin, J. de
1955 Le fossé tectonique sous le parallèle d'Ishango. Mission J. de Heinzelin l. Brussels (Inst. Parcs Natl. Congo Belge).

Holmes, A.
1937 The age of the earth. 2nd. ed. London: Nelson.

1959 A revised geologic time-scale. Trans. Edinb. Geol. Soc. 17:183–216.

King, B. C., and G. R. Chapman
1972 Volcanism of the Kenya Rift Valley. Philosophical Transactions of the Royal Society, London. A271:185–208.

Kulp, J. L.
1961 Geologic time scale. Science 133:105–114.

Maglio, V. J.
1972 Vertebrate fauna and chronology of hominid-bearing sediments east of Lake Rudolf, Kenya. Nature 239:379–385.

Martyn, J. E.
1967 Pleistocene deposits and new fossil localities in Kenya (with a note on the hominid remains by P. V. Tobias). Nature 215:476–479.

1969 Notes on the geology of the Kapthurin Beds. *In* An Acheulian industry with prepared core technique and the discovery of a contemporary hominid mandible at Lake Baringo, Kenya. Margaret Leakey. Proc. Prehist. Soc. 3:48–70.

Miller, J. A., and F. J. Fitch
1970 Radioisotopic age determination of Lake Rudolf artefact site. Nature 226:226–228.

Patterson, B.
1966 A new locality for early Pleistocene fossils in north-western Kenya. Nature 212:577–578.

Patterson, B., A. K. Behrensmeyer and W. D. Sill
1970 Geology and fauna of a new Pliocene locality in north-western Kenya. Nature 226:918.

Solomon, J. D.
1939 The Pleistocene succession in Uganda. *In* The Prehistory of the Uganda Protectorate. T. P. O'Brien. Cambridge: Cambridge Univ. Press.

Wayland, E. J.
1926 The geology and palaeontology of the Kaiso Bone-beds. Geol. Surv. Uganda, Occ. Paper 2, 5.

1934 Rifts, rivers, rains and early man in Uganda. J. Roy. Anthrop. Inst. 64:333.

1940 Desert versus forest in Eastern Africa. Geogr. Journal XCVI 5:329–341.

A scene from Miocene times in East Africa, reconstructed by artist-naturalist Jay H. Matternes working in consultation with Louis Leakey and John Napier. In the forest trees is the extinct variety of ape *Dryopithecus (Proconsul) africanus*. Beyond is a patch of open savannah that interfingered with the forest. The plate is from the volume *Animals of East Africa* by L. S. B. Leakey. © 1969 National Geographic Society.

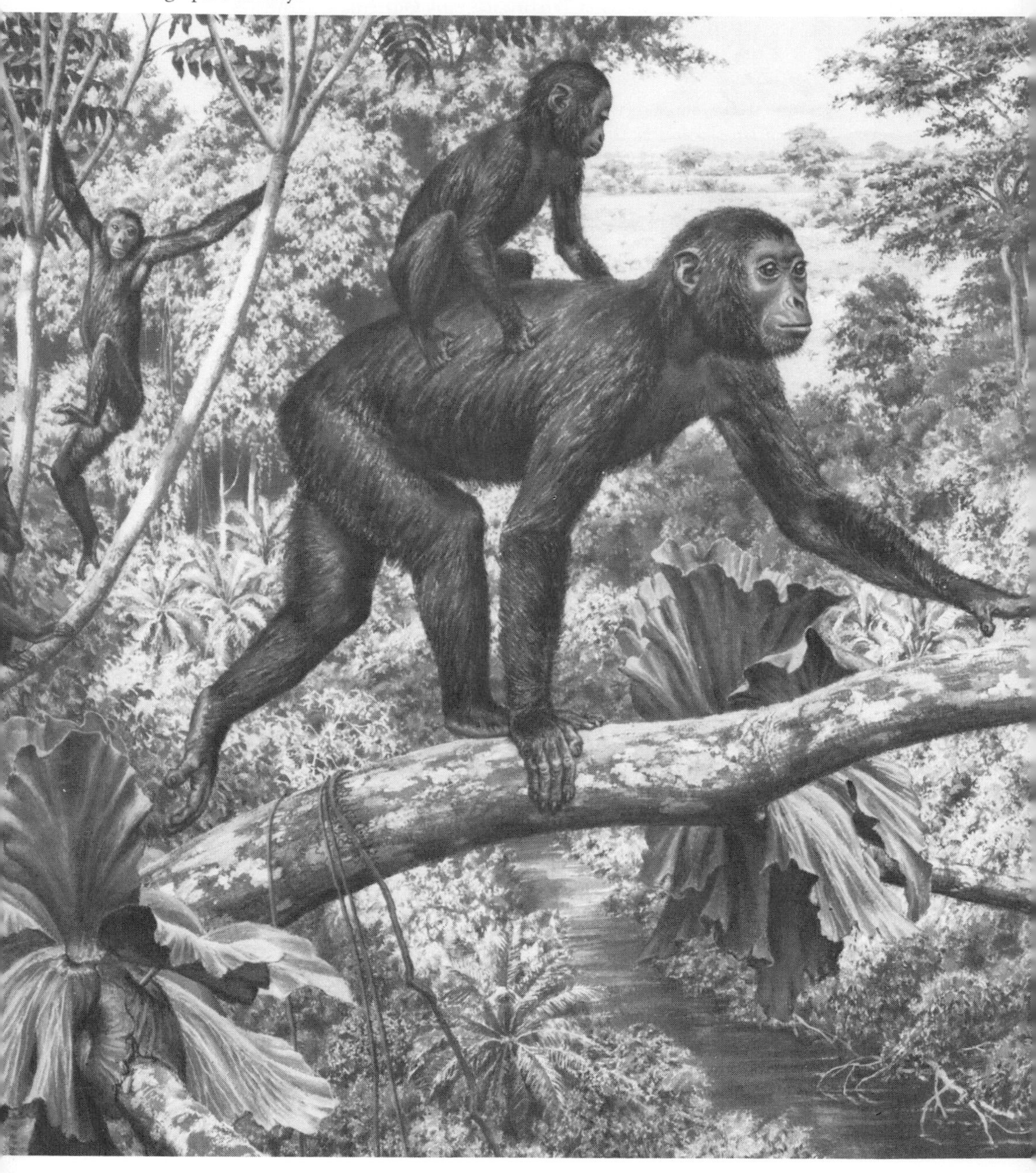

Judith A. H. and
John A. Van Couvering:

Early Miocene Mammal Fossils from East Africa:

Aspects of Geology, Faunistics and Paleoecology

Men and apes together comprise the zoological group known as the Hominoidea, whereas mankind and its closest fossil relatives are classified in the family Hominidae. Prior to the Middle Miocene, fossil hominoids are known only from Africa, where they were a diversified and successful set of animals. What were the biogeographic relationships of East Africa at this time? What other kinds of mammals were the contemporaries of these prehuman, ape-like creatures? Judith and John Van Couvering provide a wide-ranging, and necessarily somewhat technical summary of the geological and biological characteristics of East Africa's crucial Early Miocene paleontological record.

Two enormously important fossil sequences have been found in East Africa. One is the Plio-Pleistocene succession that spans the time range from six million years to half a million years ago. The evidence for the evolution of the Hominidae, that has been obtained from such sites as Olduvai Gorge, East Rudolf, and the Omo has made them justly famous. The second important fossil sequence is that obtained from the Miocene sediments of East Africa—which have yielded a record of hominoid evolution that spans the interval between 23 million years and 16 million years ago. Louis Leakey began his pioneer researches into both successions in the same year, 1931, and our knowledge of both now stands as a monument to his vision and energy.

At present the East African Early Miocene mammalian fauna, with over 110 recognized species, is one of the most varied ever to be described from the fossil record, and its seven species of hominoids represent all that is known of this group during the period between the Middle Oligocene and the Middle Miocene—an interval of about 20 million years. In addition, the quantity of early hominoid material from East Africa completely dominates the record of this group with more than 200 individuals represented in the holdings of the Kenya National Museums alone.

The descriptive study of East African Miocene mammals and the geology of the deposits in which they are found, which began many years ago, is now all but complete. New research aimed at answering more pointed questions about the fossil animals and their environments is currently building on this foundation. It seems appropriate at this time, and in this volume, to summarize the descriptive work and incorporate the current studies in a general review. We will compare the different sites (Figure 1) on the basis of what is now known about them, and we will also attempt to make an environmental or ecological comparison between the Early Miocene and present-day conditions in East Africa.

DEFINITIONS

Our use of the term "Miocene" conforms to the current practice of marine biostratigraphers in that it is based on the sequence of molluscan and echinoderm faunas in southwest Europe to which Lyell himself referred in formulating the original definition of the term. The age limits of the Miocene are presently calibrated at 23.5 and 5 m.y. (Berggren and Van Couvering 1974, Appendix). However, "Early," "Middle," and "Late" Miocene are not standardized time terms. In East Africa we use Early Miocene for the time when the mammal fauna remained remarkably stable and uniform, which has K-Ar dates ranging from approximately 23 to 16 m.y. The transition to the East African Middle Miocene, as represented at the Fort Ternan site, dated 14 m.y., is marked by a strong change in the mammal fauna. The age of the transition, 16 to 15 m.y., is more or less coincidental with the dates applied to the conventional Early/Middle Miocene boundaries in Europe—in marine sections the base of the Langhian Stage, and in continental sections, the base of the "Vindobonian" mammal age (Van Couvering and Berggren, in press). In European continental biochronology, the Late Miocene is now generally considered to begin with the appearance of *Hipparion,* about 12 m.y. (see Van Couvering *et al.,* in press); the first *Hipparion* in East Africa appears to be dated by occurrences in the Upper Ngorora Beds of about the same age (Bishop and Chapman 1970; Hooijer and Maglio 1974). This important datum event may conveniently provide a basis for the beginning of the Late Miocene in East Africa as well.

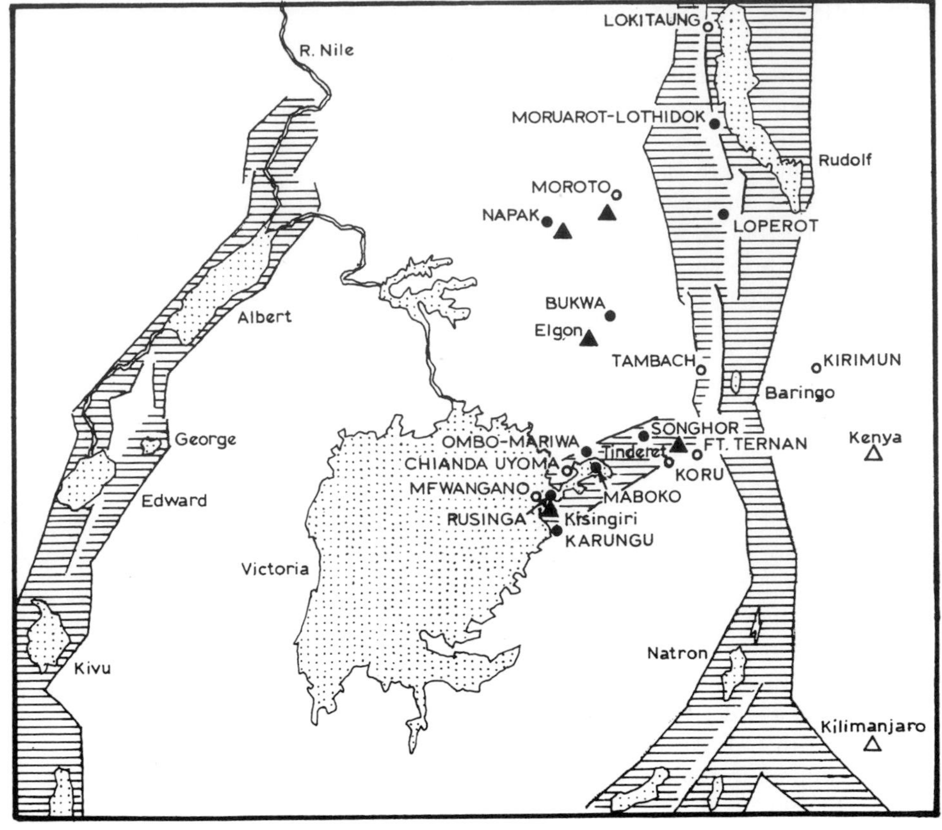

FIGURE 1
Index map of Early Miocene collecting localities in Kenya and Uganda. (Congo sites, east of Lake Albert and Lake Edward, are not shown.) Solid circles represent the localities referred to in statistical analyses and in the Appendix; open circles are other Early–Middle Miocene sites. Triangles indicate major volcanic centers.

We have used the term "savanna" very broadly here, in place of the more cumbersome term "non-forest," to include woodland, bushland, and grassland. This usage is intended to avoid using the more precise terms where precision is unwarranted. For a discussion of the meaning of the terms used for African vegetation associations see Greenway (1943); Pratt *et al.* (1966); and Andrews and Van Couvering (1975).

In this article, "species diversity" refers to the total number of species in a given area and does not take into account the relative abundance of different species. This concept has been called "species density" by some (Simpson, G. G., 1964; Armstrong 1972). Knowledge of modern African faunas is not yet adequate to attempt a diversity study in terms of relative abundance (cf. Simpson, E. H., 1949).

HISTORY OF STUDY

Figure 2 summarizes the history of collecting in African fossil mammal sites of Early and Middle Miocene age by indicating the years in which the major sites were visited. The first African Miocene fossils known to science were isolated fragments of proboscidean and anthracothere reported from North Africa in the last century (cf. Savage 1967, for details), but it was not until the second decade of the 1900s that prolific Early Miocene sites came to light, first in Kenya and shortly thereafter in Southwest Africa and in Egypt. Despite this very nearly even start, collecting in East Africa has almost completely eclipsed the early work at the far ends of the continent, due in large part to the activities of Louis Leakey and his colleagues, as well as to the stimulus provided by the discovery of abundant hominoid fossils. It should be noted, however, that political and logistical difficulties have always hindered the scope of exploration in North Africa, and access to the fossiliferous deposits in the diamond fields of Southwest Africa has been equally limited. Recently, however, significant new finds in Morocco (Lavocat 1961), Tunisia (Robinson and Black 1969), Libya (Savage and Hamilton 1973) and Israel (Savage and Tchernov 1968) indicate that African Early and Middle Miocene mammals may indeed be present in some abundance outside of East Africa, and it is to be hoped that work in such areas will eventually provide science with a wider view of Miocene hominoids.

In East Africa, work went forward on the Miocene fossils with little support from outside for almost thirty years after the initial discoveries. In 1947 Leakey and W. E. LeGros Clark organized the first well-funded campaign—the British-Kenya Miocene Expedition. This was an outstandingly productive effort (Clark and Leakey 1951) and was followed by important new discoveries in Uganda (Bishop 1964; Walker 1969) and in Kenya (Leakey 1968; Bishop 1972; Patterson *et al.* 1970).

GEOLOGICAL SETTING

The Early Miocene (23–16 m.y. old) sites of East Africa are in three groups: Eastern Rift, Western Rift, and Inter-Rift. The Inter-Rift sites, which have so far produced most of the fossils, are associated with the eroded remnants of great Miocene volcanoes in Uganda (Elgon, Napak, Moroto) and in the Nyanza (Kavirondo) Gulf region of Western Kenya. The Nyanza volcanoes of Tinderet and Kisingiri are the locations of the Songhor and Rusinga Island sites respectively, the two most prolific sources of Early Miocene hominoids. Tinderet is also the location of Koru, where the first Miocene fossils were found south of the Sahara, and Fort Ternan, which has provided the earliest dated specimens of *Ramapithecus*

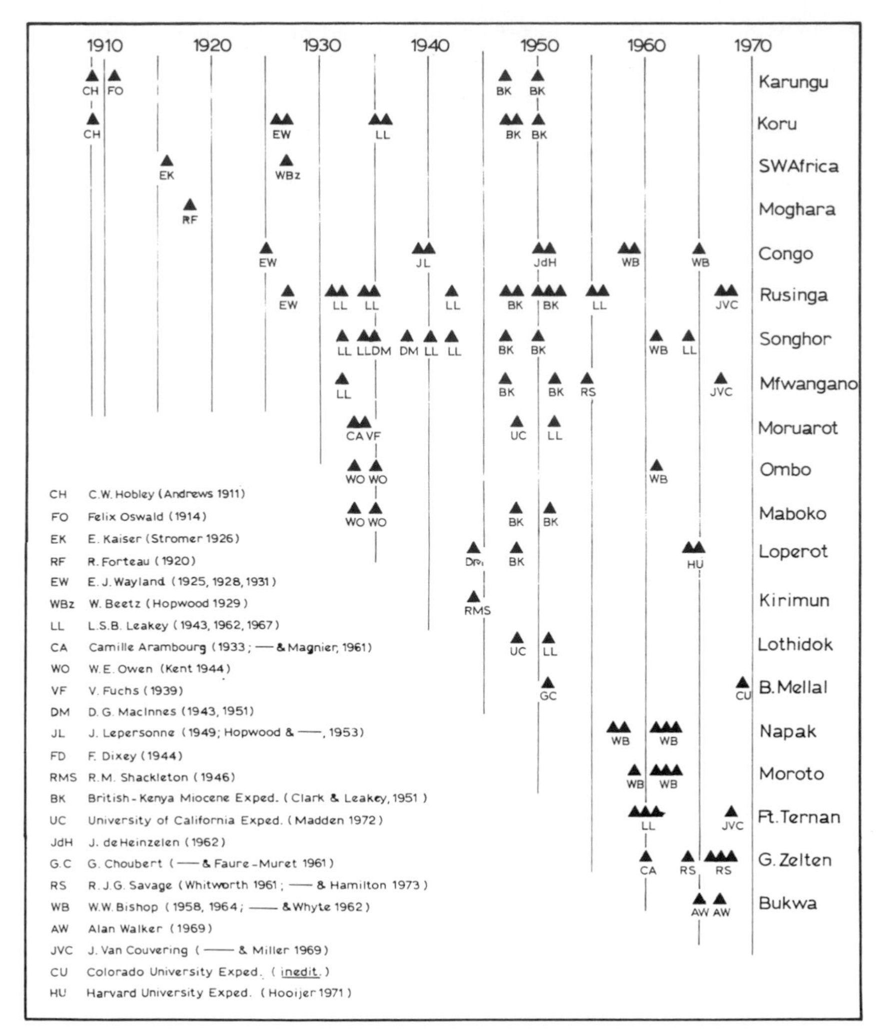

FIGURE 2

History of field work on African Early Miocene fossil mammal localities. Except for the largest expeditions, the names of the principal investigators are listed. References are to publications which describe the relevant field work, stratigraphy and general setting if not necessarily the fossil collections (cf. Appendix), and only the productive visits are indicated. Among the localities not shown, Chianda Uyoma (Figures 1 and 3) north of Kisingiri, discovered by W. E. Owen in the 1930s, but never published, is the only one with an important collection.

(although older specimens may be present at Maboko). The Western or Congo Rift Valley deposits are non-volcanic and have not been radiometrically dated, but those of the Inter-Rift and Eastern Rift group are generally associated with dated lavas and tuffs. A great range of depositional situations can be recognized in East African Neogene sediments, and

Early Miocene fossils are preserved in such sedimentary environments as soda lakes, barren floodplains, swamps, riverbeds and soil-forming surfaces. These sediments are almost invariably associated with volcanoes or lava-plateaus. (See Figure 3.)

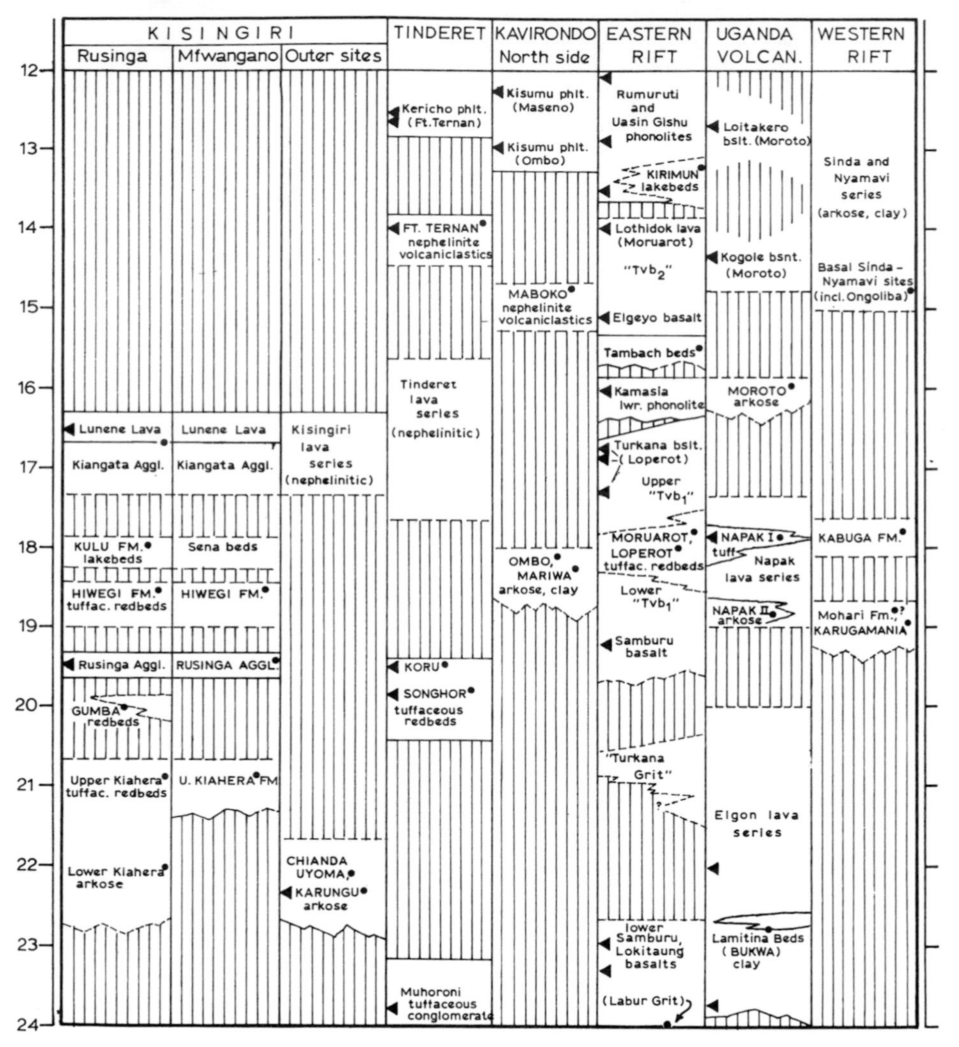

FIGURE 3

Stratigraphic succession and radiochronology in the East African Early Miocene. Filled triangles indicate selected K-Ar ages (analytical precision limits not shown) in their stratigraphic context; inconsistent and redundant values have been omitted. The collecting localities, indicated by filled circles, are also shown in stratigraphical context, and major fossil beds are capitalized. The *Kisingiri* and *Tinderet* columns show actual superpositional relationships, whereas the others are partly composite (see text). All chronological limits to the depicted stratigraphic units are estimated, and dashed boundaries indicate stratigraphical uncertainties as well.

In the earlier literature, it was commonplace to read interpretations of East African Early Miocene environments which emphasized evidence for semi-arid savanna and brushland ecologies much like those of the same region today (i.e., Chesters 1957; Clark and Leakey 1951). The present East African climate is dominantly influenced, however, by the rain shadow of the Rift Valley highlands, which block both Atlantic and Indian Ocean rainfall from reaching the interior. Because these highlands were first uplifted in the Miocene during the early stages of rifting, we prefer to accept other evidence—and other interpretations of the evidence referred to earlier—which strongly suggests that the Early Miocene began with heavily forested lowlands extending from the Congo Basin up to and probably across a low continental divide in central Kenya (cf. Andrews and Van Couvering 1975). Above the lowlands rose updomed eruptive centers which slowly grew into great volcanoes comparable in size to present-day Kilimanjaro and Mount Kenya; it was on these peaks that the first African equatorial montane forests may have developed, but it was mainly beneath their lavas (and in a few instances between them) at the base of the volcanoes that the major fossil beds were preserved. Soda-lake beds were laid down in arid basins in the Middle Miocene as the Rift Valleys began to develop and the equatorial forest belt was broken up, but earlier ash-fall eruptions also created locally barren, highly alkaline ground in poorly drained basins adjacent to the volcanoes, where forest-adapted mammal fossils were abundantly preserved (Andrews and Van Couvering 1975). During most of the Early Miocene, the climate, vegetation, and geology must have closely resembled that of the well-forested volcanic mountains and lowlands of the present-day Virunga region between lakes Kivu and Edward.

Because of the abundance of volcanic rocks in East Africa most of the fossil sites are well positioned in a K-Ar radiometric time scale (Bishop, Miller, and Fitch 1969). It is apparent from this series of dated local faunas that a distinctive East African mammalian assemblage, dominated by endemic forms, persisted without any appreciable evolutionary change from approximately 23 to 16 m.y. (cf. p. 177). The next younger dated site, at Fort Ternan (14 m.y.) has a strikingly different and more modern-looking fauna, if we can judge by the ruminants (Gentry 1970; Churcher 1970). We know that the mammal fauna of Eurasia was revolutionized by the appearance of many new forms during the "Burdigalian" land mammal age when plate movements brought the northeast corner of Afro-Arabia against southcentral Asia (Dewey *et al.* 1973), and that many of the new forms (proboscideans, chevrotains, giraffoids, certain anthracotheres and pigs) were also characteristic members of the African Early Miocene fauna. We should correlate this "Burdigalian" wave of migrations either to the beginning of the stabilized East African Early Miocene faunal interval or, as seems more likely, to its end when the dramatic changeover to the Fort Ternan-type fauna took place. This latter interpretation has been favored by one of us (Van Couvering 1972) in placing the beginning of the European "Burdigalian" land mammal age at approximately 18–17 m.y. Other workers (Evernden *et al.* 1964; Gabunia and Rubenstein 1968; Coryndon and Savage 1973) would place this time close to 20 m.y. Recently published marine-nonmarine correlations in the Tajo Basin of Portugal appear to confirm the age of the "Proboscidean Datum" at 17.5 m.y. (Van Couvering and Berggren, in press).

In either case it must be remembered that it was the Levanto-Arabian region on the south and the Armenian-Iranian region on the north which came into communication across the new land bridge (Dewey *et al.* 1973). Unfortunately Early Miocene mammal faunas from this crucial area are as yet very poorly known (Savage 1967). Once the continental plates had come to their present position, climate and topographic barriers probably restricted the distribution of immigrant mammals so that the effects of the land bridge were not felt immediately in more distant regions. The closest Early Miocene sites in Africa about which we know very much are Wadi Moghara and Gebel Zelten in North Africa. These have a fauna similar to that of the East African Early Miocene sites, although with an advanced ruminant element like that of later Early Miocene sites such as Maboko, and both are closely associated with marine Burdigalian beds (Forteau 1920; Savage and Hamilton 1973) which correlate to microplanktonic zones dated approximately 18 to 16 m.y. (Berggren 1972). At Dera Bugti in Pakistan, on the other hand, continental beds resting on marine Burdigalian strata (Eames 1970) have a peculiar and so far unique mammal fauna in which African Early Miocene and Asian Mid-Cenozoic forms are intermingled. Dera Bugti appears to represent the initial mixing across the land bridge; however, none of the Asian mammals there are as yet known from the Early Miocene of Africa, and we conclude that this is because the African Early Miocene fauna is of pre-Middle "Burdigalian" age, as the dating indicates, and that it was isolated from Eurasia up until at least 17 m.y.

Except for *Pliopithecus* (Ginsburg 1961), hominoids were apparently not part of the initial wave of African "Burdigalian" emigrants into Asia and Europe, nor have they (as yet) been found in the African Miocene outside of the equatorial region. In the Early Miocene, sites such as Gebel Zelten, Moghara, and those of Southwest Africa were close to the sea and outside of the tropical zone, as they are now, and the difference in climate might account for the absence of hominoids. The presence of hominoid-size or smaller animals in these deposits (i.e., tragulids, hyraxes, and—in southwest Africa—rodents) indicates that hydrodynamic sorting is probably not responsible. Furthermore, the first of the higher hominoids which colonized Europe and the Siwaliks after c. 15 m.y., in lower "Vindobonian" and Chinji time, are not radically different from or significantly more diverse than their equatorial predecessors, suggesting that they had not made major adaptations in order to live in the (then) largely evergreen forests of the temperate summer-rain belt in Eurasia, with its winters much milder than they were to become in the Late Miocene and Pliocene (cf. Szafer 1961).

We do not know what may have delayed the arrival of higher hominoids in Eurasia until the Middle Miocene, but the consistent absence of this group from non-equatorial African Early Miocene sites suggests that they had first to cross an ecological barrier. One would assume on present evidence that this barrier would have to be a relatively treeless zone, because (1) the known Early Miocene hominoids probably lived in a region that was mainly dense forest, (2) the known Middle Miocene hominoids also lived in regions which contained stands of dense forest, and (3) today's hominoids, except man, are still restricted to forest and dense woodland. Furthermore, it must be considered highly likely that a relatively treeless zone did in fact bound the equatorial African forest belt on the north for much the same reasons that this is true today, since configurations of continent and ocean like today's came into existence when Africa met Eurasia in the mid-Tertiary collision.

Dating and geological studies indicate that the Ethiopian highlands were already in existence, as a rifted upland volcanic region, in the Late Oligocene (Baker *et al.* 1971; Mohr 1968; Jones and Rex 1974). The existence of pre-Aquitanian freshwater deposits on the floor of the Red Sea (Abdel-Gawad 1969) suggests that the Red Sea rift valley with its flanking highlands came into existence more or less simultaneously with that of Ethiopia, with the consequence that the Arabian and Nubian lowlands were divided by rifted plateaus from Northern Kenya to Suez. The Rusinga-type fauna lived at least as far north as Moruarot-Lothidok (see "Faunal Analysis") and thus there may have already existed a continuous forest cover from Ethiopia to the eastern part of the equatorial lowland forest. The last remaining link in the chain is therefore most likely to have been in Palestine and Syria, where the initiation of updoming, anticlinal folding and erosion appears to date from a time close to the boundary between "Burdigalian" and "Vindobonian," according to our interpretation of the sparse mammalian evidence (Savage and Tchernov 1968; E. Tchernov, pers. comm. 1974) and the marine record (Derin and Reiss 1973). The formation of the Jordan Rift Valley itself appears to have begun in the Middle Miocene at the very earliest, and it is widely held to have originated only in late Miocene time (Garfunkel and Horowitz 1966).

On the present evidence, the African Rift System as we know it was not complete until Middle Miocene time. In this system, rifting produced a connected series of elongated basins in which rivers and lakes (with gallery forests?) were linked in shifting hydrological patterns, and also a parallel rim-world of bordering highlands and volcanic mountains along which rainforests could join and part in a similar way. A "high road" and a "low road" were thereby opened between equatorial Africa and the forested mountains and plains of Asia Minor, at the same time when the advanced hominoids first made their appearance outside of the tropics. This coincidence supports our opinion that this must have been the "forest bridge" that they used.

Kisingiri

The primary source of Early Miocene vertebrates in Africa, and of primitive fossil hominoids in the world, is the tuffaceous-sedimentary sequence below the lavas of the Miocene Kisingiri volcano. "Kisingiri" appears on maps as early as that of Oswald (1914) as a name for the horseshoe of mainland mountains formed by the southern half of the now-eroded cone, but local people do not recognize the word. It is probably an explorer's distortion of "Kaksingere" (Kak-*si*-ngere), the coastal plain below the mountains opposite Rusinga Island. Misapprehensions of this sort, in which a European points to a mountain peak and his African informant responds with the name of a farmland or a neighborhood in that direction, will be common until one culture or the other changes its values. Rusinga and Mfwangano islands are erosional remnants of the north flank of the mountain, and their peaks rise from Lake Victoria some 15 km. from the central vent area (Rangwa Hill, on the mainland). The sites of Karungu to the south, and Chianda Uyoma to the north, lie beneath the outermost extension of the Kisingiri lavas. Some idea of the scale of the volcano involved is given by the fact that the points are about 70 km. apart which is almost the length of a maximum diameter drawn through such huge volcanic cones as Kilimanjaro, Kenya, or Elgon.

The history of the Kisingiri volcano complex is better known than that of the other Miocene volcanoes and serves as a model for comparison in the discussion that follows. It is also possible, with this history in mind, to propose some new correlations for the Kisingiri sites. In this article we follow an interpretation based on our study of Rusinga and the previous work of many authors including Oswald (1914), Shackleton (1951), Whitworth (1953, 1961) and McCall (1958).

Briefly, we understand activity to have begun at Kisingiri in the earliest Miocene with the up-arching of a great dome of Precambrian basement rock and its intrusion by hyper-alkaline magmas (carbonatites and related types). Erosion from the dome, the apex of which may have been as much as 2000 feet (700 m.) above the surrounding land level (McCall 1958; D. Rubie, pers. comm., 1969), is represented on Rusinga Island by a very thick sequence of conglomerates and sandstones at the base of the Miocene sequence (Lower Kiahera Formation, Figure 3). This purely epiclastic deposit, the upper part of which includes fragments of the intrusive rocks but not lavas, is overlain by tuffaceous silts, sands, and conglomerates which record a long period of successive ash-cloud eruptions at the Rangwa center together with continued erosion from the basement rocks on the flanks of the dome. The tuff fragments, among which large flakes of mica are common, are derived from explosions generating the hyper-alkaline rock type *alnoite.* By contrast the Kisingiri volcanic cone itself, which began to build up over the central dome and spread out over the countryside after alnoitic tuff eruptions ceased, was composed of flows of non-micaceous *nephelinite* (Kiangata Agglomerate and Lunene Lava, Figure 3). It is in the tuffaceous series (Upper Kiahera, Hiwegi, and Kulu Formations, Figure 3) that most of the fossils are preserved, due in large part to the calcifying effect of the alkaline tuff fragments (Bishop 1963) in low-lying, poorly drained depositional areas. We believe that other evidences in the Miocene sediments of high alkalinity and resulting barren mud flats (e.g. zeolites and clay-pellet dunes) are also due to the alnoitic tuffs and not to arid climate.

Because the fossil beds at Chianda Uyoma and Karungu contain no significant amounts of pyroclastic material, and no detritus from lava flows, they are probably as old as the pre-volcanic Lower Kiahera Formation of Rusinga Island. The tuffaceous series on Rusinga and Mfwangano was evidently preserved from erosion only in the most actively downwarped part of the para-volcanic depression but not further out. This is supported by the only available K-Ar date on Kisingiri strata of the non-tuffaceous type, a very painstaking age determination of about 23 m.y. on a single mica flake obtained from the uppermost part of the Karungu exposure (Bishop, Miller, and Fitch 1969). It seems likely that the Lower Kiahera Formation is not younger than this because the Upper Kiahera tuffaceous beds are overlain by Rusinga Agglomerate dated at 19.6 m.y. (Van Couvering and Miller 1969).

If we adopt these correlations together with the most recent stratigraphy of Rusinga (Van Couvering and Miller 1969) and Mfwangano (Whitworth 1961), it is apparent that the Kisingiri sequence contains an unparalleled succession of superimposed faunal horizons, as follows:

6. Lunene horizon (top of Hiwegi Hill, Rusinga)—16.5 m.y.
5. Kulu horizon (Rusinga Island)—estimated as 18 m.y.
4. Hiwegi horizon (Rusinga and Mfwangano)—estimated as 18.5 m.y.
3. Rusinga Agglomerate horizon (Mfwangano)—19.6 m.y.
2. Upper Kiahera horizon (Mfwangano, Rusinga)—estimated as 21 m.y.
1. Lower Kiahera horizon (Rusinga, Karungu, Chianda Uyoma)—23 m.y.

It should be noted here, however, that preliminary results of a new attempt to date the pyroclastic micas from Rusinga (see Evernden *et al.*

1964; Bishop 1967; Van Couvering and Miller 1969), now in progress by G. H. Curtis and R. E. Drake at Berkeley, indicate that the entire tuffaceous series from Upper Kiahera to Upper Hiwegi formations was deposited during a relatively short interval dated approximately 18 to 17 m.y. Problems with dating these micas appear to stem from differential leaching of potassium versus radiogenic argon, and the validity of the new dates has not yet been established (G. H. Curtis, pers. comm., 1974).

It remains to be seen to what level in this sequence, if any, the local fauna of the isolated non-volcanic Gumba Redbeds of Rusinga will fit (Whitworth 1953; Van Couvering and Miller 1969), and it may be that different levels in the Lower Kiahera horizon are much more widely separated in time than those in other horizons. Unfortunately our faunal analysis of this sequence is not as complete as it should be since the collections from Mfwangano and Chianda Uyoma have yet to be properly described. All horizons in the sequence, including Gumba but possibly not the very local Lunene occurrence (Shackleton 1951:359), have yielded one or more hominoid species, not all specimens of which have been reported in the literature. It should be possible in the future to take advantage of the stratigraphic control in the Kisingiri sequence to begin the investigation of evolutionary trends during the Early Miocene of Africa.

Tinderet

Tinderet is an Early Miocene nephelinite volcano, probably with a carbonatite-intrusive complex at its still-buried core, very much like Kisingiri except for the degree to which it has been eroded. As Figure 3 shows, the succession includes the non-fossiliferous Muhoroni tuffaceous conglomerates, the Koru and Songhor fossil beds and finally the Fort Ternan fossil site. Because this succession was built up at the eastern end of the Kavirondo Rift Valley, it was partly buried by floods of phonolite that were erupted along the uprising shoulder of the Kenya Rift Valley, at the end of the early Miocene times about 16 million years ago. These phonolites underlie the Fort Ternan stratum, while 13 million-year-old phonolites overlie them (see below for details).

The oldest exposed Miocene strata in the Tinderet complex are beds of arkose and conglomerate derived from the erosion of Precambrian rocks. Known as the Muhoroni Conglomerates, these beds crop out in gullies and low foothills west of Tinderet, and form a relatively thick and locally sharply folded sequence very similar to the Lower Kiahera Formation of Kisingiri. Although they contain no other evidence of contemporary vulcanism, the Muhoroni beds yield large mica flakes dated to approximately the same age as the earliest known Kisingiri mica at Karungu (Bishop, Miller and Fitch 1969).

At the base of Tinderet Mountain, Shackleton (1951) mapped tuffaceous redbeds above Muhoroni Conglomerate, but he was not able to follow these exposures either to the lithologically similar fossil beds at Songhor to the northwest or to those of Koru to the southeast. Songhor beds lap up against the Precambrian basement rock of Songhor Hill at the fossil site and crop out in stream bluffs elsewhere in Mtete Valley, but their base is not exposed (Shackleton 1951; Bishop and Whyte 1962). On the southeast side of Legetet Hill, some 15 km. distant, the main Koru site is no longer productive (Kent 1944; Shackleton 1951), but fossils continue to turn up in other exposures of red tuffaceous silts which are uncovered during farming and quarrying operations on the east and north sides of the hill. Legetet Hill is an erosional outlier of the Tinderet sequence, and it appears to contain a sequence beginning with Muhoroni beds, seen on the southwest point of the hill, up through Koru redbeds to the nephelinite lavas which cap Legetet, and which form the main bulk of Tinderet.

Heavy vegetation prevents adequate geological mapping, but the K-Ar ages of micas from Koru (19.6 m.y.) and Songhor (19.8 m.y.) are so similar that they strongly suggest that the fossiliferous redbeds of Mtete Valley (Songhor) and Legetet Hill (Koru) are at least within the same geological formation and may even represent the same stratigraphical unit. Because of this, and because of the very marked resemblance of the collections from the two localities, we have treated the Tinderet Early Miocene sites as one horizon (SGR) in our analysis, in the same way that the different Hiwegi sites on Rusinga are treated together.

Fort Ternan is an unusual and so far almost unique Middle Miocene site. We include it in our discussion of the Kavirondo Early Miocene in order to compare it with the older levels in the same area. Because the Fort Ternan fauna is similar to that of the Ngorora Formation in the Lake Baringo region, approximately 12 m.y. old (Bishop and Chapman 1970), its 14 m.y. age based on mica flakes in the fossiliferous section has come into some doubt (F. J. Fitch, pers. comm., 1972). The general geology of the Fort Ternan area is described, somewhat inaccurately, by Binge (1962) and the detailed stratigraphy of the fossil beds has been investigated several times (Bishop and Whyte 1962; de Heinzelin, *In* Churcher 1970). Unpublished mapping of the stratigraphy of the former Wicker's Farm area, in which the site occurs, tends to confirm the value of the 14 m.y. age, although more dates should be obtained. We leave the details for another publication, but in brief, the geological observations made during the 1968 and 1974 field seasons confirm earlier conclusions (Bishop and Whyte 1962) that the Fort Ternan fossiliferous sequence itself rests upon deeply weathered phonolites and is thus younger than the (local) initiation of phonolitic vulcanism. The Fort Ternan nephelinite clastic-phonolite lava sequence is truncated by a very flat unconformity with a thick weathered zone, exposed in 1974 trenching, upon which rests a regionally mappable, undeformed pile of fresh phonolite lavas and tuffs. The basal phonolite flow of this sequence has a precisely determined age of 12.8 m.y. (Bishop, Miller and Fitch 1969). We assume here that the dated superjacent phonolite at Fort Ternan represents the "initial eruption of phonolite" event widely observed in western Kenya and placed at c. 13 m.y. (Baker *et al.* 1971), and therefore that the phonolite *beneath* the Fort Ternan fossil beds is not necessarily correlated to this event. With this assumption, and the evident maturity of the upper unconformity in mind, the 14 m.y. age attributed to the fossil fauna (Evernden *et al.* 1964; Bishop, Miller and Fitch 1969) is entirely reasonable. (For further information on Fort Ternan see Andrews and Walker, this volume.)

Kavirondo North Side

On the north side of the Nyanza Rift, three Early Miocene sites—Ombo, Mariwa, and Maboko—were discovered by Archdeacon W. E. Owen during his tenure at Maseno Mission (Figure 2). All three lie beneath the c. 12 m.y. Kisumu phonolite lava which flooded the region from the northeast (Bishop *et al.* 1969) and were first described in the scientific literature by Kent (1944), although Owen supplied accounts of his work in progress to the Nairobi newspaper.

Ombo and Mariwa sites are in erosional remnants of a granitic sand and clay deposit which filled sharply incised gullies high on the shoulder of the Nyanza Rift, whereas the Maboko Island fossil beds are bentonitic clay and volcanic pebble beds derived from the erosion of nephelinitic lavas. These differences reflect source area and not necessarily time, since Maboko would have been downstream from Tinderet on the valley floor. The Ombo-Mariwa (OMB) level is also set apart by the presence of unique species of anthracothere, rodent and suid (see Appendix) which can be ascribed to its relatively unusual location on the rift shoulder, but in other respects this is a typical Early Miocene fauna (Figure 3).

Recent studies indicate that the Maboko beds correlate with other, sparsely fossiliferous beds beneath phonolite at Majiwa Bluff on the nearby mainland and thus appear to extend over a wide area near present lake level. Saggerson (1952) indicates that on the mainland these beds rest on an irregular basement surface but the base is not exposed on Maboko Island. We note that whereas the Fort Ternan beds, dated at 14 m.y., contain abundant evidence of contemporaneous nephelinitic vulcanism just prior to the deposition of Kericho phonolites (Bishop *et al.* 1969), this is not true of Maboko except for a mica-tuff near the top of the fossiliferous sequence. Our correlation of Maboko is based partly on the assumption that it was deposited before the Fort Ternan volcanic activity began and partly on paleontological data, as yet largely unpublished, which appears to demonstrate an age between Rusinga Hiwegi and Fort Ternan. The Maboko fauna (MAB) list in the Appendix is based in large part on a 1972 letter from C. T. Madden describing the Maboko collection at the British Museum, and on new excavations in 1973.

Fossiliferous sediments at the base of the Kericho phonolite in cliffs near Sondu on the opposite side of the upper Nyanza Rift from Maboko have recently been discovered. The fauna so far is very meager and scrappy, consisting mainly of large, broken fragments deposited in high-energy depositional environments. The age of the Sondu beds is close to that of the overlying 12–13 m.y. phonolite lavas.

Uganda Volcanoes

A row of Cenozoic eruptive centers, developed from magma sources similar to those of the Nyanza rift, lies along an irregular line from Tinderet northwards to central Ethiopia parallel to the Eastern Rift axis (Nixon and Clark 1967; Williams, L. A. J., 1971; Mohr 1971). Published descriptions indicate that many of these ancient volcanoes have basal and intervolcanic sediments, but so far only Elgon, Napak, and Moroto have yielded important Early Miocene fossils (Bishop 1962, 1964, 1967; Walker 1969), together with associated Early Miocene K-Ar dates (Bishop *et al.* 1969). Fossiliferous pre-volcanic sediments at Napak and Moroto, and the very thick terrigenous pre-volcanic and early tuffaceous filling of the North Elgon Depression imply a history of early updoming and associated peripheral downwarping similar to that of the Tinderet and Kisingiri centers. However at Napak (Bishop and Trendall 1967) and Moroto (Varne 1967) the structural evidence is inconclusive and only Elgon appears to have possessed both an early dome and a true paravolcanic depression (Davies 1952; Clark 1969); unfortunately the early deposits at Elgon have been encountered only in boreholes.

The Bukwa site at Elgon is very localized, and represents preservation in pond muds interbedded with basal Elgon lavas. The dates on the lavas (Walker 1969) indicate a very early Miocene age close to 23 m.y. (cf. Berggren and Van Couvering 1974) raising the tantalizing possibility that the older strata buried in the North Elgon depression may contain Late Oligocene fossils. At Napak, Bishop (1962, 1964) sampled a number of different sedimentary environments, each with its own selection of fossils. Little has been published about diagenesis in these beds, but some of the tuffaceous fossil beds are hardened, apparently by calcite

(Bishop 1964). As pointed out above, zeolitization and calcification in alkaline tuffaceous strata need not imply arid conditions. Bishop demonstrated that Napak faunal variations were probably a taphonomical phenomenon, and the combined fauna (Table 2) suggests forested habitats, as does that of Moroto. There is little geological evidence that high mountains had already developed at these sites (the stratovolcanoes came later), but Andrews and Van Couvering (1975) give evidence that the East African region was probably covered with lowland forests in the Early Miocene, so it is not necessary to appeal to nearby elevations to account for a forest ecology.

Eastern Rift

Various stratigraphic units in the Eastern (Kenya) Rift north of Lake Baringo are grouped under this heading in Figure 3, in an arrangement according to our present view of the probable relationships between them. Direct superpositional or radiometric evidence is not yet available to support all of our conclusions, and the arguments upon which these conclusions are based are outlined below.

The conventional stratigraphy in the Miocene of this region (cf. recent examples in Joubert 1966; Walsh and Dodson 1969) is as follows:

"Pliocene" phonolites
Upper Turkana Basalt (Tvb_2)
Lower Turkana Basalt (Tvb_1)
Turkana Grit

A comparable and apparently correlative succession was described by Shackleton (1946) east of the Rift Valley, where he recognized "Samburu Basalt" overlain by "Rumuruti Phonolite," the latter associated with sparsely fossiliferous lake beds at Kirimun. Geochronological and stratigraphical studies of the northern Kenya Rift region have treated the basaltic strata as a simple but extensive unit built up by eruptions of "Miocene flood basalt" over the period from at least 23 m.y. ago to 14 m.y. ago (Baker *et al.* 1971; Williams 1971; King and Chapman 1972). Many flows are involved, and the Lower and Upper basalt members are locally distinguished by an unconformity between them. The "Turkana Grit" consists of lithified gravels and sands derived from basement rocks, and represents locally thick sedimentation in pre-volcanic basins which were subsequently buried by the earliest basalt flows in the vicinity. From the paleontological point of view, the stratigraphic system outlined above is misleadingly overgeneralized in that it implies a more or less regionally synchronous sequence of deposition.

The two important Early Miocene collecting localities in the northern Kenya Rift are Moruarot-Lothidok (with various spellings), and Loperot (Figure 1). Moruarot (originally called Lothidok, cf. Arambourg 1933; Madden 1972) and the subsequently discovered site at Lothidok Hill itself are part of a sequence of tuffaceous sediments enclosed in "Lower Turkana Basalt" or "Tvb_1" (H. B. S. Cooke, unpublished field notes, 1948; Walsh and Dodson 1969). Loperot, by contrast, is in a basal, mainly terrigenous unit which lies *beneath* "Lower Turkana Basalt" and which has therefore been called "Turkana Grit" (Joubert 1966; Hooijer 1971). Both local faunas are Early Miocene in age according to the available K-Ar dating (Figure 3) and according to faunal correlations (Table 2; Appendix) although Loperot has some very unusual elements which make comparisons difficult—including a ziphiid sea-going whale.

Nevertheless, if the Loperot pre-volcanic fossil beds are time-equivalent to the basal "Turkana Grit" further north, the Loperot local fauna is not only older than Moruarot-Lothidok, but it must also be older than the minimum acceptable age on "Lower Turkana Basalt," at least 20 m.y. and possibly more than 23 m.y. (Baker *et al.* 1971; King and Chapman 1972)—thereby the oldest Cenozoic fauna in East Africa. The evidence suggests, however, that the earliest basalts and the pre-volcanic sediments beneath them are progressively younger in age going south from the Ethiopian eruptive centers, i.e., from the Mesozoic (Arambourg and Wolff 1969) or at least pre-Miocene Labur Grit at Lokitaung near the Kenya border southward to the Tambach Beds (Shackleton 1951) beneath c. 15 m.y. Elgeyo Basalt near Eldoret. In view of this we have accepted the dating that indicates the *local* age of overlying basalt at Loperot is c. 17 m.y. (Hooijer 1971) and place the Loperot fossils at about the same 18 m.y. age level as has been estimated for Moruarot-Lothidok on paleontological grounds (Madden 1972a, 1972b). We do not intend to imply that there is a physical correlation between the two localities. It is possible nevertheless to imagine that the lower segment of "Tvb_1" beneath the Moruarot-Lothidok fossil beds is missing to the south, possibly pinched out against rising topography, and that the intervolcanic strata of Moruarot-Lothidok pass laterally into subvolcanic sediments in the Loperot area.

Congo

Neogene fossiliferous deposits in the Western Rift basins of Lake Albert and Lake Edward are locally very thick and productive but so far only the basal formations on the Congo (Zaire) side have yielded Miocene mammals. W. W. Bishop has emphasized in conversations that these remains are extremely rare and fragmentary, and the identifications by Hooijer

TABLE 1
Species diversity in modern and Early Miocene mammal faunas of equatorial Africa.

TAXON	FOREST	"SAVANNA"	RUSINGA	SONGHOR
LIPOTYPHLA				
Tenrecidae	1	1	2	2
Erinaceidae	0	2	5	3
Chrysochloridae	7	0	0	1
Soricidae	11	5	1	0
MACROSCELIDEA	1	1	4	3
CHIROPTERA	21	5	2	3
PRIMATES				
Lorisidae	5	1	4	4
Cercopithecidae	10	8	0	0
Pongidae	2	0	6	6
LAGOMORPHA	0	1	2	0
RODENTIA				
Phiomyoidea	1	2	6	6
Bathyergoidea	0	1	2	1
Anomaluridae	5	0	2	4
Pedetidae	0	1	2	1
Hystricidae	1	1	-	-
Cricetodontidae	-	-	3	3
Muridae	18	15	-	-
Sciuridae	6	5	2	1
Gliridae	2	1	-	-
CREODONTA	-	-	12	6
CARNIVORA				
Amphicyonidae	-	-	2	1
Viverridae	10	9	1	1
Felidae	2	5	2	1
Hyaenidae	0	3	-	-
Mustelidae	2	1	-	-
Canidae	0	5	-	-
PHOLIDOTA	3	1	-	-
TUBULIDENTATA	1	1	2	0
HYRACOIDEA				
Pliohyracidae	-	-	3	1
Procaviidae	1	1	0	0
PROBOSCIDEA	1	1	2	2
PERISSODACTYLA				
Chalicotheriidae	-	-	1	1
Rhinocerotidae	1	1	4	2
Equidae	0	1	-	-
ARTIODACTYLA				
Anthracotheriidae	-	-	3	0
Hippopotamidae	2	1	-	-
Suidae	2	2	5	4
Tragulidae	1	0	3	1
Gelocidae	-	-	1	0
Giraffidae	1	1	2	1
Bovidae	11	24	1	1

Dashes are entered in categories which do not have African representatives. See text for discussion.

(1963, 1970) are indeed presented with many reservations. The absence of radiometric ages in these totally non-volcanic strata also hampers correlations, and the placement of Congo units in our Figure 3 is meant only to show relative superposition and not "absolute age." Sedimentary environments representing stream beds, beaches and shallow lakes are noted by Gautier (1970).

Most of the mammals identified from the lower Congo sites are known from the Early and Middle Miocene of Kenya–Uganda. The exceptions are a possible *Macaca* from Ongoliba and *"Stegolophodon" lepersonnei* from several sites, which D. A. Hooijer (written comm., 1974) now agrees is congeneric and perhaps conspecific with *Primelephas gomphotheriodes,* the earliest true elephantid (Maglio 1970, 1973). This proboscidean has not yet been identified from a dated context older than c. 7 m.y. (Cooke and Maglio 1972). However, the long-neglected Early Miocene proboscidea of Africa are only now being restudied (Tobien 1973) and new Middle and Late Miocene sites near Lake Baringo are largely unpublished, so that it may be that the range of *Primelephas* will be extended, as well as our knowledge of Miocene cercopithecoids. Nevertheless, it is of interest to note that our stratigraphic arrangement (Figure 3) in which all the sites with these "late" forms are coincidentally placed younger than the rest is in agreement with the present geological correlations (Gautier 1970; Lepersonne 1970), if not with Hooijer's (1970) faunal groupings.

FAUNAL ANALYSIS

Comparison with Modern Faunas

General. Table 1 shows, in the two left-hand columns, the species diversities among families of extant mammals. The figures for Lipotyphla, Macroscelidea, Chiroptera, Primates, Lagomorpha, Rodentia, Carnivora, Pholidota, Tubulidentata and Hyracoidea have been compiled from reports based on material collected by the expedition of the American Museum of Natural History to the Congo in 1909–1919 under the leadership of H. Lang and J. Chapin (cf. Hollister 1916; Allen *et al.* 1917; Allen 1922a, 1924, 1925; Hatt 1934, 1936, 1940). Using their data, comparisons can be made among taxonomic groups and between many different localities. In addition, the collecting areas are similar in size to the areas sampled by the Rusinga and Songhor fossil sites (although the fossil sites may sample a slightly more ecologically diverse area due to their apparent proximity to "highlands"). Data for the Proboscidea, Perissodactyla, and Artiodactyla were derived from a number of sources (Bourlière 1973; Dorst and Dandelot 1970; Williams, J. G., 1967), and represent the probably maximum diversity that can be found in local forest or savanna areas for these groups. Knowledge of the modern species diversities may be improved in the future by the current French work in western Africa (e.g., Brosset 1966; Pagès 1970) but, as yet, the entire spectrum of animals in a large number of areas has not been studied and thus the material is not comparable.

Most of the sites sampled by the AMNH Congo Expedition are forest sites. We have used a compilation of two of the more abundant sites for our "forest" column: (1) Avakubi, on the Ituri River (2°N, 25°E) and (2) Medje, between the Ituri and Uele Rivers (3°N, 25°E). Animals were also collected from a few non-forest sites. From these we have used the Faradje site (4°N, 30°E) because of the abundance and diversity of its fauna. (This site may in fact sample a savanna-forest mosaic with a greater diversity than that of a homogeneous habitat.)

From among the fossil faunas we have chosen to compare the Rusinga (Hiwegi) and Songhor faunas with the modern ones because they are the most diverse Early Miocene collections. We do not mean for them to be contrasted with one another.

Tables 1 and 2 show three main things: (1) species diversity at Rusinga and to a lesser degree at Songhor is similar to that found in the modern faunas, implying that at these fossil sites the record is singularly complete or else that in the Early Miocene the fauna was more diverse than at present; (2) the replacement of ancient taxa by more modern ones of similar trophic type, although the radiation of bovids and the replacement of browsers by grazers is a major exception; and (3) in the modern fauna, a difference in total diversity of large herbivores and carnivores between savanna and forest environments, a fact already pointed out by Bourlière (1973) and Keast (1969). These points will be explored below.

Past and Present Faunal Diversity. The mammalian fauna that has been discovered over the years from the Hiwegi Formation of Rusinga Island is comparable in diversity to that of a modern fauna of either forest or

TABLE 2
Species diversity in trophic groups of modern and of Early Miocene mammals in equatorial Africa.

TROPHIC TYPE	FOREST	"SAVANNA"	RUSINGA	SONGHOR
Insectivores	13	9	12	9
Anteaters	2	2	2	0
Small Omnivores	10	4	8	7
Large Omnivores	10	5	11	10
Small Herbivores	37	35	21	21
Medium Herbivores	8	14	10	4
Large Herbivores	5	12	10	5
Carnivores	14	23	15	9

"savanna" habitat in all groups except bats (cf. Table 1). It is unusual to find a fossil fauna in which the full complement of diversity is so well sampled, and it is therefore pertinent to ask whether the actual Early Miocene fauna was not, perhaps, even more diverse than is shown by the fossil record. A comparison of the species diversities of the extant fauna of the African equatorial region, or rain forests, with those of other equatorial areas, such as South America and Asia, may be helpful in the evaluation of this problem.

Most of the extant fauna and flora of the African rain forest is known to be depleted (cf. Meggers *et al.* 1973). This has been attributed to the Pleistocene reduction of the forest into essentially relic islands (Moreau 1966; Cahen 1954). The extant *mammalian* fauna is, however, comparable in diversity to that of the South American forest (Bourlière 1973) and, among the Rodentia, only slightly less diverse than that of India (Ellerman 1961). The limited information that we have been able to gather on Southeast Asia (Van Peenen 1969; Medway 1969) suggests that it too has approximately the same mammalian diversity as the African and South American equatorial forest. Why is it that the total species diversity among African mammals is comparable with South America and, perhaps, Asia when that of the remaining fauna is deplete? Three possibilities come to mind: (1) the mammalian fauna somehow escaped depletion; (2) the fauna was depleted and has redeveloped, by migration or evolution, to its former diversity; or (3) all of the tropical mammalian faunas are deplete. The last alternative is suggested by two independent lines of evidence: (1) we know that there was a faunal turnover in South America during the Plio-Pleistocene which drastically changed the mammalian fauna there (cf. Keast 1969); and (2) it would be extremely unusual to find a fossil sample in which all species of the contemporaneous fauna were preserved. Changing of sea level during the Pleistocene caused faunal mixing in Southeast Asia and, perhaps, local and regional extinctions and consequent reduction in diversity (cf. Hooijer's papers on Southeast Asian fossil faunas). It seems probable to us that there was a more diverse fauna in Early Miocene times, but that the Rusinga collection is not selectively sampling one group in favor of another, except to under-represent the bats (cf. p. 178 for discussion of taphonomical biases).

Faunal Replacement. Comparing Tables 1 and 2 it can be seen that, although in many cases the taxa of extant and fossil faunas are not comparable, the diversity of trophic types is comparable and similar. This is due to the complete or partial trophic replacement of ancient groups by more recently evolved groups (see Table 3). The steps in this faunal overturn are only just beginning to be uncovered. The Fort Ternan, Ngorora and

TABLE 3
Replacement in trophic types of African Neogene mammals.

TROPHIC TYPE	EARLY MIOCENE GROUPS	DEGREE OF REPLACEMENT	MODERN GROUPS
Insectivores	Chiropterans **Tenrecids Erinaceids Chrysochlorids Soricids	Partial	Chiropterans Erinaceids Chrysochlorids *Soricids*
Anteaters	Tubulidentates	Partial	Tubulidentates Pholidotids
Small Omnivores	Macroscelideans Prosimians	Partial	Macroscelideans Prosimians *Cercopithecoids*
Large Omnivores	Hominoids Suids	Almost complete	Hominoids (ex *Homo*) Suids *Cercopithecoids*
Small Herbivores	*Phiomyoids* Anomaluroids *Cricetodontids Sciuroids **Ochotonids Procaviids	Almost complete	Phiomyoids Anomaluroids Hystricids *Cricetids* *Murids* Sciuroids Leporids Procaviids
Medium Herbivores	*Pliohyracids *Tragulids* *Gelocids Bovoids	Almost complete	Suids Tragulids *Bovids*
Large Herbivores	*Deinotheriids *Gomphotheriids *Mammutids *Chalicotheriids Rhinocerotids *Anthracotheriids	Almost complete	Elephantids Rhinocerotids Equids Hippopotamids Giraffids *Bovids*
Carnivores	**Creodonts* *Amphicyonids Viverrids Felids	Almost complete	Viverrids Felids Hyaenids Mustelids Canids

Italics *mark taxa which were clearly dominant in trophic groups. Extinct groups are marked by (*) and groups which no longer live in Africa by (**). Some taxonomic groups which vary significantly in body size or feeding habits are entered twice but the table does not show minor deviates, e.g.,* Proteles *(aardwolf), the anteater hyaenid, or the rabbit-sized bovids. "Large omnivores" and "medium herbivores" are approximately the same size. The term* trophic replacement *does not necessarily mean that the modern groups competitively excluded the ancient groups, although in some instances this may have been the fact. It is simply that the fossil record is inadequate to establish the specific cause of replacement in each case.*

younger Eastern Rift faunas, when fully known, will fill the crucial gap between the Early Miocene and the nearly modern Plio-Pleistocene faunas. At the moment, however, this dramatic turnover cannot be closely analyzed. Even so, it is interesting to note that the diversities of the

TABLE 4
Comparative species diversity of modern mammals in Congo (Zaire) forest versus non-forest environments.

TAXON	FOREST	"SAVANNA"
Erinaceidae	0	2
Chrysochloridae	7	0
Soricidae	11	5
Chiroptera	21	5
Lorisidae	5	1
Pongidae	2	0
Bathyergoidea	0	1
Anomaluridae	5	0
Pedetidae	0	1
Felidae	2	5
Hyaenidae	0	3
Canidae	0	5
Equidae	0	1
Tragulidae	1	0
Bovidae	11	24

Data selected from Table 1.

trophic types remain almost identical except that, among the large herbivores, grazers come to dominate over browsers, and that this replacement begins in the Middle Miocene (cf. Fort Ternan and Ngorora).

Forest and Savanna Diversities. According to the references we have mentioned above, a few significant differences can be seen in species diversities of some families living in African forest and savanna environments today (see Table 4). Of those mammals which are more diverse in the forest than in the "savanna," or which are restricted to forest, lorisids, pongids, anomalurids, and tragulids occur at both Rusinga and Songhor in great diversity and are, in our opinion, excellent indicators of nearby forest (cf. Andrews and Van Couvering 1975). Of those which are more diverse in the savanna, or are restricted to savanna, only pedetids and bovoids occur in the fossil sites. It is possible, however, that some of the fossil taxa which have no modern analogs in Africa may be "savanna" animals (e.g., *Myohyrax,* ochotonids, and some of the ungulates and carnivores).

If we compare trophic types rather than taxa, the similarity of the modern and fossil mammalian faunas is greatly increased. Large herbivores and carnivores occur in greater diversity in savanna than in forest environments and both groups also occur in greater diversity at Rusinga than at Songhor (cf. Table 2). Although it is difficult to evaluate this evidence due to the complexity of faunal replacement among the large herbivores, the presence of a greater diversity of these types at Rusinga than at Songhor suggests that the Rusinga sites are sampling some non-forest habitats (cf. pp. 179–82 for further discussion of Rusinga paleoecology).

Comparison of Early Miocene Sites

East African Sites. General. We have compared specific and generic similarities of fossils from African Early Miocene sites (Table 5) using Simpson's Faunal Resemblance Index[1] (hereafter referred to as FRI) (Simpson, G. G., 1947, 1960). The use of FRI over such a restricted area may be of questionable value, but we feel that it does point out some general tendencies that might otherwise go unnoticed. Relative abundance figures are at the moment very spotty, but we look forward to the time when they are complete enough to be of use in a more detailed study (cf. Appendix discussion).

The following five variables are considered in our analysis of FRI: (1) sample size; (2) geographical proximity; (3) temporal proximity; (4) similarity of sedimentary setting and (5) ecology of the area in which the animals lived. Simpson (1947) used a slightly different set of variables in his comparison of Holarctic faunas through time, but he was considering a different kind of problem.

Sample Size. Simpson's FRI is designed to eliminate the error inherent in comparing a large sample with a small one by using the size of the smaller sample (N_1) as the denominator, and in this it is fairly successful (cf. Simpson 1960). Simpson (1947) points out, however, that the comparison of two small faunas by this method is often unreliable. This is because it is unlikely that in two small samples a high proportion of the same taxa will occur in both even if they were originally part of the same fauna, although our figures (and those of Simpson) show that this is not inevitably so. Fossil faunas considered here which have less than 15 taxa (Ombo, Bukwa, Loperot, and the Congo sites) have, in general, FRI below 50 when compared with all but the most diverse of the other faunas (e.g., Rusinga; cf. Table 5). For the most part the low similarity indices are in fact probably due to sampling error, as noted.

Geographical Proximity. All of the East African Early Miocene sites lie within 650 km. of each other. As most of the FRI calculated for modern Faunas have been over greater distances, and furthermore are in the Holarctic region, we have computed FRI for some of the Congo (Zaire) sites collected by the AMNH party (see p. 173). These sites are similar both in distances and ecological variety to the Early Miocene sites of East Africa. The resulting figures (Table 6) compare favorably with the figures from the fossil sites, ranging from 33 to 84 for species and 47 to 75 for genera.

---

[1] $100 (C/N_1)$ where C is the number of taxa in common between the two faunas, and N_1 is the total number of taxa in the smaller sample.

These show that the sites which are closest together and from the same habitat have the highest FRI and those which are furthest apart and from different habitats have the lowest. This is, of course, exactly what is to be expected in cases where time and sedimentary transport are not factors.[2] It is clear from this comparison that *if all else were equal* increasing distance between sites would have a small progressive negative effect on the FRI, but that similarity of habitat can swamp these differences. That geographical distance is *not* the main influence in the FRI of the fossil faunas shows up clearly when Rusinga (Hiwegi) is compared with the other sites in order of increasing distance (see Table 3 for abbreviations):

Site	RUS-KAR	RUS-MAB	RUS-OMB	RUS-SGR	RUS-BUK
Distance	45 km	55 km	60 km	125 km	210 km
Genus FRI	100	56	77	88	100

Site	RUS-NAP	RUS-LOP	RUS-MOR	RUS-CGO
Distance	290 km	350 km	445 km	450 km
Genus FRI	88	64	80	88

Other large sites are similarly irregular in terms of distance.

Temporal Proximity. The East African Early Miocene sites range in time from approximately 23 to 16 m.y. (cf. Figure 3). This period in time in East Africa is characterized by (1) an essentially unchanged fauna for the first 6 million years (23–17 m.y.) and (2) by a faunal turnover starting perhaps about 16 m.y., affecting, in the first case, the large herbivores (as reflected especially in the Fort Ternan fauna; Gentry 1970) and later other groups (Ngorora faunas; Bishop 1972). The stability of the Early Miocene, or "Rusingan," fauna can perhaps be explained by the lack of major environmental change and, perhaps even more importantly, by the lack of faunal interchange between Africa and other continents. The faunal turnover, on the other hand, can be explained both by changes in landscape, and thus local climate and vegetation, and by faunal mixing with Eurasia (cf. Van Couvering 1972 and pp. 161–64.)

[2]The values of FRI in the modern sites are lower than we anticipated, possibly because only part of the modern fauna could be used in the computations (Insectivora, Primates, Rodentia, Carnivora, Tubulidentata and Hyracoidea) due to the lack of comparative figures for the remainder. The same groups in the fossil fauna (including also Creodonta) were similarly tabulated. Although this treatment reduces the compared taxa to unreliably low totals at most sites, those that do have abundant taxa have comparable, if slightly higher FRI than the modern sites : RUS/SGR, 83; RUS/NAP, 77; SGR/NAP, 76. N_1 is 36, 26, 26, respectively.

TABLE 5
Faunal resemblance indices (FRI) of East African Early Miocene sites.

	RUS-KAR	RUS-OMB	RUS-SGR	RUS-MAB	RUS-MOR	RUS-LOP	RUS-NAP	RUS-BUK	RUS-CGO
Size of faunas	Lg/Med	Lg/Sm	Lg/Lg	Lg/Med	Lg/Med	Lg/Sm	Lg/Lg	Lg/Sm	Lg/Sm
Distance Apart	45 km	60 km	125 km	55 km	445 km	350 km	290 km	210 km	450 km
Time difference	5 Ma	?0 Ma	2 Ma	?2 Ma	?0 Ma	5 Ma	0-1 Ma	5 Ma	?
Small mammal %	46/19	46/0	46/58	46/19	46/10	46/]0	46/56	46/25	46/0
Species FRI	100	77	81	50	79	30	80	94	88
Genus FRI	100	77	88	56	80	64	88	100	88

	KAR-OMB	KAR-SGR	KAR-MAB	KAR-MOR	KAR-LOP	KAR-NAP	KAR-BUK	KAR-CGO
	Med/Sm	Med/Lg	Med/Med	Med/Med	Med/Sm	Med/Lg	Med/Sm	Med/Sm
	90 km	140 km	90 km	490 km	400 km	330 km	245 km	455 km
	5 Ma	3 Ma	?7 Ma	?6 Ma	?6 Ma	4-5 Ma	0 Ma	?
	19/0	19/58	19/19	19/10	19/10	19/56	19/25	19/0
	55	55	40	48	20	48	56	55
	55	68	43	48	45	59	57	55

	OMB-SGR	OMB-MAB	OMB-MOR	OMB-LOP	OMB-NAP	OMB-BUK	OMB-CGO
	Sm/Lg	Sm/Med	Sm/Med	Sm/Sm	Sm/Lg	Sm/Sm	Sm/Sm
	75 km	10 km	395 km	290 km	250 km	160 km	485 km
	2 Ma	?2 Ma	?0 Ma	?0 Ma	?0 Ma	?5 Ma	?
	0/58	0/19	0/10	0/10	0/56	0/25	0/0
	55	55	33	22	44	44	44
	55	55	33	22	44	44	44

	SGR-MAB	SGR-MOR	SGR-LOP	SGR-NAP	SGR-BUK	SGR-CGO
	Lg/Med	Lg/Med	Lg/Sm	Lg/Lg	Lg/Sm	Lg/Sm
	65 km	380 km	255 km	265 km	160 km	560 km
	?4 Ma	?2 Ma	?2 Ma	1–2 Ma	3 Ma	?
	25	48	10	78	63	55
	44	60	36	79	79	55

	MAB-MOR	MAB-LOP	MAB-NAP	MAB-BUK	MAB-CGO
	Med/Med	Med/Sm	Med/Lg	Med/Sm	Med/Sm
	400 km	295 km	260 km	165 km	495 km
	?2 Ma	?2 Ma	?2–3 Ma	?7 Ma	?
	[illegible]	[illegible]	[illegible]	[illegible]	[illegible]

MOR-LOP	MOR-NAP	MOR-BUK	MOR-CGO
Med/Sm	Med/Lg	Med/Sm	Med/Sm
125 km	210 km	250 km	650 km
?0 Ma	?0 Ma	?5 Ma	?
19/10	19/56	19/25	19/0
20	48	56	66
45	59	64	66

LOP-NAP	LOP-BUK	LOP-CGO
Sm/Lg	Sm/Sm	Sm/Sm
180 km	160 km	650 km
?0 Ma	?5 Ma	?
10/56	10/25	10/0
10	30	11
27	55	22

NAP-BUK	NAP-CGO
Lg/Sm	Lg/Sm
100 km	475 km
4-5 Ma	?
56/25	56/0
50	66
57	66

BUK-CGO
Sm/Sm
500 km
?
25/0
44
44

KEY	
RUS	Rusinga Hiwegi
KAR	Karungu
OMB	Ombo, Mariwa
SGR	Songhor, Koru
MAB	Maboko Island
MOR	Moruarot, Lothidok
LOP	Loperot
NAP	Napak
BUK	Bukwa
CGO	Congo

TABLE 6
Faunal resemblance indices (FRI) related to local environment in modern Congo (Zaire) mammal-collecting sites.

Compared sites	*Ava-Med*	*Ava-Nia*	*Ava-Nian*	*Ava-Far*
Local environment	for-for	for-for	for-sav	for-sav
Species FRI	84	71	51	45
Genus FRI	67	69	51	47
		Med-Nia	*Med-Nian*	*Med-Far*
		for-for	for-sav	for-sav
		78	50	43
		83	62	52
			Nia-Nian	*Nia-Far*
			for-sav	for-sav
			43	33
			50	48
				Nian-Far
				sav-sav
				62
				75

KEY
for - forest
sav - non-forest
Ava - Avakubi (N 2°, E 25°)
Far - Faradje (N 4°, E 30°)
Med - Medje (N 3°, E 25°)
Nia - Niapu (N 2°30′, E 27°)
Nian - Niangara (N 4°, E 28°)

See Allen, et al. *(1917) for index map.*

The stability of the "Rusingan" fauna is corroborated by some of the faunal resemblance indices of Table 3. For instance, the pairs of Karungu–Rusinga and Bukwa–Rusinga, which are 5–6 million years apart in time (23 m.y. and 18–17 m.y.) have a very high FRI, indicating a high degree of faunal stability over a lengthy period of time. On the other hand, Maboko–Rusinga, which may be only 2 million years or less separated in time, have a low FRI (59). We think that this low faunal similarity is most probably due to the fact that Maboko was involved in the faunal turnover, since Maboko and Fort Ternan appear to have several genera in common that are not found in the other East African Early Miocene sites (e.g., *Paradiceros, Climacoceras, Eotragus, Ramapithecus).*

Time is probably the major factor in the differences between the Congo sites and those of Eastern Africa, but the dating of these sites is unresolved (cf. pp. 171–73) and this possibility is difficult to analyze.

Similarity of Sedimentary Environments. We will treat sedimentary environmental differences in a rather simplistic manner and will be concerned here primarily with the relationship between lithology, depositional environment, and the size of the mammals preserved.

Studies of the taphonomy of certain of the East African Early Miocene sites have been begun (Bishop 1964; Hill and Walker 1972; Andrews, Van Couvering and Van Couvering 1972; Andrews and Pickford 1972) or are in progress, but the work has not yet been completed or synthesized.

The most important factor that we are concerned with in comparing faunal similarities is the extent to which bones and teeth have been water-sorted. Sites with size-sorted bones will also show taxonomic sorting between groups consisting of "small mammals" (with small bones and teeth) and others which are strictly "large mammals" (with very few, if any, small bones or teeth). Some are, of course, intermediate. In our category of "small mammals" we have included the orders Lipotyphla, Macroscelidea, Chiroptera, Primates (prosimians only), Lagomorpha and Rodentia.

Although some work has been done recently defining the velocity parameters which affect the skeletal parts of large mammals (Voorhies 1969; Behrensmeyer 1973), little work has been done on small mammals. It has been shown that the elements of small mammals group slightly differently in transport than do those of larger mammals (cf. Dodson 1973) and that they are carried by currents of very low velocity (6–35 cm./sec) whereas the *most easily moved* bones of medium to large-sized animals (e.g., sheep or larger) are not moved until current velocity reaches 20 cm./sec. Thus, in many cases, large mammals exclusively occur in coarse-grained sediments from which any small mammal bones and teeth would have been winnowed. Fine-grained sediments, on the other hand, contain predominantly small bones and teeth which were transported by currents too weak to move the large bones. This is a simplification of the true situation, but further details are not essential to this discussion.

Table 7 summarizes the relationships between lithology, depositional environment, and percentage of small mammals. An examination of this table shows that there is general agreement between the grain size of the sediments and the proportion of small mammals in the fauna, although there is no definite correlation between the depositional environment and the proportion of small mammals. All local faunas in which a high proportion of small mammals occur were found in fine-grained sediments (clays, marls, siltstones), and those in which there are *no* small mammals were found in coarser-grained sediments (coarse sandstones and gravels). Local faunas with a low percentage of small mammals occur in a rather mixed set of lithologies and are difficult to evaluate. It could thus be very misleading to compare faunas with many small mammals with those in which there are none or few, and Table 5 shows that faunal resemblance indices for these faunas are low. On the other hand, both types of faunas can be compared with that of Rusinga which samples both small and large mammals.

It is questionable whether certain fossil beds which are thought to have been formed as soils actually were formed in this manner (e.g., Napak and Songhor; Table 7). Judging from the presence of zeolites and calcite these deposits were formed in an alkaline environment although the fauna is obviously derived, at least in part, from a forest environment

TABLE 7
Depositional size-sorting in African Early Miocene mammalian fossils.

SITE	FOSSIL-BEARING LITHOLOGY	DEPOSITIONAL ENVIRONMENT	PERCENT SMALL MAMMALS*
Rusinga Hiwegi	Clayey fissile siltstone; red marl; zeolitic tuffaceous siltstone and sandstone	Floodplain	46
Karungu	Red marl; limy arkosic grits; yellow sands	Floodplain with fluviatile channels	19
Ombo Mariwa	Massive arkosic grits	Fluviatile channel	0
Songhor-Koru	Red marl and siltstone; zeolitic tuffaceous siltstone and sandstone; limy volcanic sandstone	Floodplain and ?sub-aerial or intermittently exposed mudflat	58
Maboko	Bentonitic clays; marl; volcanic pebbly sandstones	Mudflat	19
Moruarot-Lothidok	Red marl and siltstone; zeolitic volcanic sandstone and volcanic grit	Floodplain with fluviatile channels	10
Loperot	Limy coarse sandstone	Fluviatile channel	10
Napak-1	Limy tuffaceous siltstone and sandstone	?Sub-aerial	56
Napak-2	Limy coarse sandstone	Fluviatile channel	0
Bukwa	Green osctracod claystone	Small lake	25
Congo	Limy coarse sandstone, yellow sands, gravels	Fluviatile channels and littoral margins	0
North Africa	Coarse current-bedded sands, gravels	Estuarine channels and littoral margins	0
Southwest Africa	Clayey silstone; marl	?Floodplain	65

Local depositional environments are deduced from sediment lithology, and also from associated autochthonous lower vertebrates and molluscs. See van Damme and Gautier (1972), Verdcourt (1972) and relevant geological studies for the listed sites. "Sub-aerial" does not mean a soil-forming environment but rather the conditions under which lime-centered bands might develop, as in caliche or kunkar.

**"Small mammal" here means all species listed in the Appendix before Order Creodonta except for the hominoid primates.*

in which the soils would have been acid. This apparent paradox has not yet been fully resolved although it has obvious parallels with the situation at Rusinga in which forest-habitat mammals are preserved with zeolitized alkaline tuffs.

Ecology of the Source Area. The faunal resemblance indices calculated for the modern Congo local faunas are strongly affected by the ecology of the sites (cf. Table 6): those that are geographically close but from contrasting habitats have a much lower FRI than those in which both sites are situated in the same type of habitat. It is assumed that this must also be true of fossil sites.

The paleoecology of a particular site is often difficult to determine, and detailed work on many sites is necessary before we can feel secure that our reconstructions are reflecting the true paleoenvironment. Adequate and reliable relative abundance and endemicity data is needed in order to come to a more complete understanding of the ecological differences between fossil faunas.

The most complete local faunas at the present time are from Rusinga and Songhor. They are comparable in that both are large, well-collected, close in space and time, and preserve somewhat similar percentages of small mammals. They have a FRI of 86 which is fairly high but perhaps not as high as might be expected. The small mammal fauna, in particular, is indicative of microecological differences. Relative abundance data, which is moderately good for these two sites, shows that many of the small mammals which are present at both sites are far more abundant at one than the other. In addition we can be fairly confident of the negative evidence: if a species has no record from one of these sites it probably did not live nearby. The lipotyphlan insectivores provide a good example of differential distribution: (1) the tenrec, *Protenrec tricuspis* is known from 10 individuals at Songhor and only 1 at Rusinga; (2) the hedgehogs, *Gymnurechinus leakeyi* and *G. camptolophus* occur in abundance at Rusinga but not at all at Songhor while *G. songhorensis* occurs in abundance at Songhor but not at all at Rusinga; (3) the golden mole-like *Prochrysochloris miocaenicus* occurs in moderate abundance at Songhor but not at all at Rusinga. These must reflect different microenvironmental conditions at the two sites, but it is difficult to judge from these insectivores just what those differences might be. (However, see Andrews and Van Couvering 1975 for discussion of probable preferred habitat of *P. miocaenicus.*)

Other species which occur either exclusively, or in much greater abundance, at Rusinga are suggestive of a non-forest habitat, although none of them conclusively so: the pika-like *Kenyalagomys rusingae* and *K. minor,* the elephant-shrew *Myohyrax oswaldi,* the spring hare-like *Megapedetes pentadactylus* (but see Andrews and Van Couvering 1975), the large hyrax *Pachyhyrax championi* (and hyraxes in general) and a number of large herbivores. The abundance of large herbivores at Rusinga in comparison with Songhor may be a taphonomical phenomenon, however. In addition there are four species of gliding, forest-dwelling anomalurid rodents at Songhor and only one at Rusinga. Thus although both sites have a forest-facies strongly represented, Rusinga also seems to have species derived from a more open, non-forest habitat, be it woodland, bushland or grassland. The presence of low-crowned browsing herbivores rather than high-crowned grazers suggests that, at this time, large areas of non-forest had not yet developed. Nonetheless the Rusinga local fauna does seem to have been derived from a setting in which there was more non-forest area than there was at Songhor.

Loperot is the only local fauna in our study which has a species resemblance index of 33 or less compared to *all* other African Early Miocene sites (Table 5). The generic resemblance index, however, is not anomalous, and factors such as sample size, geologic age, distance and sedimentary setting simply do not account for the low species FRI. A brief examination of the collected fauna (see Appendix) and recent geological reports suggest that this difference is due to ecology.

Foremost among the unique features of the Loperot local fauna is a ziphiid whale (V. J. Maglio, written comm., 1972; B. Patterson, written comm., 1972). Modern ziphiids, the beaked whales, are specialized deep-sea odontocetes, but primitive Early Miocene forms closely resembled generalized squalodonts (Romer 1966) and might have been less restricted in habitat. The absence of large estuarine forms in the Loperot collection (e.g., skates, sharks, sirenians) suggests that the whale was actually in fresh water when it died, although it is reasonable to assume from its presence that the site was close to sea level in a large river system. Some other large mammals at Loperot belong mainly to genera which are thought to prefer waterside habitats; it may be significant that the anthracotheriids, of which this is particularly the case, appear to be clearly different from those of the more "inland" sites.

Recent petroleum exploration in northeast Kenya has demonstrated that Lower Miocene marine deposits extend inland to the edge of the Rift Valley Swell in the latitude of Loperot (Walters and Linton 1973). The former inland limits of the Miocene marine embayment were postulated according to contours on the "sub-Miocene surface" of Saggerson and Baker (1965), but these contours only demonstrate a broad uplift immediately prior to the *local* beginning of basaltic eruptions (c. 17 m.y.?) and may thus be irrelevant to paleotypography of the earliest part of the Miocene. In any event, the Early Miocene "Indian Ocean" was in fact not far from Loperot, and it seems likely that the mammalian fauna may be derived from a coastal riverplain or delta habitat insofar as it differs significantly from the other local faunas of the region. Other evidence (Andrews and Van Couvering 1975) also suggests that the Early Miocene land in Kenya was much closer to sea level than it is today.

North and South West African Sites. A rough comparison between the modern faunas of East, North and Southwest Africa, and also those of the same regions in the Early Miocene, can be seen in the table below. The modern figures were derived primarily from Dorst and Dandelot (1970) and supplemented by those of Harrison (1964, 1968, 1972) and are not complete, but they do provide an estimate of the true situation. All three areas include vast regions of open country varying from woodland to desert (and in East Africa, some areas of forest). The similarity between the East African and Southwest African sites has been explained by the past existence of an "arid corridor" connecting the two regions (cf. van Zinderen Bakker 1967, 1969), but the Early Miocene figures are parallel.

	EAf/NAf	EAf/SWA	SWA/NAf
Modern species FRI	25	77	25
Early Miocene species FRI	25	52	9

The lower indices between SWA and NAf fossil faunas is most likely attributable to sedimentological factors: in North Africa the collections are exclusively large mammals (Gebel Zelten and Wadi Moghara are the two faunas considered here) while 61% of the Southwest African Early Miocene fauna is small mammals (cf. Stromer 1926). The lower FRI between East and Southwest Africa in the Early Miocene compared to the present may be attributable to ecological differences since a relatively large percentage of the genera in the Southwest African fauna are thought to be associated with a more open, non-forest environment: (1) the elephant-shrew, *Myohyrax;* (2) the pika-like lagomorph, *Australogomys,* a hyracoid *Prohyrax,* and two bovids *Protragoceras* and *Eotragus.* Furthermore, there are no forms from Southwest Africa that are known to be exclusively forest. The Gebel Zelten and Wadi Moghara faunas are made up almost exclusively of carnivores and large herbivores. Although it is impossible to determine their *exact* habitat, both types being ubiquitous today, the lack of forest animals *may* be indicative of open conditions. On the other hand, most known forest mammals are small and this may be merely a sedimentological bias. However, Savage and Hamilton (1973) conclude that at Gebel Zelten there is open country ("savanna") surrounding a riverine forest. It is probably fair to conclude that both North and Southwest Africa were, on the whole, drier than East Africa in the Early Miocene. This is also the case today, but the FRI between East Africa and Southwest Africa suggests that there was an even greater dissimilarity in the past. This would fit with our conception of a predominantly forested land in the Early Miocene of East Africa.

SUMMARY

Collection of Early Miocene fossil mammals in East Africa over the last forty years has led to the description of over 110 species, one of the most varied faunas ever to be described. This work was initiated and, to a great extent, accomplished by Louis Leakey.

Early Miocene sites can be classified geographically into three major groups: Eastern Rift, Western Rift and Inter-Rift. The Inter-Rift sites are, in general, associated with volcanoes and have produced the major collections. Perhaps the best known of these volcanoes is Kisingiri which lies today, in a highly eroded state, at the mouth of the Nyanza (Kavirondo) Rift. Karungu, Chianda Uyoma, Mfwangano and the Rusinga sites are all associated with it. Taken together they provide an unparalleled succession of faunal horizons ranging from 23 m.y. to 16.5 m.y. Tinderet, which lies at the head of the Nyanza Rift, is another of these large Miocene volcanoes. It has been partially buried by younger flood phonolites which erupted along the shoulder of the Eastern Rift. Songhor, Koru and Fort Ternan, which are associated with it, are all extremely productive and important sites. The Uganda Inter-Rift volcanoes, Napak, Moroto and Elgon, also have Early Miocene sites associated with them, the most productive of which has been Napak. All of these Early Miocene volcanoes lay just to the west of the continental divide of that time. The animals buried in these deposits were, in great part, forest dwellers, and lowland rain forest probably stretched all across this area broken by small areas of glade, woodland, and grassland.

The Early Miocene sediments of the Eastern Rift were deposited in down-faulted blocks which preceded major rifting. The local faunas from this area are rather meager but, for the most part, contain no true forest species. Whether this area was actually predominantly non-forest or whether this is a collection bias cannot now be resolved. The Miocene fossils from the Western Rift are from several different sites and are few in number. Their stratigraphic relationships are not well understood and it is not known if they were contemporaneous.

Comparisons of East African fossil faunas with recent occurrences of mammals in forest and non-forest areas shows that (1) the species diversity at some of these fossil sites is equivalent to that of the modern fauna suggesting a greater diversity in the Early Miocene; (2) ancient taxa have been replaced by modern taxa of similar trophic types with the major exception of the replacement of browsers by grazers among the ungulates. The apparent greater diversity of the Early Miocene mammals suggests that the fauna of the African equatorial lowland rainforest is depleted, like that of South America and Southeast Asia.

Faunal resemblance indices (FRI) are used to compare the fossil sites. These indices must be viewed in terms of sample size, geographical and temporal proximity, similarity of sedimentary setting, and the past ecology of the areas in which the animals lived.

Some general trends in the East African Early Miocene and its faunas have come to light. The single most important observation may be the stability of the Early Miocene fauna which remained essentially unchanged for 6 million years. This stability can perhaps be explained by the absence of major environmental change and by the lack of faunal interchange between Africa and other continents. The period of stability was terminated by a faunal turnover starting 17–16 million years ago which appears to have affected the large herbivores first, but which in the long run affected the entire fauna. This faunal change can be correlated to increased rifting which changed the landscape, and thus the local climate and vegetation, as well as to faunal mixing with Eurasia.

Acknowledgments

The authors wish to express their appreciation for generous financial support from the Wenner-Gren Foundation for Anthropological Research and to the L. S. B. Leakey Foundation. In addition support from the Boise Fund of Oxford University and the Royal Society of London (grants to P. J. Andrews for field work at Rusinga Island), the National Science Foundation (grant to D. Pilbeam for field work at Maboko Island, Kenya) and the Wenner-Gren Foundation (grant to A. Walker for field work at Fort Ternan, Kenya) made much of our field work and research possible. Our thanks to P. J. Andrews, G. Meyer, D. Pilbeam and A. C. Walker for collaboration in the field and free use of their data. Conversations and correspondence with G. H. Curtis, C. T. Madden, P. Robinson and particularly W. W. Bishop have been of great benefit, but the degree to which we have allowed ourselves to be influenced by their good advice is our own responsibility. The most is owed, however, to Louis Leakey, who introduced us to East African Miocene paleontology and stratigraphy.

References Cited

Abdel-Gawad, M.

1969 New evidence of transcurrent movements in Red Sea area and petroleum indications. Bull. Am. Assoc. Petrol. Geol. 53:1466–1499.

Allen, J. A.

1922a The American Museum Congo Expedition Collection of Insectivora. Bull., Amer. Mus. Nat. Hist. 47:1–38.

1922b Sciuridae, Anomaluridae and Idiuridae collected by the American Museum Congo Expedition. Bull., Amer. Nat. Hist. 47:39–72.

1924 Carnivora collected by the American Museum Congo Expedition. Bull., Amer. Mus. Nat. Hist. 47:73–281.

1925 Primates collected by the American Museum Congo Expedition. Bull., Amer. Mus. Nat. Hist. 47:283–499.

Allen, J. A., H. Lang and J. P. Chapin

1917 The American Museum Congo Expedition collection of bats. Bull., Amer. Mus. Nat. Hist. 37:405–563.

Andrews, P., and M. Pickford

1972 Report on the Songhor 1972 field season. Unpublished report.

Andrews, P., and J. H. Van Couvering

1975 Paleoenvironments in the East African Miocene. *In* Approaches to Primate Paleobiology, 5. F. Szalay, ed. Basel: S. Karger Publ. AG. pp. 62–103.

Andrews, P., J. H. Van Couvering and J. A. Van Couvering
1972 Rusinga 1971 expedition: associations and environment in the Miocene of Kaswanga Point. Unpublished report.

Arambourg, C.
1933 Mammifères Miocènes du Turkana. Annal. Paléontol. 22:121–146.

Arambourg, C., and R. G. Wolff
1969 Nouvelles données paléontologiques sur l'âge des "grès du Lubur" (Turkana Grits) à l'ouest du lac Rudolphe. C. R. Somm. Soc. Géol. France 6:190–191.

Armstrong, D. M.
1972 Distribution of mammals in Colorado. Monogr., Univ. Kansas Mus. Nat. Hist. 3:x + 1–415.

Baker, B. H., L. A. Williams, J. A. Miller and F. J. Fitch
1971 Sequence and geochronology of the Kenya rift volcanics. Tectonophysics 11:191–215.

Behrensmeyer, A. K.
1973 The taphonomy and paleoecology of Plio-Pleistocene vertebrate assemblages east of Lake Rudolf, Kenya. Ph.D. Thesis, Harvard University, Cambridge.

Berggren, W. A.
1972 A Cenozoic time-scale: some implications for regional geology and paleobiogeography. Lethhaia 5:195–215.

Berggren, W. A., and J. A. Van Couvering
1974 The Late Neogene: biostratigraphy, geochronology, and paleoclimatology of the last 15 million years in marine and continental sequences. Paleogeogr. Paleoclimat. Paleoecol. 16, nos. 1–2.

Binge, F. W.
1962 Geology of the Kericho District. Kenya Geol. Survey Report 50. 67 pp.

Bishop, W. W.
1962 The mammalian fauna and geomorphological relations of the Napak Volcanics, Karamoja. Rec. Geol. Surv. Uganda 1957–58:1–18.

1963 The later Tertiary and Pleistocene in Eastern Equatorial Africa. *In* African Ecology and Human Evolution. F. C. Howell and F. Bourlière, eds. Chicago: Aldine.

1964 More fossil primates and other Miocene mammals from North-East Uganda. Nature 203:1327–1331.

1967 The later Tertiary in East Africa—volcanics, sediments and faunal inventory. *In* Background to Evolution in Africa. W. W. Bishop and J. D. Clark, eds. Chicago: Chicago University Press. pp. 31–56.

1972 Stratigraphic succession 'versus' calibration in East Africa. *In* Calibration of Hominoid Evolution. W. W. Bishop and J. A. Miller, eds. Edinburgh: Scottish Academic Press. pp. 219–246.

Bishop, W. W., and G. R. Chapman
1970 Early Pliocene sediments and fossils from the Northern Kenya Rift Valley. Nature 226:914–918.

Bishop, W. W., J. A. Miller and F. J. Fitch
1969 New potassium-argon age determinations relevant to the Miocene fossil mammal sequence in East Africa. Am. J. Sci. 267:669–699.

Bishop, W. W., and A. F. Trendall
1967 Erosion surfaces, tectionics and volcanic activity in Uganda. Quart. J. Geol. Soc. London 122:385–420.

Bishop, W. W. and F. Whyte
1962 Tertiary mammalian faunas and sediments in Karamoja and Kavirondo, East Africa. Nature 196:1283–1287.

Bourlière, F.
1973 The comparative ecology of rain forest mammals in Africa and tropical America: some introductory remarks. *In* Tropical Forest Ecosystems in Africa and South America: A Comparative Review. B. J. Meggers, E. S. Ayensu and W. D. Duckworth, eds. Washington: Smithsonian Inst. Press. pp. 279–292.

Brosset, A.
1966 Recherches sur la composition qualitative et quantitative dans la fôret primaire du Gabon. Biologia Gabonica 2:163–177.

Cahen, L.
1954 Géologie du Congo Belge. Liège: Vaillant-Carmanne.

Chesters, K. I. M.
1957 The Miocene flora of Rusinga Island, Lake Victoria, Kenya. Palaeontographica 101:30–67.

Churcher, C. S.
1970 Two new Upper Miocene Giraffids from Fort Ternan, Kenya, East Africa: *Palaeotragus primaevus* n. sp. and *Samotherium africanum* n. sp. *In* Fossil Vertebrates of Africa, Volume 2. L. S. B. Leakey and R. J. G. Savage, eds. London: Academic Press.

Clark, L.
1969 The North Elgon Depression, eastern Uganda. (Abstract and discussions by D. L. Searle, K. A. Davies, A. F. Trendall.) Geol. Soc. London Proc. 1959: 310–313.

Clark, W. E. LeGros, and L. S. B. Leakey
1951 The Miocene Hominoidea of East Africa. Br. Mus. (Nat. Hist.), Fossil Mammals of Africa, No. 1. 117 pp.

Cooke, H. B. S., and V. J. Maglio
1972 Plio-Pleistocene stratigraphy in relation to proboscidean and suid evolution. *In* Calibration of Hominoid Evolution. W. W. Bishop and J. A. Miller, eds. Edinburgh: Scottish Academic Press. pp. 303–329.

Coryndon, S. C. and R. J. G. Savage
1973 The origin and affinities of African mammal faunas. *In* Organisms and continents through time. A. Hallam, ed. London: Paleontological Association. pp. 121–135.

Damme, D. van, and A. Gautier
1972 Some fossil molluscs from Moruarot Hill (Turkana District, Kenya). J. Conch. 27:423–426.

Davies, K. A.
1952 The Building of Mount Elgon. Mem. Geol. Surv. Uganda VII. 62 pp.

Derin, B., and Z. Reiss
1973 Revision of marine Neogene stratigraphy in Israel. Isr. J. Earth-Sci. 22:199–210.

Dewey, J. F., W. C. Pitman III, W. B. F. Ryan and J. Bonnin
1973 Plate tectonics and the evolution of the Alpine System. Bull., Geol. Soc. Amer. 84:3137–3180.

Dodson, P.
1973 The significance of small bones in paleoecological interpretation. Contrib. to Geol. 12:15–19. University of Wyoming.

Dorst, J. and P. Dandelot
1970 A Field Guide to the Larger Mammals of Africa. London: Collins.

Eames, F. E.
1970 Some thoughts on the Neogene/Palaeogene boundary. Paleogeogr. Paleoclimat. Paleoecol. 8:37–48.

Ellerman, J. R.
1961 The Fauna of British India . . . Mammalia, Vol. 3 (parts 1–2), Rodentia. 2nd ed. Calcutta. pp. 1–884.

Evernden, J. F., D. E. Savage, G. H. Curtis and G. James
1964 K-Ar dates and Cenozoic mammalian chronology of North America. Am. J. Sci. 262:145–198.

Forteau, R.
1920 Contribution à l'étude des vertébrés Miocènes de l'Egypte. Cairo: Govt. Printer.

Gabunia, L. and M. Rubenstein
1968 On the correlation of the Cenozoic deposits of Eurasia and North America based on the fossil mammals and absolute age data. International Geological Congress, Report of the Twenty-Third Session, Czechoslovakia 1968. Proc. Sect. 10:9–17.

Garfunkel, Z., and A. Horowitz
1966 The Upper Tertiary and Quaternary morphology of the Negev, Israel. Isr. J. Earth-Sci. 15:101–117.

Gautier, A.
1970 Fossil fresh water Mollusca of the Lake Albert–Lake Edward Rift. Ann. Mus. Roy. Afr. Centr. No. 67:3–165.

Gentry, A. W.
1970 The Bovidae (Mammalia) of the Fort Ternan fossil fauna. *In* Fossil Vertebrates of Africa, Vol. 2. L. S. B. Leakey and R. J. G. Savage, eds. London: Academic Press. pp. 243–324.

Ginsberg, L.
1961 Découverte de *Pliopithecus antiquus* Bl. dans le falum sauvignéen de Noyant-sous-le-Lude (Maine-et-Loire). Acad. Sci. Fr., C. R. Ser. D. 252:585–587.

Greenway, P. J.
1943 Second draft report on vegetation classification for the approval of the Vegetation Committee, Pasture Research Conference (typed script). Nairobi.

Harrison, D. L.
1964, 1968, 1972 The Mammals of Arabia. Vols. 1, 2, 3. London: Ernest Benn Ltd.

Hatt, R. T.
1934 The pangolins and aard-varks collected by the American Museum Congo

Expedition. Bull., Amer. Mus. Nat. Hist. 66:643–672.

1936 Hyraxes collected by the American Museum Congo Expedition. Bull., Amer. Mus. Nat. Hist. 72:117–141.

1940 Lagomorpha and Rodentia other than Sciuridae, Anomaluridae and Idioridae, collected by the American Museum Congo Expedition. Bull., Amer. Mus. Nat. Hist. 76:457–604.

Hill, A., and A. Walker
1972 Procedures in vertebrate taphonomy: notes on a Uganda Miocene fossil locality. J. Geol. Soc. London 128:399–406.

Hollister, N.
1916 Shrews collected by the Congo Expedition of the American Museum. Bull., Amer. Mus. Nat. Hist. 35:663–680.

Hooijer, D. A.
1963 Miocene mammalia of Congo. Mus. Roy. Afr. Centr. Tervuren, Ann. Sci. Géol. 46:1–77.

1970 Miocene mammalia of Congo, a correction. Mus. Roy. Afr. Centr. Tervuren, Ann. Sci. Géol. 67:161–167.

1971 A new rhinoceros from the late Miocene of Loperot, Turkana District, Kenya. Bull., Mus. Comp. Zool. 142:339–392.

Hooijer, D. A., and V. J. Maglio
1974 Hipparions from the Late Miocene and Pliocene of Northwestern Kenya. Zool. Verhandel. No. 134, 34 pp.

Howell, F. C.
1972 Pliocene/Pleistocene Hominidae in eastern Africa—absolute and relative ages. *In* Calibration of Hominoid Evolution. W. W. Bishop and J. A. Miller, eds. Edinburgh: Scottish Academic Press. pp. 331–368.

Jones, P. W., and D. C. Rex
1974 New dates for the Ethiopian plateau volcanics. Nature 252:218–219.

Joubert, P.
1966 Geology of the Loperot area. Kenya Geol. Surv. Report 74.

Keast, A.
1969 Evolution of mammals on southern continents. VII. Comparisons of the contemporary mammalian faunas of the southern continents. Quart. Rev. Biol. 44:121–167.

Kent, P. E.
1944 The age and tectonic relationships of East African volcanic rocks. Geol. Mag. 81:15–27.

King, B. D., and G. R. Chapman
1972 Volcanism of the Kenya Rift Valley. Phil. Trans. R. Soc. London (A), 271:185–208.

Lavocat, R.
1961 Etude systématique de la faune de mammifères. Notes Mem. Serv. Géol. Maroc 155:29–94.

Leakey, L. S. B.
1968 Upper Miocene primates from Kenya. Nature 218:527–528.

Lepersonne, J.
1970 Revision of the fauna and the stratigraphy of the fossiliferous localities of

the Lake Albert–Lake Edward Rift (Congo). Mus. Roy. Afr. Centr., Tervuren, Ann. Sci. Géol. 67:169–207.

Madden, Cary T.
1972a Miocene mammals, stratigraphy and environment of Moruarot Hill, Kenya. PaleoBios No. 14. 12 pp.

1972b Miocene mammalian fauna from Lothidok Hill, Kenya. M.A. thesis, University of California.

Maglio, V. J.
1970 Early Elephantidae of Africa and a tentative correlation of African Plio-Pleistocene deposits. Nature 225:328–332.

1973 Origin and evolution of the Elephantidae. Trans. Amer. Philos. Soc. 63:149 pp.

McCall, G. J. H.
1958 The geology of the Gwasi area. Kenya Geol. Survey Report 45. 88 pp.

Medway, Lord
1969 The Wild Mammals of Malaya. London: Oxford Univ. Press. 127 pp.

Meggers, B. J., E. S. Ayensu and W. D. Duckworth, eds.
1973 Tropical Forest Ecosystems in Africa and South America: A Comparative Review. Washington: Smithsonian Inst. Press.

Mohr, P. A.
1968 The Cainozoic volcanic succession in Ethiopia. Bull. Volcan. 32:5–14.

1971 Ethiopian Rift and Plateaus: some volcanic petro-chemical differences. J. Geophys. Res. 76:1967–1984.

Moreau, R. E.
1966 The Bird Faunas of Africa and Its Islands. London: Academic Press. 424 pp.

Nixon, P. H., and L. Clark
1967 The alkaline centre of Yelele and its bearing on the petrogenesis of other Eastern Uganda volcanics. Geol. Mag. 104:455–472.

Oswald, F.
1914 The Miocene beds of the Victoria Nyanza and the geology of the country between the lake and the Kisii Highlands. Quart. J. Geol. Soc. London. 70:128–162.

Pagès, E.
1970 Sur l'écologie et les adaptations de l'oryctérope et des pangolins sympatriques du Gabon. Biologia Gabonica 6:27–92.

Patterson, B., A. K. Behrensmeyer and W. D. Sill
1970 Geology and fauna of a new Pliocene locality in north-western Kenya. Nature 226:918–921.

Pratt, D. J., P. J. Greenway and M. D. Gwynne
1966 A classification of East African rangeland, with an Appendix on terminology. J. Appl. Ecol. 3:369–382.

Robinson, P., and C. C. Black
1969 Note Préliminaire sur les vertébrés fossiles du Vindobonien (Formation Beglia) du Bled Douarah, Gouvernerat de Gafsa, Tunisie. Travaux de Géologie Tunisienne, II. Notes Serv. Géol. Tunis. 31:67–70.

Romer, A. S.
1966 Vertebrate Paleontology. Chicago: University of Chicago Press.

Saggerson, E. P.
1952 Geology of the Kisumu area. Geol. Surv. Kenya, Report 21.

Saggerson, E. P., and B. H. Baker
1965 Post-Jurassic erosion-surfaces in eastern Kenya and their deformation in relation to rift structure. Quart. J. Geol. Soc. London 121:51–72.

Savage, R. J. G.
1967 Early Miocene mammal faunas of the Tethyan region. Systematics Association Publication No. 7, Aspects of the Tethyan Biogeography. London, pp. 247–282.

Savage, R. J. G., and W. R. Hamilton
1973 Introduction to the Miocene mammal faunas of Gebel Zelten, Libya. Bull. Brit. Mus. (Nat. Hist.), Geol. 22:516–527.

Savage, R. J. G., and E. Tchernov
1968 Miocene mammals of Israel. Proc. Geol. Soc. London 1648:98–101.

Shackleton, R. M.
1946 Geology of the country between Nanyuki and Maralal. Geol. Survey Kenya, Report 11:1–54.
1951 A contribution to the geology of the Kavirondo Rift Valley. Quart J. Geol. Soc. London 106:345–383.

Simpson, E. H.
1949 Measurement of diversity. Nature 163:688.

Simpson, G. G.
1947 Holarctic mammalian faunas and continental relationships during the Cenozoic. Bull., Geol. Soc. Amer. 58:613–688.
1960 Notes on the measurement of faunal resemblance. Amer. J. Sci. 258-A:300–311.
1964 Species density of North American recent mammals. Syst. Zool. 13:57–73.

Stromer, E.
1926 Reste Land- und Süsswasser-bewohnender Wirbeltiere aus den Diamantenfeldern deutsch-südwest-afrikas. *In* Die Diamantenwüste südwestafrikas, Band II. E. Kaiser, ed. Berlin, 1926. pp. 107–153.

Szafer, W.
1961 Miocene flora from Stare Gliwice in Upper Silesia. Inst. Geologiczny, Prace 33:162–195 (English summary).

Tobien, H.
1973 On the evolution of the Mastodonts (Proboscidea, Mammalia). Part 1: the bunodont trilophodont group. Hess. L.-Amt Bodenforsch. Notizbl. 101:202–276.

Van Couvering, J. A.
1972 Radiometric calibration of the European Neogene. *In* Calibration of Hominoid Evolution, W. W. Bishop and J. A. Miller, eds. Edinburgh: Scottish Academic Press. pp. 247–272.

Van Couvering, J. A., and W. A. Berggren
In press The biostratigraphic basis of the Neogene. *In* Concepts in Biostratig-

raphy, J. E. Hazel and E. H. Kauffman, eds. Paleontological Society, Lawrence, Kansas.

Van Couvering, J. A., and J. A. Miller
1969 Miocene stratigraphy and age determinations, Rusinga Island, Kenya. Nature 221:628–832.

Van Couvering, J. A., P. Robinson and C. C. Black
In press Geochronology of the Hipparion datum. Geological Society of America, Special Paper.

Van Peenen, P. F. D.
1969 Preliminary Identification Manual for Mammals of South Vietnam. Washington: U.S. Nat. Mus. Smiths. Inst. 310 pp.

Van Valen, L.
1966 Deltatheridia, a new order of mammals. Bull., Am. Mus. Nat. Hist. 132:1–126.

Varne, R.
1967 The growth of the Moroto volcano, Eastern Uganda Bull. Volcan. 31:163–174.

Verdcourt, B.
1972 The zoogeography of the non-marine mollusca of East Africa. J. Conch. 27:291–348.

Voorhies, M. R.
1969 Taphonomy and population dynamics of an Early Pliocene vertebrate fauna. Knox County, Nebraska. Univ. Wyoming Contrib. to Geol., Spec. Paper 1.

Walker, A.
1969 Lower Miocene fossils from Mount Elgon, Uganda. Nature 223:591–593.

Walsh, J. and R. G. Dodson
1969 Geology of Northern Turkana. Kenya Geol. Survey Report 82.

Walters, R., and R. E. Linton
1973 The sedimentary basin of Coastal Kenya. *In* Sedimentary Basins of the African Coasts. G. Blant, ed. Paris: Assoc. Afr. Geol. Survey: 133–158.

Whitworth, T.
1953 A contribution to the geology of Rusinga Island, Kenya. Quart. J. Geol. Soc. London 190:75–92.

1961 The geology of Mfwanganu Island, Western Kenya. Overseas Geol. Min. Res. 8:150–190.

Williams, J. G.
1967 A Field Guide to the National Parks of Africa. Boston: Houghton Mifflin Co.

Williams, L. A. J.
1971 The volcanics of the Gregory Rift Valley, East Africa. Bull. Volcan. 34:439–465.

Zinderen Bakker, E. M. van, ed.
1967 Paleoecology of Africa, Vol. II (1964–1965). Amsterdam: A. A. Balkema.

1969 Paleoecology of Africa, Vol. IV (1966–1968). Amsterdam: A. A. Balkema.

APPENDIX

Taxonomic Lists of Vertebrate Faunas from the Early Miocene of East Africa, with Discussion and References

The distribution and, where possible, the abundance of the Early Miocene mammals has been compiled from the literature and from private sources. An earlier inventory, providing a firm base against which our list should be compared, was compiled by Bishop (1967). The better descriptions allowed us to reduce the listed specimens to "minimum individuals" but in other instances the abundances we show are based on simple specimen counts adjusted as far as possible by stratigraphic and locality data, including entries in Kenya National Museum records. It is not possible to elaborate on how the abundance figures were arrived at in each instance, nor should they be taken very seriously in view of their mainly secondhand origin. The principal sources of inventory entries are in two groups, consisting of (1) systematic treatments of fossil groups and (2) descriptions of individual sites with more or less complete faunal lists. These sources are itemized below.

Systematic Treatments (Asterisk (*) indicates best abundance data.)

*Lipotyphla: Butler 1956; Butler and Hopwood 1957; Butler 1969; P. J. Andrews, pers. comm., 1974

*Macroscelidea: Butler and Hopwood 1957; Whitworth 1954; Patterson 1965

*Chiroptera: Butler and Hopwood 1957; Simpson 1967; Walker 1969

Primates—*(Prosimia): Clark and Thomas 1952; Simpson 1967; Clark 1956; Walker 1969, 1970
(Cercopithecoidea): Koenigswald 1969
*(Hominoidea): Clark and Leakey 1951; Napier and Davies 1959; Leakey 1967, 1968; Andrews 1970, in press; Andrews and Pilbeam, in press; P. J. Andrews, pers. comm., 1974

Lagomorpha: MacInnes 1953

Rodentia: Lavocat 1974; P. J. Andrews, pers. comm., 1974

Creodonta: Savage 1965

Carnivora: Savage 1965

*Tubulidentata: MacInnes 1956

Proboscidea: MacInnes 1942. See footnote 2, p. 204.

*Hyracoidea: Whitworth 1954; Meyer, in press

Sirenia: Savage and Hamilton 1973

*Cetacea: B. Patterson, pers. comm., 1972; V. J. Maglio, pers. comm., 1972

Perissodactyla—*(Chalicotheriidae): Butler 1965
(Rhinocerotidea): Hooijer 1966, 1971, 1973; Heissig 1971; Hamilton 1973a

Artiodactyla—(Anthracotheriidae): Andrews 1914; MacInnes 1951; Black, in press
(Suidae): Wilkinson, in press
(Ruminantia): Whitworth 1958; Churcher 1970; Hamilton 1973b

Locality Descriptions

Rusinga Hiwegi: Clark and Leakey 1951; S. C. Coryndon, written comm., 1969; Andrews, pers. comm., 1974

Karungu: Oswald 1914; Andrews 1914

Ombo-Mariwa: Kent 1944

Songhor-Koru: (Inventory compiled from systematic studies); Andrews, pers. comm., 1974

Maboko Island: Kent 1944; C. T. Madden, written comm., 1972; P. J. Andrews, pers. comm., 1974

Moruarot-Lothidok: Madden 1972, in press

Loperot: Hooijer 1971; V. J. Maglio, written comm., 1971

Napak: Bishop 1962, 1964

Bukwa: Walker 1969

Congo: Lepersonne 1970; Hooijer 1970

Gebel Zelten, Wadi Moghara: Forteau 1920; Savage and Hamilton 1973

Namibia: Stromer 1926; Hopwood 1929

TABLE 8
Taxonomic lists of vertebrate faunas from Early Miocene of East Africa

FAUNA	LOCALITIES											
	NYANZA RIFT					EAST RIFT		INTER-RIFT		WEST RIFT	COASTAL	
	RUS	*KAR*	*OMB*	*SGR*	*MAB*	*MOR*	*LOP*	*NAP*	*BUK*	*CGO*	*NAF*	*SWA*
LIPOTYPHLA												
Protenrec tricuspis But. & Hopwd.	X	...	...	X	...	...	...	X	...	...	...	...
Geogale aletris But. & Hopwd.	X	...	...	X	...	...	...	...	...	...	...	...
Erythrozootes chamerpes But. & Hopwd.	...	...	...	X	...	...	...	X	...	...	...	...
Lanthanotherium sp.	X	...	...	X	...	...	...	...	...	...	...	...
Galerix? africanus Butler	X	...	...	X	...	...	...	...	...	...	...	...
Gymnurechinus leakeyi Butler	XX	...	...	...	...	...	...	...	...	...	...	...
comptolophus Butler	X	...	...	...	...	...	...	...	...	...	...	...
songhorensis Butler	...	...	...	XX	...	...	...	P	...	...	...	...
Amphechinus rusingensis Butler	X	...	...	X	...	...	...	...	...	...	...	...
Prochrysochloris miocaenicus But. & Hopwd.	...	...	...	X	...	...	...	...	...	...	...	...
Crocidura sp.	X	...	...	...	...	...	...	...	...	...	...	...
MACROSCELIDEA												
Myohyrax oswaldi Andrews	XX	XXX	...	P	...	P	...	...	P	...	...	XX
Protypotheroides beetzi Stromer	...	...	...	...	...	...	...	...	...	...	...	X
Rhynchocyon clarki But. & Hopwd.	X	...	...	X	...	...	...	...	...	...	...	...
rusingae Butler	X	...	...	P	...	...	...	...	...	...	...	...
Rhynchocyonid sp. I	...	...	...	X	...	...	...	...	...	...	...	...
Rhynchocyonid sp. II	X	...	...	...	...	...	...	...	...	...	...	...
CHIROPTERA												
Taphozous incognita But. & Hopwd.	X	...	...	X	...	...	...	...	...	...	...	...
Megadermid	X	...	...	...	...	...	...	...	...	...	...	...
Hipposideras sp.	...	...	...	X	...	...	...	...	...	...	...	...
Propotto leakeyi Simpson	X	...	...	X	...	...	...	...	...	...	...	...
PRIMATES												
Progalago dorae MacInnes	X	...	...	X	...	...	...	...	...	...	...	...
songhorensis Simpson	X	...	...	X	...	...	...	...	...	...	...	...
Komba robustus (Clark & Thomas)	X	...	...	X	...	...	...	...	...	...	...	...
minor (Clark & Thomas)	X	...	...	X	...	...	...	...	...	...	...	...
Mioeuoticus bishopi Leakey	X	...	...	...	...	...	...	X	...	...	...	...
Prohylobates tandyi Forteau	...	...	...	...	...	...	...	...	...	...	P	...

	NYANZA RIFT					EAST RIFT		INTER-RIFT		WEST RIFT	COASTAL	
	RUS	*KAR*	*OMB*	*SGR*	*MAB*	*MOR*	*LOP*	*NAP*	*BUK*	*CGO*	*NAF*	*SWA*
Victoriapithecus macinnesi von K.	?	. . .	P	. . .	X	. . .	. . .	. . .	. . .	. . .	. . .	. . .
leakeyi von K.	. . .	. . .	. . .	. . .	X	. . .	. . .	. . .	. . .	. . .	. . .	. . .
Victoriapithecus sp.	. . .	. . .	. . .	. . .	. . .	. . .	P	. . .	. . .	. . .	. . .	. . .
Cercopithecoid aff. *Mesopithecus*	. . .	. . .	. . .	. . .	?	P	. . .	P	. . .	P	. . .	. . .
Limnopithecus legetet Hopwd.	X	. . .	X	XX	X	. . .	. . .	P	P	. . .	. . .	. . .
Dendropithecus macinnesi (Clark & Leakey)	XX	X	. . .	X	. . .	P	. . .	P	. . .	. . .	. . .	. . .
Dryopithecus africanus (Hopwd.)	XX	P	. . .	P	. . .	X	. . .	. . .	. . .	. . .	. . .	. . .
nyanzae (Clark & Leakey)	XX	P	. . .	. . .	X	X	. . .	. . .	. . .	. . .	. . .	. . .
major (Clark & Leakey)	. . .	. . .	. . .	P	. . .	?	. . .	X	. . .	. . .	. . .	. . .
gordoni Andrews	X	. . .	. . .	XX	. . .	. . .	. . .	. . .	. . .	. . .	. . .	. . .
vancouveringi Andrews	X	. . .	. . .	X	X	. . .	. . .	. . .	. . .	. . .	. . .	. . .
Dryopithecus sp.	. . .	. . .	. . .	. . .	X	. . .	X	. . .	. . .	. . .	. . .	. . .
Ramapithecus cf. *R. wickeri* (Leakey)	?	. . .	. . .	. . .	X	. . .	. . .	. . .	. . .	. . .	. . .	. . .
CREODONTA												
Kelba quadaemae Savage	X	. . .	. . .	X	. . .	. . .	. . .	X	. . .	. . .	. . .	. . .
Teratodon spekei Savage	. . .	. . .	. . .	X	. . .	. . .	. . .	. . .	. . .	. . .	. . .	. . .
enigmae Savage	. . .	. . .	. . .	X	. . .	. . .	. . .	. . .	. . .	. . .	. . .	. . .
Anasinopa leakeyi Savage	XX	XX	. . .	. . .	X	X	. . .	. . .	. . .	. . .	. . .	. . .
Anasinopa sp. nov. Savage	X	. . .	. . .	. . .	. . .	. . .	. . .	. . .	. . .	. . .	. . .	. . .
Anasinopa sp.	. . .	. . .	. . .	. . .	. . .	. . .	. . .	. . .	. . .	. . .	P	. . .
Metasinopa napaki Savage	. . .	. . .	. . .	. . .	. . .	. . .	. . .	X	. . .	. . .	. . .	. . .
Dissopsalis pyroclasticus Savage	P	. . .	. . .	P	P	. . .	. . .	. . .	. . .	. . .	. . .	. . .
Metapterodon kaiseri Stromer	X	X	. . .	. . .	. . .	. . .	. . .	. . .	. . .	. . .	. . .	X
zadoki Savage	X	. . .	. . .	. . .	. . .	. . .	. . .	. . .	. . .	. . .	. . .	. . .
Pterodon africanus Andrews[1]	X	X	. . .	. . .	. . .	. . .	. . .	. . .	. . .	. . .	. . .	. . .
Hyainailourous forteaui von Koenigswald	. . .	. . .	. . .	. . .	. . .	. . .	. . .	. . .	. . .	. . .	P	. . .
Leakitherium hiwegi Savage	X	. . .	. . .	. . .	. . .	?	. . .	. . .	. . .	. . .	. . .	. . .
Hyaenodon (Isohyaenodon) andrewsi Savage	X	. . .	. . .	X	. . .	X	. . .	. . .	. . .	. . .	. . .	. . .
metthewi Savage	X	. . .	. . .	X	. . .	. . .	. . .	. . .	. . .	. . .	. . .	. . .
pilgrimi Savage	X	. . .	. . .	. . .	. . .	. . .	. . .	. . .	. . .	. . .	. . .	. . .
Megistotherium osteothalastes Savage	X	. . .	. . .	. . .	. . .	. . .	. . .	. . .	. . .	. . .	X	. . .
Creodont	. . .	. . .	. . .	. . .	. . .	. . .	P	. . .	. . .	. . .	. . .	. . .
CARNIVORA												
Hecubides euryodon Savage	X	. . .	. . .	X	. . .	. . .	. . .	P	. . .	. . .	. . .	. . .
macrodon Savage	X	. . .	. . .	. . .	. . .	. . .	. . .	. . .	. . .	. . .	. . .	. . .

Afrocyon buroletti Arambourg	...	...	...	...	...	...	...	...	...	...	X	...
Amphicyonid	...	...	...	...	...	...	...	...	...	...	X	...
Kichechia zamanae Savage	XX	...	...	X	...	X	...	P	...	...	...	...
Metailurus africanus (Andrews)	X	X	...	X	...	...	...	...	...	...	...	...
cf. *Metailurus*	...	...	...	...	...	...	...	...	...	...	X	...
Nimravine (non *Metailurus*)	X	...	...	...	...	...	...	...	...	...	...	...
TUBULIDENTATA												
Myorycteropus africanus MacInnes	X	...	...	...	...	...	...	...	...	...	...	...
Tubulidentate (large)	X	...	...	...	...	...	...	...	...	...	...	...
PROBOSCIDEA[2]												
Prodeinotherium hobleyi (Andrews)	XX	P	P	P	P	P	P	...	P	P	P	...
Platybelodon kisumuensis (MacInnes)	XX	P	P	P	P	P	...	...	P	P	P	...
Platybelodon sp. cf. *P. grangeri* (Osborn)	...	...	...	...	...	...	P	...	...	...	...	...
Gomphotherium angustidens (Cuvier)	...	...	...	...	...	...	...	...	...	...	P	...
pygmaeus (Deperet)	...	...	...	...	...	...	...	...	...	...	P	...
Gomphotherium sp.	?	?	...	?	?	?	?	...	?	...	...	...
Mammutid aff. *Zygolophodon*	?	...	...	...	...	X	...	X	...	...	X	...
Primelephhas lepersonnei (Hooijer)	...	?	...	...	cf.	...	...	...	...	X	...	...
LAGOMORPHA												
Austrolagomys inexpectatus Stromer	...	...	...	...	...	...	...	...	...	...	...	X
simpsoni Hopwd.	...	...	...	...	...	...	...	...	...	...	...	X
Kenyalagomys rusingae MacInnes	XXX	XX	...	...	cf.	...	...	P	...	...	...	...
minor MacInnes	XX	...	...	...	...	...	...	...	...	...	...	...
RODENTIA												
Phiomys andrewsi Stromer	X	...	...	XX	...	...	...	...	...	...	...	X
Phiomyid	...	...	...	...	X	...	...	...	...	...	...	...
Paraphiomys stromeri (Hopwd.)	XX	X	...	XX	...	?	...	P	P	...	...	X
pigotti Lavocat	XXX	XX	...	XX	...	X	...	P	P	...	...	X
Epiphiomys coryndoni Lavocat	X	...	...	X	...	...	...	X	...	...	...	...
Diamantomys leuderitzi Stromer	XXX	XXX	...	XXX	...	...	...	P	...	...	...	X
Pomonomys dubius Stromer	...	...	...	...	...	...	...	...	...	...	...	X
Kenyamys mariae Lavocat	X	...	...	X	...	...	...	...	...	...	...	...
Simonomys genovefae Lavocat	X	...	...	XX	...	...	...	X	...	...	...	...
rarus Lavocat	(Mfwango only)											
moreli Lavocat	...	...	...	...	...	...	...	P	...	...	...	...
Myophiomys arambourgi Lavocat	X	...	...	XX	...	...	...	X	...	...	...	...
Elmerimys woodi Lavocat	X	...	...	P	...	...	...	...	...	...	...	...
Phiomyoides humilis Stromer	...	...	...	...	...	...	...	...	...	...	...	X

		NYANZA RIFT				EAST RIFT		INTER-RIFT		WEST RIFT	COASTAL	
	RUS	*KAR*	*OMB*	*SGR*	*MAB*	*MOR*	*LOP*	*NAP*	*BUK*	*CGO*	*NAF*	*SWA*
Bathyergoides neotertiarus Stromer	X	...	...	XX	...	...	...	X	...	...	...	X
Proheliophobus leakeyi Lavocat	X	...	...	?	...	...	...	X	...	...	...	...
Paracryptomys mckennae Lavocat	...	...	...	...	...	...	...	...	...	...	...	X
Paranomalurus bishopi Lavocat	X	...	...	XX	...	...	...	XX	...	...	...	...
soniae Lavocat	XX	...	...	XX	...	...	...	X	...	...	...	...
walkeri Lavocat	...	...	...	XX	...	...	...	X	...	...	...	...
Zenkerella wintoni Lavocat	...	...	...	X	X	...	...	...	...	...	...	...
Parapedetes namaquensis Stromer	...	...	...	...	...	...	...	...	...	...	...	X
Megapedetes pentadactylus MacInnes	XX	...	...	XX	...	...	...	X	P	...	...	...
Megapedetes sp. nov.	X	...	...	...	P	...	...	...	...	...	...	?
Afrocricetodon songhori Lavocat	X	...	...	XX	...	...	...	X	...	...	...	...
Afrocricetodon sp.	...	...	...	...	...	...	...	X	...	...	...	...
Protarsomys macinnesi Lavocat	X	...	...	X	...	...	...	...	...	...	...	...
Notocricetodon petteri Lavocat	X	...	...	XX	...	...	...	...	...	...	...	...
Notocricetodon sp.	...	...	...	XX	...	...	...	...	...	...	...	...
Afrocricetodontine	...	...	...	X	...	...	...	X	...	...	...	...
Vulcanisciurus africanus Lavocat & Mein	XX	...	...	X	...	...	...	X	...	...	...	...
Sciurid	X	...	...	...	...	...	...	...	...	...	...	...
HYRACOIDEA												
Pachyhyrax championi (Arambourg)	XX	P	...	P	P	P	...	P	P	...	...	...
Pachyhyrax cf. *P. pygmaeus* (Matsumoto)[3]	X	X	...	...	...	...	...	...	...	...	...	...
Meroehyrax bateae Whitworth	X	...	...	...	...	...	...	...	X	...	...	...
Pliohyracine aff. *Meroehyrax*	...	...	...	...	...	...	P	...	...	...	...	...
Prohyrax tertiarus Stromer	...	...	...	...	...	cf.	...	...	...	...	...	X
Procaviid aff. *Prohyrax*	...	...	...	...	...	...	...	...	...	...	P	...
SIRENIA												
Sirenia gen. et sp. nov.	...	...	...	...	...	...	...	...	...	...	P	...
CETACEA												
Ziphiid gen. et sp. nov. Meade	...	...	...	...	...	...	X	...	...	...	...	...
PERISSODACTYLA												
Chalicotherium rusingense Butler	XX	P	...	P	...	...	...	P	...	...	...	...
Aceratherium acutirostratrum (Deraniyagala)	XX	X	?	...	...	X	...	P	...	P	...	...
Aceratherium sp.	...	...	...	...	...	...	...	...	...	...	P	...
Chilotheridium pattersoni Hooijer	X	...	...	...	?	...	XX	...	P	...	...	...
Dicerorhinus leakeyi Hooijer[4]	XX	X	P	X	...	...	...	X	...	...	...	...

Dicerorhinus sp.[4]	. . .	. . .	. . .	. . .	?	?	. . .	. . .	P	?	P	. . .
Brachypotherium heinzelini Hooijer	XX	X	P	X	X	. . .	. . .	X	. . .	X	. . .	X
snowi Forteau	. . .	. . .	. . .	. . .	. . .	. . .	. . .	. . .	. . .	. . .	P	. . .
Paradiceros cf. *P. mukirii* Hooijer[4]	. . .	. . .	. . .	. . .	X	. . .	. . .	. . .	. . .	. . .	. . .	. . .
ARTIODACTYLA												
Masritherium aequitorialis (MacInnes)	XX	XX	. . .	. . .	cf.	P	. . .	P	P	P	. . .	. . .
depereti Forteau	. . .	. . .	. . .	. . .	. . .	. . .	. . .	. . .	. . .	. . .	P	. . .
Masritherium sp. nov. Black	?	. . .	. . .	. . .	. . .	. . .	P	. . .	. . .	. . .	. . .	. . .
Gelasmodon sp.	. . .	. . .	. . .	. . .	. . .	. . .	P	. . .	. . .	. . .	. . .	. . .
Brachyodus africanus Andrews	X	X	. . .	. . .	P	. . .	. . .	. . .	?	. . .	P	. . .
moneyi Forteau	cf.	. . .	cf.	. . .	. . .	. . .	. . .	. . .	. . .	?	P	. . .
Hyoboops aff. *H. paleindicus* (Lydekker)	. . .	. . .	. . .	. . .	. . .	. . .	. . .	. . .	. . .	. . .	P	. . .
Hyotherium dartevillei (Hooijer)	XX	X	. . .	XX	. . .	X	. . .	X	. . .	X	. . .	. . .
Hyotherium sp. nov. Wilkinson	XX	. . .	. . .	X	. . .	X	. . .	X	. . .	. . .	. . .	. . .
Bunolistriodon jeanneli (Arambourg)	XX	. . .	. . .	X	X	X	. . .	X	P	X	. . .	. . .
massai Arambourg	. . .	. . .	. . .	. . .	. . .	. . .	. . .	. . .	. . .	. . .	XX	. . .
Bunolistriodon aff. *B. gigas* Pearson	. . .	. . .	. . .	. . .	X	. . .	. . .	. . .	. . .	. . .	XX	. . .
Hyotheriine aff. *Propalaeochoerus*	. . .	. . .	. . .	. . .	. . .	X	. . .	. . .	. . .	. . .	. . .	X
Listriodon sp. nov. Wilkinson (large)	. . .	. . .	. . .	. . .	X	. . .	. . .	. . .	. . .	. . .	. . .	. . .
Listriodon sp. nov. Wilkinson (small)	X	. . .	. . .	. . .	. . .	X	. . .	. . .	. . .	. . .	. . .	. . .
Xenochoerus africanus (Stromer)	X	X	. . .	X	. . .	. . .	. . .	. . .	X	. . .	X	X
Xenochoerus sp. nov. Wilkinson	. . .	. . .	. . .	. . .	. . .	X	. . .	. . .	. . .	. . .	. . .	. . .
Sanitherium sp. nov. Wilkinson	. . .	. . .	X	. . .	. . .	. . .	. . .	. . .	. . .	. . .	. . .	. . .
Dorcatherium crassum Kaup	XX	X	. . .	. . .	X	X	. . .	. . .	. . .	. . .	. . .	. . .
pigotti Whitworth	XX	X	X	. . .	X	X	P	P	P	. . .	. . .	. . .
parvun Whitworth	XX	X	. . .	. . .	X	X	. . .	. . .	P	. . .	. . .	. . .
songhorensis Whitworth	. . .	. . .	. . .	XX	. . .	. . .	. . .	P	. . .	. . .	. . .	. . .
libiensis Hamilton	. . .	. . .	. . .	. . .	. . .	. . .	. . .	. . .	. . .	. . .	X	. . .
Dorcatherium sp.	. . .	. . .	. . .	. . .	. . .	. . .	P	. . .	. . .	. . .	. . .	X
Gelocus whitworthi Hamilton	X	. . .	. . .	X	. . .	. . .	. . .	. . .	. . .	. . .	. . .	. . .
Propalaeoryx austroafricanus Stromer	. . .	. . .	. . .	. . .	. . .	. . .	. . .	. . .	. . .	. . .	. . .	X
nyanzae Whitworth	X	. . .	. . .	P	. . .	P	. . .	. . .	. . .	. . .	. . .	. . .
Canthumeryx sirtensis Hamilton	X	?	. . .	. . .	?	P	. . .	. . .	. . .	. . .	XX	. . .
Climacoceras africanus MacInnes	. . .	. . .	. . .	. . .	XX	. . .	. . .	. . .	. . .	. . .	. . .	. . .
Prolibytherium magnieri Arambourg	. . .	. . .	. . .	. . .	. . .	. . .	. . .	. . .	. . .	. . .	X	. . .
Palaeotragus primaevus Churcher	?	. . .	. . .	. . .	X	X	. . .	. . .	. . .	. . .	. . .	. . .
Zarafa zelteni Hamilton	. . .	. . .	. . .	. . .	. . .	X	. . .	. . .	. . .	. . .	X	. . .
Protragoceras sp.	. . .	. . .	. . .	. . .	. . .	. . .	. . .	. . .	. . .	. . .	. . .	X

	NYANZA RIFT					EAST RIFT		INTER-RIFT		WEST RIFT	COASTAL	
	RUS	*KAR*	*OMB*	*SGR*	*MAB*	*MOR*	*LOP*	*NAP*	*BUK*	*CGO*	*NAF*	*SWA*
Walangania africanus (Whitworth)	XX	?	. . .	XX	. . .	X	. . .	P	. . .	. . .	. . .	. . .
Eotragus sp.	?	. . .	. . .	. . .	?	. . .	. . .	. . .	. . .	. . .	P	X
Gazella sp.	. . .	. . .	. . .	. . .	?	. . .	. . .	. . .	. . .	. . .	P	. . .
	RUS	*KAR*	*OMB*	*SGR*	*MAB*	*MOR*	*LOP*	*NAP*	*BUK*	*CGO*	*NAF*	*SWA*
TOTAL SPECIES	92	27	9	69	32	29	10	41	16	9	29	22
TOTAL GENERA	67	22	9	56	25	24	11	34	14	9	25	21

[1]*Van Valen (1966) questions whether this is actually the species named by Andrews from the Fayum Oligocene, and suggests a Creodont taxonomy different from the one shown here.*

[2]*Proboscidean systematics are currently in a state of flux. Our usage follows Tobien (1973), modified by personal communication from J. T. Harris (1972), D. A. Hooijer (1974) and C. T. Madden (1974–75). Madden points out that molar teeth of primitive gomphotheres and ambelodonts are very likely to be confused if both are present; pending further study we have referred all East African "gomphothere" material reported in the literature to* Platybelodon kisumuensis, *fide Tobien (1973), with true gomphothere species queried to indicate the possibility that the material may be mixed.*

[3]*G. H. Meyer (written comm., 1974) suggests that these specimens may be juvenile* P. championi.

[4]*Heissig (1973) proposes a revision of rhinocerotid taxonomy such that* Dicerorhinus *and* Paradiceros, *as used here, would be replaced by* Didermocerus *and* Diceros, *respectively.*

KEY						
X	Abundance estimated 1 - 10 individuals.	RUS	Rusinga Hiwegi	LOP	Loperot
XX	Abundance estimated 11 - 100 individuals.	KAR	Karungu	NAP	Napak
XXX	Abundance estimated more than 100 individuals.	OMB	Ombo-Mariwa	BUK	Bukwa
P	Present; abundance unknown.	SGR	Songhor-Koru	CGO	Congo
?	Presence uncertain.	MAB	Maboko Island	NAF	G. Zelten, Moghara
cf.	Present on basis of undiagnostic material.	MOR	Moruarot-Lothidok	SWA	Southwest Africa

References Cited in the Appendix

(For references which are here only indicated by author see the main references.)

Andrews, C. W.
1914 On the Lower Miocene vertebrates from British East Africa, collected by Dr. Felix Oswald. Quart. J. Geol. Soc. London 70:163–186.

Andrews, P.
1970 Two new fossil primates from the Lower Miocene of Kenya. Nature 228:537–540.
1974 New species of *Dryopithecus* from Kenya. Nature 249:188–190.

Andrews, P., and D. R. Pilbeam
In press Hominoidea. *In* Mammalian Evolution in Africa. V. J. Maglio, ed. Princeton Univ. Press.

Bishop, W. W.
1962
1964
1967

Black, C.
In press Anthracotheriidae. *In* Mammalian Evolution in Africa. V. J. Maglio, ed. Princeton Univ. Press.

Butler, P. M.
1956 Erinaceidae from the Miocene of East Africa. Brit. Museum (Nat. Hist.), Fossil Mammals of Africa No. 11:1–75.
1965 East African Miocene and Pleistocene chalicotheres. Fossil Mammals of Africa No. 18. Brit. Museum (Nat. Hist.) Bull. Geol. 10:163–237.
1969 Insectivores and bats from the Miocene of East Africa: new material. *In* Fossil Vertebrates of Africa, Volume I. L. S. B. Leakey, ed. London: Academic Press.

Butler, P. M., and A. T. Hopwood
1957 Insectivora and Chiroptera from the Miocene rocks of Kenya Colony. Brit. Museum (Nat. Hist.), Fossil Mammals of Africa No. 13:1–35.

Churcher, C. S.
1970

Clark, W. E. LeGros
1956 A Miocene lemuroid skull from East Africa. Brit. Museum (Nat. Hist.), Fossil Mammals of Africa No. 9, 6 pp.

Clark, W. E. LeGros, and L. S. B. Leakey
1951

Clark, W. E. LeGros, and D. P. Thomas
1952 The Miocene Lemuroids of East Africa. Brit. Museum (Nat. Hist.), Fossil Mammals of Africa No. 5, 20 pp.

Forteau, R.
1920

Hamilton, W. R.
1973a North African Lower Miocene rhinoceroses. Bull., Brit. Museum (Nat. Hist.) Geol. 24:351–395.
1973b Lower Miocene ruminants of Gebel Zelten, Libya. Bull. Brit. Museum (Nat. Hist.) Geol. 21:75–150.

Heissig, K.
1971 *Brachypotherium* aus dem Miozän von Südwestafrika. Mitt. Bayer. Staatssamml. Paläont. Hist. Geol. 11:125–128.

1973 Die Unterfamilien und Tribus der rezenten und fossilen Rhinocerotidae (Mammalia). Säuget. Mitteil. Jhg. 21:25–30.

Hooijer, D. A.
1966 Fossil Mammals of Africa, No. 21: Miocene rhinoceroses of East Africa. Bull. Brit. Mus. (Nat. Hist.) Geol. 13:120–190.

1970

1971

1973 Additional Miocene to Pliocene rhinoceroses of Africa. Zool. Mededel. 46:149–178.

Hopwood, A. T.
1929 New and little-known mammals from the Miocene of Africa. Am. Mus. Nat. Hist. Novit. 344. 9 pp.

Kent, P. E.
1944

Koenigswald, G. H. R. von
1969 Miocene Cercopithecoidea and Oreopithecoidea of East Africa. *In* Fossil Vertebrates of Africa, Volume I. L. S. B. Leakey, ed. London: Academic Press. pp. 39–54.

Lavocat, R.
1973 Les rongeurs du Miocène d'Afrique orientale. I. Miocène inférieur. Mém. Trav. E.P.H.E., Inst. Montpellier, 1. 284 pp.

Leakey, L. S. B.
1967 An Early Miocene member of Hominidae. Nature 213:155–163.

1968 Lower Dentition of *Kenyapithecus africanus.* Nature 217:827–830.

Lepersonne, J.
1970

MacInnes, D. G.
1942 Miocene and post-Miocene Proboscidea from East Africa. Trans. Zool. Soc. London 25:33–106.

MacInnes, D. G.
1951 Miocene Anthracotheriidae from East Africa. Brit. Mus. (Nat. Hist.), Fossil Mammals of Africa No. 4.

1953 Miocene and Pleistocene Lagomorpha of East Africa. Brit. Mus. (Nat. Hist.), Fossil Mammals of Africa No. 6.

1956 Fossil Tubulidentata from East Africa. Brit. Mus. (Nat. Hist.), Fossil Mammals of Africa No. 10.

Madden, Cary T.
1972

In press

Meyer, G.
In press Hyracoidea. *In* Mammalian Evolution in Africa. V. J. Maglio, ed. Princeton Univ. Press.

Napier, J. R., and P. R. Davies
1959 The fore-limb skeleton and associated remains of *Proconsul africanus*. Brit. Mus. (Nat. Hist.), Fossil Mammals of Africa 16:1–69.

Oswald, F.
1914

Patterson, B.
1965 The fossil elephant shrews (family Macroscelididae). Bull. Mus. Comp. Zool. 133:295–350.

Savage, R. J. G.
1965 Fossil Mammals of Africa, 19: The Miocene carnivora of East Africa. Bull. Brit. Mus. (Nat. Hist.) Geol. 10:239–316.

Savage, R. J. G., and W. R. Hamilton
1973

Simpson, G. G.
1967 The Tertiary lorisiform primates of Africa. Bull. Mus. Comp. Zool. 136:39–62.

Stromer, E.
1926

Walker, Alan
1969 True affinities of *Propotto leakeyi* Simpson 1967. Nature 223:647–648.

1970 Post-cranial remains of the Miocene Lorisidae of East Africa. Am. J. Phys. Anthrop. 33:249–261.

Whitworth, T.
1954 The Miocene hyracoids of East Africa. Brit. Mus. (Nat. Hist.), Fossil Mammals of Africa 7:1–58.

1958 Miocene ruminants of East Africa. Brit. Mus. (Nat. Hist.), Fossil Mammals of Africa No. 15.

Wilkinson, E.
In press Suidae. *In* Mammalian Evolution in Africa. V. J. Maglio, ed. Princeton Univ. Press.

Olduvai Gorge, Tanzania. (Photo by R. L. Hay.)

Richard L. Hay:

Environmental Setting of Hominid Activities in Bed I, Olduvai Gorge

What kind of terrain did the early hominids occupy? Can we discover the kinds of situations in which they preferred to camp? Can we ascertain what variety of habitats and resources occurred in the regions where we find their fossil bones? Richard Hay's account of his work at Olduvai illustrates the contribution that careful and imaginative geological research can make towards the reconstruction of the setting of early human life.

APPROACH TO ENVIRONMENTAL ANALYSIS

The environmental analysis of Bed I is based on its subdivision into lithologic facies, or lithologically different but laterally equivalent rock assemblages, each of which was deposited in the same environment or a closely related series of environments. The facies are analyzed from the standpoint of environmental information, and the data from individual facies are then integrated into an environmental synthesis and geologic history for the entire basin.

Several features make Bed I particularly well suited for environmental analysis. It was deposited in a small basin and encompasses several different lithologic facies. Sediment sources vary considerably in their composition, and thus the detrital sediments of Bed I can generally be assigned to their source areas. With the sediment sources known, stream-channel alignments can be used to determine the drainage pattern. Several tuffs are widespread, allowing lithologic comparison of age-equivalent deposits over a wide area. Chemically precipitated minerals in the lake deposits provide a wealth of information about the chemistry of the lake water. Finally, the Olduvai fauna is rich, varied, and well studied, and some elements of the fauna are diagnostic of rather specific environmental conditions.

GEOLOGIC AND CLIMATIC SETTING

Olduvai Gorge is a steep-sided valley in the Serengeti Plain near the western margin of the Eastern Rift Valley in northern Tanzania (Figure 1). Where cut by the gorge, the plain has an elevation of 1350–1500 m. The gorge is generally 45–90 m. deep in the lower 26 km. of its course, where Bed I is exposed. About 9 km. upstream from its mouth the gorge divides into two branches—a smaller, southern branch, the Side Gorge, and a larger, northern branch, the Main Gorge. Rainfall varies widely but averages 61 cm./year over the period 1966–72. The temperature ranges from 14 to 32°C, averaging about 22°C over parts of the year 1972 in which records were kept. The Serengeti Plain is a grassland with *Commiphora* scrub and scattered *Acacia* (thorn trees).

Metamorphic rocks of Precambrian age, principally gneisses and quartzites, are exposed in the upstream part of the gorge, in inselbergs in the vicinity of the gorge, and in highlands to the north. A trachytic welded tuff, the Naabi Ignimbrite, overlies the metamorphic basement in the western part of the gorge. It is overlain by the Olduvai Beds: Beds I through IV, the Masek Beds, the Ndutu Beds, and the Naisiusiu Beds (Hay 1971, in press). Beds I through IV accumulated in a small, shallow basin along the northwestern foot of the volcanic highlands as represented by Ngorongoro, Lemagrut, and Olmoti.

All units of the Olduvai sequence except the Naisiusiu Beds are cut by numerous faults. Reck (1951) numbered the larger of the faults from east to west as the First, Second, Third, Fourth, and Fifth faults. At the Second Fault is exposed almost the entire stratigraphic thickness of the Olduvai Beds (Figure 2). Faulting resulted in the Balbal Depression, a graben into which the gorge presently drains. The faulting began early during the deposition of Bed II and continued intermittently through deposition of the Ndutu Beds.

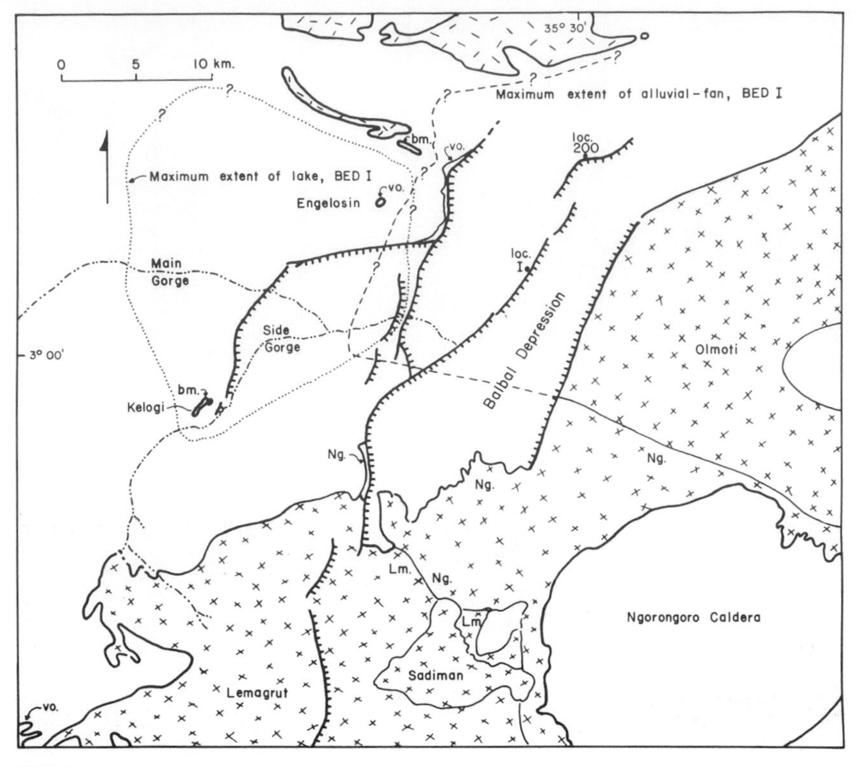

FIGURE 1
The Olduvai region showing principal faults, volcanoes, and outcroppings of metamorphic basement rock. Abbreviations used for volcanoes are Ng = Ngorongoro and Lm = Lemagrut. Dotted line indicates the inferred maximum extent of the Bed I lake; dashed line shows the maximum extent of the alluvial fan of Bed I. Courtesy, Quarternary Research.

Bed I is the lowest unit of the Olduvai Beds, and where fully exposed in the western part of the gorge, it is generally 30–43 m. thick. The only complete section in the eastern part of the gorge, near the Third Fault, is 54 m. Here Bed I is underlain by a semi-welded tuff, or ignimbrite, probably erupted from Ngorongoro, which very likely interfingers westward into the lower part of Bed I.

Bed I accumulated along the northwestern foot of the volcanic highlands, here comprising Olmoti, Ngorongoro, Lemagrut, and Sadiman (Figure 1). A lake occupied the lowest part of a small, shallow basin be-

FIGURE 2
Southside of the Main Gorge at the Second Fault. Here the gorge is about 75 m. deep. Copyright © 1976 by *The Regents of the University of California;* reprinted by permission of the University of California Press.

tween the volcanic highlands on the southeast and basement highlands to the north. The margin of the lake basin can be located accurately on the east, west, and southwest within exposures in the gorge, but the northern limit can only be inferred indirectly. The maximum east-west diameter of the lake was about 20 km., and its north–south diameter may have been somewhat more than this (Figure 1).

Tuffs provide the principal basis for correlating in Bed I, and the six tuffs most useful in correlating have been designated Tuffs IA, IB, IC, ID, IE, and IF (Hay 1971). Tuffs IA, IC, ID, and IF are ash-fall tuffs; tuff IB comprises an ash-flow tuff, both primary and reworked, and a widespread ash-fall tuff; and tuff IE is an ash-flow tuff. The oldest of the artifacts and hominid remains lie a short distance beneath Tuff IB; the others lie at varying levels between Tuffs IB and IF.

Potassium-argon dating and geomagnetic polarity measurements give the upper part of Bed I one of the firmest dates of any hominid-bearing lower Pleistocene stratigraphic unit. Potassium-argon dates clearly indicate an age on the order of 1.7–1.8 m.y. for the fossiliferous deposits. The most satisfactory dates are from Tuff IB, which has an age of 1.79 ± .03 m.y. (Curtis and Hay 1972). The magnetic stratigraphy has been worked out, at least roughly, for Beds I and II, and this gives a more satisfactory basis for estimating the duration of Bed I than do the K/Ar dates. Briefly, a normal polarity event, the Olduvai event (~1.86–1.71 m.y.), is recorded by the lava flows and overlying deposits of Bed I and the lower part of Bed II, including the lowermost part of the Lemuta Member (Grommé and Hay 1971). Both older and younger rocks have reversed polarity. On the basis of magnetic stratigraphy, stratal thickness, and development of paleosols, I estimate that the fossiliferous, artifact-bearing part of Bed I represents a time span of 50,000 to 100,000 years.

Lithologic Facies

Bed I can be subdivided into five lithologically different but laterally equivalent rock assemblages, or lithologic facies, which are the primary basis for interpreting Bed I in terms of environment. Four of the facies are named for the environment in which they were deposited: lake deposits, lake-margin deposits, alluvial-fan deposits, and alluvial-plain deposits. Lava flows constitute the fifth lithologic facies. It should be emphasized that these facies names oversimplify the environmental interpretation, as some of the facies contain sediments deposited in two or more environments.

Facies relationships are shown in Figure 3, which is an east–west cross section through Bed I at the end of its deposition. The lake deposits accumulated in a perennial lake in the lowest part of the basin. Lake-margin sediments were laid down on low-lying terrain intermittently flooded by the lake. Lava flows and alluvial-fan deposits are along the eastern margin of the basin, and they interfinger westward with lake-margin deposits. Alluvial-plain deposits are exposed only in the western half of the basin, where they underlie lake- and lake-margin deposits. Lake-margin deposits of the eastern and western parts of the basin differ in lithology as well as in fossils and archaeologic content, and the two assemblages of lake-margin sediments will be described separately under the headings *Eastern lake-margin deposits* and *Western lake-margin deposits.* The following sections will describe briefly the lithologic facies of Bed I and interpret them in terms of depositional environment.

Lake Deposits

The lake deposits are exposed westward about 6.5 km. from the Fifth Fault, and extend an undetermined distance to the east beneath younger

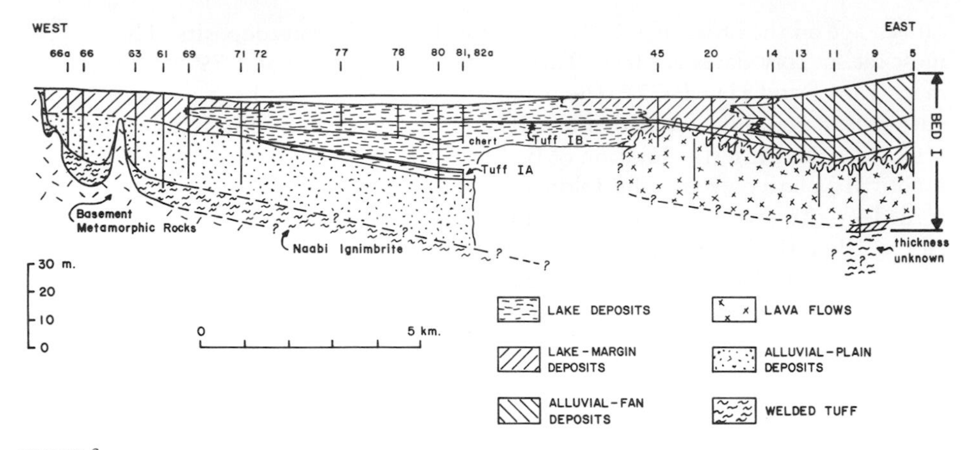

FIGURE 3
East–west cross section of Bed I, showing its lithofacies as exposed in the Main Gorge. Numbers refer to localities of measured sections, and vertical lines indicate thickness of section measured (see Figure 4; also location map in Leakey, M. D., 1971). Cross section is reconstructed with the top of Bed I approximately as it would have appeared when Tuff IF was deposited. Courtesy, Quaternary Research.

sediments; they have a maximum thickness of 26 m. near the Fifth Fault (e.g., loc. 80) and thin to about 10 m. near the western margin. The lake deposits which interfinger and intergrade westward with lake-margin deposits over a zone about 2 km. wide, are chiefly claystone (80%) but include tuff (15%), sandstone (2%), limestone and dolomite (3%), and a very small amount of chert and conglomerate.

They also contain substantial amounts of authigenic (i.e., chemically precipitated) minerals which indicate an alkaline lake of moderate to high salinity. Examples of the mineralogic evidence for lake-water chemistry include casts of soluble salts, chert, dolomite, K-feldspar (Hay 1968, 1970; O'Neil and Hay 1973) and the isotopic composition of calcite (Hay 1973). Several lines of evidence point to fluctuations in level and extent (Hay 1973).

Western Lake-Margin Deposits

The western lake-margin deposits are 8–12 m. thick and extend westward 2 km. from the zone of interfingering with the lake deposits. These sediments are chiefly sandstone (42%) and claystone (40%) but include a substantial amount (14%) of tuff. There are a few steep-sided stream channels filled with tuff or conglomerate which have an average orientation of N70°E. The deepest channel, at locality 60, has a depth of 3.4 m. The detritus in this facies is from westerly sources, hence the channels were cut by streams flowing northeast (Figure 4). An antelope skull was

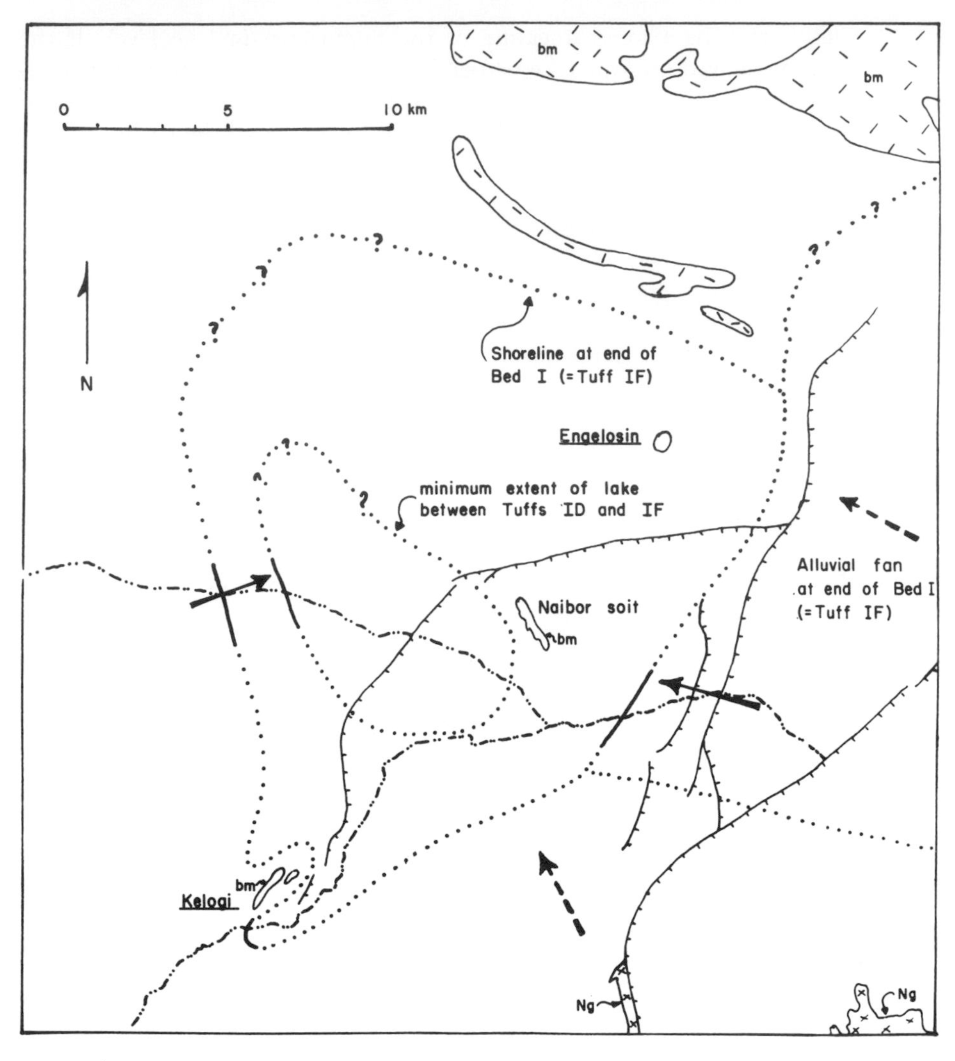

FIGURE 4
Paleogeographic diagram showing the outline of the lake and western extent of the alluvial fan at the time Tuff IF was deposited. Also shown is the inferred minimum extent of the lake between the deposition of Tuffs ID and IF. Solid arrows represent flow directions of streams as based on channel measurements; dashed arrows indicate flow directions as inferred on other grounds. Lithologic symbols are the same as in Figure 1. Courtesy, Quaternary Research.

found in locality 63 either at the base of the lake-margin deposits or in the topmost alluvial-plain deposits.

These deposits have both lacustrine and fluviatile features, indicating that they accumulated on terrain intermittently flooded by the lake. The stream channel, 3.4 m. deep, was near the margin of the lake, and its depth may be a rough measure of the fluctuation in lake level. If the streams flowed into the lake perpendicular to its margin, then the shoreline was oriented about N20°W. The lake water flooding the lake-margin terrain was overall less saline than that in the center of the basin.

Eastern Lake-Margin Deposits

Lake-margin deposits to the east of the lake beds are exposed almost continuously in the Main Gorge above the lavas over an east–west distance of 5.4 km. These are as much as 17 m. thick in the Main Gorge and include 5 meters of beds near Kelogi to the southwest (locs. 99, 100) and 7 meters of beds to the north of the gorge (locs. 201, 202).

The vertical sequence of lake-margin deposits above the lavas is subdivided at the base of Tuff IB into lower and upper units. The lower unit extends from the Second Fault west to FLK NN, a distance of 5.4 km. (Figures 5 and 6); the upper unit is restricted by alluvial-fan deposits to the western half of this area. The lower unit, except for its easternmost

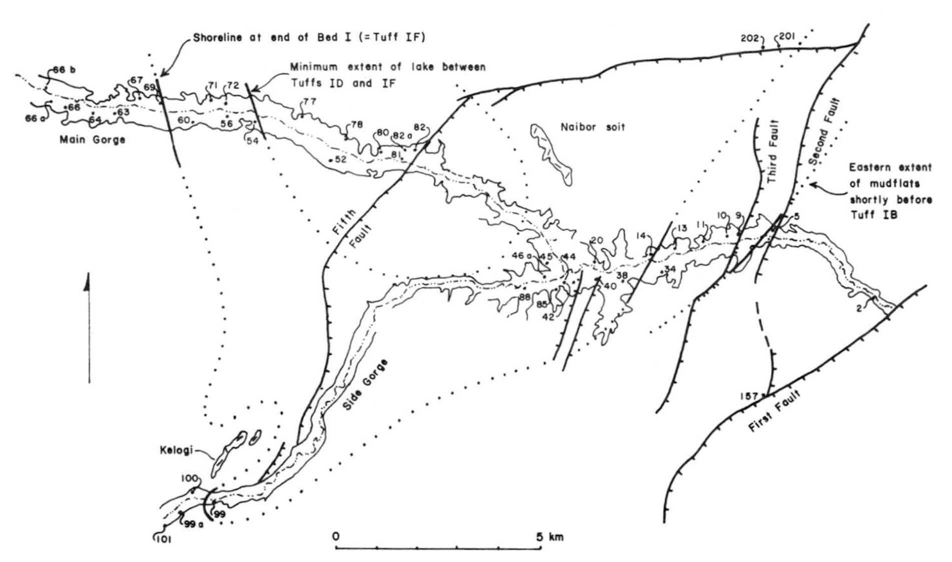

FIGURE 5
Outline of Olduvai Gorge, showing principal faults, and localities referred to in text. Also indicated are the shoreline at the end of Bed I (= Tuff IF), the minimum extent of the lake between Tuffs ID and IF, and the eastern extent of lake-margin mudflats shortly prior to eruption of Tuff IB. These paleogeographic features are shown as solid lines where based on exposures in the gorge and as dotted lines where inferred from overall paleogeographic considerations. Localities 11 (=MK), 13 (=DK), and 45 (=FLK) are sites of excavations for hominid remains and archaeologic materials. Not indicated on map are localities 45a (=FLK-N) and 45b (=FLK-NN), which lie, respectively, about 70 and 180 m. northwest of FLK (see Leakey, M. D., 1971). Courtesy, Quaternary Research.

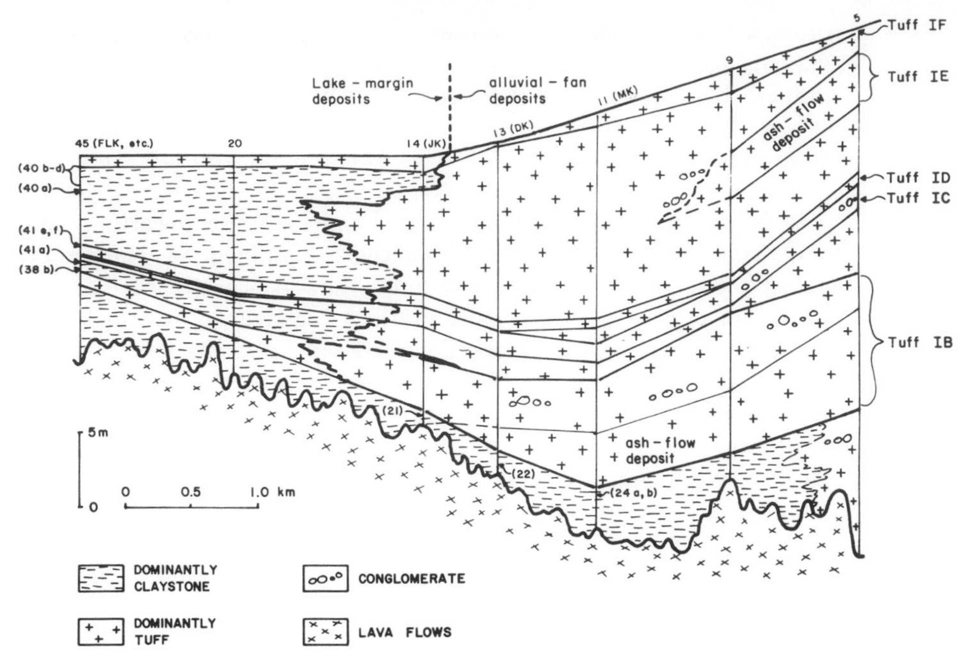

FIGURE 6
Stratigraphic diagram showing lithofacies and marker tuffs in Bed I between the Second Fault (loc. 5) and FLK (loc. 45). Numbers in parentheses refer to archaeologic and hominid sites (see Leakey, M. D., 1971, Appendix A). Diagram is drawn with the top of Bed I approximately as it would have appeared after deposition of Tuff IF. Courtesy, Quaternary Research.

exposures, is 0 to 6 m. thick, this variation reflecting the uneven surface of the lavas. It consists largely of claystone but contains a few trachytic tuffs. These sediments grade eastward into a section that is dominantly tuff and is as much as 11 m. thick. These easternmost deposits have some lithological similarity to the overlying alluvial-fan deposits and some mineralogical similarity to the underlying lavas, but they are however included within the lake-margin deposits in view of their limited exposure and uncertain genetic relationship. The upper unit of lake-margin deposits is about two-thirds tuff and one-third claystone and also includes limestone, pumice-pebble conglomerates, and siliceous earth.

Overall, the eastern lake-margin deposits are about 98 percent tuff and claystone, which are in roughly equal proportions. There is perhaps 2 percent of limestone and far less than a percent each of conglomerate and siliceous earth. Paleosols are weakly developed at many horizons over both tuffs and claystones. Where developed on claystones, they are crumbly, brown or brownish-gray, rootmarked zones. Paleosols on tuffs are rootmarked zones as much as 30 cm. thick in which much of the finest-grained vitric ash is weathered to montmorillonite.

Lithologic Descriptions. Claystones are both tuffaceous and non-tuffaceous and can be sandy or relatively free of sand. Rootmarkings are common, particularly above the level of Tuff IC. Calcite concretions are widespread and abundant in the claystones and calcite replacements of gypsum rosettes ("desert roses") have been found beneath Tuffs IB and IF.

Tuffs vary considerably in the degree and nature of reworking and provide considerable information about the environment in which they were deposited. A few tuffs represent single showers of ash that fell on the land surface (e.g., Tuff IB at FLK NN). A massive 1.2-m.-thick bed above Tuff ID, which is rootmarked and contains diatoms and ostracods in addition to pyroclastic materials, is a product of several eruptions in which the ash was deposited either in marshland or on a land surface intermittently flooded by lake water. A few tuffs (e.g., Tuff IF) are for the most part evenly laminated and are dominantly lacustrine.

Limestones are almost entirely of two types—either coarsely crystalline concretions cemented together or fine-grained beds of irregular shape—which both occur within claystones and were probably precipitated from ground water in clay at shallow depth.

Siliceous earth is used here to refer to a white or cream-colored silica-rich deposit that is friable, porous, and has an earthy luster. The silica is in the form of fine-grained particles of biogenic opal (e.g., diatoms, phytoliths, and silicified plant remains). A 60-cm. thickness of siliceous earth was found at one place in the widespread rootmarked massive tuff that overlies Tuff ID. In a few other places the tuff is siliceous and earthy.

Flora and Fauna. Fossilized leaves are rare, and little is known of the pollen, but swamp vegetation is indicated by abundant coarse, generally unbranching vertical root channels and casts which are commonly 10–30 cm. long, 2–6 mm. wide, and could have been made by *Typha* and some kinds of reeds. Fossil rhizomes of *Papyrus* (?) point to marshland or shallow water and silicified remains of a water plant (cf. *Potamageton*) were identified by Howard Schorn (1971, personal communication). Diatoms and encysting cases of chrysophyte algae are in the siliceous earth and siliceous, earthy tuff that overlies Tuff ID (e.g., at loc. 85); they point to a fluctuating lake level and periodically saline, alkaline conditions (Hay 1973).

Ostracods are in a few samples of the siliceous tuff unit above Tuff ID, and fresh-water snails were collected from clays of the *"Zinjanthropus"* level (Leakey, M. D., 1971). Fossil remains of urocyclid slugs have been found at 2 levels below Tuff IC (ibid.). This evidence suggests damp conditions in evergreen forests "where the rainfall exceeds 35 inches per year or where damp conditions are maintained by regular mists" (Verdcourt 1963).

Vertebrate remains are both varied and abundant, and the various forms represent a wide variety of habitats (see, Leakey, L. S. B., 1965; Leakey, M. D., 1971; Hay 1973). Remains of large mammals are found at many levels: proboscideans (principally elephants), rhinoceros, equids, suids, hippos, bovids, and giraffids. M. D. Leakey (1971) gives the number and proportions of specimens of identifiable larger mammals from excavated sites. In this tabulation the bovid remains, principally of gazelles and other antelopes, are much more abundant between the "Zinjanthropus" level and Tuff IF than they are in the underlying deposits. This difference reflects either a real difference in the fauna or the increasing importance of bovids in the hominid diet. Small mammals (e.g., rodents, gerbils, carnivores) are abundant in some levels of the excavations at FLK and FLK N, and the rodents are of particular ecological interest (see Lavocat, p. 17–19 in Leakey, L. S. B., 1965). The fossil avifauna is one of the largest heretofore known, and a short summary by P. Brodkorb is given in Hay 1973. The fish fauna, described by Greenwood and Todd (1970), comprises cichlids (e.g., tilapia) and clariids (e.g., catfish). All of this faunal material suggests widespread marshland and the frequent occurrence of standing water up through deposition of the siliceous earthy tuff above Tuff ID. Drier conditions are indicated by fish, birds, and rodents for excavated levels in the uppermost 1.2 m. below Tuff IF at FLK N (loc. 45b).

Depositional Environment. These deposits accumulated on relatively flat terrain that was intermittently flooded and dried in response to changes in level of the lake (Hay 1973). Claystones and widespread water-laid tuffs were deposited at times of high water level, whereas paleosols, hominid occupation sites, eroded surfaces, and extensive land-laid tuffs represent periods of exposure. Prior to Tuff IB, the zone of marginal terrain along the eastern margin of the lake was at least 5.4 km. (Figure 6); after Tuff IB, the zone was narrowed by about 3.4 km. in westward growth of an alluvial fan. Most of the fluctuations may have been relatively short—either seasonal or involving no more than a few tens of years. A longer-term paleogeographic or climatic fluctuation is probably indicated by the fauna collected at several horizons through 1.2 m. of claystone beneath Tuff IF at FLK N. This fauna reflects substantially drier conditions than prevailed at lower horizons for which the fauna is known.

The pattern of chemical alteration in lacustrine tuffs points to a salinity gradient at times of flooding, with the highest salinities on the northwest and the freshest water on the southeast, nearest the stream inlets.

Initially, the lava surface had a local relief of at least 6 m. which was reduced as the lavas were buried by sediments. The lake-margin terrain had a relief of about 2 m. just prior to Tuff IF, which filled in topographic irregularites.

Alluvial-Fan Deposits

Alluvial-fan deposits lie along the east side of the basin and interfinger westward with lake-margin deposits. They are exposed in the gorge over an east–west distance of 7.5 km. and crop out intermittently for a distance of 5 km. to the northeast along the west side of the Balbal depression. The uppermost part of the facies is exposed 15 km. northeast of the gorge (loc. 200).

The alluvial-fan deposits are almost entirely of eruptive origin and range from 21 to 28 m. in thickness. Approximately 15 percent of the facies consists of ash-fall and ash-flow deposits that have not been reworked; the remainder was deposited chiefly by streams but includes some layers emplaced as mud flows. Brown, rootmarked paleosols are widely developed, and rootmarkings are through most of the section near the western margin of the facies. The paleosols have weakly developed profiles, and only a small amount of clay has been formed by weathering. Stream channels are numerous with an average alignment of N73°W, indicating northwestward-flowing streams. Remains of a suid and *Parapapio* sp. have been collected from alluvial-fan deposits at DK (loc. 13) (Leakey, M. D., 1971, personal communication), but no other faunal remains have been found.

These deposits were erupted from Olmoti and accumulated on the lower part of an alluvial fan sloping northwestward to the lake-margin terrain. An average slope of 1:300 (3.3 m./km.) is used in the cross sections of Figures 2 and 5. As shown by the rootmarked paleosols, vegetation grew periodically on the fan, but its surface must have been relatively barren for much of the time. The scarcity of mammal remains probably reflects the sparseness of the vegetation.

Lava Flows

Lava flows which are exposed almost continuously in the gorge from the Second Fault west to FĹK NN, a distance of 5.4 km., have a total thickness of 20 m. where they are fully exposed upstream from the Third Fault. Lavas of Bed I also crop out along the western margin of the Balbal Depression, and 45 to 60 m. of flows are exposed in fault scarps 5 to 12 km. north of the gorge. In the gorge near the Second and Third Faults, the lava sequence comprises a basal-thick flow of trachyandesite overlain by several thin flows of olivine basalt with ropy or pahoehoe surfaces. Perfect preservation of pahoehoe crusts on the basalts shows that they spilled out in rapid succession, almost certainly from a vent nearby rather than from a summit crater of one of the large volcanoes. Directional features in the olivine basalts indicate a vent to the south of the gorge. The trachyandesites were discharged from Olmoti.

Alluvial-Plain Deposits

Alluvial-plain deposits constitute the lower part of Bed I to the west of the Fifth Fault, where they crop out discontinuously over an east–west distance of 9 km., and range in thickness from 12.4 to 21 m. except where they thin and pinch out over the top of a buried inselberg (loc. 64). These deposits are chiefly tuff (56%) and claystone (28%) but include limestone, conglomerate, and sandstone. Tuffs form most of the lower two-thirds of the facies, and nearly all of the claystones are in the upper one-third. Both tuffs and claystones are highly rootmarked in most places. Many of the limestones are a type of surface limestone or calcrete that, as it accumulated, was contaminated with detrital sediment, in part eolian. The only faunal remains known from these deposits are suid and elephas teeth, noted in claystone near the top of the facies in locality 71.

Clayey tuffs in the lower part of the facies probably accumulated in a steppe environment where the vegetation was at least seasonally insufficient to prevent considerable reworking of sediment by wind. The degree of weathering and clay formation suggests a rather slow rate of accumulation. Rootmarkings are similar to those produced by grass, and claystone pellets and clay-coated grains indicate eolian transport (Hay, in press). Conglomerates and sandstones were deposited by streams flowing northeast. The source of most of the tuffs is probably Ngorongoro.

ENVIRONMENTAL SYNTHESIS AND GEOLOGIC HISTORY

The overall paleogeography of Bed I is clear, although several questions remain unanswered. Bed I was deposited in a lake basin at the western foot of the volcanic highlands. The lake either did not have an outlet, or it overflowed infrequently, hence fluctuating in level, alternately flooding and exposing a broad marginal zone (Figure 4). The basin was filled by pyroclastic and detrital sediments and by lava flows—the pyroclastic deposits were erupted from Olmoti, and the lavas are both from Olmoti and from a vent to the south of the gorge.

The basin is at least partly tectonic in origin. If the Naabi Ignimbrite was erupted from Ngorongoro, as concluded here, then faulting or warping must have reversed the slope of the land surface from westward to eastward prior to the deposition of Bed I. The initial paleogeography of the lake basin remains obscure because of the very limited exposure below the lavas in the eastern part of the gorge. The lavas must have displaced the eastern shoreline, and presumably the overall position of the lake, a substantial distance to the west, where it remained during the deposition of Bed I.

Warping or faulting continued during the deposition of Bed I, as indicated by anomalous paleogeographic relationships at the level of Tuffs IB and IC compared to Tuff IF. Tuff IF must have been very nearly horizontal when deposited in the eastern lake-margin terrain, but a reconstructed cross section of Bed I drawn on this basis places lacustrine clays between Tuffs IB and IC at FLK, topographically higher than stratigraphically equivalent stream-laid sediment to the east (Figure 6). Moreover, if any reasonable water depth is taken for deposition of Tuff IF in the center of the lake, then Tuff IB, here deposited in lake water, lies either at the same or at a higher elevation than it does at the eastern margin of the lake, where it is a subaerial ash-fall deposit (Figure 3). The basin appears to have been tilted slightly downward to the east, and the area occupied by the lavas may in addition have been warped into a shallow syncline (Figures 3 and 6).

The average diameter of the lake fluctuated between limits of about 7 and 25 km. during all but the most extreme periods of desiccation. The lake was widest prior to the ash flow of Tuff IB, which spread eastward and narrowed the area subject to flooding. It appears to have been smallest, on the average, between deposition of Tuffs ID and IF, when, for most of the time, its east–west diameter was probably between 5 and 8 km. (Figure 4). When smallest, it was probably an elongate body of water oriented northwest–southeast (Hay, in press). The lake expanded rather abruptly at the time Tuff IF was deposited, at which time the shoreline is known with greater accuracy than any earlier period in the deposition of Bed I (Figure 4).

The perennial part of the lake was comparatively shallow at times of low level as indicated by evidence of wave or current action at several horizons and by desiccation cracks at one level in the central part of the basin. The wide marginal zone of intermittent flooding and drying also fits with a shallow lake. Water-level fluctuations on the order of 1.5 to 3.5 m. can account for most or all of the features noted in the lake-margin deposits.

The perennial part of the lake was alkaline and moderately to highly saline, and water flooding the lake-margin terrain varied widely in salinity, with fresh water near stream inlets along the eastern margin of the lake. Frequent fluctuations in level and high salinity of the lake suggest that its level was controlled by the balance of inflow and evaporation. Whether or not the lake ever overflowed is not known. The only possible direction of overflow is to the northwest, toward Lake Victoria.

The climate was probably semi-arid, though wetter than that in the Olduvai region today. High salinity and fluctuations in lake level are the best evidence that the climate was relatively dry over a long period.

The fauna includes water- and swamp-dwelling elements as well as plains animals, but these are compatible with a semi-arid climate.

Urocyclid slugs found below the level of Tuff ID appear to be the best evidence which clearly indicates a climate appreciably moister than that of today in the same area. The fauna collected beneath Tuff IF seem to show that the climate was considerably drier than that prevailing prior to Tuff IC. Geologic evidence, although not conclusive, supports the climatic change inferred from the fauna.

PALEOGEOGRAPHY OF HOMINID ACTIVITIES

During the known period of hominid occupation in Bed I, the Olduvai region can be visualized as a lake basin about 30 km. in diameter and bounded on the south, east, and north by forested highlands. The lake had no outlet and fluctuated seasonally in level and extent. At times of low level, the lake was enclosed by broad mudflats. Several inselbergs, or steep-sided hills, were located near the margin of the lake at times of low level and formed islands in the lake at times of flooding. The principal streams entered the lake on the east, and along this side of the lake were concentrated areas of marshland and *Acacia* forest. Savanna grassland, probably with areas of bush and scrub, extended from the lake-margin zone to the forested highlands.

Evidence of hominid activities in Bed I is restricted to the eastern lake-margin deposits, which accumulated on relatively flat terrain intermittently flooded by the lake (see p. 220). This area supported a great variety of animal life and a large amount of vegetation. Lack of game, fresh water, and appropriate vegetation can account for the apparent absence of evidence for hominid activities in other time-equivalent lithologic facies of Bed I. Within the eastern lake-margin deposits, the known distribution of hominid materials differs considerably between upper and lower units (Figure 6). In the lower unit, hominid remains and archaeologic materials are confined to the north side of the gorge between localities 10 and 14. These sites lay in the eastern or inland half of the terrain intermittently flooded by the lake, and most of them come from the uppermost 2 meters of sediment beneath Tuff IB between localities 10 and 14. The excavation of DK (loc. 13) provided some detailed information of paleogeographic significance (Leakey, M. D., 1971). The stone circle and occupation site overlie a paleosol developed partly on the eroded surface of a tuff and partly on the basalt where it rose above the level of the tuff. The surface of the tuff showed a number of narrow, steep-sided channels 45 to 60 cm. deep that resemble game trails leading to the edge of the lake in Ngorongoro (ibid., p. 12–23). Fossil *Papyrus (?)* rhizomes and large quantities of crocodile remains testify to permanent water nearby.

Above the level of Tuff IB, hominid remains and artifacts are found only to the west in the vicinity of FLK and HWK. These sites appear to have positions paleogeographically more or less comparable to the site at DK. The occupation site at FLK NN (site 38b, level 3) is on the weathered surface of a thin claystone above Tuff IB. "The many rootlet holes and reed casts in Tuff IB and occurrence of numerous fish and amphibian remains, together with bones of waterfowl, indicate that the site was situated near the shores of a lake or by a swamp" (ibid., p.42). The site at FLK, termed the "*Zinjanthropus* floor," lies on a paleosol developed on a lacustrine claystone. At FLK N, the topmost 1.5 m. of claystone beneath Tuff IF has yielded 5 implement-bearing levels. The lowest level (no. 6) was a butchering site where artifacts were associated with the skeleton of an elephant. The highest level (nos. 1 and 2) was an occupation site comparable in many respects to the "*Zinjanthropus* floor." Artifact concentrations at the intermediate levels are probably reworked. These claystones were exposed at the surface too briefly to develop paleosols. The shoreline probably may well have been a kilometer or so to the west of these sites for much of the time represented at these levels at FLK N (Figure 6). Scattered artifacts which have been encountered in both excavations and natural exposures at other levels between Tuff IC and level 6 at FLK N, seem to be most widespread in the massive tuff that overlies Tuff ID and which contains diatoms, ostracods, and rootmarkings of marshland vegetation (sites 41e,f).

ACKNOWLEDGMENTS

Field work at Olduvai Gorge has been supported by grants from the National Geographic Society and the National Science Foundation (G-22094); laboratory work and writing were aided by research professorships in the Miller Institute for Basic Research (Berkeley). Dr. L. S. B. Leakey provided the opportunity to study the Olduvai deposits and helped me to obtain financial support for field work. I am grateful to Dr. M. D. Leakey for use of the camp facilities at Olduvai and for generous assistance in various aspects of the geologic work between 1962 and the present. Discussions with Dr. Glynn Ll. Isaac have been a fruitful source of information and interpretation. Robert N. Jack is much to be thanked for diffractometer analyses, and I am indebted to S. J. Chebul for the thin sections used in microscopic study.

References Cited

Curtis, G. H., and R. L. Hay

1972 Further geologic studies and K-Ar dating of Olduvai Gorge and Ngorongoro Crater. *In* Calibration of Hominoid Evolution. W. W. Bishop and J. A. Miller, eds. Edinburgh: Scottish Academic Press. pp. 289–301.

Greenwood, P. H., and E. J. Todd

1970 Fish remains from Olduvai. *In* Fossil Vertebrates of Africa, Volume 2. L. S. B. Leakey and R. J. G. Savage, eds. London: Academic Press. pp. 225–241.

Grommé, C. S., and R. L. Hay

1971 Geomagnetic polarity epochs: age and duration of the Olduvai normal polarity epoch. Earth and Planetary Science Letters, Vol. 10:179–185.

Hay, R. L.

1966 Zeolites and zeolitic reactions in sedimentary rocks. Geol. Soc. America Special Paper 85.

1968 Chert and its sodium-silicate precursors in sodium-carbonate lakes of East Africa. Contr. Mineral. Petrol. 17:255–274.

1970 Silicate reactions in three lithofacies of a semi-arid basin, Olduvai Gorge, Tanzania. Mineral Soc. America Special Paper 3:237–255.

1971 Geologic background of Beds I and II. *In* Olduvai Gorge. Volume III. Excavations in Beds I and II, 1960–1963. Cambridge: The University Press. pp. 9–18.

1973 Lithofacies and environments of Bed I, Olduvai Gorge, Tanzania. Quaternary Research 3:541–560.

In Press Geology of the Olduvai Gorge, University of California Press.

Leakey, L. S. B.

1965 Olduvai Gorge 1951–1961. Volume I. A Preliminary Report on the Geology and Fauna. Cambridge: The University Press.

Leakey, M. D.

1971 Olduvai Gorge. Volume III. Excavations in Beds I and II, 1960–1963. Cambridge: The University Press.

O'Neil, J. R. and R. L. Hay

1973 O^{18}/O^{16} ratios in cherts associated with the saline lakes of East Africa. Earth and Planetary Science Letters 19:257–266.

Reck, H.

1951 A preliminary survey of the tectonics and stratigraphy of Olduvai. *In* Olduvai Gorge, a report on the evolution of the hand-axe culture in beds I–IV. L. S. B. Leakey. Cambridge: The University Press. pp. 5–19.

Verdcourt, B.

1963 The Miocene nonmarine mollusca of Rusinga Island, Lake Victoria, and other localities in Kenya. Palaeontographica 121:1–37.

Exposures of the Shungura Formation in the Lower Omo Valley. (Photo by F. C. Howell.)

F. Clark Howell:

Overview of the Pliocene and Earlier Pleistocene of the Lower Omo Basin, Southern Ethiopia*

What should one do if one finds a stack of fossiliferous, hominid-bearing sediments, over half a mile thick? Clark Howell and Yves Coppens worked for nine years with internationally sponsored teams of scientists and have uncovered a vast wealth of data on early men and the environment in which they lived. This summary deals with the hominid fossils, but more especially it deals with the geologic, geographic, and biologic circumstances which molded human evolution in this part of the world between one and more than three million years ago.

INTRODUCTION

Since the fortuitous discovery and insightful recognition of *Australopithecus africanus* a half century ago, the African continent has proved to be of primary importance for investigations into hominid origins and

*This is an abridged version of a paper which was presented simultaneously for publication in this volume and in the volume *Hominidae of the African Plio-Pleistocene*, edited by C. J. Jolly (Duckworths, London, 1976). Some of the more technical geological and paleontological sections of this paper have been cut down so as to render it more readily usable by students and general readers. Passages thus condensed are marked by cross references (Howell 1975).

the earlier evolution of the Hominidae. The initial twenty-five years thereafter witnessed devoted, episodic, and highly fruitful efforts to obtain further remains of such early Hominidae in the fossil-rich cavern infillings of South Africa. As a consequence of that research, the reality and the apparent complexity of hominid origins was forcefully demonstrated.

In the past fifteen years, after the remarkable discovery in 1959 of *Australopithecus* at Olduvai Gorge, Tanzania, sustained field programs have been mounted in eastern Africa to seek early Hominidae and to investigate their world in situations in and adjacent to the Rift Valley System. The presistent field researches at Olduvai Gorge have been the catalyst for such interests. Subsequently, small-scale explorations were mounted in the Kenya section of the Rift Valley, in the Baringo area, and in the Kerio drainage area south of the Rudolf basin. Slightly later, extensive and intensive field programs were developed in the more northern portions of the Rudolf basin itself.

A major interdisciplinary effort was initiated in 1966 in the northern reaches of the greater Rudolf basin, in the valley of the lower Omo river, in southern Ethiopia. The research has been highly rewarding and has afforded an unexpected and diverse array of data bearing on early Hominidae, their age, and the world of their time. The scope of these researches is briefly set out here.

The lower Omo basin was first visited by the European explorers, Count Samuel Teleki von Szék and Ludwig Ritter von Höhnel, in 1888, after their discovery of Lake Rudolf (von Höhnel *et al.* 1891; von Höhnel 1938). In 1896 Maurizio Sacchi, geographer with the ill-starred (second) Bottego Expedition (Vannutelli and Citerni 1897, 1899; cf. Sclater 1899), was the first to note the existence there of flat-lying, undeformed sediments with freshwater mollusca exposed to the north of the Omo delta. He lost his life in southern Ethiopia the next year, but his collections were subsequently studied and published by de Angelis d'Ossat and Millosevich (1900). These deposits have recently been investigated by the Omo Research Expedition (Butzer, Brown and Thurber 1969; Butzer and Thurber 1969) and defined as the Kibish Formation (Butzer 1971, 1975), considered now to be of largely later Pleistocene to Holocene age.

Emil Brumpt, naturalist with the Bourg de Bozas expedition (Bourg de Bozas 1903, 1906), was the first (in 1902) to recover vertebrate fossils from older, tectonically disturbed deposits underlying these largely horizontal sediments. Their importance for the study of the Cenozoic of sub-Saharan Africa, then practically unknown, was first noted by Haug (1912:1727), who recorded the occurrence of silurid fish, crocodilians (two species), chelonians, a hipparionine, rhinoceros, hippopotamus, a suid, various bovids, and deinothere, and elephant. Subsequently the remains of the aquatic reptile *Euthecodon (=Tomistoma) brumpti* (Joleaud

1920b, 1930; Boulenger 1920), a deinothere and *Elephas* (Joleaud 1928), hippopotamus (Joleaud 1920a), and a hipparionine and *Equus* (Joleaud 1933) were described from Brumpt's collections, deposited at the Muséum National d'Histoire Naturelle, Paris.

The first geological reconnaissance and extensive paleontological prospection of these older fossiliferous deposits was undertaken by the late Camille Arambourg between January 30 and March 13, 1933 as part of the activities of the Mission Scientifique de l'Omo (Arambourg and Jeannel 1933; Jeannel 1934). Arambourg (1943) subsequently reported on the geological results of this work and on the substantial collection of fossil vertebrates, including fish (9 species), reptiles (6 species), and mammals (29 species) (Arambourg 1947). These deposits, which Arambourg (1943) considered to extend some 90 km. up the lower Omo valley, were informally termed "depôts anciens du Lac Rodolphe" or "depôts fluvio-lacustres de la vallée de l'Omo" or simply "Omo Beds."

During World War II this area was occupied by Allied military forces from Kenya. Some fossils were collected and sent to the Coryndon Museum in Nairobi, of which L. S. B. Leakey was then Honorary Curator. In early 1942 he dispatched his assistant, Heselon Mukiri, to the lower Omo valley who spent three weeks there collecting a substantial number of mammalian fossils (now housed in the National Museums of Kenya, Nairobi). These collections, the precise geological provenience of which are of course unknown, represent a diversity of taxa. Except for some suids, which Leakey (1943) designated a new genus *(Pronotochoerus jacksoni)* and two new species *(Gerontochoerus scotti, Mesochoerus heseloni),* they remain undescribed.

A further reconnaissance of these fossiliferous sediments was made by the author in July–August, 1959, which resulted in the recognition of the geological complexity of the deposits and of their vertebrate fauna (Howell 1968). Subsequently in 1966 the formation of an international Omo Research Expedition was authorized by the Imperial Ethiopian Government. During that year, an extensive geological reconnaissance of the area was made by F. H. Brown (1969), after which various participants of the expedition have worked annually in the lower Omo basin, during the summer months, since 1967.

GEOLOGIC SETTING

The lower Omo basin, like the Rudolf basin to the south, is related to the Turkana depression of the Eastern Rift Valley (Baker, Mohr and Williams 1972). It appears to represent a northward continuation of the depression into the southwestern margin of the Ethiopian plateau. The Omo basin is considered a tectonic depression (Merla 1963), downwarped and down-faulted after episodes of planation of the adjacent highlands during the late Mesozoic and Paleogene. Although the structure of the highly folded and faulted area to the southwest is known (Walsh and Dodson 1969), that of the lower Omo basin and the highlands to the northwest and to the east remain to be investigated (Butzer 1970). Projected gravity surveys in the lower Omo valley will help in part to resolve this problem. The Korath volcanics (Nakua), a low range of hills to the northeast of Sanderson's Gulf, are a series of basalt flows and coalescent tuff cones (Brown and Carmichael 1969). They were extruded in late Pleistocene times along a line of tensional faulting, several fractures of which are recognized to transect the Plio/Pleistocene deposits farther east.

STRATIGRAPHY

The lower Omo basin preserves and exposes the thickest, most continuous fossiliferous record of Pliocene and earlier Pleistocene sedimentation ever discovered in relation to the East African Rift System. The Plio/Pleistocene succession exposed in the lower Omo basin is now known to have an aggregate measured thickness in excess of one kilometer (actually 1093 meters). The sediments outcrop discontinuously over an area of some 200 km.^{2} in four sectors of the basin (Figure 1). These respective exposures have been given formational status, on a geographical basis, and together comprise the Omo Group (de Heinzelin, Brown and Howell 1970). All the formations have been surveyed and mapped on aerial photographs (taken for the expedition by Robert Campbell in 1967 and 1970) at a scale of 1:10,000, or smaller. These Omo Group formations can be defined as a set of somewhat consolidated sediments (clays, silts, sands) and pyroclastics (tuffs and, rarely, extrusive lavas), usually tectonically deformed by tilting and subsequent faulting, which discordantly underlie horizontal sediments comprising the Kibish Formation. Their radiometric age (Brown and Lajoie 1971; Brown 1972; Brown and Nash, 1975), paleomagnetic record (Shuey, Brown and Croes 1974; Brown and Shuey 1975) and vertebrate fauna (Howell and Coppens 1974; Coppens and Howell 1974), indicate a Pliocene and earlier Pleistocene age for their accumulation.

Minerals contained in the Mursi and Usno basalts, and in pumices and tuffs of the Usno and Shungura Formations afford a means of radiometric age determination by potassium-argon. Thirty determinations have been made thus far. All are stratigraphically and biostratigraphically consistent and, with two exceptions, all are also consistent with the paleomagnetostratigraphy established for the Shungura and the Usno Formations (see below). The results indicate that: (a) the Mursi Formation sediments have an age of 4.1 m.y.; (b) the Nkalabong Formation has an

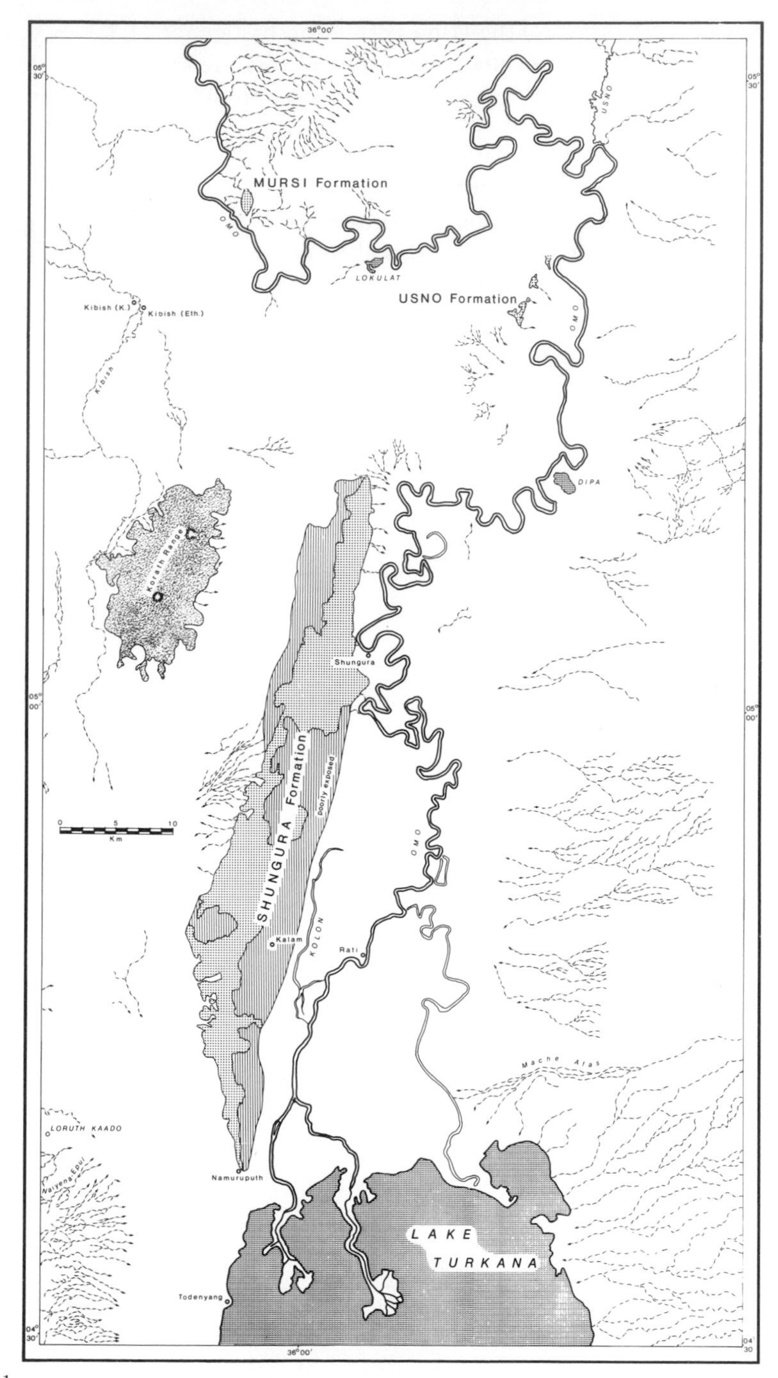

FIGURE 1
The lower Omo basin, southern Ethiopia, and situation of formations comprising the Omo Group.

age of 3.95 m.y.; (c) the Usno Formation has an age of 3.3 m.y. (basalt) to 2.97 m.y. (main tuffs), the principal fossiliferous horizon closely approximating the latter figure; and (d) the Shungura Formation encompasses a time span of about 2 m.y. (from approximately 3.3 to 0.8 m.y.). Figure 2 shows the relationship between dates and stratigraphy in the Mursi, Usno, and Shungura Formations.

(Individual descriptions of formations other than the Shungura Formation have been replaced in this abridged version by a tabular summary of the constituents of the Omo Group. See Table 1.)

SHUNGURA FORMATION

The main source of fossils and paleoanthropological evidence of the Omo has been a huge body of sediments for which the name Shungura Formation was first suggested by de Heinzelin and Brown (1969; cf. Butzer and

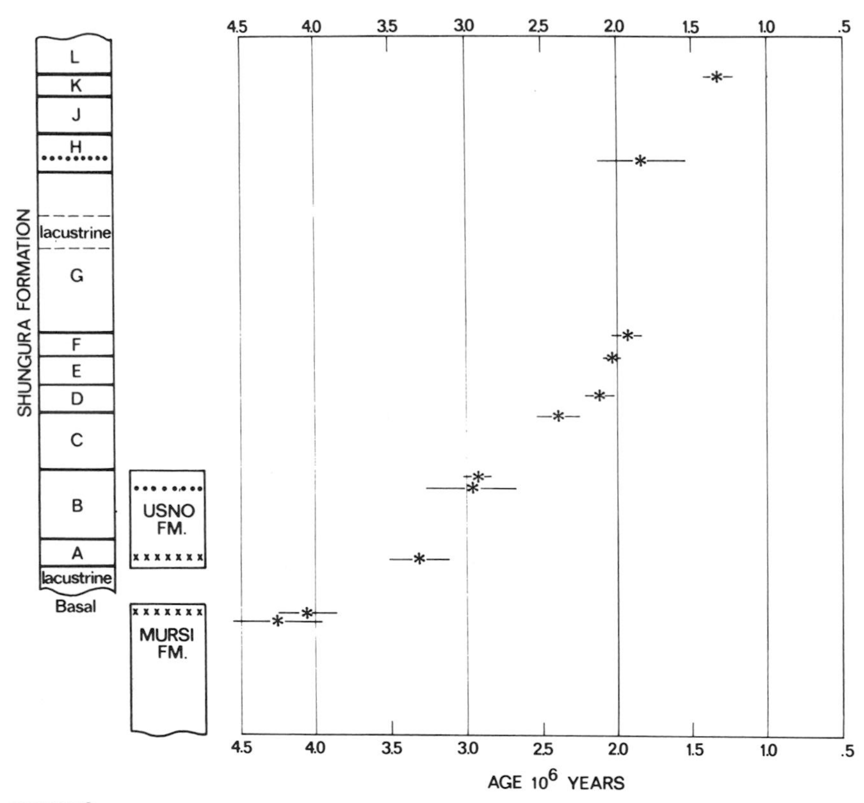

FIGURE 2
Pliocene-Pleistocene formations of the Omo group, southern Ethiopia, and their radiometric datation.

TABLE 1
The constituent formations of the Omo group.

M.Y.A.	FORMATIONS				
1.0					
1.5					
2.0	SHUNGURA*+				
2.5	FM.				
3.0		USNO FM.*	?Loruth Kaodo Fm.?		
3.5					
4.0				Nkalabong Fm.	
4.5					MURSI FM.

Mursi Formation: about 140 m. of deltaic and fluvio-littoral sediments with vertebrate and mollusc fossils.

Nkalabong Formation: 90 m. of fluvio-lacustrine sediments; not known to be fossiliferous.

Loruth Kaado Formation: little known exposures at the northern end of the Labur massif of tuffs, tuffites, fine-grained sediments and piedmont clastics. Mollusc fossils only, which is perhaps correlative with basal Shungura.

Usno Formation: more than 170 m. of fluvio-lacustrine sediments with vertebrate and molluscan fossils.

Shungura Formation: the "classic" Omo Beds; more than 770 m. of fluvial and lacustrine sediments with fossils. See text for more detailed information.

Capitalized = yielding a significant mammal fossil assemblage
** = yielding hominid fossils*
+ = yielding artifacts

Thurber 1969). It was formally defined by de Heinzelin, Brown, and Howell (1970). It includes the ancient fluvio-lacustrine beds from which E. Brumpt, and subsequently C. Arambourg, collected some classic (and type) Omo fossil vertebrates. Its complexity, substantial thickness, and the significance of its tuffs for radiometric dating and for extensive mapping were first demonstrated by Brown (1969). The formation outcrops only on the west side of the Omo river between 5° and 5°10′N and comprises a series of fluvial, deltaic and lacustrine sands, silts, and clays and a set of 12 principal widespread tuffs (labelled A to L), as well as numerous subsidiary intercalated tuffs (de Heinzelin, Haesaerts and Howell, 1975). These deposits are faulted and tilted to the west and are of late Neogene to earlier Pleistocene age. The measured aggregate thickness of the formation (base unknown and top unconformably overlain by Kibish Formation) is some 770 meters. The thickness of the principal exposures east of Korath (Nakua) Hills is 670 meters (extending from the Basal Member through middle Member J-4). The thickness of the exposures south of Korath Hills and west of Kalam police post (from upper Member C to an uppermost shellbed of Member L) is 387 meters. The differences in thickness are presumably related to differential subsidence (see below).

This sequence of deposits has been subdivided into a number of members through the (convenient) occurrence of a series of widespread, usually readily distinguishable volcanic ash horizons (designated A to L upwards) (de Heinzelin 1971). Each member (except the Basal Member) is represented by a major or principal tuff and its *overlying* sediments (de Heinzelin 1971; de Heinzelin *et al.* 1970).[1] These marker horizons at the base of the principal volcanic tuffs are true chronohorizons and are essentially isochronous over the whole lower Omo basin. (Essentially the same interpretation and procedure was followed by R. L. Hay at Olduvai Gorge and has been utilized by Ian Findlater in East Rudolf.)

There is no single, simple east-to-west type section, due to tectonic disturbance. However, exposures of about a half-kilometer in width, extending west from the vicinity of Shungura village (on the river) toward the Korath hills, includes all members through Member J. Members C upwards, including extensive outcrops of Members G through L, crop out south of the Korath hills, and west of Kalam police post, where the type sections of Members K and L have been drawn. Mapping in 1972 and 1973 clearly demonstrated the interrelationships of these sets of sediments in the two major areas of exposure of the Shungura Formation and permitted their correlation.

A particular lithologic sequence occurs repeatedly through much of the Shungura Formation. The sequence consists of silt and clay overlain successively by coarse sand, medium sand, fine sand, silt and clay. In general the sands are yellowish gray to light brown, poorly cemented, cross-bedded, and composed of fragments of quartz, feldspar, lithic fragments and bits of chalcedony. The clayey silts are brown to reddish brown and may contain gypsum or halite. Aside from the major tuffs which define the members, there are also many minor tuffs in the succession. These minor tuffs occur only discontinuously along strike, but their position within a member is constant. All tuffs are composed predominantly of volcanic glass and have only a small admixture of foreign fragments. In general the major tuffs are less contaminated than the minor tuffs. Pebbles and cobbles of pumice occur in many of the tuffs, but most of the material is of sand and silt size.

These repetitive sequences of coarser to finer sediments are basically similar, but may vary in thickness and completeness. They are termed cyclic units and evidently represent sheets of meandering stream deposits

[1]This procedure (cf. Bonnefille *et al.* 1973), adopted for ease of mapping since tuffs could be readily mapped from the prominent morphology of the tuff cuestas, replaced an initial informal terminology, that of a "series" of sedimentary units *below* a capping tuff (Arambourg, Chavaillon and Coppens 1967, 1969). Any initial designation of "series" is thus placed into this formal terminology by a lower letter designation (thus Mb. G = "series" H; Mb. F = "series" G, etc.).

(Allen 1965a, 1970). They are defined as formal units in the type sections of members and are numbered successively upward (B-1, B-2, etc.) above the relevant tuff (Tuff B, etc.); the exception is the Basal Member which comprises those sediments *below* Tuff A. They may be traced over kilometers, although surficial cover or faulting may effect disjunction between outcrops. Correlation between more distant outcrops have been effected through the enumeration of cycles between known marker horizons (customarily the major tuffs A–L), and through the recognition of similar or identical sequences of deposits. The principle is of course well established and in numerous instances has led to the recognition of reliable chronosequences. These have led to the definition of local chronozones (see Hedberg 1971:21–22), as the greater the complexity of events the more diagnostic is the resultant fit. The units of the Shungura Formation can be considered chronozones, although they are somewhat diachronous, given their formation by the lateral displacement of a meander belt. Thus, in a strict sense they represent para-chronostratigraphic units.

The general outlines of the lower Omo basin were apparently delineated prior to the accumulation of Omo Group sediments (Butzer 1970; Butzer and Thurber 1969). The Mursi Formation sediments are similar to those of the Shungura Formation and represent deposits of an ancestral Omo river and Lake Rudolf. The river bed and the lake shore (as evidenced subsequently in the Ileret and Koobi Fora areas of East Rudolf) probably shifted substantially through time. The basic geographical features of the lower Omo basin are thought to have been essentially as they are now, except that the Korath Range was probably absent, and there is the possibility that the adjacent highlands had a different elevation. Certainly there was a large river flowing over a broad plain from north to south, though its ancient course is not certainly known, and emptying into a lake. There is the possibility that the river's course was more direct and lacked the extensive easterly deviation past the Nkalabong highlands (F. H. Brown, pers. comm.). At any rate this was the general paleogeographic setting.

The principal characteristics of the members of the Shungura Formation are summarized in Table 2 (after de Heinzelin, Haesaerts and Howell, 1975). These include their thicknesses, number of units and subunits, number of major and minor tuffs, number of structured paleosoils, number of vertebrate fossiliferous horizons, and invertebrate fossil occurrences.

The Shungura Formation represents three episodes of lacustrine incursions into the basin during an otherwise protracted period of largely fluviatile (floodplain and delta plain) environments. Brief lacustrine depositional situations are recorded at the very base (lowest Basal Member)

TABLE 2
Principal characteristics of members of the Shungura Formation

SHUNGURA FORMATION MEMBERS	THICKNESS IN METERS	NUMBER OF UNITS	NUMBER OF SUBUNITS	MAJOR AND MINOR TUFFS	STRUCTURED PALEOSOILS	VERTEBRATE FOSSIL HORIZONS	MOLLUSCAN ASSOCIATIONS	ETHERIA BEDS	OSTRACOD BEDS
Mb. L	>48,60	9	15	6	5	4	4	1	4
Tuff L	0,40	1	1	1					
Mb. K	26,40	4	>5	6	2	1			
Tuff K	3,60	1	2	2					
Mb. J	43,00	7	12	10	2	5	2		
Tuff J	6,60	1	1	1					
Mb. H	48,00	7	7	5		7	3	1	
Tuff H	10,00	1	1	1					
Upp.Mb. G	97,60	16	20	4		27	2		3
Low.Mb. G	112,60	13	28	11	2	16		3	
Tuff G	6,00	1	2	1					
Mb. F	35,50	5	9	2		5			
Tuff F	7,20	1	4	1		3			
Mb. E	35,00	5	13	5	1	6		2	
Tuff E	2,00	1	2	1		1			
Mb. D	37,00	5	11	5	1	5			
Tuff D	4,00	1	2	2		1			
Mb. C	78,00	9	32	12	3	12	1	4	
Tuff C	1,80	1	3	2		1			
Mb. B	84,40	12	44	13	8	16	1	2	
Tuff B	13,00	1	4	4		2			
Mb. A	31,60	4	10	5	2	2			
Tuff A	3,20	1	1	1					
Basal Mb.	>32,00	5	10	3		5	1		
Total	>767,50	112	>239	104	26	119	14	13	7

and the top (uppermost Member L) of the sequence. Lagoonal estuarine prodeltaic followed by shallow, fully lacustrine depositional situations predominated during the accumulation of the middle third of Member G.

The detailed nature of paleoenvironmental situations associated with these principal depositional regimes has been elucidated through the recognition, mapping, and analysis of the components of fluvial sedimentary cycles in type sections and along strike, and through study of the varied expressions of land surfaces and attendant weathering horizons.

The members and units of the formation and their respective land surface criteria are summarized in Table 2. From this diagram the sequence can be divided into successive sedimentary periods, of which seventeen have been distinguished (de Heinzelin, Haesaerts and Howell, 1975). In this respect the main tuffs, which form the basis for the stratigraphic subdivision of the sequence are extraneous and incidental, as are, too, the numerous minor tuffs which are incorporated in almost every unit.

GEOMAGNETIC POLARITY MEASUREMENTS

Continental sediments and lavas, as well as marine sediments, are now well known to register the orientation of the earth's magnetic field at the time of their deposition. The unusually thick, essentially continuous record of deposition represented in the Omo Formations affords a unique opportunity to examine the sequence of polarity changes over a period of several million years. This record can then be compared with the well-defined epochs and events of the known Magnetic Polarity Time Scale.

The Shungura Formation affords a long, very complete, and also complex magnetostratigraphy for the Matuyama Epoch and the preceding Gauss Normal Epoch (Shuey, Brown and Croes 1974; Brown and Shuey, 1975). At present this is the most complete continental record of polarity changes known (Figure 3) and is comparable to that recorded in sediments from deep sea cores. The results are in large part concordant with the results of K/Ar age measurements (except those on Tuff B toward the base of the sequence). The sequence records essentially all of the Matuyama Reversed Epoch, including the Jaramillo, Gilsa/Olduvai, and three earlier normal events, and essentially all of the preceding Gauss Normal Epoch, with the Kaena and Mammoth Reversed Events.

DEPOSITIONAL ENVIRONMENTS AND TAPHOCOENOSES[2]

The sediments of Omo Group formations largely reflect lacustrine and fluviatile depositional environments. Lacustrine environments are less commonly represented and occur in the Mursi Formation, the Upper Member of the Nkalabong Formation, and in several Members of the Shungura Formation; namely the Basal Member, the middle portion of Member G, Upper Member L.

Thus, the lower Omo Valley was an embayment of Lake Rudolf intermittently at the time of deposition of the early strata of the Omo Group, that is, 3–4.5 million years ago, with subsequent short-lived transgressions again at about 1.85 m.y. ago and between ~1.6 and ~1.4 m.y. ago. It is not yet clear as to whether the episodes of lake expansion were due to tectonic subsidence, to the infilling of the trough, or to climatic fluctuation. It is probable that some combination of these factors

[2]A taphocoenosis is the set of fossil traces that becomes buried in a particular stratum. It invariably differs in qualitative and quantitative composition from the living biota (biocoenosis) of the area. Taphonomy is the study of the processes whereby selective parts of a living community of plants and animals comes to be represented by a fossil assemblage.

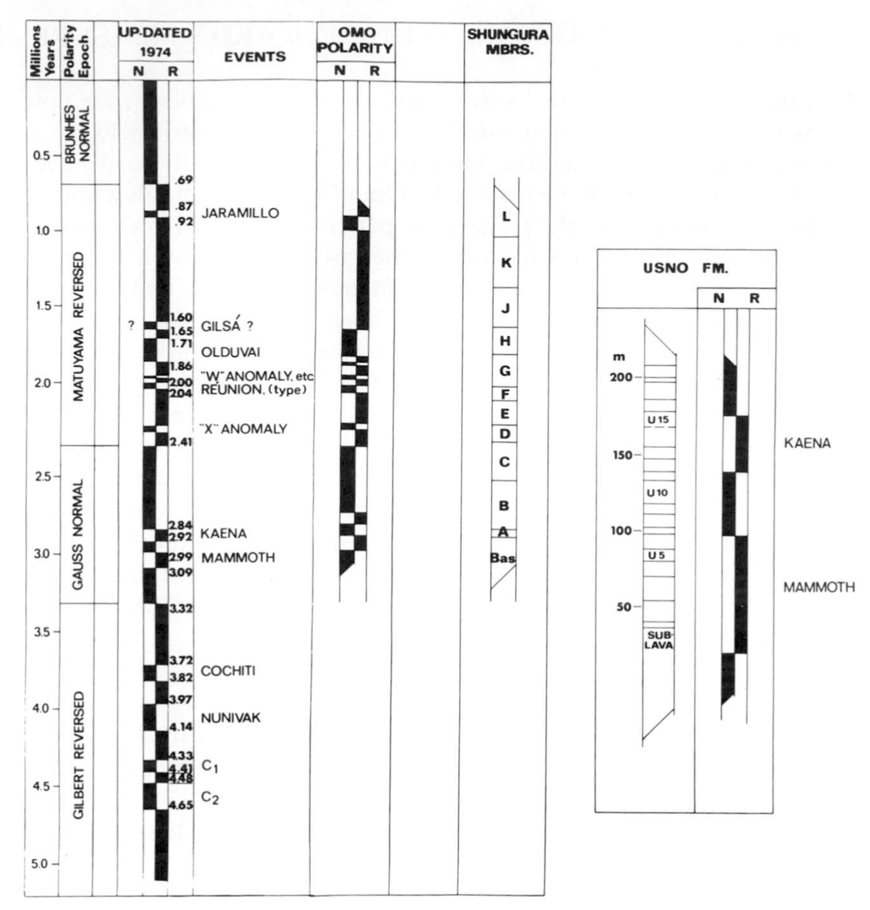

FIGURE 3
Magnetostratigraphy of the Shungura and Usno Formations.

was involved. (For more detailed treatment of the lacustrine units see Howell 1975, and de Heinzelin *et al.*, 1975.)

Hominid fossils have been recovered from lake margin deposits in Member G only. With these exceptions most of the Plio/Pleistocene sedimentation and related fossil occurrences in the lower Omo basin reflect alluvial landscapes. Fluviatile settings predominate in the lower portions (Members I, II) of the Nkalabong Formation (Butzer, 1975), in the Usno Formation, and essentially all of the Shungura Formation (through lower Member G). There are, however, important differences and variations within this general setting.

The Usno Formation comprises 19 sedimentary units above the extrusive basalt (unit U-1). After the brief episode of shallow lacustrine conditions (Unit U-3-1) the succession reveals a prolonged period of fluviatile and fluvio-deltaic sedimentation, including marsh and swamp situations comparable in many respects to that recorded in the lower three members of the Shungura Formation (Table 2). Significant vitric tuff accumulations occur at three horizons in the succession (Units U-6; U-10 and U-11); they are the counterparts of Shungura Formation Tuffs A and B. Subsidiary tuffs associated with other fine sediments occur higher in the succession (Units U-12, U-17 and U-20) and these, at least in part (like U-20), have their equivalents also in the Shungura Formation (eg., the B-10-1 tuff). The successive accumulations of coarse clastic products, which generally incorporate vertebrate fossils at four localities, all reflect to a greater or a lesser extent the proximity of piedmont alluvial fans, drained by the Usno and related streams, along the eastern margin of the basin and affording quantities of basement complex detritus.

The Gravel Sands, Brown Sands/White Sands, and Flat Sands deposits represent three successively younger concentrations of vertebrate fossils in the Usno Formation. Gravel Sands (in U-6 and -7), which equates with the upper part of the Basal Member, Shungura Formation, is not particularly fossilferous and only a dozen identifiable specimens have been collected there (mostly bovid, a few hippo, *Giraffa,* cercopithecoids, and *Protopterus*). Flat Sands, the youngest occurrence (in U-19 and -20) which equates with the upper part (probably about unit B-9) of Member B, Shungura Formation, is substantially more productive of vertebrates from fluviatile sands interbedded with floodplain silts and with clays. Over 50 identifiable specimens have been collected, about half bovid and hippo, some cercopithecoids and suids, and a few proboscideans, including deinothere and an hipparionine.

By comparison the Brown Sands and White Sands occurrences are extremely productive of vertebrates, yielding in excess of 1500 identifiable specimens (Howell, Fichter and Eck 1969). White Sands has yielded the largest assemblage. At these sites two subunits of Unit U-12, a light gravel and medium sands, and an underlying gravelly sand, are interbedded with clay silts, a laminated tuff, and stratified sands with silt and clay lenses. In some instances the finer sands and silts yield more complete specimens and even associated skeletal elements of individuals may occur (eg. bovids, suids, elephant, cercopithecoids). In the coarser sediments at both localities specimens are usually disarticulated, incomplete or fragmentary. Both localities have yielded a substantial diversity of vertebrate taxa including fish, aquatic and other reptiles, and nearly fifty species of mammals, including Hominidae.

Most members of the Shungura Formation represent deposits of a large meandering river system. Its sinuosity was undoubtedly substantial, perhaps even on the order of some meander belts of the present lower Omo River. However, the details of the ancient hydrographic system have still to be elucidated. In many respects the present lower Omo River can serve as an appropriate model for much of the Plio/Pleistocene situation in the lower Omo basin.

The fining-upwards sedimentary cycles (Allen 1965b, 1970) characteristic of the Shungura Formation merit consideration. The basal coarse sands are moderately sorted, frequently cross-bedded, composed predominantly of quartz, feldspar and fragments of chalcedony and chert. The measured attitudes of cross-beds generally indicate north to south flowing currents. Reefs of the river oyster *Etheria* suggest a stable bed with a persistent supply of fresh flowing water, as these molluscs require such a habitat. Such conditions only obtain in a fluvial situation along the river channel below the usual water mark. The coarse sands at the base of these units are thus thought to represent fairly well sorted, coarse sediments accumulated at or near the bottom of the channel and on the point bars of meander bends.

The overlying medium and fine sands are similar mineralogically, have similar bedding features, and are moderately sorted. They may contain silt interbeds, and they could represent the higher levels of point bar deposits formed by hydraulic sorting. They might also be the consequence of rises in lake level leading to coarse sand and pebble gravel deposition upstream and finer sands in their former place.

The finer silts and clays presumably represent levee and overbank deposits on the river floodplain. The clays are reduced in places and show rootcasts and calcareous concretions suggestive of situations with standing water in river edge flood basins.

Paleosoils are recognizable at the top of many clayey silt beds. Soil development is generally weak and the grade of oxidation moderate. Reduced clays, with rootcasts and calcareous concretions, have been interpreted as swamp soils. The porous tops of tuffs sometimes contain concretions and abundant rootcasts, even mammalian remains, and weak soil profiles are developed.

Most of the Shungura sediments are fluviatile or shallow-water deposits, and as they comprise more than 750 meters of section, they must have been deposited as the infilling of part of a steadily subsiding trough in which sedimentation kept pace with subsidence. In this situation the river could migrate back and forth across its floodplain and fail to completely remove deposits accumulated previously. Truncated cyclic units are known in many instances, and well-preserved channels are evident (Tuff F shows several such instances, with channels infilled with fine pebble gravel and coarse sands overlain by medium to fine sand, the distinctive tuff lithology making the channel relations particularly apparent.)

Extensive channelling occurred during two main intervals in this fluviatile succession. Three major episodes of cut-and-fill in upper Member B attended the accumulation of Unit B-10 which attains a cumulative thickness nearly double (about 40 meters) that of its actual thickness (20 meters). Subsequently, the sediments of Member D were largely accumulated in a series of substantial channels with dissections of several meters.

Such episodes of channeling could have been induced, among other things, by tectonic subsidence of remote portions of the main lake basin, by slight tectonic uplift of the lower Omo area, or by a climatically determined drop in lake level.

The meandering river situation was in part replaced by braided stream situations on at least two occassions. This occurred during the deposition of Member F (unit F-1) and early in Member G times (units G-3 to G-5). During these intervals channels were diverted and rejoined around intervening channel islands or braid bars. The riverine floodplain was less continuous and permanent than previously and channel banks were shallower and broader. There were doubtless attendant transformations in distribution and structure of the fringing riverine forest, in the vegetation of the floodplain(s), and in the nature and diversification of microhabitats and ecotones in those portions of the basin. Such transformations may be reflected in the palynological record and in the micromammal record as well.

A substantial diversity of depositional environments is represented within such a riverine system (Leopold, Wolman and Miller 1964; Allen 1965a)—channels with point bars enclosed between meander loops; floodplains and their abandoned and weathered terraces; levees with crevasses and crevasse splays; alluvial ridges with abandoned channels and meander loops, and associated ox-bow lakes, backswamps, marshes and mudflats; and flood basins with variably expressed dry, poorly vegetated, or more permanently watered situations with denser, more established vegetation patterns. Microstratigraphic studies at the unit and intraunit level in members of the Shungura Formation reveal the varied microsedimentary situations with which the abundant Omo vertebrate assemblages are associated. They afford the requisite data for taphonomical studies, both for surface occurrences and for a series of excavated situations (Johanson, Boaz and Splingaer, 1975).

Some of these varied fossil occurrences include (de Heinzelin, Haesaerts and Howell, 1975):

(a) those on or in paleosoils, incorporated in calcic concretions which have afforded protection from cracking and dissolution (with various land mammals, including cercopithecoids and hominids);

(b) those in porous tops of major tuffs (particularly D, E, F, and G) (with land mammals, including cercopithecoids and hominids);

(c) those on reduced horizons constituting ancient swamp soils (with land mammals, including cercopithecoids, hominids and artifact occurrences, as well as hippopotamus, aquatic reptiles and fish);

(d) those in transitional sand/silt situations attendant upon floodplain construction and evidently constituting accumulations in small ponds or other ephemeral water bodies (usually rich and varied land mammal assemblages, often exceptionally well-preserved *in situ,* but extremely friable and quickly dispersed and destroyed upon surface exposure);

(e) those derived by reworking from paleosoils and gentle washouts, including erosion of shallow embankments and levees (abundant concentrations of smaller skeletal parts, and including micromammals—particularly lowermost and upper Member B and in upper Member C); in other situations incipient floodplains have been alluviated and buried under sheets of fluviatile sands from divergent channels resulting in extensive overall dispersal of coarse elements (fossils) and small derived clastics (particularly the case for a number of artifactual occurrences in lower Member F);

(f) those in small pebble gravels where sorting has produced extensive vertebrate tooth and fragmentary bone concentrations (rich examples, with a great diversity of vertebrates, including some hominids, are known in upper Member B and lower Member G);

(g) those in erosional channels, in cross-bedded sands, and sometimes even in coarse gravels (with concentrated, mixed faunal assemblages, including some autochthonous aquatic reptiles and allochthonous skeletal parts of both small and large land vertebrates);

(h) those in tuff-choked erosional channels, as in the tuff facies D′, E′, F′, and G′ (extraordinarily dense, rich and diverse vertebrate assemblages, including usually teeth of Hominidae, associated usually with baked clay pebbles and pumices—two large excavations in lower F′ expose such situations).

VOLCANIC TUFFS

The repeated inflow into the lower Omo basin of vast quantities of volcanic ashes was surely of significance for the sedimentary regime of this ancient river system. Over a hundred intercalations of volcanic ashes have been identified in the Shungura Formation, suggesting a frequency of on the average of one tuff influx every 20,000 years or 1 tuff per 7.5 meters of sediment. These tuffs represent episodic, but repetitive and immediate, introductions of foreign pyroclastic products into the sedimentary regime throughout the lithologic sequence. Except for their lateral continuity there are no significant differences between the major tuffs (A to L) and the minor tuffs. The tuffs are composed almost exclusively of volcanic glass shards; the shards are fresh and angular, often with small glass bubbles or elongate vesicles. Some tuffs contain cinders or pumices, sometimes in remarkable quantity and of large size (up to 60–70 cm. diameter in the case of Tuff D), and they also show vesiculation (Brown 1972; Brown and Nash, 1975).

The source of these volcanic products is still uncertain, but a source in the middle reaches of the Omo drainage basin, possibly Mt. Damota (near Soddu, Sidamo province), now appears likely. It is some 275 kilometers distant, and composed of silica-rich and peralkaline rocks (Brown and Nash, 1975). Pantelleritic ignimbrites do occur extensively, covering an area of some 150 thousand square kilometers, over much of western and southern Ethiopia, including the upper Omo basin (Mohr 1968). These ignimbrites have an (inferred) late Tertiary age and hence would fall, at least in part, within the time span of the Omo Group Formation.

The difficulty of identifying the precise source could well be the consequence of the type of eruption. It could indicate either that there was no cone formation, that there was subsequent caldera collapse consequent upon emptying of the magma reservoir, or that the sources were fissures.

The agencies responsible for the deposition of the tuffs within the lower Omo basin are still inadequately understood. It is more than likely that several different depositional processes are responsible for their accumulation.

Some tuffs may be primary air fall tuffs, for example Butzer's tuff IIf in the Mursi Formation, the tuffs of the Upper Member (III) of the Nkalabong Formation, Tuffs A, H, J and K-β and various minor tuffs of the Shungura Formation (see de Heinzelin *et al.,* 1975; Howell 1975; and Butzer, 1975 for further information). Other tuffs are medium to coarse grained, have large current cross beds, climbing ripples, etc., and may include pebbles, cobbles, and even boulders of pumice. These tuffs have presumably been transported by the proto-Omo river from a fairly distant source and deposited rapidly on its floodplains and delta. The purity of the tuffs may simply be a consequence of the dilution of the normal sediment load by vast quantities of ash, or as a consequence of preferential transport and deposition of the glass owing to its lower specific gravity.

However, there is some evidence to suggest that in the case of certain major tuffs, particularly D, E, F, G, K-α and L, that emplacement occurred through agencies that were not merely fluviatile in nature (see Howell 1975).

Investigation of these volcanic products, and their source area, is obviously of importance not only because of their usefulness as stratigraphic marker horizons, their suitability for radiometric age determination, and their rapid emplacement and significance as isochronous surfaces. The catastrophic intensity and protracted activity of the volcanic source(s) produced a series of temporally delimited paleoenvironmental disturbances. These doubtless had their effects on the nature and distribution of habitats in and about the basin. The attendant effects on natural communities may well have been substantial and of considerable paleoecological importance, with implication for hominid adjustments over this protracted interval of time.

PALYNOLOGY AND VEGETATIONAL HISTORY

The Omo succession is exceptional[3] in sub-Saharan Africa in preserving both macrofloral and microfloral fossils essential to the reconstruction of Pliocene plant communities. There is every expectation that the vegetational history of the basin, and relationships with the adjacent highlands, can ultimately be determined for a time span of some 3 m.y. from silicified wood, seeds, and pollen. Such a vegetational history would be the first such sequence ever established in Africa, or the whole of the Old World tropics for that matter, and is unique as well in its association with extensive fossil vertebrate assemblages and early Hominidae.

Pollen samples processed from eight members of the Shungura Formation provide indications of the nature and diversity of plant associations and of their transformations over a period of about a million years (3–2 m.y. ago) (Bonnefille 1972; also 1970). They have been compared with the current local atmospheric pollen rain and the pollen content of riverine sediments, as well as with the composition and diversity of plant communities in the basin and the adjacent highlands. The principal associations are those of bush, wooded grassland, riverine and montane forests. The montane forest elements, currently restricted to higher elevations (above 2300–2500 meters), evidently result from long-distance pollen transport.

The available evidence indicates that montane forest was present in the adjacent highlands and that the lower Omo basin supported substantial wooded savanna as well as riverine forest communities during this million-year interval. All spectra contain relatively higher percentages of montane (allochthonous) elements than do present Omo sediments. Further details of the composition of montane forest marginal to the basin, and of fringing riverine forest within the basin, will accrue from macrofloral studies of silicified woods currently in progress (by Roger Dechamps, Tervuren).

There is evidence for changes in the nature, composition and extent of plant communities between the lower (2.4 to 3 m.y.) and the middle (2 m.y.) portions of the Shungura succession (Bonnefille, 1975). A more humid montane forest was replaced by associations of more drought-resistant trees. Overall, the representation of arboreal species show increased percentages of taxa with reduced water requirements, and there is also a marked overall decrease (by some 50%) of arboreal taxa. There is an accompanying increase in non-arboreal pollen, although the consistent representation of nearly 50% grasses indicates well-developed savanna

[3]Samples from certain horizons of early Pleistocene age at Olduvai Gorge now demonstrate that, with appropriate palynological techniques, microfloral assemblages will be recovered there as well (R. Bonnefille, pers. comm.).

conditions throughout the sequence. A change in the composition of grasses, especially those of Andropogonaeae, demonstrates, however, that the savanna came to assume a different structure and aspect. This is also confirmed by the presence of certain herbaceous species. A marked decrease in *Typha* (cattails) and Cyperaceae (e.g., *Papyrus*) also denotes reduction in the amount and distribution of local water sources and bodies in the course of this interval. These changes apparently reflect decreased rainfall.

These transformations in plant communities, their composition and distribution, are being investigated further, particularly with regard to earlier and later portions of the Omo succession. The results to date are in accord with the sedimentary and microfaunal evidence. They are profoundly important for the full-scale interpretation of paleoenvironments.

PALEONTOLOGY AND FAUNAL HISTORY

The Omo succession affords rich and diverse assemblages of invertebrate and vertebrate fossils which document the history of life in this portion of the Rift Valley System between ~ 4.5 and 1.0 m.y. ago. This essentially continuous record of deposition, fossiliferous throughout, for Pliocene and early Pleistocene times, is exceptional for eastern Africa. As a consequence of many radiometric age determinations and an extended paleomagnetic stratigraphy it now constitutes the basis of comparison for this time range for other known fossiliferous situations in eastern Africa.

Invertebrates

Three major invertebrate assemblages are known from the Pliocene and earlier Pleistocene time range of the lower Omo basin (Gautier, 1975; Van Damme and Gautier 1972) which show distinct differences in composition, in extinction of species, in the appearance of new species, in adaptations to particular biotopes, and in biogeographic affinities. The assemblage from the later Shungura Formation (upper Member G and Member H) is distinctive in the number of endemic forms, the new immigrants, the replacement of older species, and its overall endemic character. An assemblage from the latest Pleistocene is also known and affords important evidence for the origin and evolution of invertebrate life in recent East African lakes, and particularly the Rudolf basin.

Fish and Reptiles

Fish and chelonian fossils are abundant in appropriate deposits. Detailed taxonomic studies are still in progress, but it can be anticipated that when

TABLE 3

Representation of mammalian species (exclusive of Chiroptera, Insectivora, Rodentia, Lagomorpha, Hyracoidea, and Tubulidenta) in the Mursi and Usno Formations and in members of the Shungura Formation.

	MURSI	USNO	BASAL	SHUNGURA FORMATION								
TAXA	FM.	FM.	MB.	A	B	C	D	E	F	G	H	J-K-L
Proboscidea	3	4	?	3	4	3	3	2	2	2	2	3
Rhinocerotidae	2	2	?	?	2	2	2	1	2	2		
Equidae	1	1	?	1	1	1	1	1	2	4	3	2
Chalicotheriidae			?			1	1					
Hippopotamidae	1	2	?	1	1	3	2	1	2	3	3	3-4
Suidae	2	4-5	?	4	6	6	3+	3	5	6	4	4+
Giraffidae		4	?	1	4	4	3		3	4	1	3
Camelidae			?		1		1		1	1		
Bovidae			?									
Tragelaphini	1	1			2	3	3	3	2+	3+	1	1
Reduncini	1				4	6	3+	5	3	5+		3+
Hippotragini						1+		1		?		
Caprini							1		1+			
Neotragini								1	1	1		
Alcelaphini		1+			1+	1+	1	1	1	1+		2
Aepycerotini	1	1			1	1	1	1	1	1	1	1
Bovini		1+			1+	2+	1	1+	1+	3+	1	1
Antilopini		1+			1+	1	2		2+	2+	1	1+
Carnivora			?									
Hyaenidae		4			1	2		1	2	2		1
Felidae	?	5-6		1	5	6	2	2	6+	6	1	1
Mustelidae		2			1	1		2	2			
Viverridae		1			2	5	1	3	1	3		
Primates			?									
Lorisidae					2-3							
Colobinae		1		1	2	3	2	2	2	2	1	2
Cercopithecinae		5		2	5	4	2	2	4	5	3	4
Total	12	45-47	?	14	47-48	56	35	33	48	56	22	32-33

these have been completed they will contribute greatly to our understanding of past changes in hydrographic systems, climate and biogeography. Nine species of fish and three genera of chelonians have been described by Arambourg (1947).

Aquatic crocodilians are abundant and ubiquitous throughout the Omo Group Formations. They are of interest since they show affinities in part with other hydrographic systems within Africa, still broader temporal and biogeographic affinities outside Africa, as well as autochthonous distributions within this portion of the Rift System. At least four taxa are represented (Tchernov, 1975). The narrow-snouted *Euthecodon brumpti,* a tomistomid derivative, occurs throughout the sequence, perhaps as more than one form; related species are known from older (Lothagam) as well as contemporaneous (East Rudolf; Kanapoi) fossiliferous localities around the Rudolf basin. The nilotic crocodilian *(Crocodilus niloticus)* is represented also throughout the sequence, and is of course present at most late Neogene and Pleistocene localities in eastern Africa. The slender-snouted crocodile, *C. cataphractus,* restricted to the Congo basin now, is recorded from the Omo (cf. Arambourg 1947), from an unconfirmed provenience (probably later rather than earlier in the succession), and is also recorded at East Rudolf; it is unknown elsewhere. A new species of broad, short-snouted crocodile is also represented, both early (Members B and C) and late (Member K) in the Omo succession; it is also recorded at East Rudolf and in lower Olduvai.

Mammals

Fossils of this class are extremely abundant, often well preserved, and represented by an extraordinary diversity of taxa in Omo Group formations. Some 50,000 identifiable fossil specimens have been collected, from surface survey and from controlled excavations, and catalogued by the recent expedition.

Eleven orders and 33 families (of 50 living families) are documented in the mammalian fossil record of the Omo succession. Ninety-nine genera and over 140 species, exclusive of Hominidae, have been identified thus far (cf. Coppens and Howell 1974, and Howell and Coppens 1974). The representation of species in the Mursi and Usno Formations, and in the successive members of the Shungura Formation, is set out in Table 3.

Small mammals (such as rodents, insectivores, bats, lemurs, galagos, etc.) are known from various members of the Shungura Formation. Two species of Muridae (mice, etc.) are known in Members C and G, Hystricidae (porcupines) are known in the Usno Formation and Members C, E, and H, J, and L, and Hyracoidea *(Gigantohyrax,* a Makapan Limeworks taxon) is known in Member C. However, only two varied assemblages have been recovered from washing and sorting operations at two hominid-

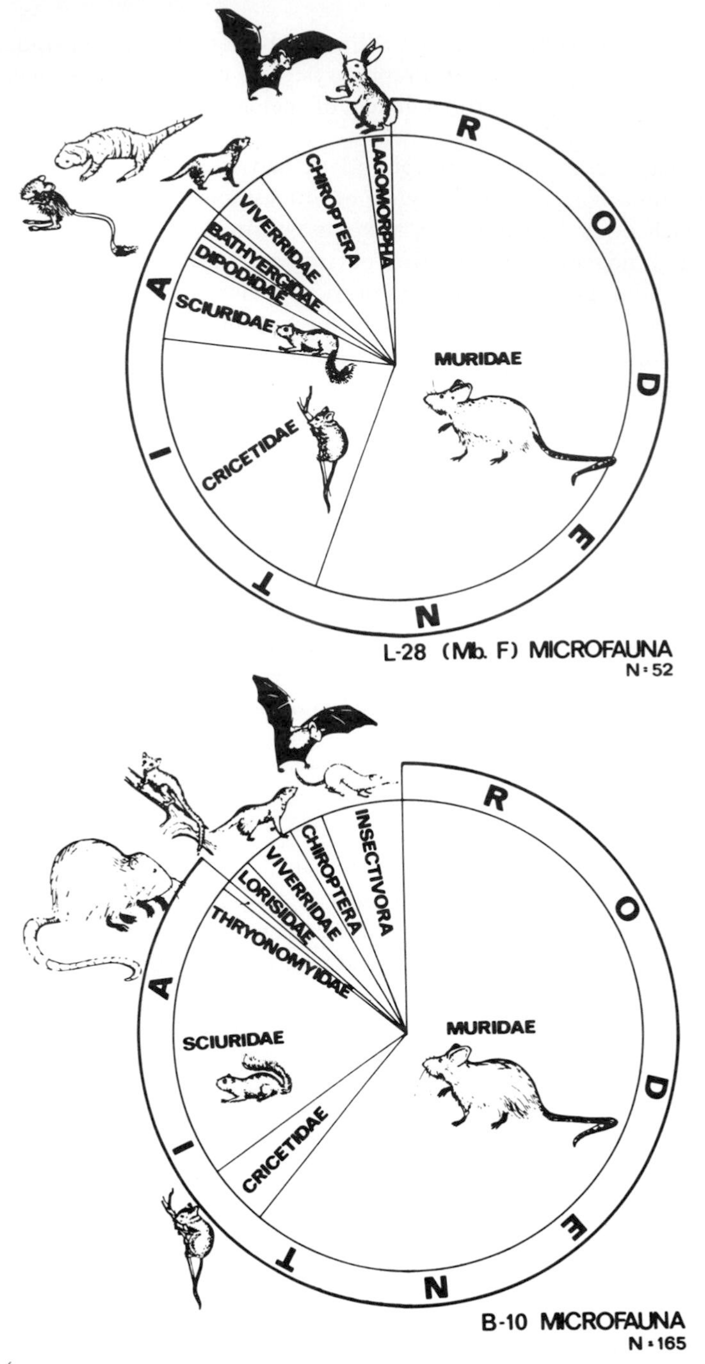

FIGURE 4
Micro-mammalian assemblages from upper Member B and lower Member F, Shungura Formation.

bearing localities, and include Chiroptera (bats), Insectivora, Rodentia, Prosimii, and small Carnivora (Viverridae). These assemblages (Figure 4) from upper Member B and lower Member F show interesting and significant differences in composition (Jaeger and Wesselman 1975). That from Member B is the most substantial and overall the most diverse, having 6 species of Muridae (3 in F), 1 species of Cricetidae (gerbils) (3 in F), 2 species of Sciuridae (2 in F), 1 species of Thryonomyidae (1 in F), 1 species of Hystricidae (2 in F), 3 species of Chiroptera (0 in F), 3 species of Insectivora (0 in F), 2–3 species of Lorisidae (0 in F), 1 species of Hyracoidea (0 in F), and 1 species of Viverridae (1 in F). Noteworthy are differences in the frequency and differing species representation of Muridae (particularly those characteristic of "closed" habitats), the markedly increased frequency of Gerbillinae, and the appearance of Dipodidae and of Bathyergidae, all indicative of substantially more "open" habitat situations in Member F times. The older assemblage is indicative of wooded savanna conditions including, at least locally, well-developed forest. This is confirmed by sedimentary and palynological evidence. Ultimately the Omo succession will afford a microvertebrate biostratigraphy of primary importance for biogeographic and biostratigraphic comparisons with other African hominid-bearing localities as well as with sequences in northern Africa and in Eurasia. The rodents already recognized show resemblances both to species known from early Olduvai (Bed I) as well as to some from australopithecine sites in southern Africa.

Some aspects of species diversity, extinctions, phyletic evolution, and immigration among other mammalian groups are worthy of brief mention.

PROBOSCIDEA (Elephants and relatives). Five proboscideans are represented early in the succession (Beden, 1975). The gomphothere *Anancus* persists only into the Mursi Formation. *Stegodon* persists into upper Mb. B. The early loxodont, *L. adaurora,* persists into Mb. C (but is apparently absent from the Usno Formation, perhaps for ecological reasons). *Loxodonta,* represented by another species, is present in Mb. D. The extant species of *Loxodonta, L. africana,* apparently appears about Mb. J times. *Deinotherium* occurs throughout the Omo succession and seemingly becomes extinct by Mb. H times (as is the case in Olduvai, upper Bed II). The extinct *Elephas recki* appears at the base of the sequence, persists throughout, and has a complex phyletic evolution through four stages. It is a species of particular significance in biozonation of the African Pliocene and Pleistocene (Maglio 1973;1970).

RHINOCEROTIDAE (Rhinoceroses). An extinct, primitive species of white rhino, *ceratotherium praecox,* is present only in the Mursi Fm. The extant white rhino *(C. simum)* and black rhino *(D. Diceros bicornis)* occur throughout the sequence, from the Usno Fm. through Mb. G at least, their presence subsequently being uncertain (either due to sampling and/or ecological factors). (Hooijer, 1975; Guerin, 1975).

CHALICOTHERIIDAE (without living counterparts). This extinct perissodactyl is everywhere rare, but is documented in Mbs. D and G of the Shungura Fm. (Hooijer, 1975; Guerin, 1975).

EQUIDAE (Horses, zebras and relatives). Two genera are represented, *Hipparion* and *Equus* (Eisenmann, 1975; Hooijer, 1975). The diversity and evolution of *Hipparion* is complex, with four species probably represented. The primitive *H. turkanense,* first known at Lothagam, is known only from the Mursi Fm. In the Usno Fm. and the lowermost members (A and B, perhaps C), another large and perhaps descendant species, *H.* cf. *albertense,* is represented. The characteristic gracile, hyposodont species *H. ethiopicum* is present thereafter, from Mb. F through Mb. L. Another, dwarf form, referred to *H.* cf. *sitifense,* seems to be present in Mbs. F and G, and is the characteristic species in this time range in northern Africa.

Equus appears suddenly in Mb. G as a new immigrant. Three species are represented. A large species, referred to *E. oldowayensis,* is most common, and occurs from Mbs. G through L. Another very large, new species is also present, but ill-known in lower Mb. G; it is also recorded in lower and upper Koobi Fora, East Rudolf. Another small species, the size of extant *E. burchelli granti,* apparently is also present.

HIPPOPOTAMIDAE. This family reveals an unexpected diversity in this time range. Five, perhaps six, species are represented in Omo Group Formations (Coryndon and Coppens 1973; Coryndon, 1975). A large hexaprotodont species, *H.* sp. nov. "D," occurs in the earliest part of the succession (through Mb. B), the same species first well-documented at Lothagam and Kanapoi. This species ultimately becomes the tetraprotodont, slender-limbed species, *H. protamphibius,* which is so characteristic and unique to the Omo succession and which persists into Mb. H. A small hexaprotodont species, *H.* sp. nov. "A," perhaps related to the former, appears sporadically from Mbs. C through G. The remaining species are tetraprotodont. A pygmy tetraprotodont *(H. aethiopicus),* which also occurs in upper East Rudolf and in lower Olduvai, occurs sporadically in Mbs. C, G, H, and finally in K and L. The large tetraprotodont, *H. gorgops,* so characteristic of Olduvai (and now also known at East Rudolf), appears first in Mb. G and is present thereafter.

The extant species, *H. amphibius,* the African evolutionary history of which is extremely poorly known, may also be present in the uppermost several members of the Shungura Fm. The distinctive diprotodont hippo of East Rudolf, *H.* sp. nov. "C," if represented at all in the Omo succession, appears only at the end of the Shungura Fm. (perhaps Mb. K). It is difficult to conceive of this unusual diversity as other than a series of complex, still ill-appreciated ecological adjustments.

SUIDAE (Pigs, warthogs, etc.). This family epitomizes diversity, extinctions, and phyletic evolution through this 3 m.y. time range. Six genera and some 12 species are documented (Cooke 1975, 1976). Two species of *Nyanzachoerus, N. pattersoni* and *N. jaegeri* (ex-*N. plicatus*) are variably represented in the Mursi and Usno Formations and the lower members of the

Shungura Fm.; both were first documented at Kanapoi. Three or four species of the phacochoerine genus *Notochoerus* are known. The most primitive species *N. capensis* is restricted to the Usno Fm. (and is also known at Kanapoi and Kubi Algi, E. Rudolf). *N. euilus* is present in the Usno Fm. and the lower members (A through C) of the Shungura Fm.; a related, presumably derivative species seems to occur through Mb. G. The other, hypsodont species, *N. "scotti,"* occurs from Mbs. B through H and demonstrates an interesting phyletic evolution of the molar dentition. The suine genus *Mesochoerus* is represented by *M. limnetes* in the Usno Fm. and Mbs. A through G of the Shungura Fm., after which it evolves into *M. olduvaiensis,* the characteristic species at Olduvai.

The aberrant phacochoerine genus *Metridiochoerus* is represented by the small species *M. jacksoni* which occurs regularly and essentially unchanged through the Shungura Fm. from Mbs. B through K (and has a long evolutionary history at Olduvai as well). The related genus, *Stylochoerus,* is represented by *S. nicoli* (ex-*Afrochoerus*), a species with peculiar elephantine-like canines, is represented in the two uppermost members of the Shungura Fm. The extant genus *Phacochoerus* is represented by two species, *P. antiquus,* present in Mbs. G and H, and probably the recently extinct species *P. aethiopicus,* which seemingly appears from Mb. G upwards.

GIRAFFIDAE. *Sivatherium* and three species of *Giraffa* have a continuous documentation from the Usno Fm. and the lower members (B and C) of the Shungura Fm. through Mb. G and, apparently, also into the uppermost members of that formation. All species are documented at the same horizon in several instances. The giraffine species include the distinctive slender-limbed *G. gracilis,* the large species *G. jumae,* and a new diminutive, okapi-sized species (also now known at a number of localities in eastern Africa).

CAMELIDAE. An extinct species of *Camelus,* perhaps with affinities to the extinct *C. thomasi* of North Africa, is documented in four members of the Shungura Fm., last in Mb. G (Grattard, Howell and Coppens, 1975). It is another and unexpectedly early immigrant into eastern Africa, and has a subsequent documentation at Olduvai Gorge (upper Bed II) as well as at Marsabit (Gentry and Gentry 1969).

BOVIDAE. Nine tribes are represented in the Omo formations (Gentry 1975, 1976). Of the smallest species only Neotragini are known, and only by fragmentary remains in the middle members of the Shungura Formation. Hippotragini are unknown except for a miniscule record in Mb. G. Antilopini (gazelles and relatives) are always rare and their diversity is surely poorly recorded. The extinct springbuck, *Antidorcas recki,* occurs repeatedly from early (Mb. B) to late (Mb. K). The Usno Fm. and Mbs. D and F thru K all yield few indeterminate antilopines, with *Gazella* sp. definitely recorded in D, F and H. The extinct *G. praethomsoni* is certainly known only in Mb. B. The occurrence of an extinct blackbuck species, *Antilope subtorta,* affords an interesting link with southern Asia (Pinjor zone, Upper Siwaliks). Caprini are still unknown. Oviborini occur in C, D, and G.

The remaining five tribes are the most common bovids, although quite unequally represented. Alcelaphini (gnu, hartebeeste, topi, etc.) are always infrequent and usually have the poorest representation, although they occur in the Usno Fm. and throughout the Shungura Fm. (increasing in frequency in Mbs. D, E, and F). The extinct hartebeest, *Megalotragus,* occurs from Mb. G onwards (but may also have an earlier history in the Omo). Another wildebeest, whether *Connochaetes* or the extinct *Oreonagor* is uncertain, is documented in Mb. B. The extinct *Parmularius altidens* occurs late in the Shungura Fm., and a related form may occur as early as Mb. C. An extinct species of herola, *Beatragus antiquus,* is definitely documented in Mb. H. Except in Mbs. B and C bovines are always in very low frequency in the Shungura as well as the Usno Fms. An ancient, short-horned species of the buffalo *Syncerus* is in Mbs. B, C, and G. The extinct genus *Pelorovis* is documented as early as Mb. C and again in Mb. F, but is better known in Mb. G, as a species ancestral to, or identical with *P. oldowayensis* (of Bed II Olduvai). An extinct genus of Asiatic waterbuffalo, *Hemibos,* first known from the Upper Siwaliks (Pinjor zone), is present in Mb. G, and perhaps E, and affords another indication of extra-African faunal affinities.

Tragelaphini (kudus, bushbucks, etc.) and Reduncini (reedbuck, waterbuck and relatives) are most abundant and varied and with *Aepyceros* (impala) form the bulk of the bovid fauna. *Aepyceros* is ubiquitous and represents from 30 to nearly 50% of the Usno and Shungura Fm. bovid fauna. An early form is also known from the Mursi Fm. Four or five species of Tragelaphini are represented. The extant greater kudu *(T. strepsiceros)* appears first in Mb. G. The common kudu in the middle members (C through G) is the extinct *T. gaudryi.* Another, indeterminate species is also present in Mbs. C, D, and E. A primitive bushbuck, *T. pricei,* is present in Mb. C. The most common and persistant species is *T. nakuae,* first recognized in the Omo by Arambourg (1947), which occurs in the Usno Fm. and in the Shungura Fm. through Mb. H, in strongly decreasing frequency after Mb. E. This species preserves markedly boselaphine features and is strongly reminiscent of *Selenoportax vexillarius* from the Middle Siwaliks.

Reduncines are more diverse than tragelaphines, and customarily show reversed frequencies with the former group. Reduncines are essentially unknown from the Mursi Fm. and are undocumented in the Usno Fm. At least seven taxa are recorded in the Shungura Fm., with particularly high frequencies in Mbs. B and G. Three taxa are particularly characteristic of the Omo succession—*Menelikia lyrocera* (recorded as late as Mb. K), *Kobus sigmoidalis* (recorded as late as Mb. J), and *Kobus ancystrocera* (recorded as late as Mb. J). The former two species in particular show an interesting phyletic evolution through the Shungura Fm. in respect to cranial and horn core morphology. *K. sigmoidalis* is considered to be ancestral to both the waterbuck, *K. ellipsiprymnus* (which appears in Mb. G) and the lechwe, *K. leche. Kobus kob* may be present in Mb. G, is known subsequently from the uppermost members of the Shungura Fm, and is represented by an ancestral form in Mb. B. In Mbs. B and C another *Kobus* sp. is now known. A species of *Redunca* is also present in several members of the Shungura Fm.

CARNIVORA. Carnivores are never common fossils and are usually, and not unexpectedly fragmentary and incomplete. Larger taxa are more adequately represented than smaller taxa, particularly viverrids and the lesser felids. Over 25 carnivores are known thus far from the Usno and Shungura Formations, and include 9 viverrids (mongeese, civets and genets), at least 3 mustelids (otters, ratels and zorillas), 5 hyaenids, 2 machairodontines (sabre-tooth forms), and more than 6 felines (Howell and Petter 1975). There is still no record of Canidae from Omo Group formations, surely a sampling and/or ecological bias. The stratigraphic distribution of these carnivores, and their presence/absence, must in part reflect this latter bias, but in some instances repetitive occurrences and first appearances must have significances which supersede these biases.

Hyaenids are next to felids the most common carnivores in the Omo succession. Four genera and five species appear to be represented. *Percrocuta* appears to be represented in the Usno Fm., and *Euryboas* (Mb. F) in the Shungura Fm. A small hyaena comparable to *H. hyaena makapani* is apparently represented in the Usno Fm. and Mbs. B and C, Shungura Fm. A larger *Hyaena hyaena* occurs in the Usno Fm. and throughout (Mbs. B through K) the Shungura Fm. *Crocuta* appears first in Mb. G, Shungura Fm., in accordance with its belated appearance in stratigraphically and/or radiometrically well-documented presence elsewhere throughout Africa.

Felidae are the most common carnivores. The machairodontines *Megantereon* and *Homotherium* are associated in the Usno Fm. and in the Shungura Fm. from Mbs. B through G. The specific attributions of their remains is still uncertain, the remains being fragmentary. The "false sabretooth" feline *Dinofelis* is persistently present in the Usno Fm. and through Shungura Fm. Mbs. A through G. *P. Leo* is in Mb. L.

PRIMATES EXCLUSIVE OF HOMINIDAE. Prosimians, at least two species of galagines, are recorded only in the Mb. B microfaunal assemblages.

Cercopithecoidea are common and very diversified in both the Usno and the Shungura Formations (Eck, 1975). Three and probably four taxa of Colobinae (colobus monkeys) are represented, but they represent less than 10% of the total cercopithecoid fauna. A small *Colobus* sp. is known from Mb. K, and presumably another, about the size of *C. polykomos,* is known from Mb. L. Colobinae sp. indet. A, rather larger than *C. polykomos,* is known from Mbs. B and C, but its affinities are unclear; they may be with *Cercopithecoides williamsi,* a South African taxon. The most abundant species is Colobinae gen. et. sp. nov. which occurs in the Usno Fm. and throughout the Shungura Fm. It is a large form, comparable in size to *Paracolobus* and *Cercopithecoides molletti,* but is distinctive in its cranial and dental morphology. Finally Colobinae sp. indet. C is present in Mbs. C through G and is an exceptionally large form, comparable in size to an anubis baboon. Both sp. indet. A and C may ultimately prove to be unique to the Omo and otherwise unrepresented in the fossil record.

Representatives of the Cercopithecinae (monkeys) are much more common and at least four taxa can be distinguished. A small *Cercopithecus* is represented in the Usno Fm. and in the lower (Mb. B) and the upper (Mbs. G and J) members of the Shungura Fm. (Eck and Howell 1972). Two small Papionini (baboons) are known, but their affinities are unclear—gen. et sp. indet. A is restricted to the Usno and the lower (Mbs. B, C) Shungura Fm., is smaller than *Parapapio jonesi* and shows *Cercocebus* resemblances; gen. et sp. indet. B, which occurs in the Usno Fm. and Mbs. B through J of the Shungura Fm., is similar in both size and morphology to *P. jonesi* or *P. broomi*. *Papio* occurs in the Usno Fm. and throughout the Shungura Fm. (at least to Mb. K) and although broadly similar to modern representatives of the genus may prove to be distinctive in facial morphology.

Theropithecus is by far the most abundant cercopithecine, about 90% of all specimens recovered (and over 80% of all the Cercopithecoidea). It is represented in the Usno Fm. and throughout the Shungura Fm. (to Mb. K). The characteristic and dominant form (through Mb. G) is *T. brumpti,* restricted to the Omo except for one other occurrence from Kubi Algi, East Rudolf. This species is theropithecine in overall dental, mandibular and cranial morphology, but has a most distinctive facial morphology unknown in any other cercopithecoid primate. The more widespread species *T. oswaldi* was probably also present, however, and perhaps made its appearance by (at least) Mb. F times and subsequently may have become the dominant species. Of this genus only the gelada baboon now survives *(T. gelada).*

This account of the fossils is a slightly abridged version of Howell 1975. For further information see that paper and the various paleontological contributions to the Lake Rudolf basin symposium volume (Coppens *et al.* 1975).

Measures of change in the taxonomic composition of the fauna

The extraordinary fossil record preserved in the Omo Group Formations affords a basis for assessing overall resemblances between successive faunal assemblages. Various methods of comparison may be employed, and various approaches which utilize the aggregate body of stratigraphic, radiometric and paleomagnetic evidence will ultimately be appropriate. A useful and expedient measure is Simpson's (1960) index of faunal resemblance which measures the number of taxa in common as a function of the smaller of any two assemblages. Figure 5 shows the values obtained in such comparisons for the larger elements in the mammalian fauna plotted against differences in age between the samples compared. The relationship between the similarity index and age difference is clearly demonstrated, regardless of whatever other factors might be operative in effecting faunal resemblance. The Omo succession thus affords a useful comparative baseline for evaluating other faunal assemblages for which stratigraphic and age control may be less clearly defined.

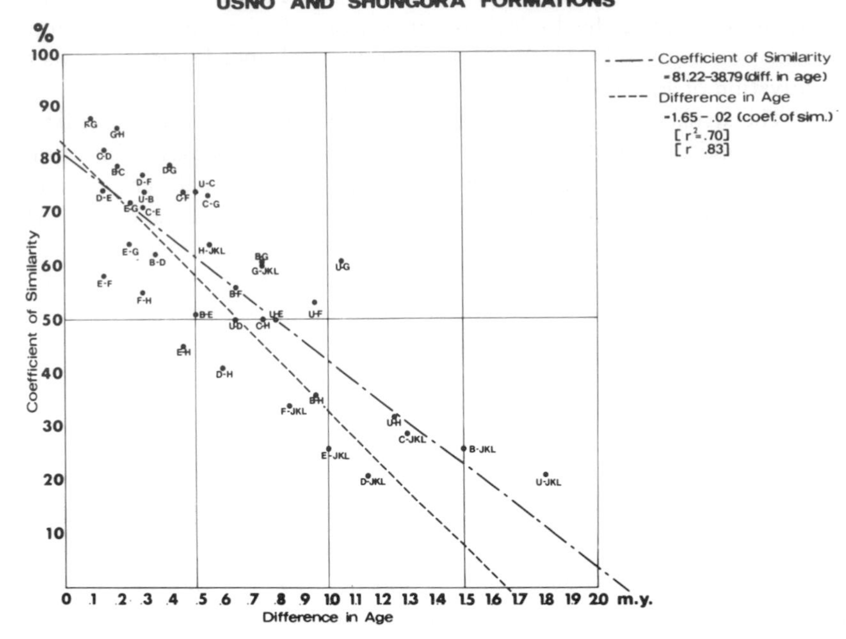

FIGURE 5
Indices of faunal resemblance between mammal assemblages from the Usno Formation and members of the Shungura Formation.

HOMINIDAE

Cultural Evidence

Occurrences of hominid artifactual materials have been surely demonstrated in three members of the Shungura Formation. They are suspected, from survey, to occur in other members, but occurrences suitable for excavation have still to be found and tested, so they remain unconfirmed. The oldest occurrence is in Upper Mb. C.

The youngest occurrences are in fossiliferous gravels of channels in lower Member G (units G4/5). They are situated above Tuff G dated 1.93 m.y. in sediments with normal remanent magnetism which must represent a Reunion Normal Event. Several out-crops were tested and yielded fresh and abraded small quartz artifacts. Concentrations in association with fine sediments suitable for excavation have still to be discovered.

Numerous artifact occurrences are now known in a number of areas of exposures of Member F (Merrick *et al.* 1973; Merrick and Merrick 1975;

Merrick 1976; Chavaillon 1970, 1971, 1975; Coppens, Chavaillon and Beden 1973). These situations are above Tuff F, dated at 2.04 m.y. A full account of archeological investigation of sites in the Shungura Formation is given by H. V. Merrick in this volume. Figure 6 shows the excavation of an archeological locality of Member F.

The artifact occurrences are certainly relevant to an understanding of the cultural capabilities and technological practices of early Hominidae, their dietary habits, and the nature and variety of paleoenvironments which they exploited. The seemingly "sudden" appearance of lithic artifacts in the Omo succession, long after the first documentation of the presence of Hominidae, is unexpected and puzzling. It still requires explanation in respect to paleoenvironmental factors as well as the paleobiology and adaptive adjustments of the hominids of that time.

Skeletal Remains

Two localities in the Usno Formation and 85 in the Shungura Formation have afforded skeletal parts of Hominidae (Howell, Coppens and de Heinzelin, 1974; Howell and Coppens, 1975). In the Shungura Formation, nine of twelve members have yielded Hominid fossils. On the

FIGURE 6
Excavation of archaeological locality FtJi-2, Member F, Shungura Formation.

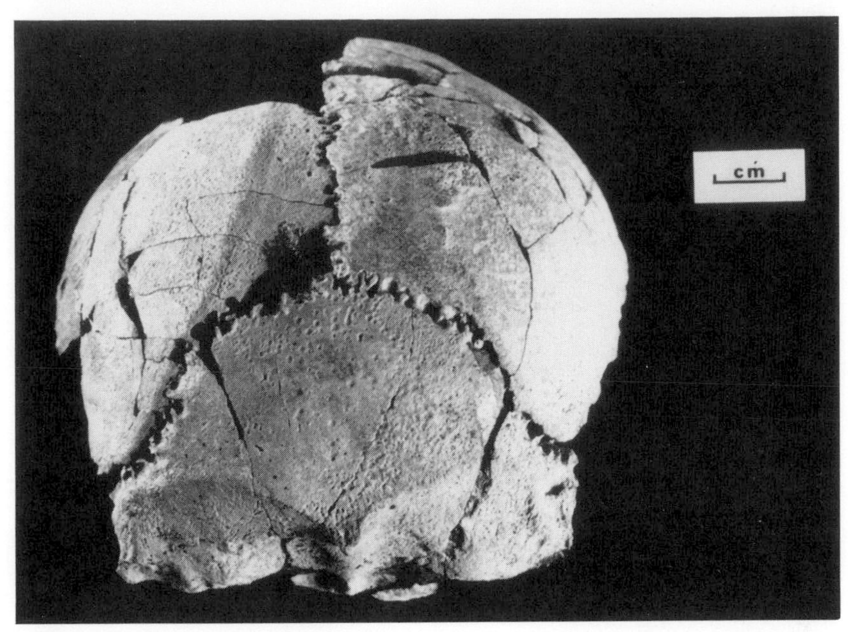

FIGURE 7
Posterior view of a partial hominid cranium referred to *Australopithecus* species, from Locality 338, Member E, Shungura Formation.

basis of conventional potassium-argon age determinations and paleomagnetic measurements these fall within a time span of 2.9 to 1 m.y. ago. The provenience and nature of the specimens are set out in Table 4. These remains are frequently fragmentary, and most often consist of teeth. However, they are informative in regard to hominid diversity and taxonomy in this important time range.

An overview of Omo Hominidae is given elsewhere by Howell and Coppens (1975). Some of the material has been discussed, and illustrated, by Arambourg and Coppens (1967, 1968), Coppens (1970a,b, 1971, 1973a,b), Howell (1969a,b), Howell and Coppens (1973), and Howell and Wood (1974). Four hominid taxa appear to be represented in this time range in the lower Omo basin:

(a) Remains from the Usno Formation and from Members B, C, D, E, and F and the lower units of G of the Shungura Formation are attributed to *Australopithecus* aff. *africanus*. The oldest specimens are generally small, with simple dental morphology, and might ultimately prove with additional more complete material to represent a distinctive, though related lower taxonomic category.

(b) Some specimens from units of Members E, F and G, and perhaps from L, of the Shungura Formation are attributed to a robust australopithecine, *Australopithecus* aff. *boisei*.

TABLE 4
Inventory by geological formation (and member) of skeletal parts of Hominidae, Omo succession.

	LOCALITIES	CRANIA	MANDIBLES	TEETH	POSTCRANIALS
Shungura Formation:					
Mb. L	2			2	
Mb. K	2	1i		1	
Mb. H	1			1	
Mb. G	23	1i;f	5	52	2
Mb. F	13		1	44	
Mb. E	11	1i	2	16	1
Mb. D	10			12	1
Mb. C	18	f	1	39	2
Mb. B	4			13	1
Mb. A	1			1	
Usno Formation:	2			21	
Totals:	87	3i;ff	9	202	7
Total Specimens: 221					

(i = incomplete; f = fragment)

(c) A very few teeth from localities in Members G (lower), H, and a partial cranium with upper teeth from Mb. G-28 are remarkably similar to those attributed to *Homo habilis* (Hominids 7 and 13) from Olduvai Gorge.

(d) From (uppermost) Member K are cranial fragments with features diagnostic of *Homo erectus*.

Thus, the Omo succession appears to reveal the presence of *A. africanus,* or an allied diminutive species between ~3.0 and 2.5 m.y. ago; the persistence of this, or a derivative species until ~1.9 m.y. ago; the presence of a robust australopithecine, *A.* aff. *boisei* between ~2.1 and nearly 1.0 m.y. ago; the appearance, by ~1.85 m.y. ago, of a hominid dentally like specimens assigned to *Homo habilis;* and, by ~1.1 m.y. ago (at least) *Homo erectus* is present.

CONCLUSIONS

In recent years very substantial progress has been made toward understanding various aspects of the Pliocene and the Pleistocene of the lower Omo basin in southern Ethiopia. The study of the late Pleistocene and recent sedimentation and depositional environments, as well as the natural communities in the basin, afford important data for the evaluation and interpretation of Pliocene and earlier Pleistocene conditions. The principal late Cenozoic geologic formations are now mapped, their stratigraphy measured in detail, and their depositional environments ascertained.

Conventional potassium-argon dating has afforded an internally consistent and sound chronological framework. This framework has been confirmed and extended through rigorous and intensive paleomagnetic sampling, resulting in one of the most complete magnetostratigraphic records ever obtained in continental sediments. A major gap in current knowledge is the absence of geophysical data for this area as neither gravity nor seismic surveys have been undertaken.

The history of life preserved in the Pliocene/Pleistocene sediments is rich and varied. Molluscan assemblages have afforded an important biostratigraphy as well as useful indications of paleoenvironments. The prolonged and diverse vertebrate fossil record, in stratigraphic contexts the ages of which are well established radiometrically and magnetostratigraphically, affords not only an exceptional biostratigraphy, but an unusual opportunity to investigate important, unresolved questions of phyletic evolution, species diversity, extinctions, and faunal exchange. These and other matters of paleobiological interest can also be related, through microstratigraphic, pedological, palynological and microvertebrate evidence, to other aspects of paleoecology in the basin. And the occurrences of Hominidae now affords another opportunity for intensive and comparative investigations into the earlier biological evolution and diversification of Hominidae and the development of capabilities for culturally patterned behavior.

Acknowledgments

The work of the many participants of the Omo Research Expedition has only been realized through the authorization and encouragement of the Imperial Ethiopian Government and, in particular, its Antiquities Administration. The cooperation of the Kenya government has enabled the expedition to function effectively across national frontiers. To both, and to the many colleagues participating in the expedition, the author is deeply grateful.

For Louis Leakey—colleague, an inspiration, and closest friend. This is a measure of what we have sought to do.

References Cited

Allen, J. R. L.

1965a A review of the origin and characteristics of alluvial sediments. Sedimentology 5:89–191.

1965b Fining-upwards cycles in alluvial succession. Liverpool Manchester Geol. J. 4:229–246.

1970 Studies in fluvial sedimentation: a comparison of fining-upwards cyclothems, with special reference to coarse-member composition and interpretation. J. Sed. Petrol. 40:298–323.

Angelis d'Ossat, G. and F. Millosevich
1900 Studio geologico sul materiale raccolto da M. Sacchi. Secondo spedizione Bòttego (Afrique orientale). 212 pages. Rome: Società Geografica Italiana.

Arambourg, C.
1943 Mission scientifique de l'Omo (1932–1933). Géologie-Anthropologie. Tome 1, fasc. 2, pp. 60–230. Mémoire, Muséum national d'histoire naturelle, Paris.

1947 Mission scientifique de l'Omo (1932–1933). Tome 1, fasc. 3. Paléontologie. pp. 231–562. Mémoire, Muséum national d'histoire naturelle, Paris.

Arambourg, C., J. Chavaillon and Y. Coppens
1967 Premiers résultats de la nouvelle mission de l'Omo (1967). C. R. Acad. Sci., Paris 265-D:1891–1896.

1969 Résultats de la nouvelle mission de l'Omo (2e campagne 1968). C. R. Acad. Sci., Paris 268-D:759–762.

Arambourg, C. and Y. Coppens
1967 Sur la découverte dans de Pléistocène inférieur de la vallée de l'Omo (Ethiopie) d'une mandibule d'australopithecien. C. R. Acad. Sciences, Paris 265-D:589–590.

1968 Découverte d'un australopithecien nouveau dans les gisements de l'Omo (Ethiopie). So. Afr. J. Science 64:58–59.

Arambourg, C. and R. Jeannel
1933 La mission scientifique de l'Omo. C. R. Acad. Sci., Paris 196:1902–1904.

Baker, B. H., P. A. Mohr, and L. A. J. Williams
1972 Geology of the Eastern Rift System of Africa. Geological Society of America, Special Paper 136:67 pages.

Beden, M.
1975 Proboscideans from Omo Group formations. *In* Earliest Man and Environments in the Lake Rudolf basin. Stratigraphy, Paleoecology and Evolution. Y. Coppens, F. C. Howell, G. L. Isaac and R. E. F. Leakey, eds. Chicago: University of Chicago Press.

Bonnefille, R.
1970 Premiers résultats concernant l'analyse pollinique d'échantillons du Pléistocène inférieur de l'Omo (Ethiope). C. R. Acad. Sci., Paris 270-D:2430–2433.

1972 Considérations sur la composition d'une microflore pollinique des formations plio-pléistocènes de la basse vallée de l'Omo (Ethiopie). Palaeoecology of Africa. E. M. van Zinderen Bakker, ed. 7:22–27.

1975 Palynological evidence for an important change in the vegetation of the Omo Basin between 2.5 and 2 million years. *In* Earliest Man and Environments in the Lake Rudolf basin. Stratigraphy, Paleoecology and Evolution. Y. Coppens, F. C. Howell, G. L. Isaac and R. E. F. Leakey, eds. Chicago: University of Chicago Press.

Bonnefille, R., F. H. Brown, J. Chavaillon, Y. Coppens, P. Haesaerts, J. de Heinzelin and F. C. Howell
1973 Situation stratigraphique des localités à hominidés des gisements Plio-Pléistocènes de l'Omo en Ethiopie. C. R. Acad. Sci., Paris 276-D:2781–2784, 2879–2882.

Boulenger, G. -A.
1920 Sur le gavial fossile de l'Omo. C. R. Acad. Sci., Paris 170:913–914.

Bourg de Bozas, R.
1903 D'Addis-Ababa au Nil par le lac Rodolphe. La Géographie 7:91–112.

1906 Mission scientifique du Bourg de Bozas de la Mer Rouge à l'Atlantique à travers l'Afrique tropicale, Octobre 1900, Mars 1903. 442 pages. Paris: F. R. de Rudeval.

Brown, F. H.
1969 Observations on the stratigraphy and radiometric age of the "Omo Beds," lower Omo basin, southern Ethiopia. Quaternaria 11:7–14.

1972 Radiometric dating of sedimentary formations in the lower Omo valley, southern Ethiopia. *In* Calibration of Hominoid Evolution. W. W. Bishop and J. A. Miller, eds. Edinburgh: Scottish Academic Press. pp. 273–287.

Brown, F. H. and I. S. E. Carmichael
1969 Quarternary volcanoes of the Lake Rudolf regions: I. The basanite-tephrite series of the Korath range. Lithos 2:239–260.

Brown, F. H. and K. R. Lajoie
1971 Radiometric age determinations on Pliocene/Pleistocene formations in the lower Omo basin, southern Ethiopia. Nature 229:483–485.

Brown, F. H. and W. P. Nash
1975 Radiometric Dating and Tuff Mineralogy of Omo Group deposits. *In* Earliest Man and Environments in the Lake Rudolf basin. Stratigraphy, Paleoecology and Evolution. Y. Coppens, F. C. Howell, G. L. Isaac and R. E. F. Leakey, eds. Chicago: University of Chicago Press.

Brown, F. H. and R. T. Shuey
1975 Magnetostratigraphy of the Shungura and Usno Formations, lower Omo Valley, Ethiopia. *In* Earliest Man and Environments in the Lake Rudolf basin. Stratigraphy, Paleoecology and Evolution. Y. Coppens, F. C. Howell, G. L. Isaac and R. E. F. Leakey, eds. Chicago: University of Chicago Press.

Butzer, K. W.
1970 Geomorphological observations in the lower Omo basin, southwestern Ethiopia. *In* Argumenta Geographica, Carl Troll Festschrift. W. Lauer, ed. Colloquium Geographicum 12:177–192.

1971 The lower Omo basin: geology, fauna and hominids of Plio-Pleistocene age. Naturwissenschaften 55:7–16.

1975 The Mursi, Nkalabong and Kibish Formations, lower Omo basin, Ethiopia. *In* Earliest Man and Environments in the Lake Rudolf basin. Stratigraphy, Paleoecology and Evolution. Y. Coppens, F. C. Howell, G. L. Isaac and R. E. F. Leakey, eds. Chicago: University of Chicago Press.

Butzer, K. W., F. H. Brown, and D. L. Thurber
1969 Horizontal sediments of the lower Omo basin: the Kibish Formation. Quaternaria 11:15–29.

Butzer, K. W. and D. L. Thurber
1969 Some late Cenozoic sedimentary formations of the lower Omo basin. Nature 222:1132–1137.

Chavaillon, J.
1970 Découverte d'un niveau Olduwayen dans la basse vallée de l'Omo (Ethiopie). C. R. séances Soc. préhist. franc. 1:7–11.

1971 État actuel de la préhistoire ancienne dans la vallée de l'Omo (Ethiopie). Archeologia 38:33–43.

1975 Evidence for the technical practices of early Pleistocene hominids. Shungura Formation, lower valley of the Omo, Ethiopia. *In* Earliest Man and Environments in the Lake Rudolf basin. Stratigraphy, Paleoecology and Evolution. Y. Coppens, F. C. Howell, G. L. Isaac and R. E. F. Leakey, eds. Chicago: University of Chicago Press.

Cooke, H. B. S.
1975 Suidae from Pliocene/Pleistocene strata of the Rudolf basin. *In* Earliest Man and Environments in the Lake Rudolf basin. Stratigraphy, Paleoecology and Evolution. Y. Coppens, F. C. Howell, G. L. Isaac and R. E. F. Leakey, eds. Chicago: University of Chicago Press.

1976 The Pliocene-Pleistocene Suidae. *In* Mammalian Evolution in Africa. V. J. Maglio, ed. Princeton: Princeton University Press.

Cooke, H. B. S. and V. J. Maglio
1972 Plio-Pleistocene stratigraphy in East Africa in relation to proboscidean and suid evolution. *In* Calibration of Hominoid Evolution. W. W. Bishop and J. A. Miller, eds. Edinburgh: Scottish Academic Press. pp. 303–329.

Coppens, Y.
1970a Localisations dans le temps et dans l'espace des restes d'Hominidés des formations plio-pléistocènes de l'Omo (Ethiopie). C. R. Acad. Sci., Paris 271-D:1968–1971.

1970b Les restes d'Hominidés des séries inférieures et moyennes des formations Plio-Villafranchiennes de l'Omo en Ethiopie. C. R. Acad. Sciences 271-D:2286–2289.

1971 Les restes d'Hominidés des séries supérieures des formations Plio-Villafranchiennes de l'Omo en Ethiopie. C. R. Acad. Sciences, Paris 272-D:36–39.

1973a Les restes d'Hominidés des séries inférieures et moyennes des formations Plio-Villafranchiennes de l'Omo en Ethiopie (récoltes 1970, 1971, et 1972). C. R. Acad. Sciences, Paris 276:1823–1826.

1973b Les restes d'Hominidés des séries supérieures des formations Plio-Villafranchiennes de l'Omo en Ethiopie (récoltes 1970, 1971, et 1972). C. R. Acad. Sciences, Paris 276:1981–1984.

Coppens, Y., J. Chavaillon and M. Beden
1973 Résultats de la nouvelle mission de l'Omo (campagne 1972). Découverte de restes des Hominidés et d'une industrie sur éclats. C. R. Acad. Sciences, Paris 276:161–164.

Coppens, Y. and F. C. Howell
1974 Les faunes de mammiferes fossiles des formations Plio-Pléistocènes de l'Omo en Ethiopie (Proboscidea, Perissodactyla, Artiodactyla). C. R. Acad. Sci., Paris 278:2275–2278.

Coppens, Y., F. C. Howell, G. L. Isaac and R. E. F. Leakey, eds.
1975 Stratigraphy, Paleoecology and Evolution. Chicago: University of Chicago Press.

Coryndon, S. C.
1975 Fossil Hippopotamidae from Pliocene/Pleistocene successions of the Rudolf basin. *In* Earliest Man and Environments in the Lake Rudolf basin. Stratigraphy, Paleoecology and Evolution. Y. Coppens, F. C. Howell, G. L. Isaac and R. E. F. Leakey, eds. Chicago: University of Chicago Press.

Coryndon, S. C. and Y. Coppens
1973 Preliminary report on Hippopotamidae (Mammalia, Artiodactyla) from the Plio/Pleistocene of the lower Omo basin, Ethiopia. Fossil Vertebrates of Africa 3:139–157.

Curtis, G. H.
1968 The stratigraphy of the ejecta from the 1912 eruption of Mount Katmai and Novarupta, Alaska. Geol. Soc. of Amer., Memoir 116:153–210.

Eck, G. G.
1975 Cercopithecoidea from Omo Group deposits. *In* Earliest Man and Environments in the Lake Rudolf basin. Stratigraphy, Paleoecology and Evolution. Y. Coppens, F. C. Howell, G. L. Isaac and R. E. F. Leakey, eds. Chicago: University of Chicago Press.

Eck, G. G., and F. C. Howell
1972 New fossil *Cercopithecus* material from the lower Omo basin, Ethiopia. Folia Primatologia 8:325–355.

Eisenmann, V.
1975 Equidae from the Shungura Formation. *In* Earliest Man and Environments in the Lake Rudolf basin. Stratigraphy, Paleoecology and Evolution. Y. Coppens, F. C. Howell, G. L. Isaac and R. E. F. Leakey, eds. Chicago: University of Chicago Press.

Fitch, F. J. and J. A. Miller
1969 Age determinations on feldspar from the lower Omo basin. Nature 222:1143.

1975 Conventional Potassium-Argon and Argon-40/Argon-39 dating of volcanic rocks from East Rudolf. *In* Earliest Man and Environments in the Lake Rudolf basin. Stratigraphy, Paleoecology and Evolution. Y. Coppens, F. C. Howell, G. L. Isaac and R. E. F. Leakey, eds. Chicago: University of Chicago Press.

Gautier, A.
1975 Assemblages of fossil freshwater mollusks from the Omo Group and related deposits in the Lake Rudolf basin. *In* Earliest Man and Environments in the Lake Rudolf basin. Stratigraphy, Paleoecology and Evolution. Y. Coppens, F. C. Howell, G. L. Isaac and R. E. F. Leakey, eds. Chicago: University of Chicago Press.

Gentry, A. W.
1975 Bovidae of the Omo Group deposits. *In* Earliest Man and Environments in the Lake Rudolf basin. Stratigraphy, Paleoecology and Evolution. Y. Coppens, F. C. Howell, G. L. Isaac and R. E. F. Leakey, eds. Chicago: University of Chicago Press.

1976 Artiodactyla. Camelidae. Tragulidae. Bovidae. *In* Mammalian Evolution in Africa. V. J. Maglio, ed. Princeton: Princeton University Press.

Gentry, A. W. and A. Gentry
1969 Fossil camels in Kenya and Tanzania. Nature 222:898.

Grattard, J. -L., F. C. Howell and Y. Coppens
1975 Remains of *Camelus* from the Shungura Formation, lower Omo valley. *In* Earliest Man and Environments in the Lake Rudolf basin. Stratigraphy, Paleoecology and Evolution. Y. Coppens, F. C. Howell, G. L. Isaac and R. E. F. Leakey, eds. Chicago: University of Chicago Press.

Guerin, C.
1975 Rhinocerotidae and Chalicotheriidae (Mammalia, Perissodactyla) from the Shungura Formation, lower Omo basin. *In* Earliest Man and Environments in the Lake Rudolf basin. Stratigraphy, Paleoecology and Evolution. Y. Coppens, F. C. Howell, G. L. Isaac and R. E. F. Leakey, eds. Chicago: University of Chicago Press.

Haug, E.
1912 Traité de Géologie. II. Les Périodes géologiques. Paris: Armand Colin.

Hedberg, H. D., ed.
1971 Preliminary report on chronostratigraphic units. (International Subcommission on Stratigraphic Classification, Report No. 6, 39 pages). International Geological Congress, 24th Session, Montreal, 1971.

de Heinzelin, J.
1971 Observations sur la formation de Shungura (vallée de l'Omo, Ethiopie). C. R. Acad. Sci., Paris 272-D:2409–2411.

de Heinzelin, J. and F. H. Brown
1969 Some early Pleistocene deposits of the lower Omo valley: the Usno Formation. Quaternaria 11:31–46.

de Heinzelin, J., F. H. Brown and F. C. Howell
1970 Pliocene/Pleistocene formations in the lower Omo basin, southern Ethiopia. Quaternaria 13:247–268.

de Heinzelin, J., P. Haesaerts and F. C. Howell
1975 Plio-Pleistocene formations of the lower Omo basin, with particular reference to the Shungura formation. *In* Earliest Man and Environments in the Lake Rudolf basin. Stratigraphy, Paleoecology and Evolution. Y. Coppens, F. C. Howell, G. L. Isaac and R. E. F. Leakey, eds. Chicago: University of Chicago Press.

von Höhnel, L.
1938 The Lake Rudolf region. Its discovery and subsequent exploration (1888–1909). J. Royal Afr. Soc. 37:21–45, 206–226.

von Höhnel, R. L., A. Rosiwal, F. Toula and E. Suess
1891 Beiträge zur geologischen Kenntnis des östlichen Afrika. Denkschr. d. Akad. d. Wiss. Wien, Math.-naturw. Kl. 58:140 pages.

Hooijer, D. A.
1975 Evolution of the Perissodactyla of the Omo Group deposits. *In* Earliest Man and Environments in the Lake Rudolf basin. Stratigraphy, Paleoecology and Evolution. Y. Coppens, F. C. Howell, G. L. Isaac and R. E. F. Leakey, eds. Chicago: University of Chicago Press.

Howell, F. C.
1968 Omo Research Expedition. Nature 219:567–572.

1969a Remains of Hominidae from Pliocene/Pleistocene formations in the lower Omo basin, Ethiopia. Nature 223:1234–1239.

1969b Hominid teeth from White Sands and Brown Sands localities, lower Omo basin (Ethiopia). Quaternaria 11:47–64.

1975 Overview of the Pliocene and earlier Pleistocene of the lower Omo basin, Southern, Ethiopia. *In* African Hominidae of the Plio-Pleistocene. C. J. Jolly, ed. London: Duckworths.

Howell, F. C. and Y. Coppens
1973 Deciduous teeth of Hominidae from the Pliocene/Pleistocene of the lower Omo basin, Ethiopia. Journal of Human Evolution (R. A. Dart Memorial Issue) 2:461–472.

1974 Les faunes de mammifères fossiles des formations Plio/Pléistocènes de l'Omo en Ethiopie (Tubulidentata, Hyracoidea, Lagomorpha, Rodentia, Chiroptera, Insectivora, Carnivora, Primates). C. R. Acad. Sci., Paris 278:2421–2424.

1975 Hominidae from the Usno and Shungura Formations, lower Omo valley. *In* Earliest Man and Environments in the Lake Rudolf basin. Stratigraphy, Paleoecology and Evolution. Y. Coppens, F. C. Howell, G. L. Isaac and R. E. F. Leakey, eds. Chicago: University of Chicago Press.

Howell, F. C., Y. Coppens and J. de Heinzelin
1974 Inventory of remains of Hominidae from Pliocene/Pleistocene formations of the lower Omo basin, Ethiopia (1967–1972). Am. J. Phys. Anthrop. 40:1:1–16.

Howell, F. C., L. S. Fichter and G. Eck
1969 Vertebrate assemblages from the Usno Formation, White Sands and Brown Sands localities, lower Omo basin, Ethiopia. Quaternaria 11:65–88.

Howell, F. C., L. S. Fichter and R. Wolff
1969 Fossil camels in the Omo Beds, southern Ethiopia. Nature 223:150–152.

Howell, F. C. and G. Petter
1975 Carnivora from Omo Group formations, southern Ethiopia. *In* Earliest Man and Environments in the Lake Rudolf basin. Stratigraphy, Paleoecology and Evolution. Y. Coppens, F. C. Howell, G. L. Isaac and R. E. F. Leakey, eds. Chicago: University of Chicago Press.

Howell, F. C. and B. A. Wood
1974 An early hominid ulna from the Omo basin, Ethiopia. Nature 249:174–176.

Jaeger, J. -J. and H. B. Wesselman
1975 Fossil remains of micro-mammals from the Omo Group deposits. *In* Ear-

liest Man and Environments in the Lake Rudolf basin. Stratigraphy, Paleoecology and Evolution. Y. Coppens, F. C. Howell, G. L. Isaac and R. E. F. Leakey, eds. Chicago: University of Chicago Press.

Jeannel, R.

1934 Un cimetière d'éléphants. 159 pages. Paris: Societé des Amis du Muséum national d'histoire naturelle.

Johanson, D. C., N. Boaz and M. Splingaer

1975 Paleontological excavations in the Shungura Formation, lower Omo basin, 1969–1973. *In* Earliest Man and Environments in the Lake Rudolf basin. Stratigraphy, Paleoecology and Evolution. Y. Coppens, F. C. Howell, G. L. Isaac and R. E. F. Leakey, eds. Chicago: University of Chicago Press.

Joleaud, L.

1920a Contribution à l'étude des hippopotames fossiles. Bull. Soc. Géol. de France, sér. 4, 20:13–26.

1920b Sur la présence d'un gavialidé du genre *Tomistoma* dans le Pliocène d'eau douce de l'Ethiopie. C. R. Acad. Sci., Paris 170:816–818.

1928 Éléphants et Dinothériums Pliocènes de l'Ethiopie: contribution à l'étude paléogéographique des proboscidiens africains. Intern. Geol. Congress, Madrid, 14th Session, 3:1001–1007.

1930 Les crocodiliens du Pliocène d'eau douce de l'Omo (Ethiopie). Contribution à l'étude paléobiogéographique des *Tomistoma* et des crocodiles à museau de gavial. Soc. Géol. de France, Livre Jubilaire, 1830–1930, 2:411–423.

1933 Un nouveau genre d'Equidé quaternaire de l'Omo (Abyssinie): *Libyhipparion ethiopicum*. Bull. Soc. Géol. de France, sér. 5, 3:7–28.

Leakey, L. S. B.

1943 New fossil Suidae from Shungura, Omo. J. E. Afr. Uganda Nat. Hist. Soc., Nairobi 17:45–61.

Leakey, M. G.

1975 Cercopithecoidea of the E. Rudolf succession. *In* Earliest Man and Environments in the Lake Rudolf basin. Stratigraphy, Paleoecology and Evolution. Y. Coppens, F. C. Howell, G. L. Isaac and R. E. F. Leakey, eds. Chicago: University of Chicago Press.

Leakey, M. G. and R. E. F. Leakey

1973 New large Pleistocene Colobinae (Mammalia, Primates) from East Africa. Fossil Vertebrates of Africa 3:121–138.

Leopold, L. B., M. G. Wolman and J. P. Miller

1964 Fluvial processes in geomorphology. 522 pages. San Francisco and London: W. H. Freeman and Co.

Maglio, V. J.

1970 Early Elephantidae of Africa and a tentative correlation of Plio-Pleistocene deposits. Nature 225:328–332.

1973 Origin and evolution of the Elephantidae. Trans. Am. Phil. Soc., n.s. 63:3:1–149.

Merla, G.

1963 Missione geologica nell'Etiopia meridionale del Consiglio nazionale delle ricerche 1959–1960. Notizie geo-morfologiche e geologiche. Giornale di Geologia, ser. 2, 31:1–56.

Merrick, H. V.
1976 Recent archaeological research in the Plio-Pleistocene deposits of the lower Omo valley, southwestern Ethiopia. In this volume.

Merrick, H. V., J. de Heinzelin, P. Haesaerts and F. C. Howell
1973 Archaeological occurrences of early Pleistocene age from the Shungura Formation, lower Omo valley, Ethiopia. Nature 242:572–575.

Merrick, H. V. and J. P. S. Merrick
1975 Archaeological occurrences of earlier Pleistocene age from the Shungura Formation. *In* Earliest Man and Environments in the Lake Rudolf basin. Stratigraphy, Paleoecology and Evolution. Y. Coppens, F. C. Howell, G. L. Isaac and R. E. F. Leakey, eds. Chicago: University of Chicago Press.

Mohr, P. A.
1968 The Cainozoic volcanic succession in Ethiopia. Bull. Volcanologique 32:5–14.

Petter, G.
1973 Carnivores Pléistocènes du ravin d'Olduvai (Tanzanie). Fossil Vertebrates of Africa 3:43–100.

Reilly, T. A., A. E. Musset, P. R. S. Raja, R. L. Grasty and J. Walsh
1966 Age and polarity of the Turkana lavas, northwest Kenya. Nature 210:1145–1146.

Ross, C. S. and R. L. Smith
1961 Ash-flow tuffs: their origin, geologic relations and identification. U.S. Geol. Survey. Professional Paper 366.

Sclater, P. L.
1899 Results of the second Bòttego expedition into eastern Africa. Science 10:951–955.

Shuey, R. T., F. H. Brown and M. K. Croes
1974 Magnetostratigraphy of the Shungura Formation, southwestern Ethiopia: fine structure of the lower Matuyama polarity epoch. Earth and Planetary Science Letters 23:249–260.

Simpson, G. G.
1960 Notes on the measurement of faunal resemblance. Am. J. Sci. 258-A:300–311.

Smith, R. L.
1960 Ash flows. Bull. Geol. Soc. America 71:985–842.

Sparks, R. S. J. and G. P. L. Walker
1973 The ground surge deposit: a third type of pyroclastic rock. Nature-Phys. Sci. 241:62–64.

Tchernov, E.
1975 Crocodilians from the late Cenozoic of the Rudolf basin, *In* Earliest Man and Environments in the Lake Rudolf basin. Stratigraphy, Paleoecology and Evolution. Y. Coppens, F. C. Howell, G. L. Isaac and R. E. F. Leakey, eds. Chicago: University of Chicago Press.

Van Damme, D. and A. Gautier
1972 Molluscan assemblages from the later Cenozoic of the lower Omo basin, Ethiopia. Quaternary Research 2:1:25–37.

Vannutelli, L. and C. Citerni

1897 Relazione preliminare sui risultats geografici della seconda spedizone Bòttego. Bolletino Società Geografica Italiana, ser. 3, 10:320–330.

1899 Seconda spedizione Bòttego. L'Omo. Viaggio di esplorazione nell'Africa orientale. 650 pages. Milano: Vottoepli.

Walsh, J. and R. G. Dodson

1969 Geology of North Turkana. Kenya Geol. Survey Rept. 82:42 pages. (1:500,000 map.)

PART THREE

The Interpretation of Hominid and Hominoid Fossils

Introduction to Part Three

Recognition of the East African Rift as one of the world's richest sources of early hominid fossils began in 1959 with the discovery of *"Zinjanthropus"* by Mary and Louis Leakey. Since that time several hundred specimens have come to light from points all along the 600 kilometers of Rift Valley that lies between Olduvai and Lake Rudolf. Indeed recent finds in the Afar depression in Ethiopia, where the Rift approaches the Red Sea, reinforce one's belief that the area to be searched extends over the whole length of this tectonic gash in the earth's crust.

The East African fossil beds contain the world's longest known palaeontological sequence of evolving apes and men. The record starts in Miocene deposits of about 20 million years ago and extends to the present. The fossils are unevenly spaced along the thread of time, and from the interval between about 14 m.y.a. (million years ago) and about 5 m.y.a., we have as yet only tantalizing fragments of teeth (see Bishop's essay in Part Two of this volume). From the so-called Plio-Pleistocene, a time range from about 3 m.y.a. to 1 m.y.a., there

How a group of Australopithecus (Zinjanthropus) boisei *may have appeared. An oil painting by Jay H. Matternes reproduced by his kind permission and with the help of Aubrey Buxton of Survival Anglia Ltd., who hold the copyright.*

is a particularly rich and diverse series of fossil specimens. These include skulls, mandibles, long bones, hand and foot bones, and in a few cases associated elements from several parts of a single skeleton. A discontinuous string of fossils connects this early series to the present: there are early Middle Pleistocene fossils from the Olduvai region (Leakey, M. D., 1971a, b; Day 1971), late Middle to early Upper Pleistocene fossils from the Kapthurin Beds (Leakey, M. D. *et al.* 1969) and the Kibish Formation at the Omo (Leakey, R. E. F., Butzer and Day 1969). These latter are said to include specimens that are among the oldest known examples of *Homo sapiens.*

The interpretation of human fossils has always been a contentious business and each segment of the hominoid record has its own set of fierce arguments. Concerning the 20- to 14-million-year-old Miocene fossils, there is keen debate on the question of whether any taxon among the diverse hominoidea can specifically be picked out as being exclusively ancestral to man, that is, as being the root stock of the family Hominidae. Louis Leakey identified fossils from the 14-million-year-old beds at Fort Ternan as very early hominids and named the taxon *Kenyapithecus wickeri* (Leakey, L. S. B., 1962). Later Simons and Pilbeam (1965) and Pilbeam (1969) argued that the *Kenyapithecus* specimens belonged to the same genus as the Asian fossils named *Ramapithecus.* Most authorities have accepted this revision, and the real debate has come to turn on whether these fossils really prove that evolutionary divergence between the separate lineages leading to men and to apes had occurred by 15 or more million years ago. An alternative interpretation may be that an adaptive radiation of Miocene apes produced a spectrum of forms that included some with masticatory mechanics and dental morphology paralleling that which later became characteristic of the hominidae. The main reasons for being cautious over the phylogenetic interpretation of the existing fragmentary fossils is that various comparative biochemical evidences can be interpreted as indicating that the divergence between man and the African apes did not take place more than 5 to 10 million years ago. This line of evidence has been reviewed by Sarich (1968) and by Washburn and Moore (1974).

This volume offers two contributions to the discussion of the earlier part of the hominoid fossil record: in the geology section (Part Two) the Van Couverings have provided a general account of the Miocene fossil beds, their fauna and paleoenvironmental setting; while in this section Peter Andrews and Alan Walker present a very thorough report on both the anatomy and the context of the Fort Ternan *Ramapithecus* material. Andrews and Walker reach interesting conclusions regarding the distinctiveness of the creature represented and the adaptive pattern that may be indicated by the overall features of the masticatory apparatus. They decide that an assignment to a definite family within the Hominoidea is not warranted on the basis of the available fragmentary material.

For the Plio-Pleistocene time range (3–1 m.y.a.) the most active current discussion concerns how many contemporary species of hominid co-existed in Africa. In Southern Africa two contrasting morphologies have long been recognized: the "gracile" form, *Australopithecus africanus* (Dart) and the "robust" forms, *Australopithecus (Paranthropus) crassidens* and *robustus* (Broom). For the most part these seemed * to occur separately in the deposits of different caves, so that the two forms could have been successive rather than contemporary. However, a different situation has been found in East Africa. It has emerged that a wider morphological range than that comprised by both the Transvaal taxa together is commonly found within the fossiliferous strata of one locality. How is this diversity to be interpreted? Perhaps the majority of comparative anatomists familiar with the material favor the view that at least two separate, co-existing species of hominid are represented through the time range from 3 million to 1 million years ago. However, the view has also been held by some, that the so-called principle of competitive exclusion dictates that there can only have been one species of hominid. Under this view, the diversity encountered is to be accounted for by a combination of sexual dimorphism and subspecific variation (cf. Clark, LeGros, 1967, Wolpoff 1971). Other authorities, Louis Leakey among them, have suggested that there may have been more than two contemporary species—perhaps even four or five.

If one accepts that more than one genetic system (lineage) is represented among the East African hominid fossils, then other questions follow: How are the phylogenetic lines connecting specimens of different ages to be drawn? When did a

*Many authorities now believe that the segregation of taxa between sites in Southern Africa may be less clear-cut than was formerly supposed (see Tobias this volume).

Mr. Kamoya Kimeu points to the 1½-million-year-old australopithecine mandible that he had just found near Lake Natron (1964). Kimeu, who was first trained by Louis and Mary Leakey, is now deputy leader of the East Lake Rudolf expeditions. He probably holds the world record for number of stock with the particular evolutionary trends characteristic of the genus *Homo,* become differentiated? And above all, what were the mechanisms of speciation and ecological separation among the various co-existing forms?

There are as yet no definite answers to these questions, but four papers in this volume contribute information and views relating directly to current research on these problems. The first, by F. Clark Howell, presented in the geological section, provides a comprehensive summary of the Omo investigation, which among other things, serves well to show the value of a long, well-dated hominid bearing stratigraphic sequence. Howell's article also illustrates how much modern paleoanthropology is concerned with context rather than with minutiae of morphology.

The second, a composite contribution from Richard Leakey and Glynn Isaac, is a brief summary plus three reprints from *Nature* which clearly show how rapid the rate of discovery has been since 1968 at East Rudolf. The second of these reprints is a report to *Nature* on the 1972 discovery of evidence for an unexpectedly large-brained and fully bipedal Plio-Pleistocene hominid—the cranium now widely known as "1470." This has been one of the most dramatic and controversy-provoking finds yet made in East Africa. The third reprint is the report to *Nature* for the following year showing that 1973 finds include additional relatively complete specimens—these occasioned more surprises and potential controversies. This report also includes a clear statement of Richard Leakey's thoughts on taxonomic relationships as of 1973. Louis Leakey loved the excitement and challenge of new discoveries and delighted in the controversies they often provoked. It was therefore fitting that his son Richard was able to show him the newly discovered 1470 a few days before his death.

In the third contribution on Plio-Pleistocene hominids, Michael Day has a valuable summary of findings made during a prolonged review of the postcranial material from Olduvai Gorge. His paper brings out very clearly the intricacy of dealing with sites containing remains from more than one closely related taxon. In attempting to sort out the tangle, Day draws heavily on the implications of the new material from East Rudolf—thereby showing again how each wave of discoveries necessitates a reexamination of earlier opinions. The article also includes particularly important revisions of anatomical data on the FLK NN I hand (OH 7).

The fourth paper is a comprehensive review by Phillip Tobias which deals with new finds in both East and South Africa. Tobias, a longtime friend and colleague of Louis and Mary Leakey, was invited by the Leakeys to prepare the detailed monograph on "*Zinjanthropus*" and on other Olduvai specimens (Tobias 1967). This article is thus written from the rare vantage point of a close firsthand knowledge of both the Southern African and the Olduvai fossils. His opinions regarding phylogenetic sequences are very clearly stated in both the text and diagrams. The essay also illustrates the importance of dating for phylogenetic interpretation. There remains considerable uncertainty over the age of the Southern African fossils, in spite of the addition of controversial geomorphological estimates to the stock of data for consideration (Partridge 1973).

The four papers in this section deal primarily with the skeletal remains of early hominids; the next section treats archaeological evidence for the activities of the hominids to whom the bones originally belonged.

References Cited

Clark, W. LeGros
1967 Man-Apes or Ape-Men. New York: Holt, Rinehart and Winston.

Day, M. H.
1971 Postcranial Remains of *Homo erectus* from Bed IV, Olduvai Gorge, Tanzania. Nature 232:383–387.

Leakey, L. S. B.
1962 A New Lower Pliocene Fossil from Kenya. Annals and Magazine of Natural History 13:689–696.

Leakey, M. D., P. V. Tobias, J. E. Martyn and R. E. F. Leakey
1969 An Acheulean Industry and Hominid Mandible, Lake Baringo, Kenya. Proceedings of the Prehistoric Society 35:48–76.

Leakey, M. D.
1971a Olduvai Gorge, Vol. 3. Cambridge: Cambridge University Press.

1971b Discovery of Postcranial Remains of *Homo erectus* and Associated Artefacts in Bed IV at Olduvai Gorge, Tanzania. Nature 232:380–383.

Leakey, R. E. F., K. W. Butzer and M. H. Day
1969 Early *Homo sapiens* Remains from the Omo River Region of South West Ethiopia. Nature 222:1132–1138.

Partridge, T. C.
1973 Geomorphological dating of cave opening at Makapansgat, Sterkfontein, Swartkrans and Taung. Nature 246:75–79.

Pilbeam, D. R.
1969 Tertiary Pongidae of East Africa. Bulletin 31, Peabody Museum of Natural History, Yale University, New Haven.

Sarich, V. M.
1968 The Origin of the Hominids: An Immunological Approach. *In* Perspectives on Human Evolution 1. S. L. Washburn and P. C. Jay, Eds. New York: Holt, Rinehart and Winston. pp. 94–121.

Simons, E. L., and D. R. Pilbeam
1965 Preliminary Revision of the Dryopithecinae (Pongidae, Anthropoidea) Folia Primatologia 3:81–152.

Tobias, P. V.
1967 The Cranium and Maxillary Dentition of *Australopithecus (Zinjanthropus) boisei. In* Olduvai Gorge, Vol. 2. Cambridge: University Press.

Washburn, S. L., and R. Moore
1974 Ape into Man. New York: Little Brown.

Wolpoff, M. H.
1971 Competitive Exclusion Among Lower Pleistocene Hominids: the Single Species Hypothesis. Man 6:601–614.

L. S. B. Leakey pointing to the caliche deposit at Fort Ternan. Courtesy, Bob Campbell.

Peter Andrews and Alan Walker:

The Primate and Other Fauna from Fort Ternan, Kenya

What are the oldest fossils belonging to the evolutionary branch which diverged from that leading to any ape and which gave rise to mankind and to his closest extinct man-like relatives? One strong contender has been a fossil found at Fort Ternan in Kenya and dated as 14 million years old. In the light of fresh evidence Peter Andrews and Alan Walker reassess the taxonomic and evolutionary status of this Fort Ternan primate and consider indicators of the kind of environment in which it lived, died and became a fossil.

The middle Miocene site at Fort Ternan, Kenya (Figure 1), was first excavated in 1961 by Dr. L. S. B. Leakey. The owner of the farm on which the site is located, Mr. Fred Wicker, had found some fossil bones weathering out of the hillside and sent them to Dr. Leakey, who quickly realized the importance of the site. His first excavation yielded a large quantity of bones, including four specimens of a hominoid primate which he named *Kenyapithecus wickeri* (Leakey 1962). These were left and right maxillary fragments, together with an upper canine which was found to fit onto the left maxilla and an isolated right lower molar that was thought to be associated.

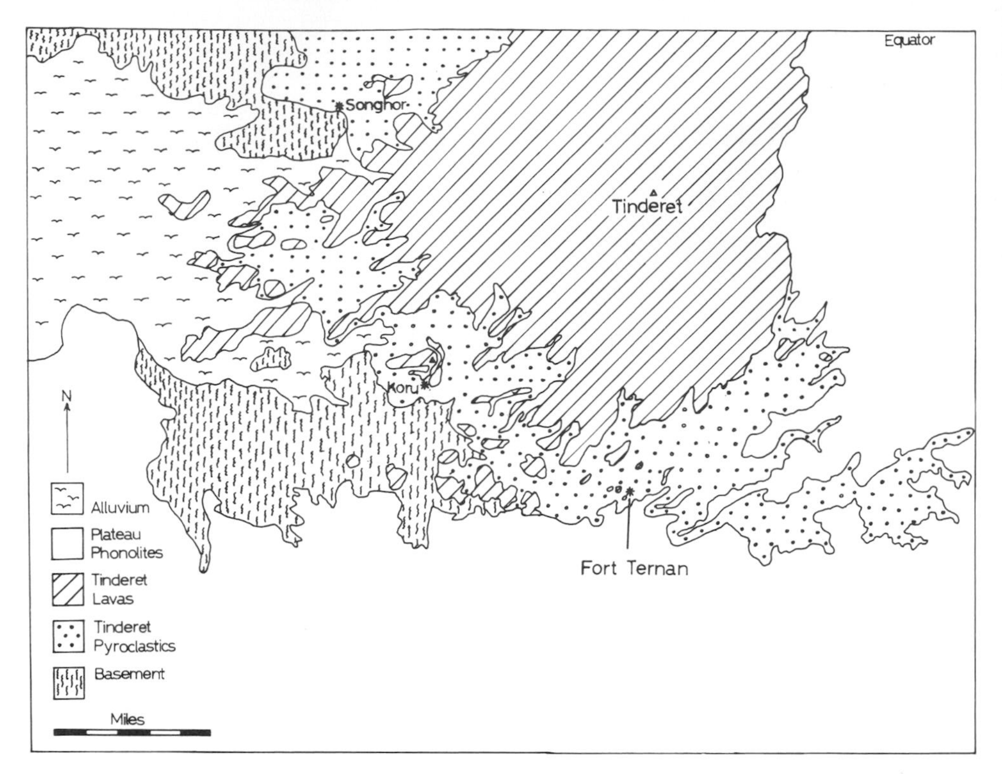

FIGURE 1
Simplified geological map showing the location of the Fort Ternan site in relation to contemporary volcanism.

The following year Leakey worked at Fort Ternan again. Very early in the season a mandibular fragment was discovered in the gully immediately below the 1961 excavation. Although it cannot be known for certain, it is probable that this specimen was almost uncovered in 1961 and that subsequent rains washed it out and on to the edge of the gully. This specimen was not described until much later (Andrews 1971) and has provided the foundation for a reconstruction of the toothrows of *Ramapithecus wickeri* (Walker and Andrews 1973).

These specimens are described again now in greater detail than has been attempted before. Some additional specimens, two of which were used in our reconstruction but which have not been described previously, are also described for the first time here. The incisor that Leakey (1967) assigned to *R. wickeri* we consider not to belong to this species but to the large pongid that is represented by a number of isolated teeth.

The descriptions of the specimens are followed by short discussions on the taxonomic and functional implications of the morphology. We do not consider that definite assignment to family is possible with the inadequate material available at present, but the functional analysis emphasizes the unusual morphology which makes this species unique among hominoid species. Finally, the geology and dating of Fort Ternan is discussed and the primate and other faunas described. Some tentative conclusions are reached concerning the paleoenvironments from which the fauna and flora in the Fort Ternan sediments were sampled, and the dental and gnathic adaptations postulated for *R. wickeri* are discussed in the light of these conclusions.

DESCRIPTIONS OF THE SPECIMENS[1]

KNM-FT 46 (Leakey 1962)

This type of specimen of *R. wickeri* (Leakey 1962) is a left maxillary fragment with the canine, P^4 and the first two molar teeth in situ, and the roots of P^3 and the mesial part of the wall of the mesiobuccal alveolus of M^3 present. The crowns of the preserved teeth are complete but for a small chip missing from the apex of the canine and another from the mesiobuccal corner of the P^4. The maxillary fragment is broken superiorly just above the level of the floor of the maxillary sinus. The zygomatic process is broken off short of the zygomaticomaxillary suture, and the palatine process is similarly broken short of the intermaxillary suture. The mesiobuccal roots of the premolars make strong, narrow juga on the alveolar processes, and although the outer alveolar bone is missing over the canine root, it too probably made a strong rounded jugum. The profile of the alveolar margin is curved, convex inferiorly, rising towards the canine neck and rising towards the neck of M^3.

The canine root is compressed with its long axis set at about 60° to the toothrow axis, and the flattened mesiolingual surface is slightly convex. The canine crown is relatively small and conical. The maximum diameters of the crown are: mesiodistal 9.6, buccolingual 8.6. There is a fairly strong lingual cingulum that in places exceeds 3.0 in height and which wraps round the mesial border of the tooth and ends rather abruptly as a mesiobuccal tubercle. The maximum height of the enamel on the buccal side is reached mesiobuccally, but on the lingual side is reached distobuccally; the enamel swings down at both mesial and distal borders. The buccal surface is mildly grooved distobuccally but is mainly characterized by very many, largely horizontal, fine perikymata. The lingual surface is more strongly ridged and grooved, the ridges ending on the cingulum. There is a fairly strong mesial groove running down the mesial border of the tooth, and wear facets are seen on either side of the

[1]All measurements are in millimeters.

groove (more developed towards the apex) and on the lowest part of the cingulum. These facets are roughly plane and set not quite at right angles to the toothrow axis. On the distal surface there is a strongly developed wear facet from the tip to the cingulum; dentine is exposed for the whole of the facet's length and the bounding enamel is raised above the dentine surface. Many, rather coarse, mostly transverse, scratches are seen on the thicker buccal enamel section. An estimated buccal height for the unworn crown is 12.0.

The P^3 root section shows that there were three roots, the lingual the largest, and the distobuccal the smallest in diameter. P^4 is a bicuspid tooth with a rather pointed buccal cusp and a blunt lingual one. Both mesial and distal buccal grooves are present. The anterior fovea is small and the posterior fovea is larger, with several tributary sulci breaking up the surrounding enamel into small tubercles. There is a single curious round central fovea. Soft grooves are seen running down the lingual surface but there is no trace of a cingulum. There is a distinct wear facet on the buccal cusp, but wear is seen as a series of transverse scratches over the whole of the lingual cusp. The dimensions are: mesiodistal diameter 7.0, buccolingual diameter 10.5, maximum buccal height 6.5.

The upper M^1 is a simple, four-cusped tooth with low conical buccal and flattened lingual cusps. The buccal cusps are set at the edge of the crown. There is a small anterior and a larger posterior fovea. Three grooves are seen on the buccal face, running on the sides of the cusps. The trigon basin is broad and well defined and there is dentine exposure on the apex of the protocone. There is a single lingual groove between the protocone and the hypocone and a possible cingular remnant is seen as a low ridge on the mesiolingual corner of the crown. Wear is seen as a series of transverse scratches that are visible to the naked eye and which have many finer transverse microscopic scratches between them (Figure 2). The dimensions are: mesiodistal diameter 10.7, buccolingual diameter 10.9, maximum buccal height 6.4. M^2 is very similar in morphology to M^1. Again the crown is complete but for a tiny chip of enamel from the paracone. The buccal grooves are less developed than in M^1 but the lingual groove is more strongly developed. The anterior fovea is almost obliterated but the features of the posterior fovea are seen clearly. There is no dentine exposure but the brown dentine shows through the grey enamel on the tip of the protocone. There is a small transversely elongated contact facet for the erupted M^3. The dimensions are: mesiodistal diameter 12.0, buccolingual diameter 12.0, maximum buccal height 6.5.

The zygomatic process of the maxilla has a plane anterior surface that swings laterally from the body at the level of M^1. As seen on the broken lateral section there is a blunt inferior margin above M^1 but the process thickens superiorly by posterior expansion. Anterior to the zygomatic process there is a well-defined fossa that is depressed into the lateral surface above the premolar roots and posterior to the longer canine root.

FIGURE 2
Enlarged view of M^1-M^2 of KNM-FT 46, showing transverse wear scratches on the occlusal surfaces.

There is no trace of the infraorbital foramina even though the preserved parts of the maxilla reach some 22.0 above the alveolar margin, but a small canal seen in the broken wall of the body may have opened as an accessory foramen just behind the canine jugum. The floor of the maxillary air sinus is preserved for a fair extent and is surprisingly high and not invaginated between the roots of the cheek teeth. The floor is triangulated in outline with one apex near the canine fossa at the level of the P^4/M^1 junction, one extending into the zygomatic process, and one posteriorly towards M^3.

KNM-FT 47 (Leakey 1962)

This is part of the right maxilla of the same individual with M^1 and M^2 preserved, together with the broken roots of M^3 and the distobuccal root of P^4. The distal wall of the distolingual alveolus of P^4 is also seen. M^1 has a small postmortem chip missing from the metacone. Both teeth are virtually identical in morphology, wear and dimensions to their antimeres on KNM-FT 46, but because P^4 is missing, the double contact facet with its parts for both dm^2 and P^4 is visible on the mesial face of M^1. The dimensions are: M^1 – mesiodistal diameter 10.5, buccolingual diameter 10.8, maximum buccal height $>$ 6.0; M^2 – mesiodistal diameter 12.0, buccolingual diameter 12.0, maximum buccal height 5.7. This last value is partly smaller than that for the left M^2 because of variation in the enamel contours near the neck and partly because of a patch of hypoplastic enamel on the tip of the paracone. M^3 was in occlusion as judged by the contact facet on the distal surface of M^2, and some estimate of its size can be gained from an examination of the broken neck of the tooth where the roof of the crown's pulp cavity is exposed. The buccolingual diameter is likely to have been close to that of the M^2, but the mesiodistal diameter was probably not as great.

The floor of the maxillary sinus is seen almost in its entirety, extending backwards over M^3. In the anterior break, cancellous bone can be seen that fills the spaces between the roots and below the sinus floor. The greater palatine foramen and the groove for the greater palatine vessels and nerves can be seen, the foramen being at the level of the mesial margin of M^3 and the grooves continuing at least as far as M^1. That part of the transverse palatine suture that extends lateral to the greater palatine foramen is seen on the broken surface superiolingual to M^3.

KNM-FT 48 (Leakey 1962)

This is an isolated right M_2, possibly of the same individual as the previous specimens, but it seems to us to be slightly too short. Basically of simple form, with a "Y5" pattern, this tooth does not have any striking features. There is a small trigonid basin, both lingual cusps are quite sharply defined, set at the edge of the crown, and both buccal cusps are of low relief. Wear scratches, both macro- and microscopic, are almost totally in a transverse direction. There is a large oval contact facet for M_1 and a smaller, round contact facet for M_3. A poorly defined buccal groove ends in a cingular remnant about halfway down the crown. The dimensions are: mesiodistal diameter 11.2, buccolingual diameter 9.3, maximum lingual height > 5.1.

KNM-FT 3318 (Walker and Andrews 1973)

This is an isolated lower right canine of simple construction that has a long, straight, buccolingually compressed root. The crown is roughly conical and has a main cingular remnant at the distal border of the neck. There is a shallow mesiolingual groove running the length of the crown and the mesial ridge is reduced. There is a strong, plane, distobuccal wear facet running from the tip to the tubercle of the cingular remnant. Chemical erosion of the surface seems to have removed any scratches that might have been present. A patch of hypoplastic enamel is seen on the buccal surface. On superficial examination it might seem that the root was still open, but the amount of wear on the crown would imply that the root tip has been broken off. The dimensions are: maximum transverse diameter 8.2, minimum transverse diameter 6.3, buccal crown height 10.3, maximum transverse diameter of the root 7.8, minimum transverse diameter of the root 5.6 (just below the neck).

KNM-FT 7 (Walker and Andrews 1973)

This is a right mandibular fragment with parts of the crowns of P_4 and M_1 preserved in situ. Only the lateral surface of the mandibular body is preserved and this is present for nearly 17.0 below the alveolar margin at P_4, at which lower point it is beginning to curve to reach the mandibular base. At the level of the distal margin of M_1 the body of the mandible is swinging laterally from the toothrow. The buccal two-thirds of P_4 are preserved. It has two well-developed anterior cusps and a large distal basin. Dentine is exposed on the buccal cusp, and on the accessory cusp buccal to the basin, brown dentine shows through the enamel. The mesial contact facet is large and concave. It undercuts the occlusal edge of the crown and runs obliquely across the mesial face, suggesting a strongly rotated P_3. Dimensions are: mesiodistal diameter 8.2, buccolingual diameter (est.) 9.0. The M_1 crown has lost enamel from the metaconid, hypoconid and protoconid and is rather heavily worn, with only the entoconid not showing dentine exposure. Two large indented islands of dentine are exposed on the two buccal cusps and the lingual cusps are almost completely flattened. Dimensions are: mesiodistal diameter 10.3, buccolingual diameter (est.) 10.0. A large contact facet for the M_2 faces slightly buccally so that the M_2 crown would be aligned with the margin of the body that is swinging slightly laterally at this point (Figure 3).

KNM-FT 45 (Andrews 1971)

This is part of a mandible, possibly from the same individual as the type specimen. It consists of the left body from the level of M_1 to the symphysis and parts of the right body adjacent to the symphysis. The crowns and roots of P_3 and P_4 and the mesial root of M_1 are preserved. Parts of the alveolar walls of the distal root of M_1, the left and right canines, the left and right central incisors and small portions of the central incisor alveoli are seen. The P_3 crown is oval in outline with the long axis directed at about 45° to the cheektooth row. There is a strong main cusp that has a minute island of dentine exposure on its tip. A cingulum is seen running round the lingual border only. The distal basin is quite well developed and seems to be produced by the distal continuation of the cingulum. A small, but clear, wear facet runs down the mesial ridge of the tooth and another large one down the grooved distal surface of the main cusp. The strong mesial root makes the jugum at the point where the body of the mandible turns to form the mental region, while the distal root indents the alveolar margin on the post-incisive planum. There is a small contact facet for the canine on the cingulum. Dimensions are: maximum diameter

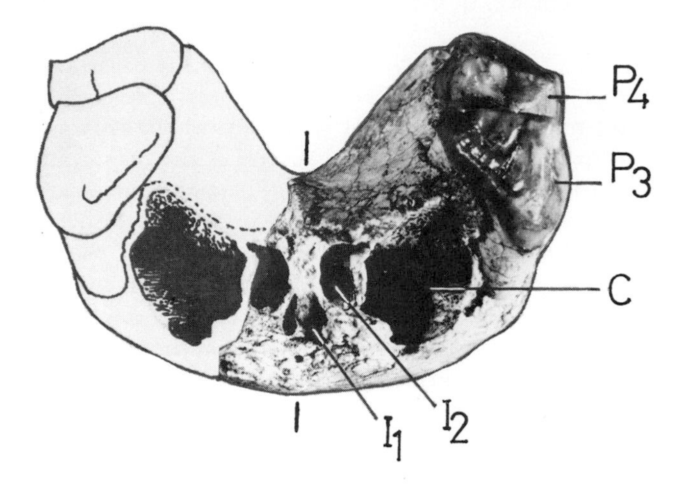

FIGURE 3
Mandible KNM-FT 7, in anterosuperior view with mirror-image right side sketched in and teeth or alveoli indicated. The midline is shown by the vertical line.

10.9, minimum diameter 6.6. The P_4 is more circular in outline and has two mesial cusps, a clear mesial fovea and a well-developed distal basin. A chip of crown is missing from the distolingual corner. Wear has mainly affected the basin, but dentine is exposed on the mesiobuccal cusp. Dimensions are: mesiodistal diameter 8.2, buccolingual diameter 8.4. From what remains of the M_1 roots it seems that they were strong and plate-like and grooved towards the septum between them.

The canine alveolus shows that the root was compressed mesiodistally and the long axis is inclined inferiorly towards the symphysis. The floor of the canine alveolus follows the contour of the sloping mental surface. The lateral incisor roots were similarly mesiodistally compressed and at the point of breakage across the symphysis they are only 2.5 apart. The tips of the canine roots passed inferior to the tips of the lateral incisors as they converged towards the midline. Part of the alveolus for the left central incisor is seen and only the basal pit of the right central is seen. These are very small alveoli at the level of the break, and the lateral incisor alveoli are slightly superior to them at this point. It is possible, even though there is slight crushing, to estimate the height of the mandibular body: at M_1 it is at least 20.0, at P_4 it is about 15.0, and at P_3 it is also about 15.0. Where preserved, the base is rounded and the outer laminar bone is thick (c. 4.0). There is a single mental foramen low on the body about 13.0 below the alveolar margin at P_4, that appears to open laterally and anteriorly. The post-incisive planum is extensive and reaches posteriorly at least as far as the mesial margin of M_1. The planum is highest near the canine but falls away more rapidly to the midline posteriorly. The very strongly developed inferior mandibular torus breaks away from the internal contour of the body at the level of the middle of M1 and curves tightly to meet its fellow from the right side. A fair approximation of the symphyseal cross section can be arrived at. Roughly oval in outline, the long axis is directed at an angle of about 25° to the mandibular base. The inferior torus is about 7.0 thick, is blunt posteriorly, and is reflected superiorly. The genial pit was deep and probably overhung by a superior torus, but this has been broken away. The mental profile was markedly retreating, and as far as can be judged from the alveoli of the anterior teeth, formed a gentle convex curve from the anterior alveolar margin to the inferior torus. The midline structures seen are: traces of the symphyseal suture on the post-incisive planum; small foramina in the genial pit; a grooved ridge on the superior surface of the inferior torus, a keel of low relief on the mental surface; a central vascular foramen in the cortical bone below the central incisor alveoli; and symmetrical cancellous bone between the lateral incisor alveoli either side of a thin median septum.

KNM-FT 8*

This is the unerupted germ of a left upper canine, together with the walls of its enclosing crypt and adjacent parts of the maxilla body. This germ, which has no root formation at this stage, is of a canine larger than that of the type specimen. The crown has all the details seen in KNM-FT 46 and shows the unworn distal crest that will be worn against the lower P3. Of interest is the way in which the floor of the maxillary sinus is also seen and this has already, by this stage, developed its anterior apex to a point just above the canine. There is a small vascular foramen opening just posterior to the bulge that represents the neck of the germ. The anterior face of the maxilla cuts across the canine at an angle of about 45 ° to the canine axis, thus implying an abbreviated snout region.

TAXONOMIC DISCUSSION

Many of the features described here for *R. wickeri* have been long taken to be characters of pongids and some have become sufficiently entrenched in people's minds for their presence in this taxon to be regarded as proof of its pongid affinities. The most notable of these characters described previously (Andrews 1971) is the presence of an inferior mandibular torus, a sectorial type of P3, and procumbant incisors. To these will probably be added the nearly straight, though diverging, toothrows shown in our reconstruction (Walker and Andrews 1973).

*This specimen has not been described or mentioned previously.

The relevance of these morphological features can be questioned from two different points of view. Firstly, assuming that they are pongid features, how does their presence in a tentative hominid ancestor affect its status? Does one expect the Hominidae to have sprung into existence fully fledged, or did the hominid features of australopithecines and later hominids evolve in a normal mammalian fashion, changing from the pongid features of Miocene dryopithecines? If the latter, how does one classify the fossils putatively intermediate between the Miocene dryopithecines and the Plio-Pleistocene australopithecines? The earlier intermediates will be more pongid-like than the later, and logically the position is reached where the earliest populations ancestral to later hominids must have been almost entirely pongid in appearance, and where only a few features, not necessarily those characteristic of modern *Homo,* distinguished them from contemporary pongids. Some would classify these hominoids as pongids on the basis of their pongid-like appearance, and others would classify them as hominids on the basis of their presumed relationship with later hominids.

This can be illustrated with one of the examples—the presence of a sectorial type of P3 in *R. wickeri.* This is a feature present in most, though not all, pongids, whether fossil or modern. It is present in the Oligocene pongids of Egypt, the early Miocene pongids of East Africa, and the middle to late Miocene pongids of East Africa and Eurasia. Since the hominids almost certainly arose from some part of this group, it must be accepted that at an early stage of their evolution hominids would have had single-cusped P3's, irrespective of the actual time of the pongid/hominid divergence. This feature in *R. wickeri,* which may have been very close to the pongid/hominid dichotomy, cannot therefore be used as evidence against its hominid status but is simply a primitive retention that tells us nothing at all about possible later development our of *R. wickeri.*

The other question that can be asked is—are these features that appear to be pongid actually so? It was shown for the mandible described in 1971 that there was a massive inferior mandibular torus. This is a condition that is often seen in living pongids but is rather uncommon in fossil pongids, especially those known from the early and middle Miocene of East Africa. It is also seen in some australopithecine mandibles, but in none of these, neither pongid nor hominid, is it developed as much as in *R. wickeri.* In particular, the symphysis is a much more massive and horizontally elongated structure than is present in living pongids. It would appear, therefore, that the symphyseal cross section and the inferior mandibular torus of *R. wickeri* are unlike both hominids and pongids, fossil and modern, and cannot therefore be said to constitute evidence for or against the inclusion of this species in either family.

For such a problematical taxon as *R. wickeri,* there seems little point in listing morphological features and then adding up the score to find which family of the Hominoidea it is closest to. Moreover, we feel that too little is known of *Ramapithecus* to come to any definite higher category taxonomic conclusion. It is still extremely instructive, however, to try to consider it as a living animal would be today, in terms of both its morphology and its adaptations. This is not easy to do with the inadequate material available, but an attempt is made and the result will be vindicated or not by more complete material.

FUNCTIONAL MORPHOLOGY

The reconstruction of the mandibular toothrows (Walker and Andrews 1973) and thereby the maxillary toothrows, shows that *R. wickeri* had a relatively narrow dental arcade (Figure 4). This is mostly a factor of the narrow incisor region, but posteriorly the toothrows do not diverge very greatly. The value of the mandibular index, intercanine breadth/inter-M_2 breadth × 100, is only 47 for *R. wickeri,* but this low value is largely due

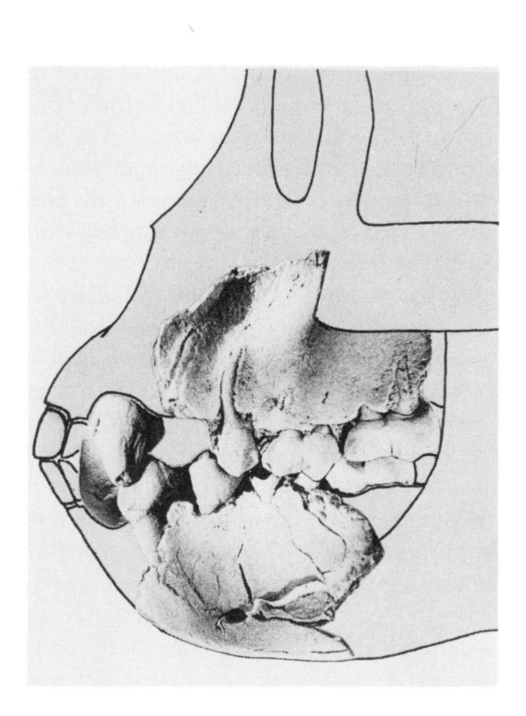

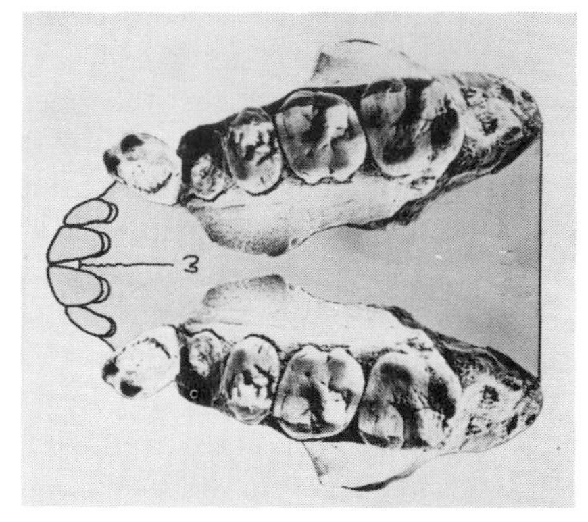

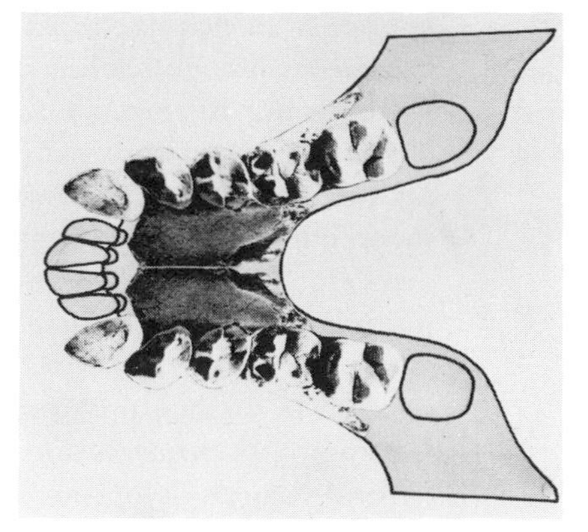

FIGURE 4
Reconstruction of the mandible, and maxilla, of *R. wickeri* in occlusal view. Reprinted from Andrews and Walker in *Nature,* 244, 1973.

to the placement of the canines lingual to the third premolars. Posterior to this, the P_3-M_3 toothrow is nearly straight, and the equivalent index comparing P_3 and M_2 mandibular breadths is 70.

Several attempts have been made to demonstrate the size of the incisors of *Ramapithecus* (Leakey 1967; Yulish 1970; Conroy 1972). Since these were based on the Fort Ternan incisor (KNM-FT 49) that is actually that of a large pongid, these arguments need not be discussed here. There is still no incisor crown of *Ramapithecus* known for sure, and the only thing we can resort to is the estimated incisor width based on our mandibular reconstruction. The index of I_2-I_2/C-M_2 × 100 is 35, which compares with a range of 42–48 for four species of early Miocene dryopithecines. This would seem to indicate relatively compressed incisors in *R. wickeri* compared with near-contemporary African pongids, which themselves have been shown to have small incisors compared with modern apes (Andrews 1973).

Upper canine size relative to first molar size has been shown to be somewhat reduced in *R. wickeri* compared with living pongids, but not significantly so (Yulish 1970). However, this form of comparison is of doubtful use, for it is open to two interpretations, one concerning the canine size relative to the molar, and the other the molar size relative to the canine. A more useful approach is that of Conroy (1972), which is to compare the complete dentition of *R. wickeri* with those of near-contemporary fossil pongids and to decide how the relative proportions of the teeth differ. Thus the canine of *R. wickeri* is similar in its length and breadth dimensions to the medium-sized specimen of *Dryopithecus africanus,* but in height it is lower crowned than all but the smallest specimens of this species. In contrast to this, the first molar is intermediate in size between the two large African dryopithecines, *D. nyanzae* and *D. major,* and is similar in size to the larger of the two Indian fossil apes, *D. indicus.* While not giving a quantitative value to the canine reduction in *R. wickeri,* it is quite clear that compared with the African dryopithecines there is a big difference in C/M^1 proportion between them and *R. wickeri.*

Correlated with this evidence for compressed incisors and small canines is the shape and size of the mandibular symphysis which is low superoinferiorly and elongated anteroposteriorly. This is an unusual condition for a primate, and an explanation given (Andrews 1971) attempted to explain it in terms of mechanical function. Thus, shearing forces set up at the symphysis by sideways movement of the mandible are best counteracted by an anteroposterior buttress of the symphysis. Such a buttress is common in primates and can be seen as the superior mandibular torus of most African dryopithecines, the inferior mandibular torus of modern pongids, and the chin of modern man. The position of the buttress, whether inferior or superior, must depend on a number of factors, among them incisor size and degree of sideways movement of the mandible. In modern apes, those species that have the incisors most strongly developed (chimpanzee and orang-utan) also tend to have the most inferior and shelf-like inferior torus. It has been shown, however, that *R. wickeri* could not have had large incisors, so we must look to another explanation. The masseter and medial pterygoid muscles attach all the way to the base of the mandible, and the force they produce during chewing is countered by the frictional drag of the occlusal surfaces of the teeth and any interposed foodstuff. The twisting effect on the mandible, transmitted through the symphysis, will be greater the deeper the mandible and symphysis, and it would be a mechanical advantage if the mandible were shallow and if the buttress at the symphysis were at the most inferior point furthest from the occlusal plane. Both these conditions are present in *R. wickeri* in greatly exaggerated form, which suggests there may have been a large sideways element in the chewing mechanism and that it may have been put to very heavy use in this species.

Further support for this hypothesis comes from a number of other morphological features in *R. wickeri.* The flare of the zygomatic process indicates a large temporal fossa and hence a massive temporalis muscle that would give a powerful bite. The anterior position of the zygomatic process leaves room for a wide mandibular ramus, and this could mean both a larger area available for attachments of the muscles of mastication as well as a greater distance between their lines of action and the point of pivot of the mandibular condyle on the articular tubercle with subsequent increase in power arms. Another feature is the thickness of the enamel on the molars; despite moderately heavy wear, enamel thickness varies from 0.8–1.9 in thickness on the occlusal surface of M^2 (measured on an X-ray photograph). The occlusal surface is relatively flat, but it is apparent that even before wear the cusps would not be prominent. The wear on the occlusal surface is almost exclusively transverse (Figure 2) with both transverse coarse scratches and many microscopic ones between them. All these factors are consistent with a masticatory system adapted for hard grinding with a strong lateral element in the movement of the mandible. This is a gnathic complex that is not found in any living or fossil pongids with the possible exception of *Gigantopithecus,* in which the canine plays a different role (Pilbeam 1970). A gnathic complex such as this would be an appropriate model for an hominid precursor, irrespective of whether the complex is found in a species that is pongid or hominid.

GEOLOGY AND DATING

The Fort Ternan sediments are within a lava and pyroclastic sequence associated with the large central volcano of Tinderet (Figure 1). The relationship of the various lavas in the area is complicated because of their derivation from at least three sources (Binge 1962). The detailed geological situation in the immediate area of the site has not yet been published (J. A. Van Couvering, in prep.), but the sediments appear to lie in a narrow fault block. Dates for the Tinderet volcanic center show that it was active from about 20 to 5.5 m.y. ago (Baker *et al.* 1971). The earliest parts of the cone are agglomerates and tuffs that overlie the early Miocene fossil sites of Koru and Songhor. Later lavas, first of nephelinite, followed by analcite basanites, cover the agglomerates over most of the mountain. The final feldsparphyric basanite gave the age of about 5.5 m.y. (Baker *et al.* 1971). Phonolites, of seemingly two ages and different sources, flowed round the Tinderet cone between 14 and 12 m.y. ago. The phonolites intercalated with the Fort Ternan sediments are stated by Baker *et al.* to be of Uasin Gishu type, but to our knowledge they have not been dated, and the correlation is based on the dates from micas within the sediments. The later phonoloites, the Kericho phonolites, cover an observed area of 1000 square miles, and Binge (1962) estimates that another 1000 square miles underlie the Mau Forest phonolitic nephelinites to the east. These phonolites lapped round the Tinderet cone during their westwards flow, and lavas of this type cap the sequence at Fort Ternan. The Kericho phonolite extrusion occurred between about 13.5 and 12.5 m.y. ago and must have been an event of great geological magnitude. The Baraget river and its tributaries have excised the junction between the plateau phonolites and the Tinderet cone and it is on the south side of this valley that the Fort Ternan fossil site lies (35°20′ E, 00°13′ S).

The sedimentary sequence at Fort Ternan was measured and described in 1961 by Bishop and Whyte (1962) when Dr. Leakey's excavations were at a relatively early stage. The section is seen (Figure 5) to consist of a sequence of tuffs that Bishop and Whyte considered to be mostly subaerial accumulations of original primary volcanic ash. Although they recognized evidence for local wash-outs and water sorting on some of the horizons, they considered that this was of local effort, probably the results of rain wash. The fossils are concentrated in two main palaeosol horizons (Beds 6 and 7), a fact that is unusual in itself, since fossils are rarely found in palaeosols that are weathering phenomena not sedimentary ones. All the tuffs are, however, highly calcareous.

We think that the evidence for more vigorous water activity, at least on some horizons, is better seen now that the excavations have been substantially enlarged. Some of the horizons (Bed 8, top of Bed 5) show fine-grained bedding and sorting with current bedding (Bed 8) and ripple marks (Bed 5). Most of the beds have sharp upper and lower junctions and one would have expected more irregularity in direct subaerial conditions, even with rain wash. The fossil bones (according to the recorded site plans) have a definite preferred orientation and are mainly concentrated in one narrow strip that wanders back and forth across the section through the various levels of the two soil horizons. The presence of crocodile and

FIGURE 5
The Fort Ternan excavation showing the fossil-bearing beds. Bed numbers from Bishop and Whyte. The staff is 14 inches high.

freshwater crabs from the palaeosols shows that water was nearby, if not actually at the spot that the sediments were deposited. The geological situation, where the phonolites lapped round a volcanic cone, would provide a sedimentary trap for any of the presumably radial drainages from Tinderet that had been base-levelled by the plateau lavas. The intercalation of plateau lavas and Tinderet ashes indicates that the site lay close to the junction of the two when the sediments were laid down. Thus it seems probable that fossils from three main habitats could be sampled in the Fort Ternan deposits: from evergreen forest on the slopes of Tinderet, from stream or lake shore and from whatever habitat (possibly open woodland to scrub) that was developing on the phonolite plateau.

Potassium-argon dating of this sequence gives a bracketing age for the fossil beds. The crystal-lithic tuff immediately underlying the palaeosols has been dated at 14.7 ± 0.7 and 14.0 ± 0.2 m.y. (Bishop *et al.* 1969) and at 14.0 m.y. (Evernden *et al.* 1964). The Losuguta-type phonolite from the top of the succession was dated first at 11.8 ± 0.2 and 11.8 ± 0.3 m.y. but was later run by the $^{40}Ar/^{39}Ar$ method and gave ages of 12.6 ± 0.7 and 12.5 ± 0.4 m.y. This further experiment was carried out because the conventional K/Ar ages were not in line with those from other

Kericho phonolites—6 ages ranging from 12.3 ± 0.2 to 13.4 ± 0.2 [an age of 12.1 ± 0.2 for the same rock that gave the 13.4 m.y. date is considered as probably too young (Bishop *et al.* 1969)]. The other Kericho phonolites have not been given the same treatment, and comparisons stand between conventional K/Ar and $^{40}Ar/^{39}Ar$ (including step heating) methods. The fossil-bearing sediments are thus bracketed between ages of about 12.5 and 14.0 m.y. Whether the fossils are nearer the older or younger date is not a question that can be resolved at present, but it has been implied that they are close to the 14.0 m.y. date on the basis of the possible primary nature of the biotites in the crystal-lithic tuff, which lies immediately below the palaeosols (Bishop 1971). It has been argued (Bishop, pers. comm.) that biotites of this size could not be reworked for any great distance and we are not in disagreement with this argument. However, there are two biotite fractions in Bed 5, one that gives the date quoted and one "thought to be derived from a preexisting rock" (Bishop *et al.* 1969) as well as pebbles of carbonatite and nephelinite. The bed is quite thin (about 15 cm.) with well-defined upper and lower interfaces and could, we believe, be reworked from a short distance (i.e., no further than, say, half the width of the present valley) without undue deterioration of the books of mica. It may prove that the fossils are close to 14.0 m.y. old, but we feel with the possibilities of reworking that it would be wiser to limit ourselves to the bracketing ages.

FLORA AND FAUNA

At least three species of higher primates other than *R. wickeri* have been found in the Fort Ternan deposits. Some of these have been mentioned briefly by Leakey (1968), but some of his identifications have not been substantiated. The three species are assigned to the following taxa:

(1) Dryopithecinae cf. *Limnopithecus legetet* (Leakey 1968: 527)
(2) Dryopithecinae *Dryopithecus* cf. *nyanzae* (Leakey 1968: 528)
(3) Dryopithecinae *Dryopithecus* cf. *africanus* (Leakey 1968: 528)

The isolated teeth assigned by Leakey (1968: 528) to the Oreopithecidae and Cercopithecoidea correctly belong to the Suidae.

Dryopithecinae, cf. *Limnopithecus legetet*

This is the most common primate in the Fort Ternan deposits and is represented by 14 specimens (KNM-FT 11–15 and 17–25). It is probable that specimens KNM-FT 19–24 belonged to one individual on the basis of similar morphology and state of wear. These specimens consist altogether of a left mandibular fragment with C-M_3, isolated right P_4-M_2, and an isolated left M^3. It is possible that two isolated incisors belong to the same individual; KNM-FT 11, left I^1 and KNM-FT 15, right I_2.

The other specimens can be assigned to three individuals. KNM-FT 14 is an immature posterior mandibular fragment with M_1 exposed in its crypt. Another individual is represented by KNM-FT 12 and 17. The former is a relatively large edentulous mandible with the symphysis and left body preserved and the roots of right I_2 to left P_4; and the latter is a left M_1 rather larger than the M_1 of the first individual. Of the remaining two teeth, one (KNM-FT 18, right P_4) cannot be fitted into any of the previous individuals because of its small size and distinctive morphology. The other (KNM-FT 15) is a left upper C that might belong with the first individual, but it is extremely similar to the canine of *R. wickeri,* and there is a slight possibility that it may be a size variant of this species.

In general size, the teeth of this taxon fall into the top half of the range of variation of *L. legetet.* The lower canine is small, and is beyond the range and 95% confidence limits for the canine of *"Limnopithecus" macinnesi.* In most dimensions the specimens are slightly smaller than the smallest specimens of *Pliopithecus antiquus* from Europe (Figure 6).

In morphology most of the teeth of this taxon are very close to *L. legetet.* The central incisor is relatively slim and broad. The canines and lower premolars are very close to those of *L. legetet* and are less specialized than in *"L." macinnesi* or *P. antiquus.* The lower molars are relatively elongated and narrow. They are slightly longer than the equivalent lower molars in *L. legetet* but are much narrower. Both upper and lower M3 are reduced in size. The latter is slightly larger than M_1 but is smaller than M_2, which is conspicuously the largest tooth in the molar toothrow.

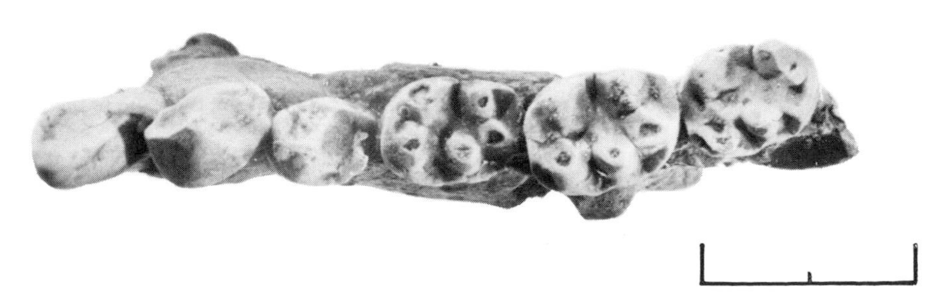

FIGURE 6
Occlusal view of reconstructed mandible, KNM-FT 19-24 with left C-M_3, attributed to *L. legetet.*

Dryopithecinae, *Dryopithecus* cf. *nyanzae*

Seven specimens represent this taxon; an upper and lower C (KNM-FT 39 and 28), an I^1 (KNM-FT 49, previously assigned to *R. wickeri*), two M_3 (KNM-FT 34 and 40), a P_3 (KNM-FT 35), and a distal humerus (KNM-FT 2751). These specimens all fall within the ranges of variation of both *D. nyanzae* and *D. fontani* which dentally are very similar to each other. No upper molars, which are the most distinctive of the teeth of this taxon, are preserved from Fort Ternan, and it has been decided to assign specimens to this species only provisionally (See Table 1).

The identification of the two third molars is doubtful. Both are heavily worn, but despite this no dentine has been exposed on the occlusal surfaces. Breaks in the enamel show enamel thicknesses ranging from 0.3 over the tips of the cusps to 1.2 on the sides of the cusps and 1.4 in the sulci between them. In addition, the general appearance of the crowns is very similar to the M_3 of *Ramapithecus punjabicus,* with the flattened occlusal surface, and the metaconid raised above the rest of the occlusal surface. None of these features is sufficiently diagnostic to indicate

TABLE 1
Measurements of Dentition of the Fort Ternan Dryopithecinae.

	MD LENGTH	BL BREADTH	$\frac{BL}{MD} \times 100$
D. africanus			
KNM-FT 16 M^1 or M^2	7.7	9.7	126.0
KNM-FT 29 P^3	5.7	9.2	161.4
D. nyanzae			
KNM-FT 28 C_1	12.5	8.8	70.4
KNM-FT 34 M_3	12.3	10.2	82.9
KNM-FT 35 P_3	13.1	7.4	56.5
KNM-FT 39 C^1	15.5	11.7	75.5
KNM-FT 40 M_3	13.0	10.7	82.3
KNM-FT 49 I^1	9.9	6.8	68.7
L. legetet			
KNM-FT 11 I^1	6.2	4.2	67.7
KNM-FT 14 M_1	6.0	4.1	68.4
KNM-FT 15 C^1	7.5	5.8	77.3
KNM-FT 17 M_1	6.3	5.2	82.5
KNM-FT 18 P_4	4.3	3.8	88.4
KNM-FT 19 M^3	4.8	6.4	133.3
KNM-FT 20-24 C_1	5.6	3.7	66.1
P_3	5.6	3.8	67.8
P_4	4.7	4.1	87.2
M_1	6.2	4.8	77.4
M_2	7.0	5.3	75.6
M_3	6.4	4.9	76.5
KNM-FT 25 I_2	3.1	4.3	138.5

MD = mesio-distal; BL = bucco-lingual

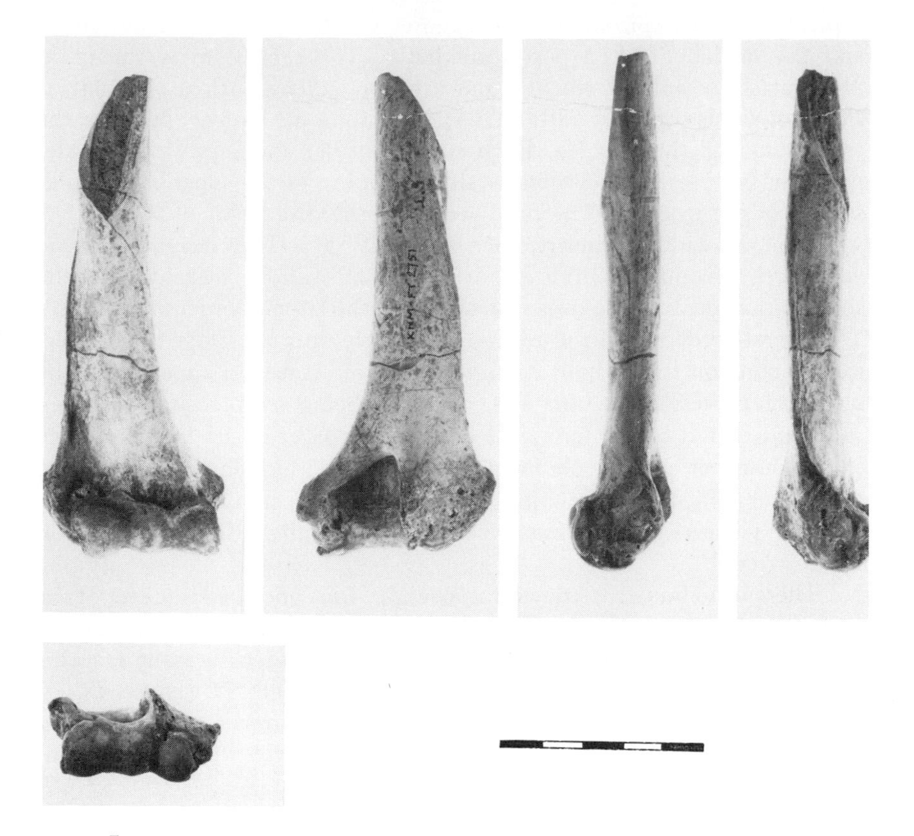

FIGURE 7
The Fort Ternan humerus. From left to right, anterior, posterior, medial and lateral views, and below, distal view.

definitely that these specimens belong to *Ramapithecus,* and the probable presence of a buccal cingulum makes it more likely that they belong to *Dryopithecus.*

Finally there is a distal humerus (KNM-FT 2751) that is important as one of the better preserved Miocene hominoid limb bones (Figure 7). It is the distal one-third of a right humerus of an adult hominoid. Apart from minor cracks and the main oblique upper fracture, the bone is remarkably well preserved. At the highest measurable point, the shaft is oval in section (16.3 × 13.7) and compressed anteroposteriorly. This compression increases distally. The brachialis flange is rather poorly developed but the brachioradialis crest is well developed and has its border turned anteriorly. This crest takes off from the shaft at about 84.0 above the most distal part of the capitulum and arises from the posterolateral surface of the bone. The shaft is bowed, concave medially, with the effect that its long axis is offset on the distal epiphysis, falling more nearly through the capitulum. Both epicondyles are rugose, the lateral reflected anteriorly and the medial directed posteromedially. The capitulum is the usual globular shape but the trochlea has a gently convex surface although it slopes upwards superolaterally. There is a strong medial keel between the capitulum and the trochlea. Both supracapitular and supratrochlear depressions are present and there is slight lipping of the superior articular surface of the trochlea. The total width of the distal part of the bone is 45.0 and the width of the articular surface is 35.3. The olecranon fossa is very deep, a rounded triangle in outline and offset from the midline of the shaft so that the midline runs just lateral to the lateral border of the fossa. The lateral border is built up into a posteriorly directed flange that has articular bone on its medially facing steep face. It would appear that the olecranon process of the ulna was capable of being accepted into the deep fossa and that it would have had an articular facet on its lateral surface that would act against the flange to resist abduction and adduction movements when the elbow joint was in full extension or even hyperextension. The articular lipping of the trochlea might indicate that flexion was restricted.

There is no humerus known for *Ramapithecus,* and the large size of the molars of *R. wickeri* does not allow sorting of the bone by a humerus/toothrow regression (Dr. H. McHenry, pers. comm.). In many features that can be compared, the bone is similar to those of cf. *D. fontani, D. africanus* and cf. *D. major,* and the most reasonable course at present is to assign it provisionally to the large *Dryopithecus* species known only from teeth at Fort Ternan.

Dryopithecinae, *Dryopithecus* cf. *africanus*

Only three specimens are referred to this species. One of these is a very characteristic molar of this species. The crown has a massive lingual cingulum and is relatively broad. The P^3 (KNM-FT 29) is also typical of *D. africanus* both in size and morphology (Figure 8). In particular it has the highly developed buccal cusp of this taxon, although in this specimen the buccal cusp is even taller than is commonly seen in *D. africanus.* These two teeth are too small to be accommodated either in the large dryopithecine just described or in *Ramapithecus wickeri,* and in addition they differ greatly from the latter in morphology. It seems unavoidable, therefore, to accept the presence of a fourth species of hominoid primate even on the slender evidence of two isolated teeth.

One other specimen is provisionally assigned to this species, a bilaterally compressed and blade-like upper canine (KNM-FT 41). This is not typically *D. africanus* in morphology, but it is the right size, and it might suggest, if correctly assigned, that this species was more specialized in Fort Ternan times than in the early Miocene.

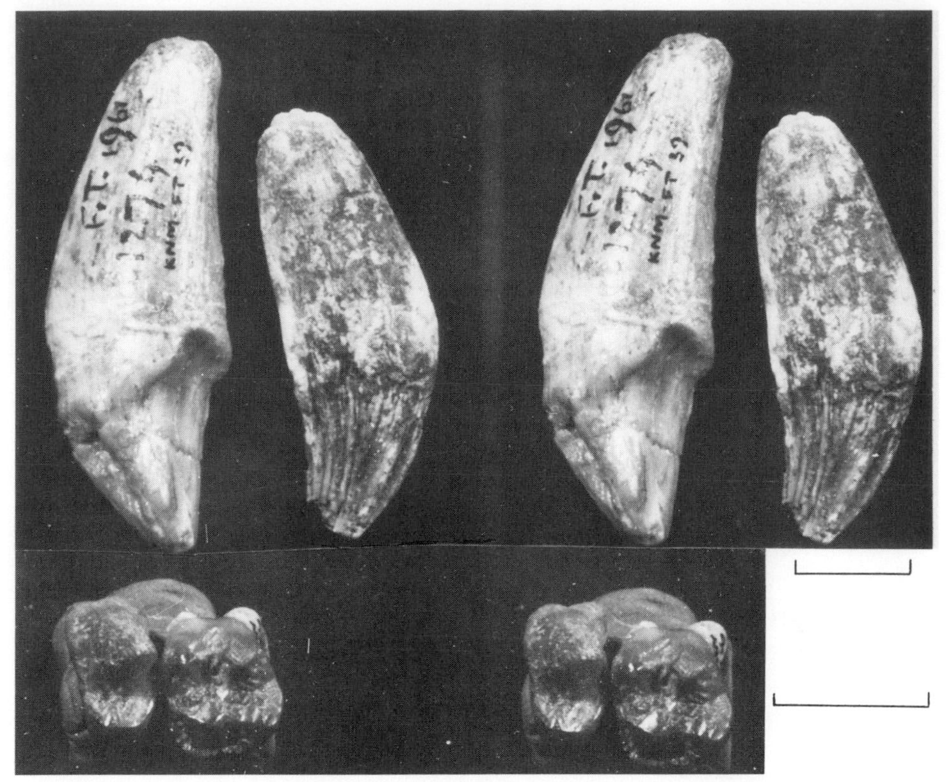

FIGURE 8
Stereo pairs of the upper canine attributed to *D. nyanzae,* KNM-FT 39, lingual view, and the two specimens attributed to *D. africanus,* KNM-FT 16 M^1 or M^2, and KNM-FT 29 P^3, occlusal views. An unidentified canine is also illustrated on the right, KNM-FT 41, lingual view: this specimen is quite unlike known dryopithecine species, but it is always possible that it is from a male individual of *R. wickeri.*

The rest of the fauna known from Fort Ternan can only be mentioned very briefly. A faunal list is given in Table 2, but since much of the collection has yet to be described it is given mostly only to family.

The few plant remains from Fort Ternan have not yet been studied. At the locality, one leaf is still in situ in a fallen block of tuff and very probably belongs to a member of the family Sterculiaceae, a family of mostly woodland trees and shrubs.

The land snails constitute a distinctive ecological group. The two most common genera, *Maizania* and *Homorus,* are equatorial African forms restricted today to evergreen forest; less common is the extant West African species, *Burtoa nilotica,* living now in woodland, and two species of *Gulella,* a genus with worldwide distribution in forest and thicket (Verdcourt 1972). The gastropod assemblage as a whole strongly suggests the presence of evergreen forest in the vicinity of the Fort Ternan deposits.

TABLE 2
Fort Ternan Faunal List.

PLANTAE
- cf. Sterculiacae
- sundry indet.

ANIMALIA

- MOLLUSCA
  - Cydophoridae
    - *Maizania*
  - Enidae
    - *Cerastua*
  - Sterogyridae
    - *Homorus*
  - Achatinidae
    - *Burtoa nilotica*
  - Heliocarionidae
    - *Trodonania*
  - Streptoxidae
    - *Gulella*
- ARTHROPODA
  - Potamidae
- REPTILIA
  - Chamaeleodontidae
  - Crocodilia
    - *Crocodylus* sp.
- AVES
  - Struthionidae
    - *Struthio* cf. *asiaticus*
  - Falconidae
- MAMMALIA
  - Insectivora
    - *Rhynchocyon*
  - Primates
    - Anthropoidea
      - *Dryopithecus* cf. *nyanzae*
      - *Dryopithecus* cf. *africanus*
      - cf. *Limnopithecus legetet*
      - *Ramapithecus wickeri*

MAMMALIA (Cont.)
- Prosimii
  - Lorisinae
- Rodentia
  - Cricetidae
    - *Leakeymys ternani* (1)
  - Anomaluridae cf. *Anomalurus*
  - Sciuridae (1)
  - Pedetidae
  - Phyomyidae (1)
- Carnivora
  - Hyainailouros (5)
  - Canidae or Amphicyonidae
  - Felidae
  - Mustelidae
  - Viveridae
- Tubulidentata
  - *Orycteropus* sp., 2 specimens (6)
- Proboscidea
  - *Gomphotherium*
  - *Deinotherium*, 1 specimen only
- Perissodactyla
  - Rhinocerotidae
    - *Paradiceros mukirii* (2)
- Artiodactyla
  - Giraffoidea
    - Giraffidae
      - *Palaeotragus primaevus* (3)
      - *Samotherium africanus* (3)
- Suidae
  - *Listriodon*
- Tragulidae
  - *Dorcatherium* (4)
- Bovidae
  - *Protragoceros labidotus* (4)
  - *Oioceros tanyceros* (4)
  - *Pseudotragus potwaricus* (4)
  - *Gazella* (4)

Identifications by present authors unless otherwise stated: (1) Lavocat 1964; (2) Hooijer 1968; (3) Churcher 1970; (4) Gentry 1970; (5) Savage 1973; (6) Pickford, in prep.

It does not rule out the possibility of more open conditions nearby, but land gastropods restricted to these environments have not been found in the sediments. Fragments of potamid crabs and a few isolated crocodile teeth show that fresh water was nearby.

The only definite bird identification so far at Fort Ternan is a species of ostrich, *Struthio* cf. *asiaticus*. This is represented by a tarsometatarsus and by eggshell fragments. The latter are important because they show that the ostrich must have been nesting nearby, for complete ostrich eggs are too big for scavengers to carry and they would not be carried far in water before being broken up and the pieces scattered. Several fragments were collected crushed together in blocks of sediment. The ostrich today lives in conditions varying from semi-desert to bushland and open woodland. Little is known of their nesting habitat preferences, but they seem to be variable.

The insectivore genus, *Rhynchocyon,* is represented by a single upper molar of a large-sized animal, larger than the living species, but similar to the largest of the early Miocene species. All the living species of this genus are restricted to forest and dense evergreen bush (where the forest has been cleared). It is impossible to be sure for the Fort Ternan species, but in view of the many adaptations of the species today to life in forest conditions (G. Rathbun, pers. comm.), and the close dental similarity between the Miocene and recent species, it is likely that the Miocene species were also forest living. The presence of a single specimen at Fort Ternan might, therefore, indicate the past presence of forest conditions somewhere within a short distance of the site.

The pongids have already been described in this paper. In addition, a single specimen of a species of lorisine has recently come to light (Walker, in prep.). All the living counterparts of these primates are exclusively forest living, and it has been shown elsewhere (Andrews and Van Couvering, 1975) that some at least of the early Miocene species were also probably forest animals.

The most common of the rodents, *Leakeymys ternani,* is of uncertain affinities, possibly a cricetid (Lavocat 1964). Three other rodent families are represented at Fort Ternan: Pedetidae by a single tooth, Anomaluridae by a single mandibular body, and Sciuridae by a single tooth. Peditids are represented today by a single genus of burrowing animals. They are restricted to areas of well-drained sandy soils ranging from semi-desert vegetation to moderately dense bushland. Anomalurids are represented today by four genera, but the affinities of the Fort Ternan specimen appears closest to *Anomalurus.* All living species of this family inhabit forest today, even the flightless ones, and the adaptation for gliding would seem to indicate an ancient forest adaptation. As with the single specimen of elephant-shrew already mentioned, the specimen of cf. *Anomalurus* may indicate the past presence of forest within close proximity to Fort Ternan.

The carnivores mostly have not been studied, but at least five families are represented in the Fort Ternan fauna. These range in size from small viverrids to a very large hyaenodontid, the canine of which is much larger than that of the present day lion (Savage 1973).

Several specimens of fossil aardvaark have been identified and will be described by M. Pickford. The material is too incomplete to record anything other than its presence. The common proboscidean is a species of *Gomphotherium,* and only one scrap definitely attributable to *Deinotherium* has been recorded so far. There is only one species of small rhinocerotid, described by Hooijer (1968).

The Fort Ternan fauna is dominated by the artidactyls, particularly the bovids and giraffids. In addition, the Maboko lagomerycid, *Climacoceras,* is represented by two ossicones, and there are several specimens each of *Dorcatherium* (Tragulidae) and a large *Listriodon* (Suidae). The bovid fauna is dominated by two species, *Protragoceros labidotus* and *Oioceros tanyceras* (Gentry 1970). *Oioceros* has cursorial adaptations in the limb bones, and the habitat deduced by Gentry (1970:310) was open or lightly wooded country. In addition there are a few specimens assigned by Gentry to *?Pseudotragus potwaricus,* and the hypsodont teeth of this species were considered to be inconsistent with a forest-living ecology. The two giraffid species described by Churcher (1970) also showed some cursorial adaptations, and Churcher concluded that the environment indicated was some kind of open steppe.

The faunal evidence thus corroborates the assumptions based on the geological settings. The remains are sampled of lake and lake-side animals, wet evergreen forest animals, and more open-country animals. On minimum numbers of known individuals, the more open-country forms predominate, and animals with presumed forest affinities are less common and more poorly preserved. *Ramapithecus wickeri,* although not the commonest primate, is represented by the best preserved primate specimens. Although this is not strong enough evidence to establish *Ramapithecus* was an open-country form, the dental and gnathic complex is certainly not what is found today in forest-living primates. It has been demonstrated that the area of western Kenya, where Fort Ternan is located, was under evergreen forest during the preceding Miocene (Andrews and Van Couvering 1975). It is intriguing that the first known dental adaptations to a grinding mastication in hominoids occurs at a time when (a) the first bovids and ostriches appeared in the region, and (b) the large-scale flows of plateau phonolites would have made great topographical and profound vegetational changes.

References Cited

Andrews, P.
1971 *Ramapithecus wickeri* mandible from Fort Ternan, Kenya. Nature 23:192–194.

1973 Miocene Primates (Pongidae, Hylobatidae) of East Africa. Ph.D. thesis. Cambridge University.

Andrews, P. and J. A. H. Van Couvering
1975 Paleoenvironments in the East African Miocene. *In* Approaches to Paleobiology, 5. F. Szalay, ed. Basel: S. Karger Publ. AG. pp. 62–103.

Baker, B. H., L. A. J. Williams, J. A. Miller and F. J. Fitch
1971 Sequence and geochronology of the Kenya rift volcanics. Tectonophysics 11:191–215.

Binge, F. W.
1962 Geology of the Kericho District. Geol. Survey Kenya, Report 50. 67 pp.

Bishop, W. W.
1971 The late Cenozoic history of East Africa in relation to hominoid evolution. *In* The Late Cenozoic Glacial Ages. K. K. Turekian, ed. New Haven: Yale University Press. pp. 493–527.

Bishop, W. W., J. A. Miller and F. J. Fitch
1969 New potassium-argon age determinations relevant to the Miocene fossil mammal sequence in East Africa. Am. J. Sci. 267:669–699.

Bishop, W. W., and F. Whyte
1962 Tertiary mammalian faunas and sediments in Karamoja and Kavirondo, East Africa. Nature 196:1283–1287.

Churcher, C. S.
1970 Two new Upper Miocene Giraffids from Fort Ternan, Kenya, East Africa: *Palaeotragus primaevus* n.sp. and *Sanotherium africanum* n.sp. *In* Fossil Vertebrates of Africa, Volume 2. L. S. B. Leakey and R. J. G. Savage, eds. London: Academic Press.

Conroy, G. C.
1972 Problems in the interpretation of *Ramapithecus:* with special reference to anterior tooth reduction. Am. J. Phys. Anthrop. 37:41–47.

Evernden, J. F., D. E. Savage, G. H. Curtis and G. James
1964 K-Ar dates and Cenozoic mammalian chronology of North America. Am. J. Sci. 262:145–198.

Gentry, A. W.
1970 The Bovidae (Mammalia) of the Fort Ternan fossil fauna. *In* Fossil Vertebrates of Africa. Volume 2. L. S. B. Leakey and R. J. G. Savage, eds. London: Academic Press. pp. 243–324.

Hooijer, D. A.
1968 A Rhinoceros from the Late Miocene of Fort Ternan, Kenya. Zoologische Mededelingen 43:77–92.

Lavocat, R.
1964 Fossil Rodents from Fort Ternan, Kenya. Nature 202:1131.

Leakey, L. S. B.
1962 A New Lower Pliocene Fossil Primate from Kenya. London: Ann. and Mag. Nat. Hist. 13:689–696.

1967 An Early Miocene Member of the Hominidae. Nature 213:155–163.

1968 Upper Miocene Primates from Kenya. Nature 218:527–528.

Pilbeam, D. R.
1970 *Gigantopithecus* and the Origins of Hominidae. Nature 225:516–519.

Savage, R. J. G., and W. R. Hamilton
1973 Introduction to the Miocene mammal faunas of Gebel Zelten, Libya. Bull. Brit. Mus. (Nat. Hist.), Geol. 22:516–527.

Simons, E. L., and D. R. Pilbeam
1965 Preliminary revision of the Dryopithecinae (Pongidae, Anthropoidea). Folia Primat. 3:81–152.

Verdcourt, B.
1972 The zoogeography of the non-marine mollusca of East Africa. J. Conch. 27:291–348.

Walker, A. C., and P. Andrews
1973 Reconstruction of the dentale arcades of *Ramapithecus wickeri.* Nature 224:313–314.

Yulish, S. M.
1970 Anterior tooth reductions in *Ramapithecus.* Primates 11:255–263.

A fossil hominid jaw has been partly exposed by erosion at Ileret, Kenya. Richard Leakey works to disengage it from the matrix that has encased it for about 1½ million years.

Richard Leakey and Glynn Isaac:

East Rudolf:

An Introduction to the Abundance of New Evidence

In the past seven years, the supply of hard facts about early man has been explosively augmented through the discoveries made at East Rudolf. These finds include more than a hundred hominid fossils and a rich archaeological record. It is also evident that the vast spatial extent of the fossil-bearing outcrops at East Rudolf provide an unusual opportunity for reconstructing a prehistoric landscape and exploring early man's environmental relationships. Richard Leakey and Glynn Isaac provide an introduction to the contribution of East Rudolf through a brief review of findings, a bibliography and through reprints of some of the most dramatic announcements of new evidence.

In 1968 Richard Leakey began the exploration of a vast area of fossiliferous sediments on the east side of Lake Rudolf.[1] In the years that followed, this area rapidly emerged as one of the world's major sources of evidence relating to human origins. More than a hundred hominid fossils have been found, including several more or less complete crania, numerous well-preserved postcranial bones, and some partial skeletons. Also, at

[1]While this paper was in press, the President of Kenya announced that the lake formerly known as "Lake Rudolf" would have the name Lake Turkana. All future publications of our research group will honor this change.

East Rudolf as at Olduvai, the hominid skeletal evidence is complemented by a well-preserved archaeological record that includes evidence of the technological and economic activities of the early hominids.

While the succession of discoveries was being made, an international team of scientists developed the East Rudolf Research Project so as to make possible the immense task of investigating the paleoanthropological traces and of reconstructing the circumstances of early hominid life. Reports on the exploratory work of this group are still in active preparation, and here we offer only a brief, provisional account of the research, in order to indicate something of its contribution to the subjects with which this volume is concerned.

The rate of discovery has been such that *Nature* has published annually, for the past several years, one or several announcements of the new fossils. In order to represent something of the actuality of the remarkable succession of finds, we are presenting reprints of three of the most important and most recent notices. These are announcements, not monographs, and should be read as documents indicative of the amazing spate of new potential knowledge, rather than as complete statements of anatomical detail or the most authoritative representation of current interpretations. The bibliography that accompanies this essay provides a guide to such follow-up papers as have appeared at the time this volume goes to press. Descriptive and comparative studies are still in progress.

GEOLOGICAL BACKGROUND

The northern half of the Lake Rudolf basin is a large asymmetrical tectonic depression that is associated with the Eastern Rift complex at a point along its length where it has a somewhat diffuse branching configuration. The deepest part of the depression is a huge trough or graben that extends down at least several kilometers into the earth's crust and which has been steadily filled with sediments as its floor subsided. The modern lake waters, which are not very deep, are more or less confined to that portion of the trough that has not been filled with sediment. The northeastern margin of the flooded graben is formed by a shelf-like strip of terrain some 20 miles wide that lies between the lake and an irregular, low escarpment formed by uplifted tertiary volcanic rocks (Figure 1).

At the present time this shelf is well drained and subject to the forces of erosion, but during the Pliocene and early Pleistocene the lake waters were at higher relative levels, so that streams and rivers were obliged to drop their loads of sand, silt and clay on the shelf. In consequence, over time, there developed a mantle of layered, fossil-bearing sediments several hundred feet thick.

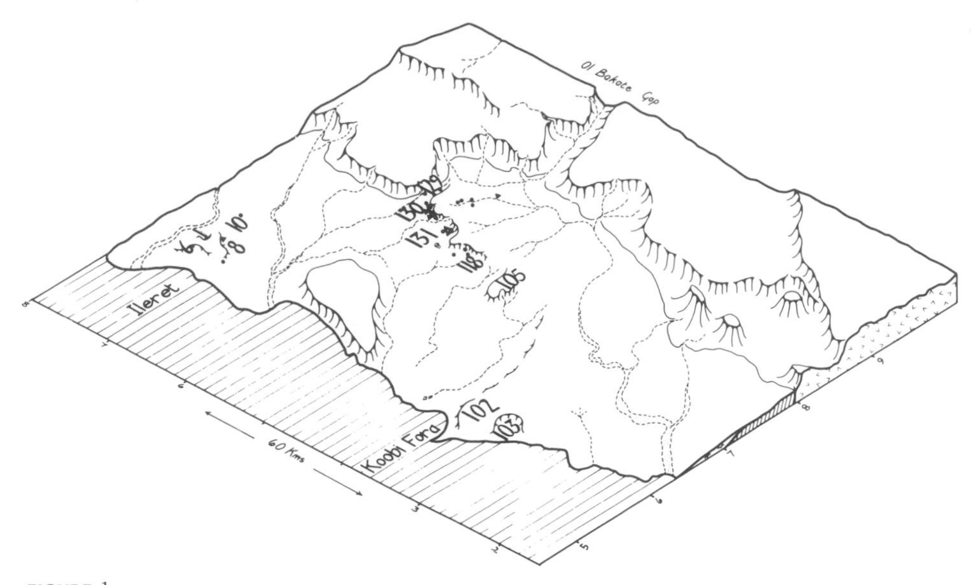

FIGURE 1
A block diagram showing the topography of East Rudolf as it is today. Each significant area has been given a reference number, and a selection of these is shown in the figure.

Subsequently, the effective height of the lake waters fell. This may have been caused by a shift to drier climatic conditions, by the disruption of rivers with larger catchment areas than those of today, or by subsidence of the central tectonic depression. Very probably all these factors were involved (Figure 2).

The fall in lake level has given streams the erosive power to cut back into the sedimentary layers on the shelf and dissect them into big tracts of badlands topography (Figure 3).

Outcrops of the sediments have been traced over an area that is more than 50 miles from north to south and some 15 to 20 miles from west to east. Part of the importance of East Rudolf lies in the fact that its huge geographic scale creates opportunities that do not exist in many other fossil-bearing localities of more restricted extent.

Sediments were deposited on the shelf during a time range of some three or four million years and can be divided into four units (Bowen and Vondra 1973):

	Age
Galana Boi Beds	5,000–11,000 years
Guomde Formation	~ 0.7 m.y.?
Koobi Fora Formation	1–3½ m.y.
Kubi Algi Formation	3½–4½m.y.

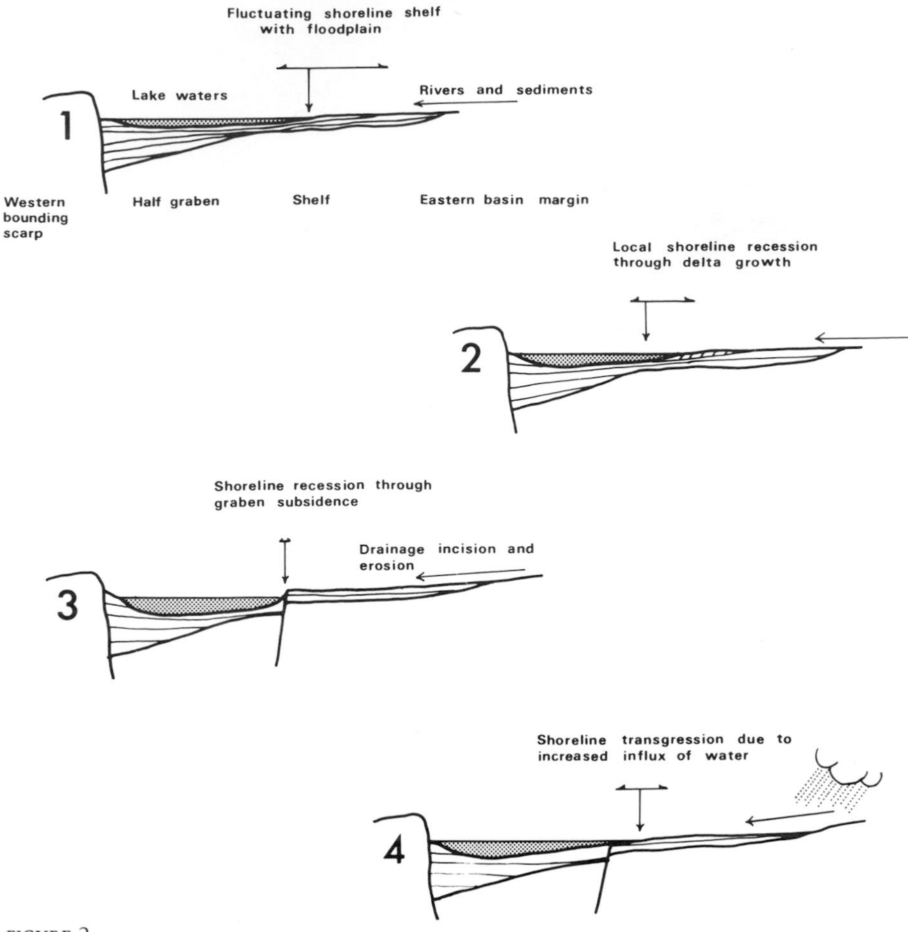

FIGURE 2
Diagrammatic representation of the kind of conditions under which sediments have accumulated in the Lake Turkana (Rudolf) graben, together with indications of factors affecting the position of the shore line.

Of these the Koobi Fora Formation is the most extensive and has yielded the most important paleoanthropological evidence, so that it is with this that the remainder of the review will be concerned.

The Koobi Fora Formation consists of a varied suite of sediments that were laid down under different local circumstances (Bowen 1974). One distinctive facies was deposited from stable lake waters which laid down fine claystones and mudstones with occasional thin interfingering beds of sandstones. The main fossils in these beds are those of aquatic creatures (fish, turtles, crocodiles and hippos, plus ostracods and molluscs) with only rare terrestrial forms that have been washed in. The facies that most

FIGURE 3
(a) In the dry conditions that prevail along the east side of Lake Rudolf, erosion dissects the ancient sedimentary rocks and the fossils that are exposed remain uncovered by vegetation or soil. (b) Vast areas are available for paleontological searches and studies.

strongly contrasts with these lacustrine sediments consists of fluviatile beds laid down along the inland edge of the shelf. These are the deposits of vigorous streams and rivers which dumped sands and coarse gravels in their shifting beds and spread broad expanses of sandy floodplain silts outside their banks. This facies is best known from the upper part of the Koobi Fora Formation where it contains a great wealth of archaeological material plus some hominid and other mammal fossils.

Between these extremes of conditions were the environments of deposition in which most of the fossil-bearing beds were formed; namely lake beaches, lakeside floodplains, and delta floodplains. These produced

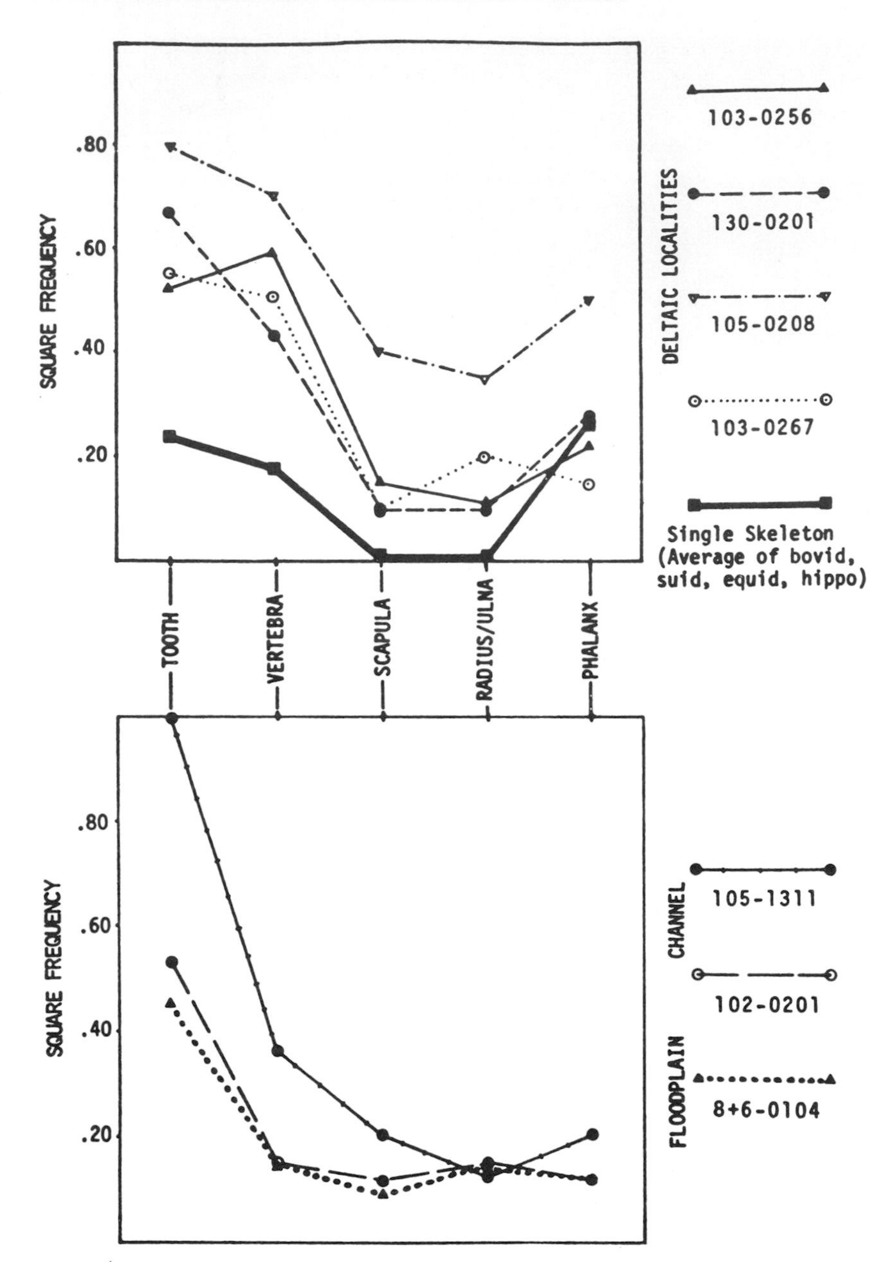

FIGURE 4
Right—The relative proportion of different bones found in sediments deposited under delta or lake margin conditions, as measured by counts of the number of 10-meter sample squares that contained the particular bone.
Left—Relative proportions of the same bones in sediments laid down by a river channel or on its flood plain. Teeth, being durable and transportable, are over-represented.

very varied suites of sediments that show oscillating depositional circumstances and include thin layers of lacustrine mud plus sands belonging to delta fronts, beaches or low energy channels. There are also extensive beds of swamp clays and silts, and there are biogenic layers of algal limestones, and shingles of mollusc shells.

In the early stages of deposition of the Koobi Fora Formation the lake waters formed a fluctuating shoreline 10 to 15 miles east of the present-day shore, but by Upper Member times the shoreline had withdrawn to a fluctuating position not far from the present margins. Subsequent changes have involved a drop in level without change in shoreline position.

Interbedded with the normal sediments are a series of beds composed largely of volcanic ash. These appear to derive from eruptions in the headwaters of the rivers that entered the basin in this area. Evidently, the eruptions so filled the catchment area with volcanic ash and pumice that for short periods the rivers delivered more or less nothing else into the lake basin. These tuff beds thus provide convenient time markers within the complex series of layers, and the pumices contained in them can be dated by means of the K/Ar technique. Each major tuff has a specific label. Figure 4 shows something of the stratigraphic framework that the tuffs help to delineate.

A general survey of the geology of the sediments was made by Bruce Bowen (Bowen and Vondra 1973; Bowen 1974), while a detailed study of the volcanic ash horizons and of changing paleogeography is currently in progress by Ian Findlater (Findlater 1975). When this has been completed we shall have a much more detailed picture of the settings in which fossils were preserved.

Volcanic rocks form the floor on which the fossil-bearing sediments rest, as well as the rim around their eastern margins. Studies of the stratigraphy, composition, and structure of these rocks are being undertaken as a contribution to understanding the geological history of the region (Fitch, Watkins and Miller 1975). Fossil-bearing sediments of Miocene age have been found enclosed by the volcanics (Harris and Watkins 1974). To the south of the East Rudolf study area, there are more fossiliferous Miocene deposits that are being investigated by a team under the leadership of R. Savage.

DATING

The age of fossil-bearing deposits such as those of the Koobi Fora Formation can be assessed in a number of different ways. The following are some of the principal methods:

Isotopic dating: especially potassium-argon dating (K/Ar and Ar^{40}/Ar^{39}) and fission-track dating, both of which can be applied to volcanic rocks (Miller 1972; Fleischer and Hart 1972; Hurford 1974).

Paleomagnetism: sequences recording changing polarity of the earth's magnetic field can be correlated with a well-dated global sequence of reversals (Cox 1969).

Paleontology: faunal assemblages change sequentially through time and the approximate position of any assemblage of fossils can be gauged by careful comparative work. However, this method can be complicated by differences in habitat and by biogeographic factors, so that it should be used only with great caution.

Ideally, the chronology of stratigraphic sequences such as those at East Rudolf should be dated primarily by geophysical techniques such as K/Ar and paleomagnetism, so that a time scale can be constructed that is independent of the evolutionary story that is being investigated. The scientists of the East Rudolf Research Project have worked extremely hard to achieve such an independent chronology, but the area poses unusual difficulties, and there have been some vigorous scientific discussions about the interpretation of different lines of evidence. The issues are not yet settled but steady progress is being made along several lines of investigation.

Frank Fitch and Jack Miller have made more potassium-argon measurements on samples from East Rudolf than have ever before been done on a fossil-bearing sequence in East Africa (Fitch and Miller 1975). However, in spite of their extensive efforts, several tuffs have yielded a very puzzling scatter of apparent ages. In order to discriminate true ages from the scattered results, they have employed a powerful new dating technique that they themselves have done much to develop, namely the $Argon^{40}/Argon^{39}$ step-heating method (Miller 1972). Table 1 shows the results available from this technique at the time of writing. It should be

TABLE 1
$Argon^{40}/Argon^{39}$ step-heating age determinations for tuffs in the Koobi Fora Formation (Fitch *et al.* 1974).

Tuff	Age
* { Chari Tuff	1.22 ± 0.01
{ Karari Tuff	1.32 ± 1.10
{ Koobi Fora Tuff	1.48 ± 0.23, 1.57 ± 0.00
{ Okote Tuff Complex (formerly BBS)	1.70 ± 0.04, 1.56 ± 0.02
{ Ileret Lower and Middle Tuff Complex	1.48
KBS Tuff Complex	2.54, 2.61 ± 0.26
Tulu Bor Tuff	3.18 ± 0.09

Where Fitch et al. *give both a maximum or a minimum apparent age, plus a specific apparent age, both figures are quoted.*

**Bracket indicates groups of tuffs which have been given separate names in different areas, but which are believed on general stratigraphic grounds to be correlated.*

pointed out that some other geochronologists have questioned the reliability of the Ar^{40}/Ar^{39} step-heating technique when it is applied to young rocks of the kind found at East Rudolf. This is a discussion involving highly technical issues, and paleoanthropologists will have to await the outcome of exchanges of information and ideas among expert geophysicists.

Additional dating measurements have recently been made by Garniss Curtis and Robert Drake using conventional K/Ar methods which have been publicly announced but not yet printed. They are currently under critical scrutiny by members of the East Rudolf Research Project and it would therefore be premature to discuss them in detail. Suffice it to say that if the results are confirmed they would imply an age for the KBS tuff complex that is substantially younger than that shown in Table 1.

Work by Hurford and Fitch is currently in progress on fission-track age determinations on zircon crystals in the volcanic tuffs. These measurements can be expected to make a significant contribution to the resolution of present uncertainties.

Paleomagnetic results from East Rudolf were first reported by Brock and Isaac (1974) and these appeared at the time to be a great help in the establishment of a geophysical chronology. However, more recent work by Brock and by Hillhouse and Ndombi working with Allen Cox at Stanford have shown that the East Rudolf sediments are much less reliable recorders of ancient polarity than at first appeared. The rather ambiguous results now available for the Lower Member could fit a number of different chronologies, and thus for the time being, it appears wise to withdraw paleomagnetism from discussions of dating at East Rudolf.

Some arguments about the dating at East Rudolf have been based on interpretations of faunal evidence. (These have mainly been presented at meetings and seminars and are not printed, but see Maglio 1972; Cooke and Maglio 1972.) However, research on details of the biostratigraphy at East Rudolf are still in active progress and there are many revisions that render obsolete arguments based on earlier approximations of the situation (see J. M. Harris 1975a, b, c; in press). Because of this and because of other complicating factors such as the possible effects of habitat differences and of biogeographic shifts, we prefer that the absolute chronology be settled through further geophysical research rather than by means of paleontological correlations.

PALEOENVIRONMENTS AND TAPHONOMY

The research team has been working towards increasingly detailed reconstructions of the environmental setting of early hominid life. This involves close collaboration between geologists involved in microstratigraphic studies and paleontologists engaged in investigating the composition of both vertebrate and invertebrate faunal communities. Preparation of reports on much of this research is still in progress so that only rather general comments can be offered.

A. K. Behrensmeyer (1975a) has done pioneering work of great significance on techniques for making quantitative assessments of the composition of vertebrate faunas in different parts of the ancient landscape. This research has made extensive use of the concepts and results of the newly fledged science of taphonomy—the study of the processes by which assemblages of bones are modified, redistributed and buried. Figure 4 illustrates Behrensmeyer's discovery that different sedimentary environments preserve rather different proportions of the parts of mammal skeletons. Under the quiet depositional conditions that prevail on the low-lying flats of a delta, the body parts occur in much the same relative frequency that obtains in the animals from which they derive. This seems to guarantee that there has been minimal disturbance and that the animals whose bones are found together actually made use of that habitat. By contrast, in the sands and gravels of river-channel deposits, dense skeletal parts like teeth and limb-bone articular ends have much higher relative frequency than they do in the skeletons from which they derive. Clearly, these assemblages have been distorted by the action of river currents and may include remains of species swept in from several different habitats.

Figure 5 shows Behrensmeyer's other results which illustrate differences in species composition from place to place. The triangle diagram shows relationships between environments of deposition and animals grouped in regard to their degree of dependence on aquatic or terrestrial habitats. The other graph illustrates the contrasting habitat associations of two sets of extinct pig species. The *Mesochoerus* forms have low-crowned teeth, perhaps most suitable for browsing; these predominated in the delta swamps and the floodplains, while *Notochoerus* and *Metridiochoerus* pigs with high-crowned teeth predominate in the channel deposits of better-drained habitats where vegetation may have been coarser. Behrensmeyer has begun to work in the same way on the taphonomic and paleoenvironmental associations of the hominid fossils (1975b).

Also at East Rudolf, for the first time, Richard and Meave Leakey have been able to collect meaningful data on the relative abundance of hominids, carnivores and nonhominid primates. John Harris, of the National Museum in Nairobi, has begun to compile lists of faunal associations that presumably reflect different animal communities making use of different environments (Harris, J. M. 1975a, b, c; in press). Craig Black and co-workers are seeking to enlarge the sample of rodent and other micromammal fossils by special recovery techniques, which are often extremely sensitive indicators of habitat characteristics.

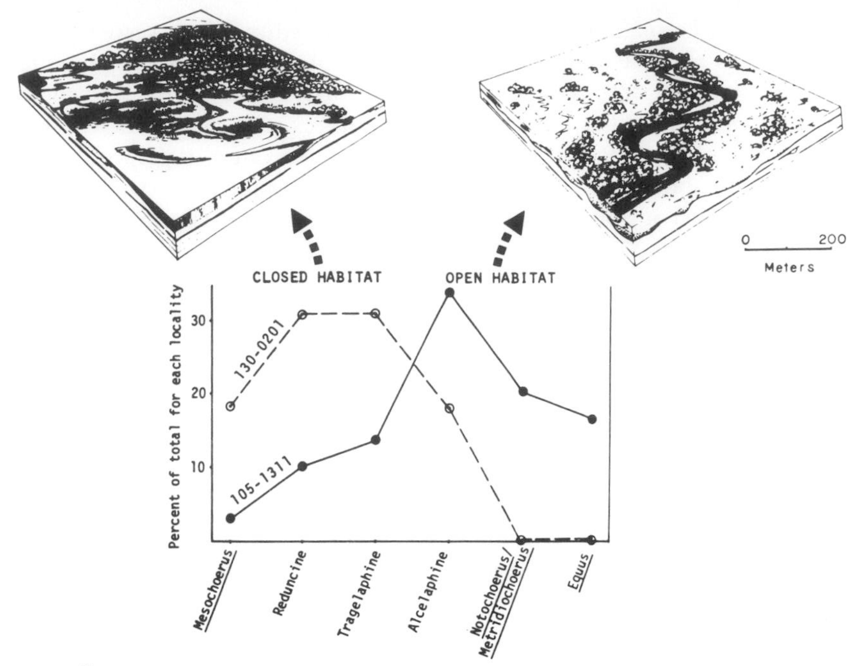

FIGURE 5
Fossils of different species of mammals show differing frequencies in sediments deposited under the different conditions of a delta, a river channel, or a flood plain. This may prove to be an important clue to the ecology of extinct animals and of hominids.

Detailed paleontological studies of the fauna of which early man was a part are being done by various participants in the Project: the bovids, giraffids and deinotheres by J. M. Harris; the carnivores and nonhominid primates by Meave Leakey; the equids by Vera Eisenmann; the suids by H. B. S. Cooke and by J. M. Harris; the elephants by V. Maglio and by M. Beden; the hippos by Shirley Coryndon and the crocodilians by E. Tchernov. Preliminary reports on all these groups appear in the proceedings of a conference held in 1973 to review research in the Rudolf basin (Coppens, Howell, Isaac and Leakey 1975), and are listed in the bibliography at the end of this review.

Knowledge of vegetation types is of course critical to an effective understanding of the ecology of the early hominids. Faunal evidence makes some contribution by giving gross indications that allow one to distinguish between swamp communities, grassland communities, gallery forest communities, etc. However, adequately detailed information can really only be obtained through the recovery of fossil pollen. Research by Raymonde Bonnefille, of the CNRS in France, has shown that the

Koobi Fora Formation, like most East African sediments preserve very little pollen. However, one sample out of 40 did prove to have a rich pollen assemblage that amplifies our knowledge of the species involved in the mosaic of riverine bush and grasslands that existed in the ancient East Rudolf landscape. Pollen samples such as these can also give otherwise unavailable information on conditions that prevailed on the hills and highlands that surround the basin. This research is continuing.

Investigations of the invertebrate fossils have also been initiated by Peter Williamson of Bristol University. His studies of the fossil shells of aquatic molluscs, in conjunction with geochemical work by Thure Cerling, should teach us something of the ancient lake—for instance, whether it was entirely fresh or slightly salty, or whether its salinity oscillated in relation to other environmental variables.

Other lines of paleoenvironmental research involve detailed reconstructions of portions of the ancient landscape. This can be done by detailed geological mapping and the painstaking recording of the minutiae of lithology, form and distribution of small-scale sedimentary units. In this way, Behrensmeyer and Isaac have been able to reconstruct the specific topography of more than a square mile of terrain that included a shoreline with a barrier bar, a lagoon and swamp complex and delta distributory channels that meandered across the swamps. The location of hominid activity within this scene can be mapped in by careful archaeological survey (Isaac *et al.* 1971, 1975). Similar detailed work has been done on the fluvial sediments of the Upper Member by the Iowa State University team working with archaeologist J. W. K. Harris.

An unusually complete and comprehensive reconstruction of the overall paleogeography is currently being worked out by Ian Findlater, and it is clear that this will provide a starting point for many years of further detailed local microstratigraphic studies.

Incomplete as they presently are, the paleontological and paleoenvironmental studies already make it clear that the early hominid inhabitants of the East Rudolf area lived under conditions that were much more lush and well watered than those that now prevail. It is also clear that animal faunas in general were more diverse in their species composition. We do not know whether the more lush conditions were due to moister climatic conditions and/or to the possible former existence of a major river bringing in water from areas that do not now drain into East Rudolf.

ARCHAEOLOGY

Scattered over the varied ancient landscapes that are documented by the sediments of the Koobi Fora Formation were traces of the cultural activities of the early hominids. These consist for the most part of camp sites

FIGURE 6
The archaeological site of KBS is now situated in the midst of an arid eroded landscape, but paleo-environmental studies show that in ancient times, when the beds were deposited, the hominids had located their camp by a watercourse that traversed a swampy delta flood plain. (Photo by H. V. Merrick.)

that are marked by concentrations of discarded stone artifacts and bone food refuse, but there are also butchery sites where sharp stone tools lie in association with a broken-up carcass.

The archaeological record is being carefully searched out by a team of researchers coordinated by one of us (G.L.I.). The Upper Member contains a particularly abundant record and the study of this has become the special responsibility of J. W. K. Harris (Isaac *et al.* 1975; Harris and Isaac, in press). The essay by Isaac in Part Four of this volume provides an account of the East Rudolf archaeology presented in the context of a general discussion of the evidence from East Africa as a whole.

With three exceptions, archaeological sites and hominid fossils are not directly associated with each other. At Ileret the archaeological site FxJj 1 occurs in close proximity to the find spots of several hominid fossils that include both *Homo sp.* and *Australopithecus* cf. *boisei,* while along the Karari escarpment in the Upper Member an *A.* cf. *boisei* mandible occurred in silts just above a major archaeological horizon at FxJj 20E (Leakey, R. E., in press). Also the cranium, 1805, is in close proximity to a site at the same horizon (FxJj 38) (Harris and Isaac, in press).

HOMINID FOSSILS

In the East Rudolf expeditions the recovery of hominid fossils has not been left to chance. Under the direction of Richard Leakey and the field leadership of Bw. Kamoya Kimeu, a team of expert paleontological prospectors has systematically searched the exposures. It is to their efforts and skill that the paleoanthropological community owes the long list of important specimens that are now available for study. When a new fossil is found it is not moved at all until Richard Leakey and, where appropriate, a geologist have examined the circumstances. Discovery sites are recorded on aerial photographs and marked on the ground with inscribed concrete posts.

The paleontological prospectors search with equal intensity for fossil remains of hominids, other primates, and carnivores. This has enabled us to make a meaningful assessment of the relative abundance of these groups. It appears that gelada-baboon-like creatures belonging to the genus *Theropithecus* were the most abundant form among these groups with as many specimens of them being found as of all the others put together (Leakey, M. G. 1975 a, b). Carnivores, cercopithecids and hominid fossils have roughly equivalent abundances. See Table 2 (from Leakey, R. E. 1975). Table 3 provides a list of the 129 hominid fossils recovered up to the end of 1974. Table 4 summarizes the representation of body parts—practically all distinctive body parts are represented.

The three papers from *Nature* that have been reprinted to follow this review provide good representation of the announcements by which news of this succession of discoveries has been made known to the scientific community. Each announcement carries a brief preliminary comment on possible interpretations. As might be expected, a spate of discoveries like this carries the potential that later finds will necessitate revisions of interpretations made on the basis of the earlier finds. This has already happened at East Rudolf and we fully expect that it will happen again!

TABLE 2
Relative numbers of specimens of Hominidae, Carnivora, and Cercopithecidae, in the East Rudolf Collection. The relative numbers of specimens attributed to *Homo*, *Australopithecus* and indeterminate, within the Hominidae, is also given.

GROUP	NUMBER OF SPECIMENS	PERCENT
Hominidae	89	33.3
Carnivora	92	32.5
Cercopithecidae (excluding *Theropithecus*)	86	32.2
Hominidae		
Australopithecuu	49	55.0
Homo	29	32.6
Genus Indeterminate	11	12.4

TABLE 3
East Rudolf hominids collected 1968–1974

KNM-ER NO.	YEAR	AREA	SPECIMEN DETAIL
164a	1969	104	Parietal fragment
164b-c	1971	104	Two phalanges, two vertebrae
403	1968	103	Right mandible
404	1968	7A	Right mandible M_2, M_3
405	1968	105	Palate lacking teeth
406	1969	10	Cranium lacking teeth
407	1969	10	Cranium lacking face
417	1968	129	Parietal fragment
725	1970	1	Left mandible
726	1970	11	Left mandible
727	1970	6A	Right mandible
728	1970	1	Right mandible
729	1970	8	Mandible with dentition
730	1970	103	Left mandible with symphysis, left M_1-M_3
731	1970	6A	Left mandible
732	1970	10	Demi-cranium, parietal P^4
733	1970	8	Right mandible, left maxilla and cranial fragments M^1, M_3
734	1970	103	Parietal fragment
736	1970	103	Left femur shaft
737	1970	103	Left femur shaft
738	1970	105	Proximal left femur
739	1970	1	Right humerus
740	1970	1	Distal left humerus fragment
741	1970	1	Proximal right tibia
801	1971	6A	Right mandible, M_3, M_2, and associated isolated teeth
802	1971	6A	Isolated teeth
803	1971	8	Associated skeletal elements
805	1971	1	Left mandible
806	1971	8	Isolated teeth
807A	1971	8A	Right maxilla fragment M^3, partial M^2
807B	1973	8A	Right maxilla fragment M^1
808	1971	8	Isolated juvenile teeth
809	1971	8	Isolated teeth
810	1971	104	Left mandible, M_3
811	1971	104	Parietal fragment
812	1971	104	Juvenile left mandible fragment
813	1971	104	Right talus and tibia fragment
814	1971	104	Cranial fragments
815	1971	10	Proximal left femur
816	1971	104	Canine and molar fragment
817	1971	124	Left mandible fragment
818	1971	6A	Left mandible P_3-M_3
819	1971	1	Left mandible
820	1971	1	Juvenile mandible with dentition
992	1971	1	Mandible with dentition
993	1971	1	Distal right femur
997	1971	104	Left metatarsal III
998	1971	104	Isolated incisor
999	1971	6A	Left femur

TABLE 3 (cont.)

KNM-ER NO.	YEAR	AREA	SPECIMEN DETAIL
1170	1971	6A	Cranial fragments
1171	1971	6A	Isolated juvenile teeth
1462	1972	130	Isolated M_3
1463	1972	1A	Right femur shaft
1464	1972	6A	Right talus
1465	1972	11	Proximal left femur fragment
1466	1972	6	Parietal fragment
1467	1972	3	Isolated M_3
1468	1972	11	Right mandible
1469	1972	131	Left mandible, M_3
1470	1972	131	Cranium
1471	1972	131	Proximal right tibia
1472	1972	131	Right femur
1473	1972	131	Proximal right humerus
1474	1972	131	Parietal fragment
1475	1972	131	Proximal left femur
1476	1972	105	Left talus and proximal left tibia
1477	1972	105	Juvenile mandible with dentition
1478	1972	105	Cranial fragments
1479	1972	105	Isolated molar fragments
1480	1972	105	Isolated molar
1481	1972	131	Left femur, proximal and distal left tibia, distal left fibula
1482	1972	131	Mandible, right P_4, left P_3-M_3
1483	1972	131	Left mandible, fragment M_2
1500	1972	130	Associated skeletal elements
1501	1972	123	Right mandible
1502	1972	123	Right mandible fragment with M_1
1503	1972	123	Proximal right femur
1504	1972	123	Distal right humerus
1505	1972	123	Proximal left femur
1506	1972	121	Right mandible, M_1, M_2, isolated P^3, P^4
1507	1972	127	Juvenile left mandible with dentition
1508	1972	127	Isolated molar
1509	1972	119	Isolated teeth C-M_3
1515	1972	103	Isolated incisor
1590	1972	12	Partial cranium with juvenile dentition
1591	1972	12	Right humerus lacking head
1592	1972	12	Distal right femur
1593	1972	12	Cranial and mandibular fragments
1648	1972	105	Parietal fragment
1686	1972	123	Cranial fragments
1800	1973	130	Cranial fragments
1801	1973	131	Left mandible, P_4, M_1, M_3
1802	1973	131	Mandible, left P_4-M_2, right P_3-M_2
1803	1973	131	Fragment right mandible
1804	1973	104	Right maxilla, P^3-M^2
1805	1973	130	Cranium and mandible with dentition
1806	1973	130	Mandible lacking teeth
1807	1973	103	Shaft right femur
1808	1973	103	Associated skeletal and cranial elements
1809	1973	121	Shaft right femur

TABLE 3 (cont.)

KNM-ER NO.	YEAR	AREA	SPECIMEN DETAIL
1810	1973	123	Proximal left tibia
1811	1973	123	Left mandible fragment
1812	1973	123	Fragment right mandible, isolated left I_1, M_1
1813	1973	123	Cranium
1814	1973	127	Maxillary fragments
1815	1973	1	Right talus
1816	1973	6A	Immature mandible with dentition
1817	1973	1	Right mandible
1818	1973	6A	Isolated I^1
1819	1973	3	Isolated M_3
1820	1973	103	Left mandible, M_1
1821	1972	123	Parietal fragment
2592	1974	6	Parietal fragment
2593	1974	6	Molar fragment
2594	1974	6A	Proximal fragment of tibia
(a + b)			Fragment tibia shaft
2595	1974	1A	Cranial fragments
2596	1974	15	Distal fragment of tibia
2597	1974	15	Molar crown
2598	1974	15	Occipital fragment
2599	1974	15	Fragment premolar crown
2600	1974	130	Partial molar crown
2601	1974	130	Molar
2602 (A-G)	1974	117	Cranial fragments
2603	1974	117	Tooth fragment, enamel
2604	1974	117	Fragment of enamel
2605	1974	117	Tooth fragment
2606	1974	117	Tooth fragment
2607	1974	105	Tooth fragment
3228	1974	102	Right innominate
3227	1974	103	Mandible with both P_3 crowns
3230	1974	131	Complete, crushed mandible with full dentition

In the commentary in the reprint which announced the 1973 finds, Richard Leakey pointed out that the Rudolf hominid collections seemed to be comprised of three different sorts of hominids, with a hint of a fourth. At the end of another season this tentative interpretation still stands.

The hominid specimen from East Rudolf which has attracted most attention—and most controversy—has been the fossil cranium KNM-ER 1470 (see the second reprint). This specimen provides unambiguous evidence of very early hominids with brain capacities markedly bigger than those of almost all the other known Plio-Pleistocene hominids. Its capacity approaches 800 cc. and is more than half the capacity of modern man, whereas most other early fossils have only a third to a quarter of the cranial volume of *Homo sapiens*.

TABLE 4
Numbers of pieces representing each body part in the East Rudolf collection (1968–1973).

SKELETAL ELEMENT		NUMBER	TOTAL
Crania	a	6	
parietal	b1	13	
maxillary	b2	5	24
Mandible	a	9	
	b	26	35
Clavicle	a		
	b		
Scapula	a		
	b	1	1
Humerus	a	2	
	b	5	7
Radius	a		
	b	2	2
Ulna	a		
	b	2	2
Pelvis	a		
	b		
Femur	a	2	
	b	14	16
Tibia	a		
	b	7	7
Fibula	a		
	b	2	2
Talus	a	3	
	b	2	5
Calcaneum	a		
	b	1	1
Phalanges	a	2	
	b	1	3
Metapodials	a		
	b	4	4
Isolated teeth			16

Where a body part from one individual is represented by more than one fragment or by both the left and right side it has been assessed as one piece only. a = relatively complete, b = half or less.

Much of the controversy over 1470 has turned on widespread skepticism over its dating. It should be stressed that the East Rudolf Research team is testing the chronological evidence as thoroughly as possible. So far none of the evidence suggests the need for major revisions to the estimate given in the *Nature* paper. The specimen comes from strata just above the projected level of the Tulu Bor Tuff which has given a reproducible series of dates that cluster around 3.2 million years (Fitch *et al.* 1974; Fitch and Miller, pers. comm.).

As stated in the reprints, postcranial remains were found at stratigraphic levels close to that of 1470. These demonstrate the existence at that time of an advanced, fully upright biped with a lower limb anatomy that contrasts in some respects with that of the robust australopithecines.

This inference has received dramatic confirmation by the recent discovery of KNM-ER 3228, a nearly complete right innominate bone (half-pelvis) from below the Tulu Bor Tuff (Leakey, R. E., 1975).

Following the announcement of each fossil it has been policy for Day, Walker and Wood, the project's specialist team of hominid anatomists, to work with Richard Leakey on the preparation of a more detailed technical description. These have been published in the *American Journal of Physical Anthropology* and a list of those reports that have already appeared is included in the bibliography that follows this review. Further detailed comparative studies are in progress and will be reported in a series of monographs now in preparation.

Kay Behrensmeyer has begun to do research on the paleoenvironmental relationships and taphonomy of the hominid fossils (Behrensmeyer 1975b). Her results hint at a significant statistical difference in the relative frequencies with which specimens of the robust *Australopithecus* cf. *boisei* fossils versus other hominid specimens were preserved in certain environments of deposition. It is too early to evaluate the significance of this pioneer study, but it is exciting that East Rudolf has provided paleoanthropology with the first opportunity to try out such investigations.

The activities of the East Rudolf Research Project during the years 1968-1974 constitute only the first phase of research, during which time the area has been merely explored. The great potential of this area as a source of paleoanthropological evidence has been demonstrated, and the team of scientists is now embarking on a second phase of work that will include detailed follow-up on the discoveries of the first phase as well as the initiation of new kinds of research that could not be accommodated in the exploratory phase.

References Cited

A complete bibliography of the publications of the East Rudolf Research Project up to June 1975 is given below. This is based on a catalogue maintained by Meave Leakey. All other references cited in the text are given in a short list that precedes the East Rudolf bibliography.

Behrensmeyer, A. K.

1975a The taphonomy and paleoecology of Plio-Pleistocene vertebrate assemblages east of Lake Rudolf, Kenya. Bulletin of the Museum of Comparative Zoology 146:473–578.

1975b Fossil assemblages in relation to sedimentary environments in the East Rudolf succession. *In* Earliest Man and Environments in the Lake Rudolf Basin: Stratigraphy, Paleoecology and Evolution. Y. Coppens *et al.*, eds. Chicago: University of Chicago Press.

Bowen, B. E.
1974 The geology of the Upper Cenozoic sediments in the East Rudolf embayment of the Lake Rudolf basin, Kenya. Ph.D. dissertation. Iowa University, Ames. 164 pages.

Bowen, B. E. and C. F. Vondra
1973 Stratigraphical relationships of the Plio-Pleistocene deposits, East Rudolf, Kenya. Nature 242:391–393.

Brock, A. and G. Ll. Isaac
1974 Palaeomagnetic stratigraphy and chronology of hominid-bearing sediments east of Lake Rudolf, Kenya. Nature 247:344–348.

Cooke, H. B. S. and V. J. Maglio
1972 Plio-Pleistocene stratigraphy in relation to proboscidean and suid evolution. *In* Calibration of Hominoid Evolution. W. W. Bishop and J. A. Miller, eds. Edinburgh: Scottish Academic Press. pp. 303–329.

Coppens, Y., F. C. Howell, G. Ll. Isaac and R. E. F. Leakey, eds.
1975 Earliest Man and Environments in the Lake Rudolf Basin: Stratigraphy, Paleoecology and Evolution. Chicago: University of Chicago Press.

Cox, A.
1969 Geomagnetic reversals. Science 163:237–244.

Findlater, I. C.
1975 Tuffs and the recognition of isochronous mapping units in the Rudolf succession. *In* Earliest Man and Environments in the Lake Rudolf Basin: Stratigraphy, Paleoecology and Evolution. Y. Coppens *et al.*, eds. Chicago: University of Chicago Press.

Fitch, F. J., I. C. Findlater, R. T. Watkins, and J. A. Miller
1974 Dating of the rock succession containing fossil hominids at East Rudolf, Kenya. Nature 251:213–215.

Fitch, F. J. and J.A. Miller
1975 Conventional Potassium-Argon and Argon-40/Argon-39 dating of volcanic rocks from East Rudolf. *In* Earliest Man and Environments in the Lake Rudolf Basin: Stratigraphy, Paleoecology and Evolution. Y. Coppens *et al.*, eds. Chicago: University of Chicago Press.

Fitch, F., R. Watkins, and J. A. Miller
1975 Age of a new carbonalite locality in northern Kenya. Nature: 254:581–583.

Fleischer, R. L. and H. R. Hart
1972 Fission track dating: techniques and problems. *In* Calibration of Hominoid Evolution. W. W. Bishop and J. A. Miller, eds. Edinburgh: Scottish Academic Press. pp. 135–170.

Harris, J. M.
1975a Bovidae from the East Rudolf succession. *In* Earliest Man and Environments in the Lake Rudolf Basin: Stratigraphy, Paleoecology and Evolution. Y. Coppens *et al.*, eds. Chicago: University of Chicago Press.

1975b Giraffidae from Omo Group formations. *In* Earliest Man and Environments in the Lake Rudolf Basin: Stratigraphy, Paleoecology and Evolution. Y. Coppens *et al.*, eds. Chicago: University of Chicago Press.

1975c Rhinocerotidae from the East Rudolf succession. *In* Earliest Man and Environments in the Lake Rudolf Basin: Stratigraphy, Paleoecology and Evolution. Y. Coppens *et al.*, eds. Chicago: University of Chicago Press.

In press New fossil mammal faunas from East Rudolf. Nature.

Harris, J. M. and R. T. Watkins
1974 A new early Miocene fossil mammal locality near Lake Rudolf, Kenya.

Harris, J. W. K. and G. Ll. Isaac
In press The Karari Industry, a distinctive series of early Pleistocene artefact assemblages from East Rudolf, Kenya. Nature.

Hurford, A. J.
1974 Fission track dating of a vitric tuff from East Rudolf, Kenya. Nature 249:236–237.

Isaac, G. Ll., J. W. K. Harris, and D. Crader
1975 Archaeological evidence from the Koobi Fora Formation. *In* Earliest Man and Environments in the Lake Rudolf Basin: Stratigraphy, Paleoecology and Evolution. Y. Coppens *et al.*, eds. Chicago: University of Chicago Press.

Isaac, G. Ll., R. E. F. Leakey, and A. K. Behrensmeyer
1971 Archaeological traces of early hominid activities, east of Lake Rudolf, Kenya. Science 173:1129–1134.

Leakey, M. G.
1975a Carnivora of the East Rudolf succession. *In* Earliest Man and Environments in the Lake Rudolf Basin: Stratigraphy, Paleoecology and Evolution. Y. Coppens *et al.*, eds. Chicago: University of Chicago Press.

1975b Cercopithecoidea of the East Rudolf succession. *In* Earliest Man and Environments in the Lake Rudolf Basin: Stratigraphy, Paleoecology and Evolution. Y. Coppens *et al.*, eds. Chicago: University of Chicago Press.

Leakey, R. E. F.
1975 An overview of the Hominidae from East Rudolf, Kenya. *In* Earliest Man and Environments in the Lake Rudolf Basin: Stratigraphy, Paleoecology and Evolution. Y. Coppens *et al.*, eds. Chicago: University of Chicago Press.

In press A partial innominate bone and additional fossil hominids from East Rudolf. Nature.

Maglio, V. J.
1972 Vertebrate faunas and chronology of hominid-bearing sediments east of Lake Rudolf, Kenya. Nature 239:379–385.

Miller, J. A.
1972 Dating Pliocene and Pleistocene strata using the Potassium-Argon and Argon-40/Argon-39 methods. *In* Calibration of Hominoid Evolution. W. W. Bishop and J. A. Miller, eds. Edinburgh: Scottish Academic Press.

Publications of the East Rudolf Research Project 1970–June 1975.

Behrensmeyer, A. K.
1970 Preliminary geological interpretation of a new hominid site in the Lake Rudolf basin. Nature 226:225–226.

1975a The taphonomy and paleoecology of Plio-Pleistocene vertebrate assemblages east of Lake Rudolf, Kenya. Bulletin of the Museum of Comparative Zoology 146:473–578.

1975b Fossil assemblages in relation to sedimentary environments in the East Rudolf succession. *In* Earliest Man and Environments in the Lake Rudolf Basin: Stratigraphy, Paleoecology and Evolution. Y. Coppens *et al.*, eds. Chicago: University of Chicago Press.

Bowen, B. E.
1974 The geology of the Upper Cenozoic sediments in the East Rudolf embayment of the Lake Rudolf Basin, Kenya. Ph.D. dissertation. Iowa University, Ames. 164 pages.

Bowen, B. E. and C. F. Vondra
1973 Stratigraphical relationships of the Plio-Pleistocene deposits, East Rudolf, Kenya. Nature 242:391–393.

Brock, A., J. Hillhouse, A. Cox, and J. Ndombi
1974 The paleomagnetism of the Koobi Fora Formation, Lake Rudolf, Kenya. Transactions American Geophysical Union 56:1109 (Abstract GP 15).

Brock, A. and G. Ll. Isaac
1974 Palaeomagnetic stratigraphy and chronology of hominid-bearing sediments east of Lake Rudolf, Kenya. Nature 247:344–348.

1975 Reversal stratigraphy and its application at East Rudolf. *In* Earliest Man and Environments in the Lake Rudolf Basin: Stratigraphy, Paleoecology and Evolution. Y. Coppens *et al.*, eds. Chicago: University of Chicago Press.

Cerling, T. E.
1975 Oxygen-isotope studies of the East Rudolf volcanoclastics. *In* Earliest Man and Environments in the Lake Rudolf Basin: Stratigraphy, Paleoecology and Evolution. Y. Coppens *et al.*, eds. Chicago: University of Chicago Press.

Cooke, H. B. S.
1975 Suidae from Pliocene/Pleistocene strata of the Rudolf basin. *In* Earliest Man and Environments in the Lake Rudolf Basin: Stratigraphy, Paleoecology and Evolution. Y. Coppens *et al.*, eds. Chicago: University of Chicago Press.

Coppens Y., F. C. Howell, G. Ll. Isaac and R. E. F. Leakey, eds.
1975 Earliest Man and Environments in the Lake Rudolf Basin: Stratigraphy, Paleoecology and Evolution. Chicago: University of Chicago Press.

Coryndon, S. C.
1975 Fossil Hippopotamidae from Pliocene/Pleistocene successions of the Rudolf basin. *In* Earliest Man and Environments in the Lake Rudolf Basin: Stratigraphy, Paleoecology and Evolution. Y. Coppens *et al.*, eds. Chicago: University of Chicago Press.

Day, M. H.
1975 Hominid postcranial remains from the East Rudolf succession. *In* Earliest Man and Environments in the Lake Rudolf Basin: Stratigraphy, Paleoecology and Evolution. Y. Coppens *et al.*, eds. Chicago: University of Chicago Press.

Day, M. H. and R. E. F. Leakey
1973 New evidence for the genus *Homo* from East Rudolf, Kenya. I. Am. J. Phys. Anthrop. 39:341–354.

1974 New evidence for the genus *Homo* from East Rudolf, Kenya. III. Am. J. Phys. Anthrop. 41:367–380.

Eisenmann, V.

1975 A preliminary note on Equidae from the Koobi Fora Formation, Kenya. *In* Earliest Man and Environments in the Lake Rudolf Basin: Stratigraphy, Paleoecology and Evolution. Y. Coppens *et al.,* eds. Chicago: University of Chicago Press.

Findlater, I. C.

1975 Tuffs and the recognition of isochronous mapping units in the Rudolf succession. *In* Earliest Man and Environments in the Lake Rudolf Basin: Stratigraphy, Paleoecology and Evolution. Y. Coppens *et al.,* eds. Chicago: University of Chicago Press.

Fitch, F. J., I. C. Findlater, R. T. Watkins, and J. A. Miller

1974 Dating of the rock succession containing fossil hominids at East Rudolf, Kenya. Nature 251:213–215.

Fitch, F. J. and J. A. Miller

1970 Radioisotopic age determinations of Lake Rudolf artefact site. Nature 226:226–228.

1975 Conventional Potassium-Argon and Argon-40/Argon-39 dating of volcanic rocks from East Rudolf. *In* Earliest Man and Environments in the Lake Rudolf Basin: Stratigraphy, Paleoecology and Evolution. Y. Coppens *et al.,* eds. Chicago: University of Chicago Press.

Fitch, F., R. Watkins, and J. A. Miller

1975 Age of a new carbonalite locality in northern Kenya. Nature: 254:581–583.

Harris, J. M.

1974 Orientation and variability of the ossicones of Sivatheriinae (Mammalia: Giraffidae). Annals of the South African Museum 65:189–198.

1975a Bovidae from the East Rudolf succession. *In* Earliest Man and Environments in the Lake Rudolf Basin: Stratigraphy, Paleoecology and Evolution. Y. Coppens *et al.,* eds. Chicago: University of Chicago Press.

1975b Giraffidae from Omo Group formations. *In* Earliest Man and Environments in the Lake Rudolf Basin: Stratigraphy, Paleoecology and Evolution. Y. Coppens *et al.,* eds. Chicago: University of Chicago Press.

1975c Rhinocerotidae from the East Rudolf succession. *In* Earliest Man and Environments in the Lake Rudolf Basin: Stratigraphy, Paleoecology and Evolution. Y. Coppens *et al.,* eds. Chicago: University of Chicago Press.

In press New fossil mammal faunas from East Rudolf. Nature.

Harris, J. M. and R. T. Watkins

1974 A new early Miocene fossil mammal locality near Lake Rudolf, Kenya. Nature 252:576–77.

Harris, J. W. K., and G. Ll. Isaac

In press The Karari Industry, a distinctive series of early Pleistocene artefact assemblages from East Rudolf, Kenya.

Hurford, A. J.

1974 Fission track dating of a vitric tuff from East Rudolf, Kenya. Nature 249:236–237.

Isaac, G. Ll.
1975 Plio-Pleistocene artefact assemblages from East Rudolf, Kenya. *In* Earliest Man and Environments in the Lake Rudolf Basin: Stratigraphy, Paleoecology and Evolution. Y. Coppens *et al.*, eds. Chicago: University of Chicago Press.

Isaac, G. Ll., J. W. K. Harris, and D. Crader
1975 Archaeological evidence from the Koobi Fora Formation. *In* Earliest Man and Environments in the Lake Rudolf Basin: Stratigraphy, Paleoecology and Evolution. Y. Coppens *et al.*, eds. Chicago: University of Chicago Press.

Isaac, G. Ll., R. E. F. Leakey, and A. K. Behrensmeyer
1971 Archaeological traces of early hominid activities, east of Lake Rudolf, Kenya. Science 173:1129–1134.

Johnson, G. D.
1974 Cainozoic lacustrine stromatolites from hominid-bearing sediments east of Lake Rudolf, Kenya. Nature 247:520–523.

Johnson, G. D. and R. G. H. Raynolds
1975 Late Cenozoic environments of the Koobi Fora Formation. *In* Earliest Man and Environments in the Lake Rudolf Basin: Stratigraphy, Paleoecology and Evolution. Y. Coppens *et al.*, eds. Chicago: University of Chicago Press.

Leakey, M. D.
1970 Early artefacts from the Koobi Fora area. Nature 226:228–230.

Leakey, M. G.
1975a Carnivora of the East Rudolf succession. *In* Earliest Man and Environments in the Lake Rudolf Basin: Stratigraphy, Paleoecology and Evolution. Y. Coppens *et al.*, eds. Chicago: University of Chicago Press.

1975b Cercopithecoidea of the East Rudolf succession. *In* Earliest Man and Environments in the Lake Rudolf Basin: Stratigraphy, Paleoecology and Evolution. Y. Coppens *et al.*, eds. Chicago: University of Chicago Press.

Leakey, M. G. and R. E. F. Leakey
1973 New large Pliestocene Colobinae (Mammalia, Primates) from East Africa. Fossil Vertebrates of Africa 3:121–138.

Leakey, R. E. F.
1970 In search of man's past at Lake Rudolf. National Geographic 137:712–732.

1970 New hominid remains and early artefacts from northern Kenya. Nature 226:223–224.

1971 Further evidence of Lower Pleistocene hominids from East Rudolf, North Kenya. Nature 231:241–245.

1972a Further evidence of Lower Pleistocene hominids from East Rudolf, North Kenya, 1971. Nature 237:264–269.

1972b New evidence for the evolution of man. Social Biology 19:99–114.

1973a Further evidence of Lower Pleistocene hominids from East Rudolf, North Kenya, 1972. Nature 242:170–173.

1973b Evidence for an advanced Plio-Pleistocene hominid from East Rudolf, Kenya. Nature 242:447–450.

1973c Skull 1470. National Geographic 143:818–829.

1973d Australopithecines and hominines: A summary of evidence from the Early Pleistocene of Eastern Africa. Symposium of the Zoological Society London 33:53–69.

1974 Further evidence of Lower Pleistocene hominids from East Rudolf, North Kenya, 1973. Nature 248:653–656.

1975 An overview of the Hominidae from East Rudolf, Kenya. *In* Earliest Man and Environments in the Lake Rudolf Basin: Stratigraphy, Paleoecology and Evolution. Y. Coppens *et al.,* eds. Chicago: University of Chicago Press.

In press A partial innominate bone and additional fossil hominids from East Rudolf. Nature.

Leakey, R. E. F. and G. Ll. Isaac

1972 Hominid fossils from the area east of Lake Rudolf, Kenya: photographs and a commentary on context. *In* Perspectives on Human Evolution 2. S. L. Washburn and P. Dolhinow, eds. New York: Holt, Rinehart, and Winston.

Leakey, R. E. F., J. M. Mungai, and A. C. Walker

1971 New australopithecines from East Rudolf, Kenya. Am. J. Phys. Anthrop. 35:175–186.

1972 New australopithecines from East Rudolf, Kenya. II. Am. J. Phys. Anthrop. 36:235–251.

Leakey, R. E. F. and A. C. Walker

1973 New australopithecines from East Rudolf, Kenya. III. Am. J. Phys. Anthrop. 39:205–222.

Leakey, R. E. F. and B. A. Wood

1973 New evidence for the genus *Homo* from East Rudolf, Kenya. II. Am. J. Phys. Anthrop. 39:355–368.

1974a New evidence for the genus *Homo* from East Rudolf, Kenya, IV. Am. J. Phys. Anthrop. 41:237–244.

1974b A hominid mandible from East Rudolf, Kenya. Am. J. Phys. Anthrop. 41:245–250.

Maglio, V. J.

1971 Vertebrate faunas from the Kubi Algi, Koobi Fora and Ileret areas, East Rudolf, Kenya. Nature 231:248–249.

1972 Vertebrate faunas and chronology of hominid-bearing sediments east of Lake Rudolf, Kenya. Nature 239:379–385.

Tchernov, E.

1975 Crocodilians from the late Cenozoic of the Rudolf basin. *In* Earliest Man and Environments in the Lake Rudolf Basin: Stratigraphy, Paleoecology and Evolution. Y. Coppens *et al.,* eds. Chicago: University of Chicago Press.

Vondra, C. F. and B. E. Bowen

1975 Plio-Pleistocene deposits and environments, East Rudolf, Kenya. *In* Earliest Man and Environments in the Lake Rudolf Basin: Stratigraphy, Paleoecology and Evolution. Y. Coppens *et al.,* eds. Chicago: University of Chicago Press.

Vondra, C. F., G. D. Johnson, B. E. Bowen, and A. K. Behrensmeyer
1971 Preliminary stratigraphical studies of the East Rudolf Basin, Kenya. Nature 231:245–248.

Walker, A. C.
1973 New *Australopithecus* femora from East Rudolf, Kenya. Journal of Human Evolution 2:545–555.

1975 Remains attributable to *Australopithecus* in the East Rudolf succession. *In* Earliest Man and Environments in the Lake Rudolf Basin: Stratigraphy, Paleoecology and Evolution. Y. Coppens *et al.*, eds. Chicago: University of Chicago Press.

Wood, B. A.
1974a Evidence on the locomotor pattern of *Homo* from early Pleistocene of Kenya. Nature 251:135–136.

1974b A *Homo* talus from East Rudolf, Kenya. Journal of Anatomy 117:203–204.

1974c Morphology of a fossil hominid mandible from East Rudolf, Kenya. Journal of Anatomy 117:652–655.

1975 Remains attributable to *Homo* in the East Rudolf succession. *In* Earliest Man and Environments in the Lake Rudolf Basin: Stratigraphy, Paleoecology and Evolution. Y. Coppens *et al.*, eds. Chicago: University of Chicago Press.

R. E. F. Leakey:

Further Evidence of Lower Pleistocene Hominids from East Rudolf, North Kenya, 1972

The following article was reprinted, with permission, from Nature, *Volume 242, pages 170–173.*

Thirty-eight fossil hominids were collected during the 1972 season at East Rudolf, and the total now known from this locality is eighty-seven. Several specimens, including some attributable to *Homo,* were recovered from deposits that are below the KBS Tuff dated at 2.61 million years.

THIS report summarizes the results of the continuation of work carried out during 1972 at East Rudolf. Preliminary reports for the years 1968–1971 have already been published.[1-3]

A further thirty-eight fossil hominids were collected during 1972, bringing the total now known from this locality to eighty-seven. The collection includes cranial and postcranial material. Several specimens were recovered from deposits that are below the KBS Tuff (2.61 m.y.[4]); these are of particular interest in view of the presence of at least two distinct forms of hominid at this early period. In this report, as before,[1-3] I have avoided specific identifications and have made provisional attributions to either *Australopithecus* or *Homo.* I believe that the final analysis of the hominids from East Rudolf should be based on as large a sample as can be reasonably available, and there is every indication that further material will be recovered with the continued investigation of the area.

The palaeontological investigations were directed by Dr J. M. Harris (Kenya National Museums) and further specimens of several taxa were collected. Some specimens collected from below the KBS Tuff should prove of considerable interest and value regarding the interpretation of the lower Pleistocene genera.[5] With the extension of exploration at East Rudolf, a unified system of locality numbers designating collecting areas has been devised by Dr V. Maglio who has provided an area map and a simplified diagrammatic representation of correlations between the principal areas.[5]

Archaeological exploration and excavations were continued under the supervision of Professor Glynn Isaac (University of California, Berkeley), assisted by graduate students. The excavations within the KBS Tuff[6] were extended and a series of sites above the KBS Tuff were examined; a large collection of artefacts was recovered from several preliminary exca-

TABLE 1
1972 Material Attributed to *Australopithecus*

KNM-ER NO.	SPECIMEN DETAIL	AREA	STRATIGRAPHIC POSITION WHERE KNOWN
1463	Right femur diaphysis	1A	Below middle tuff
1464	Right talus	6	Below lower tuff
1465	Proximal fragment left femuur	8	Below upper tuff
1467	Isolated M_3	3	Below upper tuff
1468	Right mandible	8	Below middle tuff
1469	Left mandible, M_3	131	Below KBS Tuff
1471	Proximal half right tibia	131	Below KBS Tuff
1476	Left talus, proximal tibia and fragment tibia shaft	105	At or above KBS Tuff
1477	Juvenile mandible with teeth	105	At or above KBS Tuff
1478	Cranial fragments	105	At or above KBS Tuff
1479	Fragments isolated molars	105	At or above KBS Tuff
1500	Skeletal elements	130	Below KBS Tuff
1503	Proximal right femur	123	—
1504	Distal right humerus	123	—
1505	Proximal fragment left femur and fragment of shaft	123	—
1506	Right mandible, M_1, M_2 and isolated P^4, P^3		—
1509	Isolated teeth, P_4–M_3	119	—
1592	Distal half femur	12	Below lower tuff

TABLE 2
1972 Material Attributed to *Homo*

KNM-ER NO.	SPECIMEN DETAIL	AREA	STRATIGRAPHIC POSITION WHERE KNOWN
1462	Isolated M_3	130	Below KBS Tuff
1466	Parietal fragment	1	Below upper Tuff
1470	Cranium	131	Below KBS Tuff
1472	Right femur	131	Below KBS Tuff
1475	Proximal right femur	131	Below KBS Tuff
1480	Isolated molar	105	Above KBS Tuff
1481	Left femur, proximal tibia, distal tibia and distal fibula	131	Below KBS Tuff
1483	Mandible fragments	131	Below KBS Tuff
1501	Right mandible	123	—
1502	Right mandible with molar	123	—
1507	Juvenile left mandible with teeth	127	—
1508	Isolated molar	127	—
1510	Cranial fragments	119	—
1590	Cranial fragments with juvenile dentition	12	Below KBS Tuff
1591	Humerus lacking head	12	Above KBS Tuff
1593	Cranial and mandibular fragments	12	Below KBS Tuff

vations. A detailed report on the archaeological activities will be presented elsewhere.

Mr Bruce Bowen (Iowa State University) extended the geological survey and mapping south of the Koobi Fora ridge and has established the broad continuity of the sequence from the lowest deposits at Kubi Algi, about 4.5 m.y., to the uppermost deposits at Ileret, about 1.0 m.y.

Mr Ian Findlater (Birkbeck College, London) continued the field investigation and mapping of volcanic events, and Drs J. Miller and F. Fitch are dating the volcanic horizons. In spite of meticulous collecting and rigorous laboratory analysis, however, dates for the tuffaceous horizons above the KBS Tuff continue to prove unreliable. Some intriguing technical problems indicate that depositional complications are causing the conventional techniques of radiometric age determinations to give inconclusive results. Thus there has been no advance on the situation as reported by Maglio,[5] although there is no evidence to suggest that the 2.61 ± 0.26 m.y. BP date from the KBS Tuff is unreliable (personal communication from J. A. Miller). The continued palaeomagnetic studies by Dr A. Brock (University of Nairobi) together with faunal correlations further strengthen confidence in this date.

In the summary which follows of the 1972 hominid collection a brief mention is made of the more important specimens; detailed descriptions of all the specimens will be given shortly elsewhere. The specimens attributed to *Australopithecus* are listed in Table 1; the specimens attributed to *Homo* are listed in Table 2; and the specimens which have not yet been at-

TABLE 3
1972 Material of Unclear Affinities

KNM-ER NO.	SPECIMEN DETAIL	AREA	STRATIGRAPHIC POSITION WHERE KNOWN
1473	Proximal fragment of right humerus	131	Below KBS Tuff
1474	Parietal fragment	131	Below KBS Tuff
1482	Mandible, partial dentition	131	Below KBS Tuff
1515	Isolated incisor	103	Below Koobi Fora Tuff

tributed to a particular genus, either because they are too fragmentary, or because a more detailed study is first required, are listed in Table 3. These tables also include geographical and stratigraphical data. The stratigraphical correlation of the areas 119, 121, 123, 127 south of the Koobi Fora ridge has yet to be concluded, so, for the present, this information is omitted for specimens from these areas. Specimens 1510 and 1590 to 1593 were discovered after the 1972 expedition had closed its principal research activities; these specimens, although important, are therefore only listed in the tables and not mentioned further. A separate article will be published[7] giving details of the *Homo* specimens, KNM-ER 1470, 1472, 1475 and 1481, in order to allow adequate treatment of these important finds.

AUSTRALOPITHECINE MATERIAL

Eighteen individuals, listed in Table 1, were collected. A number of specimens represent associated parts from the same individual. Several finds refer to horizons below the KBS Tuff so that there is now evidence of the species at East Rudolf during a period of more than 1 m.y.

A left half of a mandible, KNM-ER 1469, provides evidence that the large form of *Australopithecus* had developed before deposition of the KBS Tuff 2.6 m.y. ago. The mandible is cracked, although a large portion of M_3 is preserved *in situ* together with the roots and parts of the crowns of M_2 and M_1.

A mandibular fragment, KNM-ER 1506, includes M_1 and M_2 *in situ* and isolated P^3, P^4; the M_2 shows a distinct wear facet for M_3 but this tooth seems to have been lost before fossilization. The significance of the specimen is its small size; the body demonstrates the high degree of robusticity seen in other australopithecine mandibles from East Rudolf, and, for the first time, teeth of a small individual are preserved and provide crown dimensions.

A juvenile mandible, KNM-ER 1477 (see Fig. 1), discovered by Mr Musa Mbithi, a Kenyan, is beautifully preserved. The following teeth are in place: the left c, the left and right dm_1, dm_2 and erupting M_1. The

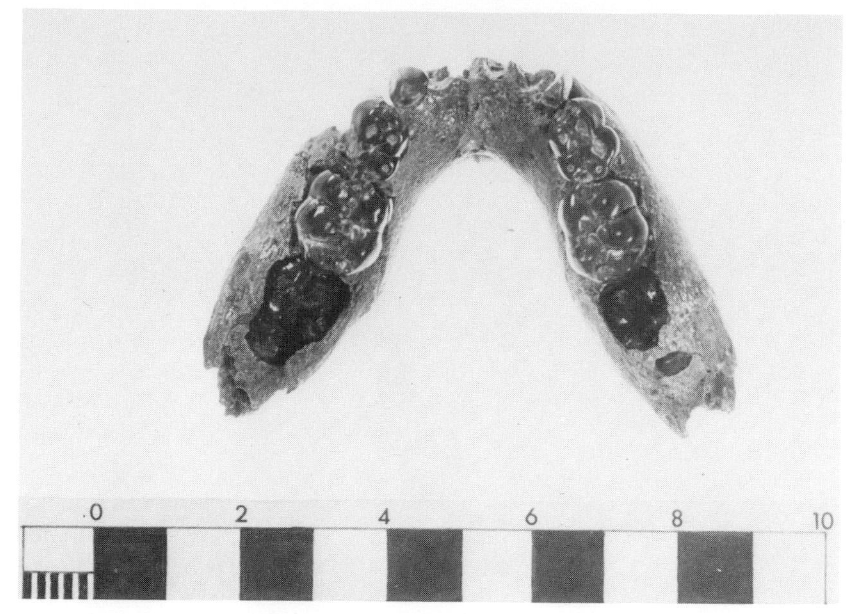

FIGURE 1
Occlusal view of juvenile mandible KNM-ER 1477.

germs of the left P3 C, and I2 were dislodged from their crypts and were recovered separately, whereas the corresponding tooth germs on the right side are preserved within their crypts.

Several postcranial specimens were recovered and a significant difference in the size of some of the specimens may provide additional evidence of sexual dimorphism in this group.[2,3] A specimen, KNM-ER 1500, which was also recovered from deposits below the KBS Tuff, includes parts of an associated skeleton. The specimen—a small individual, probably female—was discovered by a Kenyan, Mr John Kimengech, and it represents the first certain association of upper and lower limb elements of *Australopithecus* from East Africa. Unfortunately the material is badly weathered and considerable detail has been lost. The following skeletal elements have been identified: the proximal portions of the right femur, radius and ulna, and left tibia; the distal portions of the right femur, and fibula and left tibia; and scraps of the shafts of the tibia, ulna and humerus. This specimen should provide important new evidence on limb proportions in addition to morphological details of australopithecine postcranial bones that have not been available before.

A specimen, KNM-ER 1476 provides information on an associated tibia and talus, the former showing features similar to the australopithecine tibia, KNM-ER 741,[2,8] previously reported. The talus

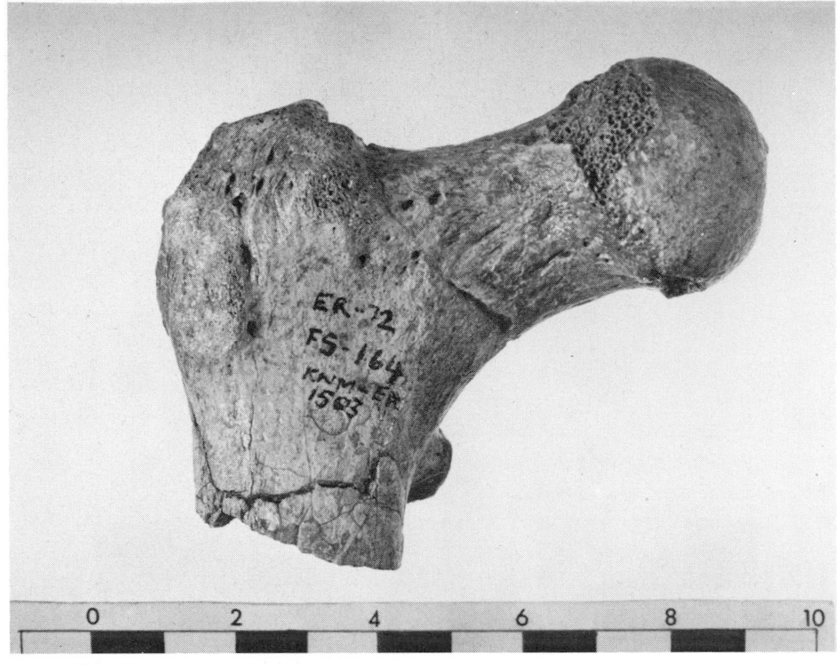

FIGURE 2
Anterior view of femur KNM-ER 1503.

exhibits a marked lateral extension for the articulation of the lateral malleolus which is also seen in a complete talus, KNM-ER 1464. This feature seems to be a characteristic of australopithecine tali and together with a number of other features makes this foot bone quite distinctive of this group.

Three specimens collected from area 123, KNM-ER 1503, 1504 and 1505 (see Fig. 2), probably represent parts of the same individual. The proximal portion of the right femur, KNM-ER 1505, is beautifully preserved and shows the characteristic long neck and small head seen in all other australopithecine femora.[9] The distal fragment of humerus is very similar to the humerus, KNM-ER 739,[2,8] but it is smaller. It shows the marked extension of the medial epicondyle, the constricted trochlea, and the relatively large capitulum.

HOMININE MATERIAL

Sixteen individuals, listed in Table 2, are represented by the material collected during 1972. Evidence of *Homo* at levels below the 2.6 m.y. KBS

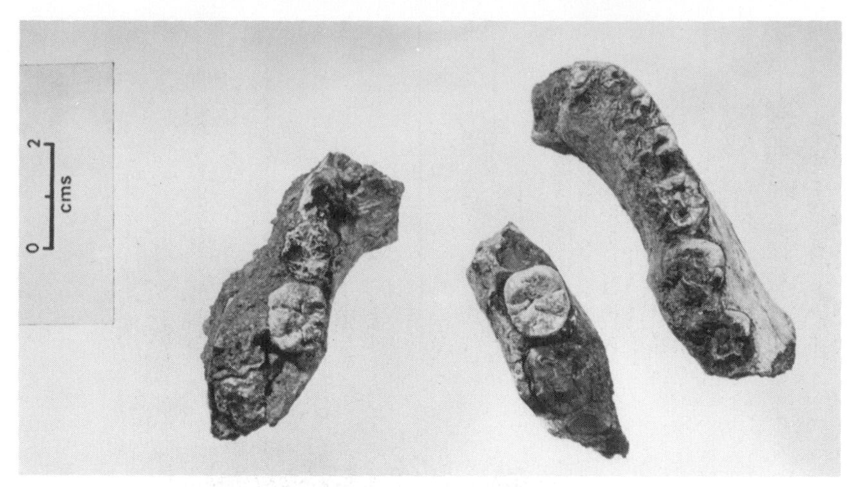

FIGURE 3
Occlusal view of mandibles (from left to right) KNM-ER 1507, 1502 and 1501.

Tuff horizon will be put forward on the basis of some remarkable material that will be discussed in a separate report.[7]

A fragment of parietal, KNM-ER 1466, is notable; it is thick boned and bears a marked temporal ridge, features which are reminiscent of the calvaria from Olduvai Gorge (OH9).[10] The specimen is unfortunately incomplete, and therefore cannot provide any conclusive evidence for the presence of this species at East Rudolf.

The mandibular fragments, KNM-ER 1501, 1502 and 1507 (see Fig. 3), from areas south of the Koobi Fora ridge are gracile and show a morphology similar to that seen on the *Homo* mandible from Ileret, KNM-ER 992.[3]

MATERIAL OF UNCLEAR AFFINITIES

Among the specimens listed in Table 3 a mandible, KNM-ER 1482 (see Fig. 4), is of particular interest; it shows characters not seen in either the contemporary *Australopithecus* or the assemblage of *Homo* mandibles from East Rudolf. The specimen was recovered from deposits below the KBS Tuff.

SIGNIFICANCE OF THE 1972 FINDS

The recovery of *Australopithecus* from levels below the KBS Tuff is particularly important in view of the close similarity of this material to that from

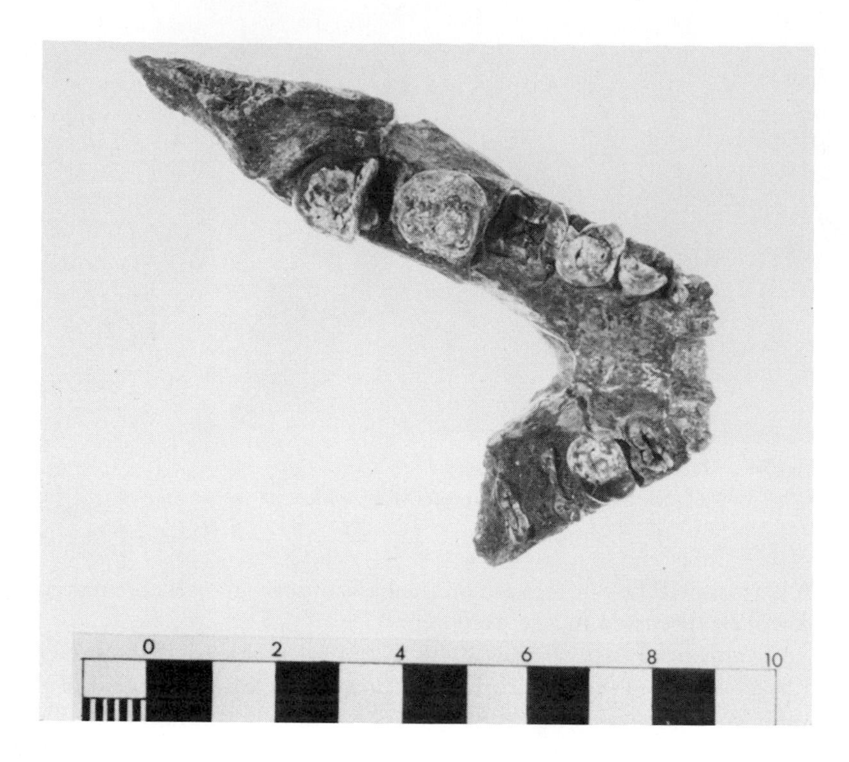

FIGURE 4
Occlusal view of mandible KNM-ER 1482.

the upper part of the East Rudolf succession. A pre-Pleistocene emergence of this genus seems reasonable although no precise data to support this have yet been put forward. Now that a large number of specimens of *Australopithecus* are known from East Africa, the taxonomic attribution of the mandibular specimen, KNM-ER 329 from Lothagam Hill, to *Australopithecus c.f. africanus*[11] might be questioned. The evidence for the presence of the "gracile" form of *Australopithecus* in East Africa is limited. The fragmentary nature of this specimen makes specific identification very difficult, particularly in the absence of other material from deposits of an equivalent age; only a few isolated dental fragments from the Usno Formation in the Omo Valley[12] are known. The generic attribution of the Lothagam specimen to *Australopithecus* might be questioned on the grounds that the specimen could equally well represent a Pliocene form such as the East African representative of the genus *Ramapithecus*. Such a conclusion would radically alter the interpretative model of hominid evolution.

The additional postcranial material will provide further evidence on which to examine the significance of the locomotory behavior of *Australopithecus.* Although there is a growing body of evidence in support of a bipedal model, it is important to recognize that the australopithecine mode of bipedality may have been quite distinctive from that of other contemporary hominids and behaviourally significant. In view of the evidence for contemporary *Homo* with morphologically distinct limb elements, some consideration must be given as to whether the australopithecine pattern of bipedal adaptation really reflects a transitional phase as has been suggested. The associated skeletal material collected during 1972 will advance the study of limb proportions within *Australopithecus;* preliminary indications point to a relatively short lower limb and a longer forelimb.

The hominine collection provides further examples of the gracile mandibles previously reported from the Ileret area.[3] Although the dating for the deposits south of Koobi Fora remains tentative, the fauna associated with these new mandibles indicates that they are earlier than those from Ileret. As such, it is possible to postulate a range of time that is greater than that represented by Ileret and the Olduvai Bed II deposits from which similar material has been reported. Mandibular material for the hominid represented by the cranium, KNM-ER 1470,[7] is not known at present but the size and shape of the palate of this specimen suggest a fairly broad big-toothed mandible, quite unlike the gracile specimens just mentioned. The affinities of the mandible, KNM-ER 1482, found in the same locality as the cranium KNM-ER 1470, are far from clear.

In conclusion, I would like to stress that the 1972 collection of hominids has raised more questions than answers; it now seems clear that the pattern of hominid evolution in eastern Africa is extremely complex. Evidence for local geographical variation in various mammalian taxa, primate and non-primate, is emerging, with significant differences occurring within forms restricted to small geographical zones such as the Omo/East Rudolf basin. Until further data on palaeogeographical barriers, environments and chronology are available the interpretation and the erection of evolutionary models should remain tentative.

The success of the 1972 programme was due to the generous support of the National Gėographic Society, the National Science Foundation, the National Museums of Kenya, the William H. Donner Foundation and various other individual donations. Many people assisted in the field to whom I am grateful; in particular I should like to thank Mr Kamoya Kimeu and his team of field collectors who are responsible for so many of the important discoveries. Dr Alan Walker, Dr Bernard Wood and my wife Meave all provided invaluable laboratory assistance.

Received January 2, 1973.

[1]Leakey, R. E. F., *Nature,* **226,** 223 (1970).
[2]Leakey, R. E. F., *Nature,* **231,** 241 (1971).
[3]Leakey, R. E. F., *Nature,* **237,** 264 (1972).
[4]Fitch, F. J., and Miller, J. A., *Nature,* **226,** 226 (1970).
[5]Maglio, V. J., *Nature,* **239,** 379 (1972).
[6]Isaac, G. Ll., Leakey, R. E. F., and Behrensmeyer, A. K., *Science,* **173,** 1129 (1972).
[7]Leakey, R. E. F., *Nature* (in press).
[8]Leakey, R. E. F., Mungai, J. M., and Walker, A. C., *Amer. J.. Phys. Anthrop.,* **36,** 235 (1972).
[9]Walker, A. C., *J. Hum. Evol.* (in the press).
[10]Leakey, L. S. B., *Nature,* **189,** 649 (1961).
[11]Patterson, B., Behrensmeyer, A. K., and Sill, W. D., *Nature,* **226,** 918 (1970).
[12]Howell, F. C., *Quarternaria,* 11, 47 (1969).

R. E. F. Leakey:

Evidence for an Advanced Plio-Pleistocene Hominid from East Rudolf, Kenya

The following article was reprinted, with permission, from Nature, *Volume 242, pages 447–450.*

Four specimens collected last year from East Rudolf are provisionally attributed to the genus *Homo*. One, a cranium KNM-ER 1470, is probably 2.9 million years old.

PRELIMINARY descriptions are presented of four specimens collected from East Rudolf during 1972. Most of the collection recovered during this field season has been reported recently in *Nature*[1]; the specimens described here are sufficiently important to be considered separately and in more detail. The collections of fossil hominids recovered from East Rudolf during earlier field seasons and detailed descriptions of some of these specimens have been published previously.[2-5]

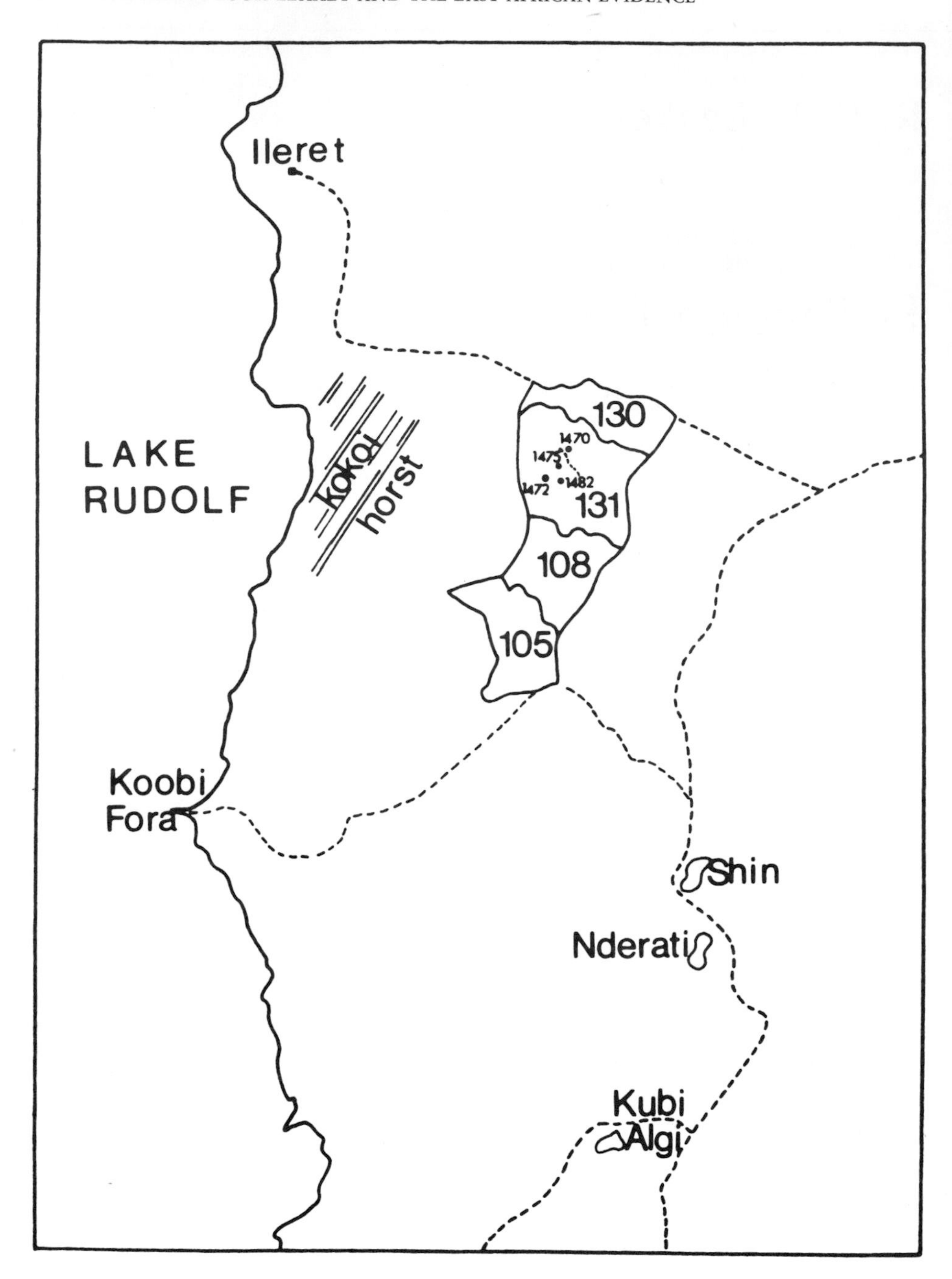

FIGURE 1
Map showing sites of discovery of fossil hominids KNM-ER 1470, 1472, 1475 and 1481 in the East Rudolf locality. Succession shown in Fig. 2 was taken from the position indicated by the dotted line.

The specimens described here are: (1) a cranium, KNM-ER 1470; (2) a right femur, KNM-ER 1472; (3) a proximal fragment of a second right femur, KNM-ER 1475; and (4) an associated left femur, distal and proximal fragments of a left tibia, and a distal left fibula, KNM-ER 1481. They were all recovered from area 131 (see Fig. 1) and from deposits below the KBS Tuff which has been securely dated at 2.6 m.y.[6]

Area 131 consists of approximately 30 km^2 of fluviatile and lacustrine sediments. The sediments are well exposed and show no evidence of significant tectonic disturbance; there is a slight westward dip of less than 3°. Several prominent marker horizons provide reference levels and have permitted physical correlation of stratigraphical units between area 131 and other areas in the East Rudolf locality.

Several tuffs occur in the vicinity of area 131. The lowest of these is the Tulu-Bor Tuff which is not exposed in the area itself but does outcrop nearby in several stream beds. Above this horizon, in a composite section, there is some 60 m of sediment capped by the prominent KBS Tuff. This latter tuff has been mapped into areas 108 and 105 (also shown in Fig. 1) from where samples have been obtained for K/Ar dates. An account of the geology is given by Vondra and Bowen.[7] A section showing the vertical position of these four hominids in relation to the KBS Tuff is given in Fig. 2.

At present, analysis of samples collected for dating from the KBS Tuff in area 131 has proved inconclusive because of the apparent alteration of the sanidine feldspars. This was not seen in the 105/108 samples from the same horizon which provided the date of 2.61 m.y. and there is no reason to suspect the validity of that date (personal communication from J. A. Miller).

Detailed palaeomagnetic investigation of the sedimentary units is being undertaken by Dr A. Brock (University of Nairobi). Systematic sampling closely spaced in the section has identified both the Mammoth and Kaena events in area 105 between the Tulu-Bor and KBS Tuffs, a result which supports the 2.61 m.y. date on the latter. The mapping of several horizons has established a physical correlation between areas 105 and 131. During the 1973 season, the area 131 succession will be sampled in detail in an attempt to confirm this correlation. Available evidence points to a probable date of 2.9 m.y. for the cranium KNM-ER 1470, and between 2.6 and 2.9 m.y. for the other specimens reported here.

Collections of vertebrate fossils recovered from below the KBS Tuff in areas 105, 108 and 131 all show the same stage of evolutionary development and this evidence supports the indicated age for this phase of deposition at East Rudolf. Maglio[8] has discussed the fossil assemblages following detailed studies of field collections from various horizons.

The cranium (KNM-ER 1470) and the postcranial remains (KNM-ER 1472, 1475 and 1481) were all recovered as a result of surface discov-

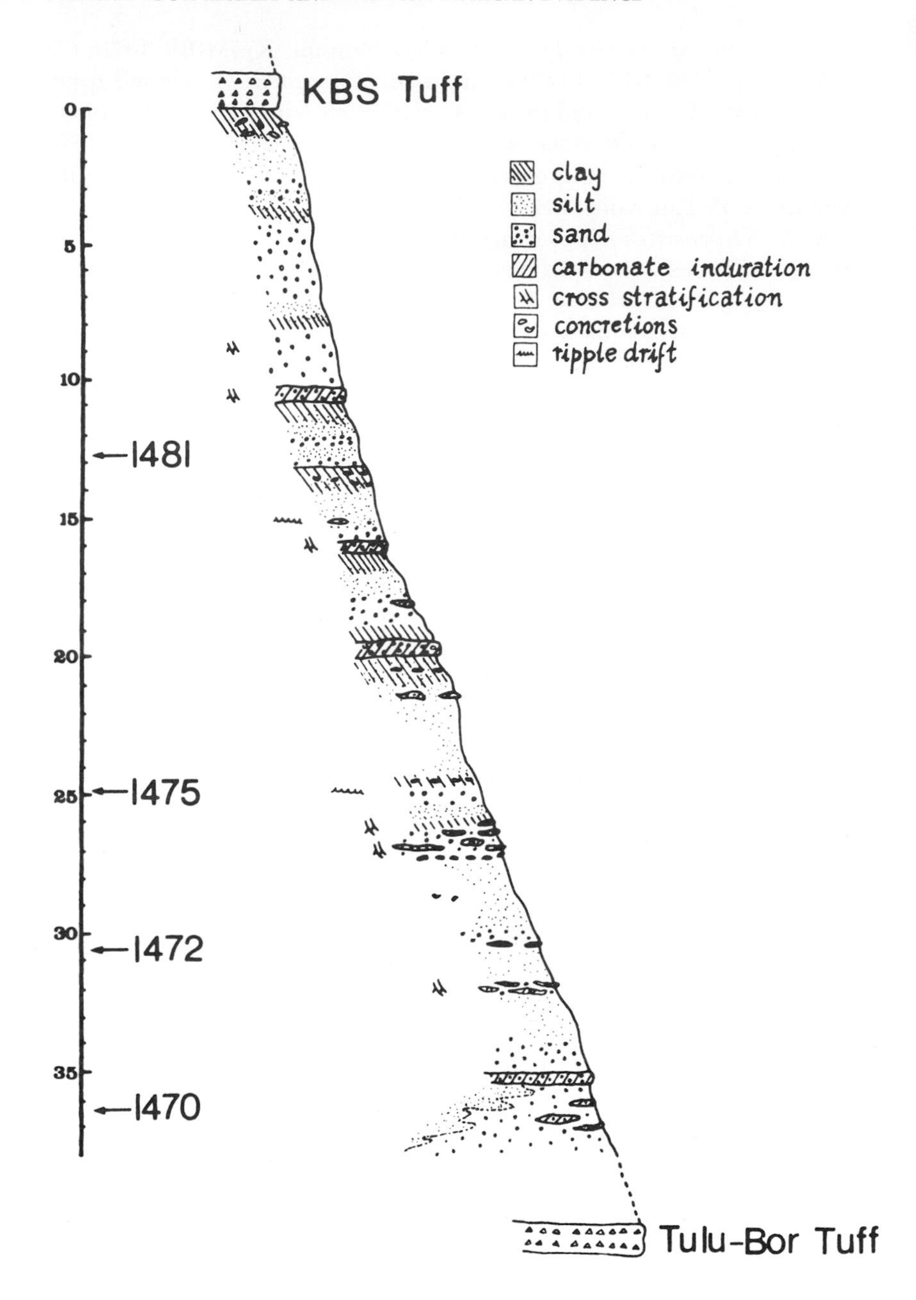

FIGURE 2
Stratigraphical succession of the sediments in area 131 and the vertical relationships of the fossil hominids KNM-ER 1470, 1472, 1475 and 1481 to the KBS Tuff. Dotted line shown in Fig. 1 marks the position at which the section was taken.

ery. The unrolled condition of the specimens and the nature of the sites rules out the possibility of secondary deposition—there is no doubt in the minds of the geologists that the provenance is as reported. All the specimens are heavily mineralized and the adhering matrix is similar to the matrix seen on other fossils from the same sites. In due course, microscopic examination of thin sections of matrix taken from the site and on the fossils might add further evidence.

CRANIUM KNM-ER 1470

Cranium KNM-ER 1470 was discovered by Mr Bernard Ngeneo, a Kenyan, who noticed a large number of bone fragments washing down a steep slope on one side of a gully. Careful examination showed that these fragments included pieces of a hominid cranium. An area of approximately 20 m × 20 m was subsequently screened and more than 150 fragments were recovered.

The skull is not fully reconstructed. Many small fragments remain to be included and it may be some time before the task is completed. At present the cranial vault is almost complete and there are good joins between the pieces. The face is less complete and although there are good contacts joining the maxilla through the face to the calvaria, many pieces are still missing. The orientation of the face is somewhat uncertain because of distortion of the frontal base by several small, matrix filled cracks. The basi-cranium shows the most damage and is the least complete region.

The cranium (see Fig. 3) shows many features of interest. The supraorbital tori are weakly developed with no continuous supratoral sulcus. The postorbital waisting is moderate and there is no evidence of either marked temporal lines or a temporal keel. The vault is domed with steeply sloping sides and parietal eminences. The glenoid fossae and external auditory meati are positioned well forward by comparison with *Australopithecus.* The occipital area is incomplete but there is no indication of a nuchal crest or other powerful muscle attachments.

In view of the completeness of the calvaria, it has been possible to prepare in modelling clay an endocranial impression which has been used to obtain minimum estimates for the endocranial volume. Six measurements of the endocast by water displacement were made by Dr A. Walker (University of Nairobi), and gave a mean value of 810 cm^3. Further work on this will be undertaken but it seems certain that a volume of greater than 800 cm^3 for KNM-ER 1470 can be expected.

The palate is shallow, broad and short with a nearly straight labial border that is reminiscent of the large *Australopithecus.* The great width in relationship to the length of the palate does contrast markedly, however,

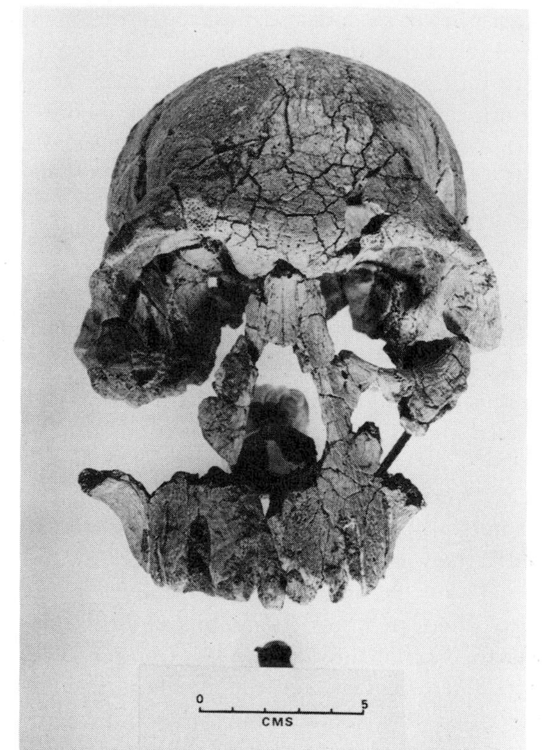

a

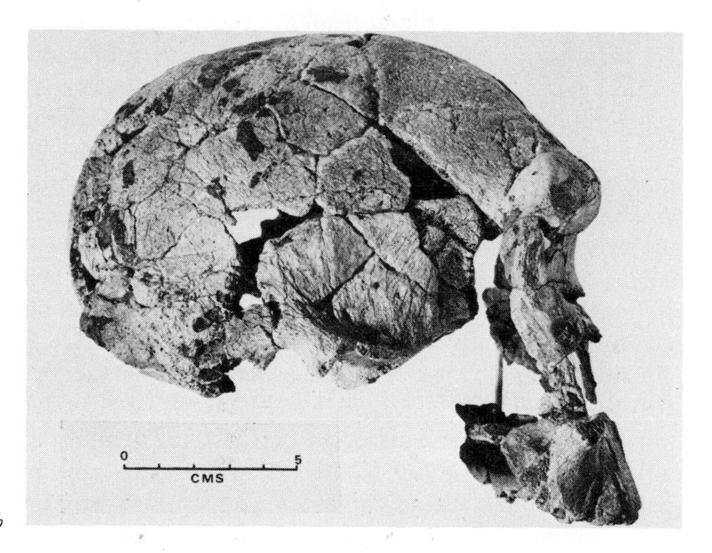

b

FIGURE 3

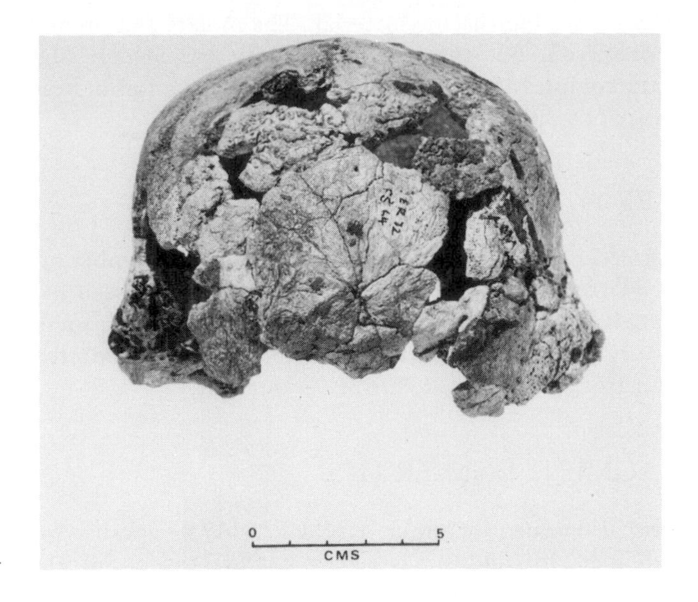

c

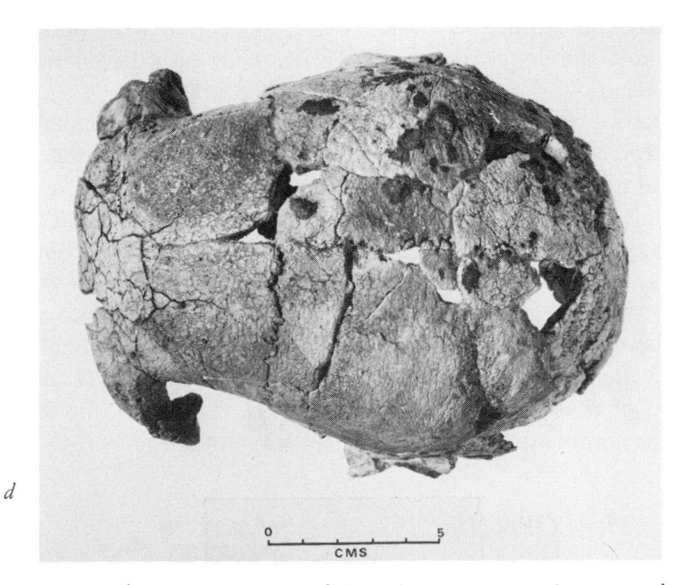

d

Cranium KNM-ER 1470. *a*, Facial aspect; *b*, lateral aspect; *c*, posterior aspect; *d*, superior aspect.

with known australopithecine material. The molars and premolar crowns are not preserved, but the remaining roots and alveoli suggest some mesiodistal compression. The large alveoli of the anterior teeth suggest the presence of substantial canines and incisors.

FEMUR KNM-ER 1472

KNM-ER 1472, a right femur, was discovered as a number of fragments by Dr J. Harris. It shows some features that are also seen in the better preserved left femur, KNM-ER 1481, but other features, such as the apparently very straight shaft and the bony process on the anterior aspect of the greater trochanter, require further evaluation.

FEMORAL FRAGMENT KNM-ER 1475

The proximal fragment of femur, KNM-ER 1475, was discovered by Mr Kamoya Kimeu. Its condition is such that a final taxonomic identification will be difficult and it is therefore included only tentatively in this report. This fragment shows some features such as a short, more nearly cylindrical neck, which are not seen in the femurs of *Australopithecus.*

ASSOCIATED SKELETON KNM-ER 1481

A complete left femur, KNM-ER 1481, associated with both ends of a left tibia and the distal end of a left fibula were also discovered by Dr J. Harris.

The femur (see Fig. 4) is characterized by a very slender shaft with relatively large epiphyses. The head of the femur is large and set on a robust cylindrical neck which takes off from the shaft at a more obtuse angle than in known *Australopithecus* femurs. There is a marked insertion for gluteus maximus and the proximal region of the shaft is slightly flattened anteroposteriorly. The femoro-condylar angle is within the range of *Homo sapiens.* When the femur is compared with a restricted sample of modern African bones, there are marked similarities in those morphological features that are widely considered characteristic of modern *H. sapiens.* The fragments of tibia and fibula also resemble *H. sapiens* and no features call for specific comment at this preliminary stage of study.

HOMO OR *AUSTRALOPITHECUS?*

The taxonomic status of the material is not absolutely clear, and detailed comparative studies which should help to clarify this problem have yet to be concluded. The endocranial capacity and the morphology of the cal-

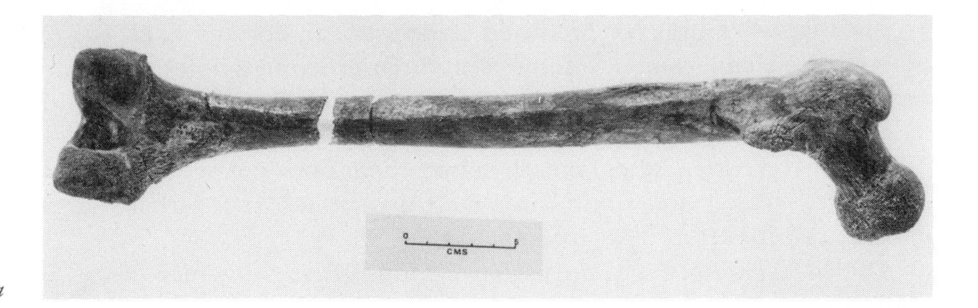

a

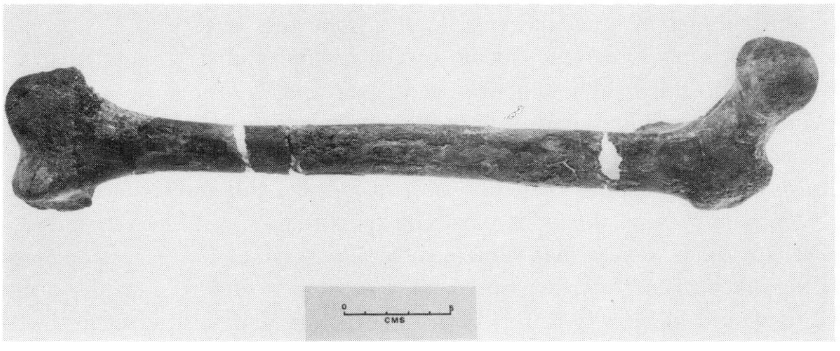

b

FIGURE 4
Left femur KNM-ER 1481. *a*, Posterior aspect; *b*, anterior aspect.

varia of KNM-ER 1470 are characters that suggest inclusion within the genus *Homo,* but the maxilla and facial region are unlike those of any known form of hominid. Only the flat fronted wide palate is suggestive of *Australopithecus,* but its extreme shortening and its shallow nature cannot be matched in existing collections representing this genus. The post-cranial elements cannot readily be distinguished from *H. sapiens* if one considers the range of variation known for this species.

The East Rudolf area has provided evidence of the robust, specialized form of *Australopithecus* from levels which span close to 2 m.y. (2.8 m.y.–1.0 m.y.)[1]; throughout this period the morphology of this hominid is distinctive in both cranial and postcranial elements. The cranial capacity of the robust australopithecine from Olduvai Gorge, *A. boisei,* has been estimated for OH 5 to be 530 cm^3 (ref. 9); this is the same value as that estimated by Holloway for the only specimen in South Africa of *A. robustus* which provides clear evidence of cranial capacity.[9] Holloway has also found the mean cranial capacity of six specimens of the small gracile *A. africanus* from South Africa[10] to be 422 cm^3. Thus, to include the 1470 cranium from East Rudolf within the genus *Australopithecus* would require an extraordinary range of variation of endocranial volume for this genus. This seems unacceptable and also other morphological considerations argue strongly against such an attribution.

The Olduvai Gorge has produced evidence of an hominine, *H. habilis;* the estimated endocranial volumes for three specimens referred to this species are 633, 652 and 684 cm^3 (ref. 10). The Olduvai material is only known from deposits that are stratigraphically above a basalt dated at 1.96 m.y. (ref. 11). At present therefore there does not seem to be any compelling reason for attributing to this species the earlier, larger brained, cranium from East Rudolf.

The 1470 cranium is quite distinctive from *H. erectus* which is not certainly known from deposits of equivalent Pleistocene age. It could be argued that the new material represents an early form of *H. erectus,* but at present there is insufficient evidence to justify this assertion.

There is no direct association of the cranial and postcranial parts at present, and until such evidence becomes available, the femora and fragment of tibia and fibula are only provisionally assigned to the same species as the cranium, KNM-ER 1470. Differences from the distinctive *Australopithecus* postcranial elements seem to support this inferred association.

For the present, I propose that the specimens should be attributed to *Homo* sp. indet. rather than remain in total suspense. There does not seem to be any basis for attribution to *Australopithecus* and to consider a new genus would be, in my mind, both unnecessary and self defeating in the endeavour to understand the origins of man.

I should like to congratulate Mr Ngeneo and Dr Harris for finding these important discoveries. Dr Bernard Wood spent many long hours at the site screening for fragments and assisted my wife, Meave, and Dr Alan Walker in the painstaking reconstruction work. I thank them all. The support of the National Geographic Society, the National Science Foundation, the W. H. Donner Foundation and the National Museum of Kenya is gratefully acknowledged.

Received January 23, 1973.

[1]Leakey, R. E. F., *Nature,* **242,** 170 (1973).
[2]Leakey, R. E. F., *Nature,* **231,** 241 (1971).
[3]Leakey, R. E. F., *Nature,* **237,** 264 (1972).
[4]Leakey, R. E. F., Mungai, J. M., and Walker, A. C., *Amer. J. Phys. Anthrop.,* **35,** 175 (1971).
[5]Leakey, R. E. F., Mungai, J. M., and Walker, A. C., *Amer. J. Phys. Anthrop.,* **36,** 235 (1972).
[6]Fitch, F. J., and Miller, J. A., *Nature,* **226,** 223 (1970).
[7]Vondra, C., and Bowen, B., *Nature,* **242,** 391 (1973).
[8]Maglio, V. J., *Nature,* **239,** 379 (1972).
[9]Tobias, P. V., *The Brain in Hominid Evolution* (Columbia University Press, New York and London, 1971).
[10]Holloway, R. L., *Science,* **168,** 966 (1970).
[11]Curtis, G. H., and Hay, R. L., in *Calibration of Hominoid Evolution* (edit. by W. W. Bishop and J. A. Miller) (Scottish Academic Press, Edinburgh, 1972).

R. E. F. Leakey:

Further Evidence of Lower Pleistocene Hominids from East Rudolf, North Kenya, 1973

The following article was reprinted, with permission, from Nature, *Volume 248, pages 653–656.*

Twenty new hominid specimens were recovered from the East Rudolf area in 1973. New evidence suggests the presence of at least three hominid lineages in the Plio-Pleistocene of East Africa.

THIS is a report of the 1973 field season at East Rudolf, Kenya, where the East Rudolf Research Project (formerly Expedition) has now concluded its sixth year of operations. Eighty-seven specimens of fossil hominid were collected[1] from the area during 1968–72; a further twenty specimens were recovered between June and September 1973 from the Upper, Lower and Ileret Members of the Koobi Fora Formation[2]. Exploration to the south of Koobi Fora was begun in 1972 and continued in 1973. No hominids have yet been found in the limited exposures of the Kubi Algi Formation. A notice of two specimens—KNM-ER 1510 and 1590—that were previously[1] mentioned only by number, is included in this report. The 1973 hominids are not here attributed to genera as there are still no clear generic diagnoses available for fossil hominids. With a few exceptions, previous attributions for the East Rudolf hominid collection remain satisfactory.

Archaeological investigation during 1973 was extended under the direction of G. Ll. Isaac, with J. C. W. Harris who conducted major excavations at several sites in the Upper Member of areas 130 and 131. Limited excavation, but extensive prospecting, in the Lower Member produced sufficient results to support further searching for artefacts below the KBS Tuff.

During the palaeontological survey, which was supervised by J. M. Harris, all identifiable fragments from certain horizons were collected; new species were recorded and some primate remains were recovered during a limited survey of the Kubi Algi Formation. A detailed account of the East Rudolf fauna will be presented upon conclusion of current studies, but there are clear indications that at times the palaeoenvironment differed from that of the lower part of the Shungura Formation of the Omo Valley in Ethiopia.

In the geological studies, emphasis was placed on microstratigraphy and palaeoenvironmental reconstruction. B. Bowen supervised a study of the Lower and Upper Members of areas 130 and 131 which included confirming the stratigraphic relationships of the cranium KNM-ER 1470. The complete section of the Koobi Fora Formation exposed in area 102 was studied by a group from Dartmouth College, New Hampshire, under G. Johnson. A. K. Behrensmeyer completed a preliminary geological investigation of the hominid sites, noting depositional environments and possible association of fauna; further studies are planned.

I. Findlater extended mapping of tuffaceous horizons to the south of Koobi Fora and collected samples for isotope dating. A series of dates has been obtained from material collected during 1972 (unpublished work of Miller, Findlater, Fitch and Watkins). Palaeomagnetic studies complement those of 1972 and there are sufficient data for internal correlations to be made.[3]

HOMINID COLLECTION

Specimen KNM-ER, 1590, reported previously,[1] consists of dental and cranial fragments which were collected from area 12, some metres below the KBS Tuff. Both parietals, fragments of frontal and other pieces of cranial vault, the left decidious $\underline{c}$ and dm^2, the left and right unerupted C,

TABLE 1
1973 hominid collection from East Rudolf

KNM-ER NO.	SPECIMEN	AREA	MEMBER
1800	Cranial fragments	130	Lower
1801	Left mandible, P_4, M_1, M_3	131	Lower
1802	Left mandible, P_4-M_2 and right P_3-M_2	131	Lower
1803	Right mandible fragment	131	Lower
1804	Right maxilla, P^3-M^2	104	Upper
1805	Cranium and mandible	130	Upper
1806	Mandible	130	Upper
1807	Right femur shaft	103	Upper
1808	Associated skeletal and cranial fragments	103	Upper
1809	Right femur shaft	127	Lower
1810	Proximal left tibia	123	?Lower
1811	Left mandible fragment	123	?Lower
1812	Right mandible fragment and left I_1 and M_1	123	Lower
1813	Cranium	123	?Lower
1814	Maxillary fragments	127	Upper
1815	Right talus	1	Upper
1816	Immature fragmented mandible	6A	Upper
1817	Right mandible	1	Upper
1818	I^1	6A	Upper
1819	M_3	3	Upper
1820	Left mandible with M_1	103	Upper

P^3 and P^4, and the erupted left and right M^1 and left M^2 were recovered. Although the cranium is immature, it was large with a cranial capacity as great as that determined for KNM-ER 1470. The parietals may show some deformation but, in any event, they suggest that the cranium was wide with a sagittal keel.

KNM-ER 1510, also reported previously,[1] includes cranial and mandibular fragments. The specimen is poorly mineralised and further geological investigation at the site indicates a Holocene rather than an early Pleistocene provenance as originally thought.

The 1973 hominids and their stratigraphical positions are listed in Table 1. Specimens from area 123 are rare, and their stratigraphical position relative to the Upper and Lower Members of the Koobi Fora Formation needs clarification.

A well preserved mandible (Fig. 1), KNM-ER 1802, was discovered by J. Harris *in situ* below the KBS Tuff in area 131. The dentition is only slightly worn, and fragments of both M_3 crowns suggest that death occurred before full eruption. The canines and incisors are represented by roots and by alveoli filled with matrix. The mandible shows some interesting features—moulding of the mandibular body, absence of a strong post-incisive planum, the development of a slight inferior mandibular torus and the distinct eversion of the mandibular body when viewed from below.

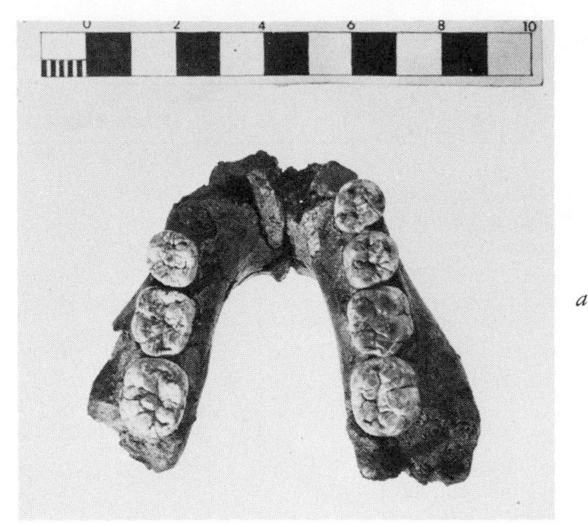

a

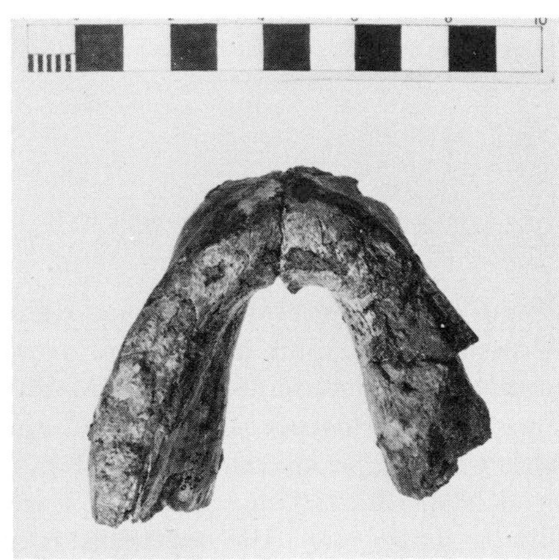

b

c

FIGURE 1
Mandible, KNM-ER 1802. *a*, Superior view; *b*, inferior view; *c*, right lateral view.

A weathered mandible, KNM-ER 1801, bears some resemblances to KNM-ER 1802, but its worn dentition and loss of surface bone prevent direct comparisons. The relative proportions of the molars and premolars may have been exaggerated by interstitial wear.

A crushed maxillary fragment, KNM-ER 1804, with P^3-M^2 preserved was discovered by R. Holloway. The teeth are complete but worn.

A skull (cranium with associated mandible), KNM-ER 1805, was discovered by P. Abell *in situ* in the BBS Tuff complex in area 130. The specimen is heavily encrusted with a hard matrix and will require careful preparation before its morphology is revealed. Comments here are thus preliminary. The cranium is in pieces which fit together. After preparation, it should be possible to determine the endocranial capacity; at present, a volume of 600–700 cm^3 is suggested. The supraorbital region, much of the face and the greater part of the basi-cranium have not been preserved. The postorbital region is preserved and the minimum breadth is approximately 90 mm. No distinct temporal lines cross the frontal area although they can be discerned and are still apart at the bregma. There are distinct parasagittal crests. The nuchal attachments are very distinctive and protrude to form a wide bony shelf. The palate is intact; all the teeth are preserved except for the right P^4 and the left I^1. The mandible, small and distinctly robust, is represented by both sides of the body but, except for the right M_2 and M_3, the tooth crowns are missing. The ascending rami are not preserved. The right M_3 and M_2 are well worn but small. The upper dentition shows wear on all the teeth, including M^3.

A large mandible, KNM-ER 1806, was discovered by Meave Leakey at the same site and horizon as was KNM-ER 1805. There are no tooth crowns preserved and the ascending rami are missing in this otherwise complete specimen. The mandible is typical of the large East Rudolf hominid that I have previously attributed to *Australopithecus.*

A fragmented specimen, KNM-ER 1808, was discovered in area 103 by Kamoya Kimeu. The specimen includes maxillary and mandibular teeth, cranial and mandibular fragments, a fragment of atlas vertebra, the distal half of a femur lacking the condyles, a large segment of humerus and other postcranial fragments. There is little doubt that the various pieces are from one individual and further sieving and excavation will be undertaken in the hope of recovering more material.

A cranium, KNM-ER 1813 (Fig. 2), was discovered *in situ* by Kamoya Kimeu in area 123. The specimen was fragmented but has been partially reconstructed. Plastic deformation is evident. The cranium is partly covered with a thin coat of matrix and considerable preparation is needed before a detailed description can be attempted. The endocranial volume is likely to be small; a figure of approximately 500 cm^3 is suggested on the basis of comparative external measurements. Other interesting features include the curvature of the frontals, a postglabella sulcus and the small dentition. The maxilla has well preserved teeth, P^3-M^3,

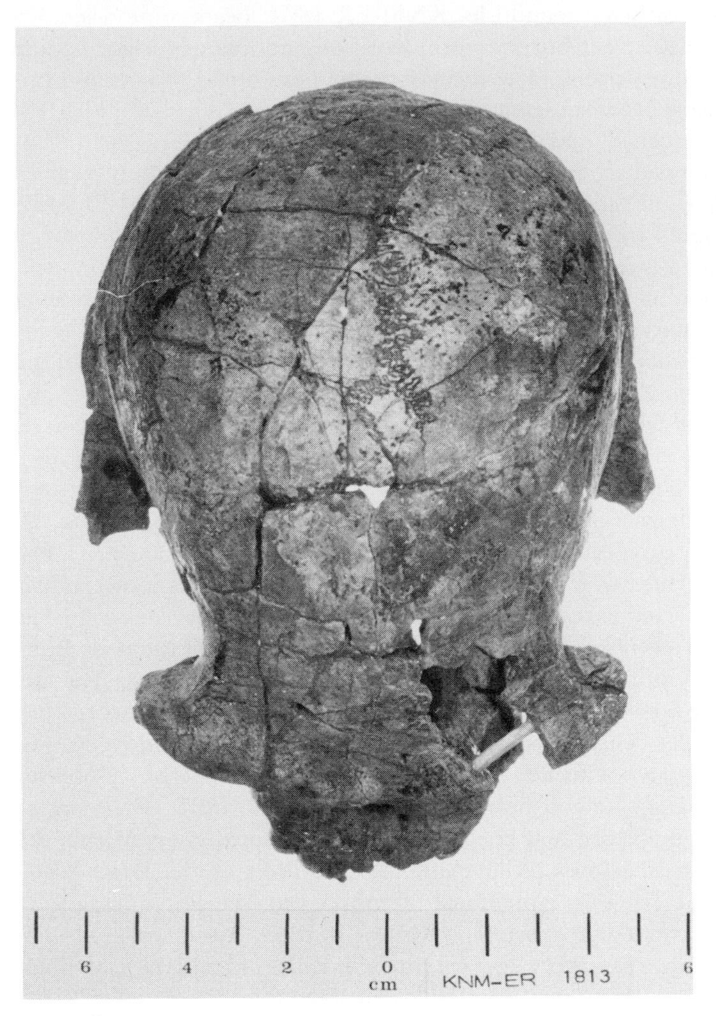

FIGURE 2
Cranium, KNM-ER 1813, superior view.

on the left side, but on the right side only the tooth roots and the complete crown of M^3 remain. Both canines and lateral incisors are present but the central incisors seem to have been lost before fossilisation. Both sides of the maxilla fit together to give the form of the dental arcade. The right maxillary fragment includes the malar region and connects with the lateral margin of the right orbit.

Other specimens recovered during 1973 are listed in Table 1 and will be described in detail after studies are completed.

SIGNIFICANCE OF THE 1973 COLLECTION

In previous reports, [1,4-7] the East Rudolf hominids were assigned to *Australopithecus, Homo* or indeterminate (the last category included both very fragmentary specimens and those of uncertain taxonomic rank.

The East Rudolf specimens that have attributed to *Australopithecus,* span a period of time from 3 million yearts to just over 1 million years with apparently little morphological change. This form is likely to be the same species as *A. boisei*[8]; it also shows similarities with *A. robustus* from southern Africa. A Pliocene origin is suggested for this specialised group.

Specimens attributed to *Homo* have been recovered from deposits covering a similar time span, but these show greater morphological variability. Those recovered from the Ileret Member seem to differ from those recovered from the Lower Member of the Koobi Fora Formation. The suggestion that a large brained, fully bipedal hominid was living at East Rudolf 3 million years ago was put forward after the 1972 discoveries.[7] This point of view is supported by the cranial fragments, KNM-ER 1590, also from below the KBS Tuff, and this specimen is provisionally attributed to *Homo.*

The 1973 collection from East Rudolf raises many questions. The new mandible, KNM-ER 1802, could be considered as belonging to the same genus and species as KNM-ER 1470 and 1590. There are striking similarities between the dental characters of KNM-ER 1802 and some specimens from Olduvai Gorge such as the type mandible of *Homo habilis,* OH 7. Although the suggested cranial capacity for *H. habilis* is appreciably smaller than that determined for KNM-ER 1470, the discrepancy may be due to the fragmentary material upon which the former estimates were made. I consider that the evidence for a "small brained" form of *Homo* during the Lower Pleistocene is tenuous.

The cranium, KNM-ER 1813, may prove to be quite distinct from the robust australopithecines and from *Homo,* as represented by KNM-ER 1470. The dentition is "hominine," yet the cranial capacity appears small. The cranium has some of the features seen in the gracile, small brained, hominid *Australopithecus africanus* Dart, from Sterkfontein.

I have previously questioned the validity of a distinct gracile species of *Australopithecus,*[6] but this new evidence reopens the possibility of its existence. Some authors[9,10] have suggested that *Homo habilis,* particularly OH 24, shows features typical of *Australopithecus africanus.* My suggestion here, that *H. habilis* may have affinities with KNM-ER 1470 and 1590, refers only to OH 7 and OH 16. Features of the calvarium of OH 24 show similarities with KNM-ER 1813. The size and morphology of the teeth of the two specimens are alike and the cranial capacities may also be comparable.[11,12]

The skull KNM-ER 1805 is undoubtedly important, but its interpretation is enigmatic. Its relatively large cranium bears sagittal and nuchal crests but has small teeth; this combination is in contrast to all the specimens previously recovered from East Rudolf.

In any consideration of the affinities of the East Rudolf hominids, the question of sexual dimorphism must not be overlooked. There does seem to be evidence for quite marked sexual dimorphism in one hominid group as demonstrated by the East Rudolf crania, KNM-ER 406 and 732.[5] Unfortunately both crania lack teeth so that the dental characteristics of the alleged female are far from clear.

The possibility of more than two contemporary hominid lineages in the Plio-Pleistocene of East Africa may now have to be recognised, whereas previously one, or at most, two forms were assumed. The attribution of isolated teeth may thus become even more difficult than it is now. Postcranial identifications likewise may be difficult, although the proximal femoral material continues to suggest a morphological dichotomy.

I suggest the following as a basis of nomenclature for Plio-Pleistocene hominids. One genus would include much of the material currently referred to *Australopithecus robustus* and *A. boisei.* A second genus would incorporate many of the gracile specimens from Sterkfontein presently referred to *A. africanus,* perhaps certain specimens from East Rudolf including KNM-ER 1813, and possibly some from Olduvai such as OH 24. A third genus, *Homo,* would incorporate specimens such as KNM-ER 1470 and 1590 from East Rudolf and possibly OH 7 and OH 16 from Olduvai. Some material from South Africa might also be considered within this last category together with later specimens from Olduvai and East Rudolf. The unusual mandible KNM-ER 1482,[1] together with the specimen from the Omo area referred to *Paraustralopithecus*[13] and certain other specimens from Omo which are contemporary with the three groups just mentioned, could be considered a fourth form—a remnant of an earlier population that disappeared during the early Pleistocene. All these forms may be traced back well beyond the Plio-Pleistocene boundary.

These remarks are necessarily speculative. A more detailed review of hominid systematics is being prepared in collaboration with B. A. Wood. The wealth of data now available presents a new era in the study of early man. The complexities of dealing with the enlarged sample are challenging, and isolated studies on specific specimens must be replaced by exhaustive studies on all the fossil hominid evidence.

I should like to express appreciation for the financial backing provided by the National Geographic Society, the National Science Foundation, the W. H. Donner Foundation and others. The support and encouragement of the National Museums of Kenya and the Kenya Government made the research possible. Members of the East Rudolf Research Project are too numerous to thank individually but all play a part in a successful

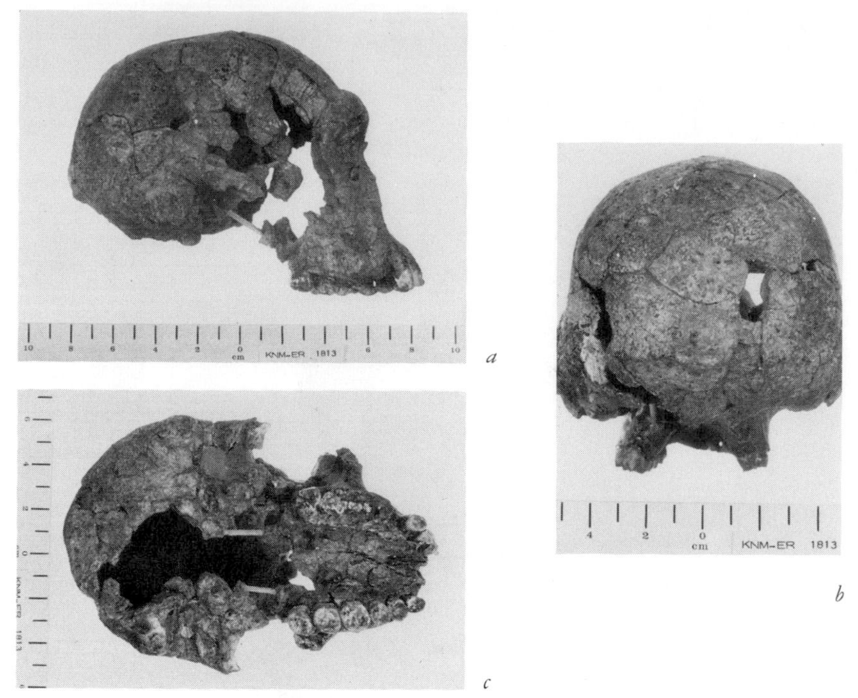

FIGURE 3
Additional photographs of the cranium, KNM-ER 1813. *a,* Right lateral view; *b,* posterior view; *c,* basal view.

field season and are thanked along with those who made important discoveries. I would also express thanks to my wife Meave who, as always, provided invaluable assistance both at the museum and in the field.

Received January 9, 1974.

[1]Leakey, R. E. F., *Nature,* **242,** 170 (1973).
[2]Bowen, B. E., and Vondra, C. F., *Nature,* **242,** 391 (1973).
[3]Brock, A., and Isaac, G. Ll., *Nature,* **257,** 344 (1974).
[4]Leakey, R. E. F., *Nature,* **226,** 223 (1970).
[5]Leakey, R. E. F., *Nature,* **231,** 241 (1971).
[6]Leakey, R. E. F., *Nature,* **237,** 264 (1972).
[7]Leakey, R. E. F., *Nature,* **242,** 447 (1973).
[8]Tobias, P. V., *Olduvai Gorge,* 2 (Cambridge University Press, 1967).
[9]Robinson, J. T., *Nature,* **205,** 121 (1965).
[10]Anon., *Nature,* **232,** 294 (1971).
[11]Leakey, M. D., Clarke, R. J., and Leakey, L. S. B., *Nature,* **232,** 308 (1971).
[12]*Nature,* **239,** 469 (1972).
[13]Arambourg, C., and Coppens, Y., *C.r. Hebd. Séanc. Acad. Sci., Paris,* **265,** 589 (1967).

Professor Michael Day at work describing the morphology of a hominid femur.

Michael Day:

Hominid Postcranial Material from Bed I, Olduvai Gorge

Most scientists now believe that there was more than one variety of hominid living in East Africa during the Plio-Pleistocene. The distinction between species is most easily made on the basis of skulls, jaws and teeth, but the question arises, can we tell to which species belong the fossil bones from the limbs and trunk? Were there significant differences in the body form, posture, and gait of the species? Michael Day takes up this problem in a study which describes the Olduvai fossils and compares them with postcranial bones from other sites.

During excavations at site FLK NN I Olduvai in 1960 Louis and Mary Leakey recovered a quantity of hominid postcranial bones from a living floor in Bed I. These important finds were associated with Oldowan stone tools and with abundant faunal remains (Leakey, L. S. B., 1960, 1961). Two groups of hominid fossil postcranial bones were present, one adult and the other juvenile. Those that were immediately recognized comprised one clavicle (two were reported in error in Leakey, L. S. B., 1960), 21 bones believed to be of the hand and 12 bones of the foot.

During the same season, a fossil hominid tibia and fibula were recovered from site FLK, the same level that produced Olduvai Hominid 5 ("Zinjanthropus"). Subsequently three other fossil hominid specimens were recovered from the Bed I/Bed II boundary zone—a terminal phalanx (O.H. 10), a femoral neck (O.H. 20) and an ulna (O.H. 36). Finally, examination of the total collection of Bed I material has revealed two more metatarsals (O.H. 43), a short section of radius shaft (O.H. 49), and part of a rib that have been attributed to the Hominidae. Initial descriptions and functional interpretations of most of this material have already been published; the remainder is in preparation and will be published shortly. A reassessment of the published material seems necessary in view of recent work on the material and the recovery of new finds from East Rudolf.

The recovery of the FLK NN I material (hand, foot, and clavicle), and its interpretation in terms of hominid locomotion, clearly played a part in influencing the decision to revise the previous definition of the genus *Homo* and to erect a new species in that genus, *Homo habilis* (Leakey, Tobias and Napier 1964). Subsequent analyses have thrown some further light on the functional implications of the finds (Day 1967; Day and Wood 1968) but the recovery of more material has been a necessary preliminary to further evaluation of the taxonomic attribution of this material. In the recent finds from East Rudolf there have been a number of specimens that seem likely to prove of assistance, indeed they are already beginning to point the way in which opinion is likely to go concerning the Olduvai material. Not all of the Olduvai Bed I postcranial material has been duplicated from East Rudolf, however, but some of the new specimens form part of associated groups of remains that each relate to one single individual—specimens such as KNM-ER 1500 and KNM-ER 803. These valuable partial skeletons allow internal comparisons between the skeleton parts of the remains of one individual representative of one species; a certainty denied those who examined the Olduvai remains other than the Hominid 8 foot bones.

THE OLDUVAI CLAVICLE (O.H. 48)

Originally reported by Louis Leakey (1960) this left clavicle was first examined and briefly described by Napier (1965). His conclusion was that its closest similarities were with *Homo sapiens,* being distinguished from those of this species only by a smooth postero-inferior groove that could indicate a steep thoracic inlet. Functional deductions were limited to the view that shoulder girdle movements were probably substantially the same as in modern man and were probably no barrier to overarm throwing or other activities requiring external rotation of the shoulder joint.

Later the angle of clavicular torsion of the Olduvai clavicle was used as an argument for suggesting "structural modifications of the shoulder girdle towards considerable ability for suspension of the body by the arms"

(Oxnard 1969). Unfortunately this angle is not available on this fossil since both ends are missing and internal cracks shown on the skiagram do not allow the certainty of longitudinal integrity in this specimen, even if the ends are reconstructed.

Anatomical comparisons of the Olduvai clavicle with collections of *Gorilla, Pan, Homo sapiens* (Bushman) and *Homo sapiens* (European) clavicles has tended to confirm Napier's original view. Similarly, comparisons with a number of fossil or sub-fossil African clavicles such as those from Willey's Kopje and Nakuru as well as that from Pin Hole Cave have not revealed any further information of particular significance. A series of multivariate analyses of ten dimensions of these clavicles and comparative series were unable to separate two alternative Olduvai reconstructions from the *Homo sapiens* groups other than to indicate some possibility that it may be male rather than female. The only conclusion that can be suggested is in line with the anatomical assessment that the Olduvai clavicle appears to have functioned in a fashion similar to that known in modern man.

THE OLDUVAI HAND (O.H. 7)

Of the original 21 "hand" bones recovered from the FLK NN I floor (Leakey, M. D., 1971), 15 were described by Napier (1962). Further examination has shown that of these 21 bones, 6 are nonhominid and one appears to be a vertebral fragment of unknown taxon (See Table 2). The fifteen hominid bones left for consideration include seven that are clearly juvenile, six that are of uncertain age and two that are adult. At this point, therefore, at least two individuals are involved. However, of the three carpal bones (which are of uncertain age since they lack the possibility of epiphyses) two are right sided (trapezium and scaphoid) and one left sided (capitate). The remaining bones of uncertain age include two distal fragments of proximal phalanges and the base of a second metacarpal (Napier's Table 1 is in error: Napier 1962). All three of these bones could be juvenile since the regions that carry epiphyses are missing. Indeed their coloration, degree of fossilization and general size is close to the remaining juvenile bones. The four undamaged but incomplete intermediate phalanges are all clearly immature since they lack epiphyses. The three terminal phalanges are also immature since they possess partially fused epiphyses.

The two adult proximal phalanges are of distinctive form and quite unlike the two possibly juvenile proximal phalangeal fragments. Further examination and comparison of these phalanges shows them to be highly curved, deeply guttered ventrally and in possession of prominent lateral and medial ridges for the attachment of fibrous flexor sheaths. In addi-

TABLE 1
Olduvai Hand Bones (O.H.7)

LETTER CODE	ALLOCATION
	HOMINID
A	Terminal thumb phalanx R., juvenile
B	Terminal phalanx of finger, juvenile
C	Terminal phalanx of finger, juvenile
D	Intermediate phalanx R. II, juvenile
E	Intermediate phalanx R. IV, juvenile
F	Intermediate phalanx R. III, juvenile
G	Intermediate phalanx R. V, juvenile
H	Proximal phalanx, juvenile?
I	Proximal phalanx, juvenile?
O	Base second metacarpal R., juvenile?
Q	R. trapezium, juvenile?
P	R. scaphoid, juvenile?
R	L. capitate, juvenile?
	NONHOMINID
J	Proximal phalanx, adult
K	Proximal phalanx, adult
T	Terminal phalanx
L	Intermediate phalanx, adult
M	Intermediate phalanx, adult
N	Proximal or intermediate phalanx, adult
S	Vertebral fragment
U	Unidentified fragment

tion, the bones appear to have disproportionately small heads. A search of the comparative collections in London and Nairobi has so far disclosed no certain identity with any known fossil or recent primate phalanges. There does seem, however, to be a clear morphological similarity between these phalanges and those of the black and white colobus monkey *(Colobus polykomos),* although the fossil bones are much larger. Unfortunately, the only giant colobus skeleton *(Paracolobus chemeroni)* known from the East African Pleistocene lacks these particular bones. The possibly arboreal character of these two bones has also been noted by Oxnard (1972a) and the suggestion tested by an ingenious comparative technique making use of stressed plastic profiles. It was concluded that the Olduvai adult proximal phalanges would have been "markedly inefficient" in a knuckle-walking context in both *Pan* and *Pongo,* "reasonably efficient" in a hanging-climbing context in *Pan* and still "efficient" in the same context in *Pongo.* Oxnard further concludes that these bones may well have been used in the "hanging-climbing" situation—a mode of locomotion frequently used by colobines. In view of this reappraisal it would seem prudent to exclude the two adult proximal phalanges from any reconstruction of the Olduvai hand since none of the remaining phalanges have arboreal features of this kind.

Thirteen hominid bones remain that could most economically come from two hands since two of the carpals and the metacarpal base are right sided, the capitate is left sided and all thirteen are juvenile or possible juvenile. Of the remaining phalanges, the thumb terminal phalanx is also considered to belong to the right side, while the other terminal phalanges cannot be sided at present. Similarly the intermediate and proximal phalanges cannot be sided, yet their digital order can be given some clear alternatives if it is assumed that they all belong to one side. The relative sizes of the largest and smallest of the four intermediate phalanges coupled with the tilts of the heads indicates that if they are a set then they belong to a right hand (Figure 1).

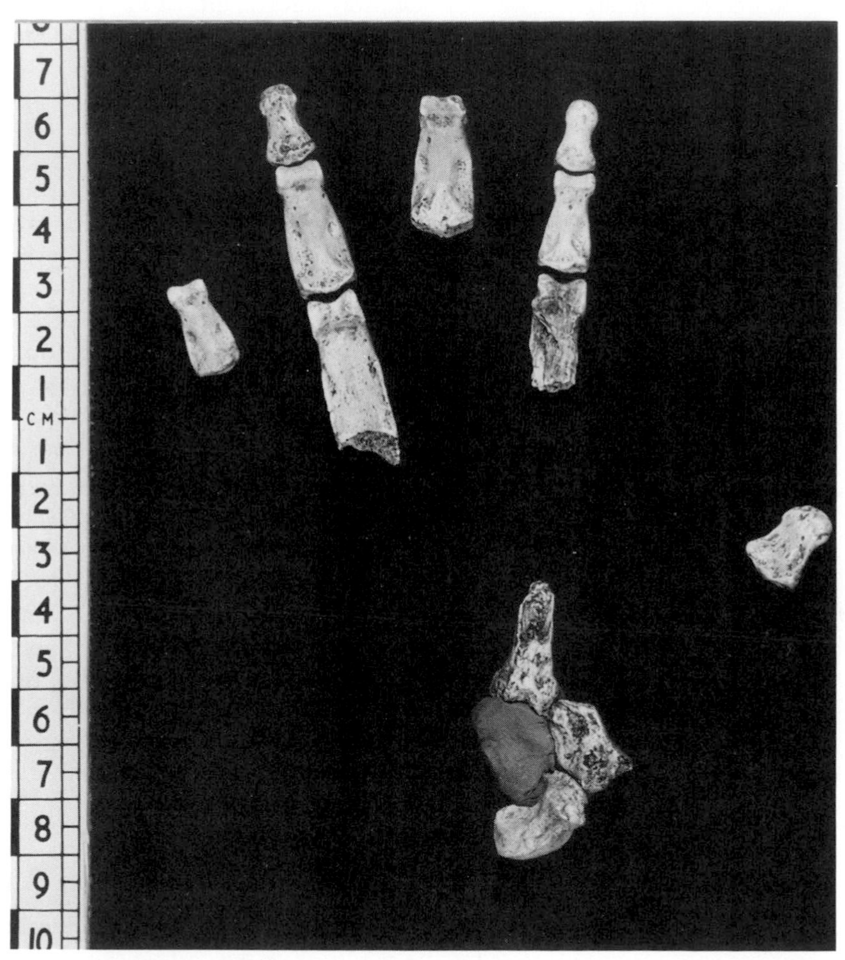

FIGURE 1
A reconstruction of the Olduvai Hand (O.H.7).

Since detailed descriptions of these bones are in preparation little more will be said at present other than to suggest that the hand was powerfully built, capable of strong finger flexion and possessed of an opposable thumb. It is of interest however, that an estimate of its maturity based upon modern human epiphyseal information (Greulich and Pyle 1950) agrees remarkably well with one given on dental grounds for the O.H. 7 mandible from the same living floor by Wolpoff (pers. comm.).

Little further hand material is available for direct comparison at present. One intermediate phalanx has been recovered from the Omo region by the French group (Omo 18-1970-1848) (Coppens 1973). From its general morphology and the relative tilts of the head and the base it would appear to be a left second intermediate phalanx that has a number of points of similarity with the Olduvai equivalent. Two proximal phalanges from high in the sequence at East Rudolf (KNM-ER 164) are very modern in form and similar to those known for *Homo sapiens.*

THE OLDUVAI TIBIA AND FIBULA (O.H. 35)

These two bones were recovered from the FLK floor which produced the O.H. 5 cranium *(Australopithecus boisei)* but they have never been firmly attributed to this species (Leakey, L. S. B., 1960). In the description (Davis 1964) it was stated that these bones could not be assigned to any given species; later, however, it was suggested as a possibility that the tibia and fibula "belonged with *Homo habilis* rather than with *Australopithecus (Zinjanthropus)*" (Leakey, Tobias and Napier 1964).

In his description and assessment of the functional implications of the tibia and fibula, Davis (1964) concluded that bony adaptations to bipedalism were well advanced at the ankle in this form but, on "slender evidence," were less well adapted at the knee joint. Further examination of these fossils confirms this view in respect to the ankle but not in relation to the knee, since it appears that Davis' opinion of the arrangements of muscular markings at the upper end of the tibia is not the only tenable interpretation. In Davis' view the markings for popliteus muscle in the fossil differ widely from both the human condition and that of the great apes; he goes on to state that the popliteal area is strongly marked and that its texture suggests that the direction of pull of the muscle was nearly vertical. It is my belief that this interpretation was based upon a misidentification of the soleal line of the tibia in this specimen and that the true soleal line skirts the broken area towards the upper end of the posterior aspect of the bone. In this event none of the area for the attachment of popliteus is available for examination. The true identification of the soleal line at the higher level appears to be confirmed by the presence of a clear vascular groove in it for the popliteal vessels, while the prominent crest below is the vertical line separating the Tibialis posterior and

Flexor digitorum longus attachment. This opinion has also been reached quite independently by Lovejoy (pers. comm.). If this interpretation is accepted then the conclusion previously put forward that the knee is less well adapted for bipedalism is not so well supported.

Recent postcranial finds from the East Rudolf area have included a number of tibial and fibular fragments such as KNM-ER 741, 803, 1471, 1476, 1481, 1500. Some of these have been previously allocated to the genus *Australopithecus* (741, 1471, 1476, 1500) and others to the genus *Homo* (803, 1481) (Leakey, R. E. F., 1971, 1972, 1973). The opportunity to examine the finds as a group rather than as a sequential series of finds each to be considered singly, leads to an appreciation of the variable nature of both the tibia and the fibula in the Hominidae, a point made by Davis (1964) concerning comparative studies made in relation to the O.H. 35 tibia and fibula. It seems likely that some of the original attributions will have to be revised, a contingency well recognized by the anatomists in the East Rudolf Research Group. However, two specimens (KNM-ER 803 and 1500) are of particular importance, since each is believed to be a collection of fossil bones ascribed to a single individual due to the homogeneity of the finds, the lack of duplications, and the isolated depositional circumstances of the find sites. In each instance there are tibial remains and femoral remains. In the case of KNM-ER 1500 there is a femoral neck of typical australopithecine nature, and the tibia shows a number of similarities with that of O.H. 35. On the other hand, the tibia of KNM-ER 803 shows a number of differences from the O.H. 35 tibia while its femur shows similarities to the Peking *Homo erectus* femora. The KNM-ER 1500 partial skeleton has been ascribed to the genus *Australopithecus* sp. (Leakey, R. E. F., 1973) while the KNM-ER 803 partial skeleton has been ascribed to the genus *Homo* sp. (Leakey, R. E. F., 1972; Day and Leakey, 1974).

In terms of dating, the australopithecine partial skeleton is from the lower member of Area 131, perhaps 2 million years B.P. while the hominine partial skeleton is from the Ileret region, Area 8A, perhaps 1.5 million years B.P. Olduvai Hominid 35 is dated at about 1.75 million years B.P.

Thus, the new material from East Rudolf would seem at the very least to raise serious doubts as to the correctness of allocating O.H. 35 to *Homo habilis,* an allocation never expressed by Davis (1964) and only raised as a possibility by Leakey, Tobias and Napier (1964).

THE OLDUVAI FOOT (O.H. 8)

The twelve bones comprising this foot were found in close association on the FLK NN I floor; when lifted they were found to articulate perfectly to form a small left tarsus and metatarsus. Their single identity is beyond question. Initial announcement and description (Leakey, L. S. B., 1960; Day and Napier 1964) has been followed by further study of the talus (Day and Wood 1968). The initial anatomical conclusions that the whole foot belonged to an upright bipedal hominid have never been seriously contested, but discussion has arisen over the interpretation of some of the results of metrical analyses of the talus based on comparative data treated by multivariate statistics (Oxnard 1972b; Day, in press).

In a study of the anthropoid hallucial tarso-metatarsal joint, Lewis (1972) has said that the Olduvai Hominid 8 foot has retained a hominoid locking mechanism of this joint, only traces of which may be seen in *Homo sapiens.* It is suggested that this locking mechanism, which "primarily evolved as a means for stabilizing the grasping hallux of arboreal primates was pre-adaptive for terrestrial life, for it provides just the right morphological requirement for producing a strong weight-bearing but arched and flexible foot." Indeed, the strongly adducted hallux with an adventitious joint between the first and second metatarsals as well as the pronounced longitudinal and transverse arches of the Olduvai foot bear out this suggestion.

In the preliminary description of the foot (Day and Napier 1964), attention was drawn to the robusticity formula of the metatarsals as being unlike that of the gorilla and similar to that of man except in one respect—the apparent dominance of metatarsal III in length and thickness. This was interpreted as either an individual variation or alternatively, an indication of the incomplete evolution of the pattern of metatarsal structure and function known in modern *Homo sapiens.* Archibald, Lovejoy and Heiple (1972), using an Amerindian sample of 50, confirmed that the commonest modern human robusticity formula was 1>5>4>3>2 (28 cases) while the Olduvai formula of 1>5>3>4>2 occurred far less frequently (3 cases); they also confirmed that "those features which are indicative of a striding gait are clearly present in O.H. 8, including a very robust M-5 and a reduced relative robusticity of M-2." They further believe that the variability that they have shown in the Amerindian feet examined indicate that the variations in relative robusticity of the median metatarsals are not critical to the gait pattern; thus favoring the first of the two alternative explanations offered previously (Day and Napier 1964).

Recent finds from East Rudolf that relate to the foot include several tali (KNM-ER 1464, 1476a, 803, 813a) and some metatarsal remains (KNM-ER 803, 997, 1500). Of these, specimens 803 and 813a are regarded as belonging to the genus *Homo* (Leakey, R. E. F., 1972), while 1476a and 1500 are regarded as belonging to the genus *Australopithecus* (Leakey, R. E. F., 1973). The KNM-ER 803 third left metatarsal and the talar fragment from part of the partial skeleton referred to previously have been described in detail (Day and Leakey 1974). The KNM-ER 813a talus has been described by Wood (1973) and confirmed as belonging to the genus *Homo.* The introduction of its dimensions into a multivariate analysis has indicated its functional similarity to those of *Homo sapiens* (Wood 1973). In addition, KNM-ER 1476a is a slightly damaged left talus whose features resemble those of the Olduvai talus in the presence of a medially placed dorsal trochlear groove, a "common-shaped" tibial facet with a dorsiflexion "stop" mechanism, a wide horizontal angle of the neck, a head set on the neck in the same way as in Olduvai Hominid 8 (Day and Wood 1968), and trochlear radii that are equal, indicating a stationary axis of the ankle joint. KNM-ER 1464 is a perfect right talus very similar to KNM-ER 813a in many respects; it has, however, two features a little reminiscent of the Olduvai talus—a grooved trochlear surface and a broad fibular facet.

Placing the Olduvai talus, O.H. 8, KNM-ER 1476a, 1464 and 813a together provides an interesting morphological comparison; the first two are comparable in size as are the last two in the series. Further studies will need to be completed before firm attributions can be made, but it appears at present that KNM-ER 1464 should also be attributed to the genus *Homo;* a viewpoint accepted by Wood (pers. comm.) and Lovejoy (pers. comm.).

The metatarsal material is very scanty. KNM-ER 1500, the australopithecine partial skeleton, has a right third metatarsal base that is comparable with the Olduvai third metatarsal while the KNM-ER 803 metatarsal is larger and more comparable with that of modern man. Similarly the KNM-ER 997 third left metatarsal shows some features of the base similar to that of the Olduvai third metatarsal, but here allocation is quite uncertain.

Once again, the new material from East Rudolf seems to be sufficient to cast doubts on the attribution of the Olduvai Hominid 8 foot to the genus *Homo* as a paratype of the species *Homo habilis* (Leakey, Tobias and Napier 1964), but nothing has been recovered so far that would suggest a radical revision of the locomotor conclusions that have been put forward in relation to the Olduvai foot material.

THE OLDUVAI TOE (O.H. 10)

This terminal phalanx was described by Day and Napier (1966) and later compared in a multivariate study (Day 1967). Criticisms of the conclusions drawn from this study (Oxnard 1972b) have been the subject of a reply (Day, in press). No terminal phalanges have been recovered from East Rudolf so far, therefore little that is new can be said other than to reiterate that the O.H. 10 toe bone is quite unlike that from any prehensile primate foot so far examined, and that its functional morphology is closest to that of man. It is of interest, however, that a terminal phalanx of a great toe has been recognized from the TM 1517 collection of hand and foot bones from Kromdraai (TM 1517k). This phalanx has some points of similarity with O.H. 10 phalanx including a degree of shaft torsion. A reconsideration of all of the TM 1517 material is in progress.

THE OLDUVAI FEMUR (O.H. 20)

This femoral neck was described following its recognition in the Olduvai collection by Dr. Mary Leakey (Day 1969). At that time the only early hominid femoral remains known were from the Transvaal collections, Sts 14, SK 82, SK 97. The striking resemblances between the two Swartkrans femora and the Olduvai specimen led to its attribution to *Australopithecus* cf. *boisei* recognized from Olduvai. Subsequent finds at East Rudolf have included 7 femora many of which show features similar to those of the Olduvai femoral neck. In a reappraisal of australopithecine femora, Olduvai Hominid 20 has been accepted as the only example of an australopithecine femur known from Olduvai (Walker 1973).

The advent of new early hominid fossil postcranial material from East Rudolf has clearly had a number of effects on opinions held about the material previously known from Olduvai. The presence at East Rudolf of material attributed to the genus *Homo* (Day and Leakey 1973; Leakey and Wood 1973; Day and Leakey 1974) that is very different in character from that known from Olduvai, yet of an age comparable to the Olduvai finds, demands a close reconsideration of the taxonomic attributions formerly made for some of the Olduvai material. While it would be premature to remove all the Olduvai postcranial material from the genus *Homo* and place it in *Australopithecus,* the possibility that the Olduvai assemblage is mixed must be taken seriously.

Acknowledgments

It is with great pleasure that I acknowledge the support of the Wenner-Gren Foundation for Anthropological Research, the Royal Society of London and the Boise Fund. I am also most grateful to my colleagues Professor J. R. N. Napier, Dr. Alan Walker, and Dr. Bernard Wood for helpful discussions and particularly to Dr. Louise Scheuer for her help with the reconstruction of the Olduvai hand. Finally I must thank the whole Leakey family for the opportunity to study their precious fossils and for so many kindnesses during successive visits to East Africa.

References Cited

Archibald, J. D., C. O. Lovejoy and K. G. Heiple

1972 Implications of relative robusticity in the Olduvai metatarsus. Am. J. Phys. Anthrop. 37:93–95.

Coppens, Y.

1973 Les restes d'Hominidés des séries inférieures et moyennes des formations plio-villafranchiennes de l'Omo en Ethiopie (recoltés 1970, 1971 et 1972). C. R. Acad. Sci. Paris 276:1823–1826.

Davis, P. R.

1964 Hominid fossils from Bed I, Olduvai Gorge Tanganyika: a tibia and fibula. Nature 201:967–968.

Day, M. H.

1967 Olduvai Hominid 10: a multivariate analysis. Nature 215:323–324.

1969 Femoral fragment of a robust australopithecine from Olduvai Gorge, Tanzania. Nature 221:230–233.

In press The interpolation of isolated fossil foot bones into a discriminant function analysis—a reply. Am. J. Phys. Anthrop.

Day, M. H., and R. E. F. Leakey

1973 New evidence of the genus *Homo* from East Rudolf, Kenya I. Am. J. Phys. Anthrop. 39:341–354.

1974 New evidence of the genus *Homo* from East Rudolf, Kenya III. Am. J. Phys. Anthrop. 41:367–380.

Day, M. H., and J. R. Napier

1964 Fossil foot bones. Nature 201:969–970.

1966 A hominid toe bone from Bed I, Olduvai Gorge, Tanzania. Nature 211:929–930.

Day, M. H., and B. A. Wood

1968 Functional affinities of the Olduvai Hominid 8 talus. Man 3:440–455.

Greulich, W. W., and S. I. Pyle

1950 Radiographic atlas of skeletal development of the hand and wrist. Stanford, Calif.: Stanford Univ. Press.

Leakey, L. S. B.

1960 Recent discoveries at Olduvai Gorge. Nature 188:1050–1052.

1961 New finds at Olduvai Gorge. Nature 189:649–650.

Leakey, L. S. B., P. V. Tobias and J. R. Napier

1964 A new species of the genus *Homo.* Nature 202:7–9.

Leakey, M. D.

1971 Olduvai Gorge. Volume 3. Excavations in Beds I and II, 1960–1963. Cambridge: The University Press.

Leakey, R. E. F.

1971 Further evidence of Lower Pleistocene hominids from East Rudolf, North Kenya. Nature 231:241–245.

1972 Further evidence of Lower Pleistocene hominids from East Rudolf, North Kenya. Nature 237:264–269.

1973 Further evidence of Lower Pleistocene hominids from East Rudolf, North Kenya. Nature 248:653–656.

Leakey, R. E. F., and B. A. Wood

1973 New evidence of the genus *Homo* from East Rudolf, Kenya II. Am. J. Phys. Anthropology 39:355–368.

Lewis, O. J.

1972 The evolution of the hallucial tarsometatarsal joint in the Anthropoidea. Am. J. Phys. Anthrop. 37:13–26.

Napier, J. R.

1962 Fossil hand bones from Olduvai Gorge. Nature 196:409–411.

1965 Current Anthropology Comment. Current Anthropology 6:402–403.

Oxnard, C. E.

1969 A note of the Olduvai clavicular fragment. Am. J. Phys. Anthrop. 29:429–432.

1972a Functional morphology of primates: some mathematical and physical methods. *In* The functional and Evolutionary Biology of Primates. R. Tuttle, ed. Chicago: Aldine Atherton Inc.

1972b Some African fossil foot bones: a note on the interpolation of fossils into a matrix of extant species. Am. J. Phys. Anthrop. 37:3–12.

Walker, A. C.

1973 New *Australopithecus* femora from East Rudolf, Kenya. J. of Human Evolution 2:529–536.

Wood, B. A.

1973 A *Homo* talus from East Rudolf, Kenya. Proc. Anat. Soc. Gt. Britain and Ireland.

Professor Phillip Tobias gives a lecture-demonstration on hominid fossils at the famous site of Sterkfontein, where he now conducts researches.

Phillip V. Tobias:

African Hominids:
Dating and Phylogeny*

Within the past fifteen years there have been more discoveries of fossil specimens that elucidate human ancestry than have been found in the entire preceding century. Phillip Tobias offers a sweeping commentary of finds in Eastern and Southern Africa and presents his views on how the new finds affect our understanding of our origins.

The fossils discovered by Dr. Leakey and his family include examples of the following groups:

(1) Miocene hominoids including *Ramapithecus wickeri* (formerly called *Kenyapithecus*), *Kenyapithecus africanus* (a probable member of *Dryopithecus*) and

---

*The major portion of this chapter represents an updated and considerably amplified version of an article which appeared in the *Annual Review of Anthropology*. Those parts of the original which are unchanged are reprinted here, with permission, from "New Developments in Hominid Paleontology in South and East Africa" by Phillip V. Tobias, *Annual Review of Anthropology,* Volume 2, pp. 311–334.

Much additional material and some 95 extra references have been added to bring the survey up to date.

species of *Dryopithecus* (formerly called *Proconsul*), from Rusinga Island, Koru, Songhor, Fort Ternan and elsewhere;

(2) Plio-Pleistocene australopithecines, from Olduvai Gorge, Peninj, the Chemeron Beds and the vast area east of Lake Rudolf;

(3) early *Homo,* including the group to which the name *Homo habilis* has been given, from Olduvai and East Rudolf:

(4) *Homo erectus* from Olduvai and perhaps Kapthurin;

(5) early *Homo sapiens* from Kanjera, Omo and perhaps elsewhere;

(6) late *Homo sapiens* from Gamble's Cave, Bromhead's Site, Naivasha Railway Rock Shelter, Willey's Kopje, Makalia and Nakuru Burial Sites.

No group of investigators in the world, other than the Leakeys, has made major contributions to every one of these levels of emergent humanity. It is fitting therefore to pause at this moment, shortly after Dr. Leakey's untimely death, to see what kind of story is crystallizing from the work of the past half-century. We need especially to take stock following the dramatic advances of the last fifteen years since the discovery of Olduvai hominid 5, that exquisite type specimen of *Australopithecus boisei.*

NEW DEVELOPMENTS IN HOMINID PALEONTOLOGY IN SOUTH AND EAST AFRICA

In recent years there has been remarkable progress in the unravelling of hominid evolution. The main advances registered—and which are now crying out for provisional synthesis—may be grouped under the following broad rubrics:

(a) the discovery of further hominid fossils in East Africa [Omo, East Rudolf, Chesowanja (Carney *et al.* 1971) and Olduvai] and in South Africa (Sterkfontein and Swartkrans), and their demographic analysis (Mann 1968, 1973, 1974; McKinley 1971; Tobias 1968a, 1974a);

(b) new studies on the morphology of many specimens, with a concomitant better appreciation of the "total morphological pattern" (Clark, LeGros, 1964) and of the ranges of variation within populations and taxa.

(c) the time scale of hominid evolution in East Africa which is becoming progressively clearer with new age determinations based upon the potassium-argon and $^{40}Ar/^{39}Ar$ methods (Brown 1972; Curtis and Hay 1972; Evernden and Curtis 1965; Fitch 1972; Miller 1972), paleomagnetism reversals (Brock and Isaac 1974; Cox 1972; Grommé and Hay 1967), fission tracking (Fleischer and Hart 1972; Hurford 1974) and faunal correlations (Butzer 1971a; Cooke and Ewer 1972; Cooke and Maglio 1972; Howell 1972a; Maglio 1971, 1972, 1973);

(d) the long uncertain and disputed time parameters of the South African australopithecine sites that are at last yielding to analysis by faunal correlations (Cooke 1970; Freedman and Brain 1972; Maglio 1973; Vrba 1974; Wells 1969) and by geomorphological methods (Butzer 1974; Partridge 1973, 1974);

(e) the resulting possibility of formulating tentative synoptic models of the chronological and phylogenetic relationships between the South and East African fossil hominids, and a single pattern of African hominid phylogeny (Tobias 1973a, 1974a);

(f) some special implications of the recent work which include the rejection of the concept that only one hominid species has existed at any one time (based upon the so-called Principle of Competitive Exclusion); the realization that hominid evolution has been cladistic in character and not predominantly phyletic; the illustration of mosaic evolution in the merging pattern of hominid phylogeny; and a greater consciousness of the extent and the limitations of sexual dimorphism in higher primates and, especially, the Hominoidea;

(g) paleoanthropologists have begun to show a healthy, critical and self-critical approach to the niceties of systematics and nomenclatural procedure, thus clearing the air for a less confused and confusing picture of hominid systematics—an essential prerequisite to the erecting of hypotheses on hominid phylogeny and on the selective mechanisms which could have operated in hominid evolution.

An attempt is made here to review the newer evidence from the major African sites under some of these headings and to limn a tentative reconstruction of hominid phylogeny over the late Tertiary and Quaternary.

NEW HOMINID DISCOVERIES AND STUDIES IN AFRICA

East Rudolf

The most spectacular set of discoveries has undoubtedly been those made by R. E. F. Leakey and his co-workers in the vast area known as East Rudolf in northern Kenya. The sequence of hominid discoveries began with four finds—two from Koobi Fora and two from Ileret—in 1968. These were followed by three specimens found in 1969 and 16 in 1970 (Day and Leakey 1973; Leakey, R. E. F., 1970, 1971; Leakey, R. E. F., *et al.* 1971). The 1971 season added 26 more hominid specimens or sets of remains (Leakey, R. E. F., 1972a, b), while the 1972 season brought to light 38 additional hominid fossils (Leakey, R. E. F., 1973a,b; Leakey and Walker 1973; Leakey and Wood 1973). The net yield over the five years was thus some 87 specimens, of which rather more than half were derived from the larger, southerly area known as Koobi Fora and the remainder from the smaller northerly zone called Ileret. Further discoveries in 1973 brought the overall total from East Rudolf to over 100 hominid specimens.

A remarkable discovery, late in 1972, was of the cranium KNM-ER-1470 (Leakey, R. E. F., 1973b). On reconstruction it proved to have an endocranial capacity of close to 800 cc., nearly twice as great as the mean for South African crania of *A. africanus* (Holloway 1970, 1972b, 1973; Robinson 1966; Tobias 1971a, 1974d) and 25% as much again as the mean for 4 Olduvai crania assigned to *Homo habilis* (Holloway 1973; Tobias 1971a, 1973b, in press c). Yet, despite its bigger brain size, the new East Rudolf specimen was found at a level *below* the KBS tuff dated to 2.6 m.y. and subsequently inferred from paleomagnetism readings to be about 2.9 m.y. (Leakey, R. E. F., 1973b). It would seem thus to be 1 m.y. or more older than the Olduvai *H. habilis* sample and would point to an appreciably earlier appearance of the *Homo* tendency to marked encephalization than had earlier been suspected.

The studies on the stratigraphy (Vondra *et al.* 1971), chronology, and vertebrate faunas led Maglio (1972) to propose four faunal zones in the East Rudolf succession. The earliest, the *Notochoerus capensis* zone, is represented in the Kubi Algi beds to the south of Koobi Fora: these beds have not so far yielded hominid fossils. The *Mesochoerus limnetes* zone has a "best fit" age of 2.3 m.y. (2.0–3.0 m.y., indicated minimum and maximum). It is represented in the Lower Member of the Koobi Fora formation and includes some of the hominids tentatively assigned to *Homo*. There follows the *Metridiochoerus andrewsi* zone with a "best fit" age of 1.7 m.y. (1.5–1.9 m.y., indicated minimum and maximum ages); this zone corresponds in age and fauna with Olduvai Bed I and lower Bed II. It is represented in the Upper Member of the Koobi Fora formation and includes many hominids, both *Australopithecus* and *Homo*. The latest of Maglio's four zones, the *Loxodonta africana* zone, has a "best fit" age of 1.3 m.y. (1.0–1.6 m.y., indicated minimum and maximum) and is represented in the Ileret Member. It includes many hominid specimens, both *Australopithecus* and *Homo*. These faunal stages are still tentative at this time.

Archaeological remains were discovered at East Rudolf by Behrensmeyer in 1969 (Behrensmeyer 1970; Isaac *et al.* 1971; Leakey, M. D., 1970; Leakey, R. E. F., 1970) associated with an outcrop of the KBS tuff dated at 2.61 ±0.26 m.y. Subsequent excavations revealed worked stone objects and manuports. At one excavation locality, FxJj1, in Koobi-Fora stratigraphic unit IIA, the stone forms a low density scatter along with small but significant quantities of broken up bones (Isaac *et al.* 1971). This occurrence is interpreted as "an occupation episode, albeit of low intensity" (Isaac 1972:406). As such, "FxJj1 may reasonably be advanced as the oldest dated occupation site" (Isaac 1972:407). R. E. F. Leakey has reported a sample of artifactual material from below this tuff, "although little detailed excavation has yet been undertaken" (1973c:58). These signs of hominid cultural activities have set the archaeological record of occupation sites back from 1.8 m.y. at Olduvai to about 2.6 m.y. or perhaps earlier at East Rudolf. This record of very ancient implemental activities is corroborated by comparably early finds at Omo.

In sum, the newest evidence from East Rudolf indicates the presence there of hominid fossils provisionally assigned to *Homo* in strata from about 2.9 m.y. and extending up to about 1.0–1.3 m.y., and of others belonging to a very robust australopithecine like *A. boisei* in strata from about 2.0 m.y. or more onwards to about 1.0–1.3 m.y.

Omo

Since 1966 the Omo Research Expedition, led by F. C. Howell of the University of California, Berkeley, and Y. Coppens of Paris, has been systematically exploring the series of sediments in the Lower Omo Valley (Arambourg 1969; Arambourg *et al.* 1969; Bonnefille *et al.* 1970; Butzer 1971a; Butzer and Thurber 1969; Coppens 1970–71, 1973a,b; Howell 1968, 1969, 1972a,b; Howell and Coppens 1973, 1974). Two of the formations in this area, the Usno (Heinzelin and Brown 1969) and the Shungura (Brown 1969; Heinzelin 1971), have yielded hominid fossils from close to 70 localities. The fossiliferous strata in the Usno Formation have an estimated age of 2.5–2.6 m.y. (Brown 1972) and have yielded some 19 isolated or associated hominid teeth (Howell and Coppens 1974). The hominid-bearing members of the Shungura Formation extend in time from 3.75 m.y. (Member B) to 1.3–1.5 m.y. (Member L): its hominid fossils are found mainly in the upper members (C-G) of the sequence, corresponding to potassium-argon ages of about 2.5–1.7 m.y. Only a few localities in the lowest part of the Shungura sequence, dated to about 3.5 m.y., have yielded hominid remains—some isolated teeth. The overwhelming majority of the Omo hominid remains are of isolated teeth; there are also some cranial and mandibular parts and a number of postcranial bones. Although detailed studies of these hominid remains are awaited, some of them, more especially the teeth, have been tentatively assigned to a very robust australopithecine like *A. boisei,* and others to one or more smaller-toothed forms, *A. aff. africanus* and/or *H. habilis* (Coppens 1970–71; Howell 1968, 1969). Howell and Coppens (1973) have published a detailed analysis of nine deciduous teeth from the Usno and Shungura Formations. Although they have attempted no taxonomic attribution, they draw attention to resemblances to the homologues of Southern Africa australopithecines. The evidence tends to suggest to them that more than one hominid is represented by the available sample from the Omo.

Artifacts have been reported by Chavaillon (1970), Bonnefille *et al.* (1970), and Howell (1972a), including sparse occurrences as low down in the sequence as Member C. They include stone artifacts and fragmentary bone remains and suggest that, in the northern Rudolf basin, "hominid object-manipulation and modification must date back to about 3 m.y." (Howell 1972a:349). This would push back the archaeological record to about 0.5 m.y. earlier than its hitherto revealed expression at East Rudolf, though it is possible that artifactual material from below the KBS tuff at East Rudolf (Leakey, R. E. F., 1973c) may be of comparable age.

Olduvai

Dr. M. D. Leakey (1971a,b) has continued to lay bare the archaeological and paleontological sequence in the Olduvai Gorge, and some 48 hominid individuals are now represented. Many of these specimens have received preliminary description and publication (Day 1971; Leakey, M. D., *et al.* 1971; Tobias 1965a, 1966, 1972a,b) while, to date, only the type specimen of the very large-toothed, extremely robust australopithecine, *A. boisei,* has been described in detail (Tobias 1967). The minute study of the other hominid cranial and dental specimens from Beds I and II is virtually complete and will be joined by Professor M. H. Day's description of most of the postcranial remains in those beds to form another volume in the *Olduvai Gorge* series. It is intended to relegate the descriptive analysis of those Bed II remains ascribed to *H. erectus,* as well as of the hominid remains from Beds III and IV, to a subsequent volume, in which also Dr. M. D. Leakey will describe the Acheulean cultural remains. Meantime, it can be recorded here that, save for Olduvai 9 and one or two other hominid fossils from the upper part of Bed II, all the identifiable hominid cranial and dental remains from Beds I and II have been assigned to a very robust australopithecine, *A. boisei,* or to the ultra-gracile *H. habilis* first described by Leakey, Tobias and Napier (1964). Thus from Beds I and II remains representing 15 individuals have been assigned to *H. habilis,* 5 individuals to *A. boisei,* while 8 further cranial or dental specimens have not been identified taxonomically. This analysis excludes the postcranial bones of some 8 individuals from Beds I and II (Archibald *et al.* 1972; Day 1967, 1969; Day and Wood 1968; Napier and Day 1964).

Studies on Olduvai 24, reconstructed by R. J. Clarke (Leakey, M. D., *et al.* 1971), have shed much further light on the structure of the cranium of *H. habilis.* In a number of respects it is nearer to later forms of *Homo* than to *A. africanus* and these include mean dental size, the position of the foramen magnum on the cranial base, and the cranial capacity. A sample of four crania attributed to *H. habilis* has a mean cranial capacity (adult values) of 640 cc. (Tobias 1971a, 1973b), with 95% population limits of 519.5–761.5 cc. (Tobias, in press c). This compares with the means for the South African *A. africanus* of 442 cc. (Holloway 1972b; Tobias 1971a), with 95% population limits of 391–491.6 cc. (Tobias, in press c).

There has been some modification in the subdivision of the Olduvai formations (Hay 1971). Bed V, as used by H. Reck and L. S. B. Leakey, has been abandoned, its place being taken by the Mesak, Ndutu and Naisiusiu stratigraphic units. In addition, Bed I has been extended downwards to include the basal lava flows and the underlying tuffs and clays; thus, there is now a Lower Member, a Basalt Member and an Upper Member of Bed I. The pioneering application by Evernden and Curtis (1965) of the potassium-argon method to the dating of Bed I has now been partly confirmed, partly modified and extended (Curtis and Hay 1972; Howell 1972a). Tuffs of the Lower Member of Bed I yielded K/Ar dates between 2.03 and 1.79 m.y.; lavas of the Basalt Member have an average age of 1.96 m.y.; while there is a most important, firm, consistent and reliable date of 1.79 m.y. for Tuff I^B within the Upper Member of Bed I. Moreover, the lapse of time from the bottom to the top of Bed I (Upper Member) is much smaller than was formerly thought and probably does not exceed 100,000 years. Howell thinks it may be of the order of 25,000 to 50,000 years (1972a:334)! These dates have received independent support from the fission-track method of Fleischer and Hart (1972) and from the studies of geomagnetic polarity (Cox 1969; Cox *et al.* 1968, Grommé and Hay 1967). Thus the lapse of time between the Bed I fossils assigned to *H. habilis* and those of Bed II attributed to the same taxon is nowhere near the figure of about 1 m.y. which had previously been adduced as evidence, militating against the possibility that these two groups of hominid fossils belonged to the same taxon.

Considerable progress has been made by M. D. Leakey (1971a) in the archaeological analysis of the Bed I-II sequence. The lithic industries into which the artifacts are classified are threefold: the Oldowan (once called "pre-Chellean"), Developed Oldowan, and Acheulean. Throughout Bed I and Lower Bed II, the characteristic tool of the Oldowan is the stone chopper. "Proto-bifaces" come into the picture from Upper Bed I to Middle Bed II and "appear to represent attempts to achieve a rudimentary handaxe by whatever means was possible" (p. 266). No true bifaces occur before the upper part of Middle Bed II, where they form an integral part of both the Developed Oldowan B and the early Acheulean industries. Yet, while tracing a strong thread of continuity from the Oldowan into the Developed Oldowan, Mary Leakey finds the Acheulean seemingly intruding as a dissonant element from Middle Bed II upwards. The Developed Oldowan and the Acheulean seem to her to "represent two distinct cultural traditions, perhaps made by two different groups of hominids" (p. 272).

Of the makers of the implements themselves, Mary Leakey is prepared in the latest *Olduvai Gorge* volume to go beyond the cautious line she followed at the Burg Wartenstein symposium in 1965 (Leakey, M. D., 1967). She has now adduced strong evidence that *H. habilis* was responsible for the Oldowan culture. From six localities, five in Bed I and one in the lower part of Middle Bed II, she has found remains of *H. habilis* directly associated with Oldowan tools. With the Acheulean remains at Olduvai, *H. erectus* is probably associated (Leakey, M. D., 1971b) although, as the late Dr. L. S. B. Leakey has pointed out in a posthumous paper (1973), evidence from elsewhere that *H. erectus* was the maker of the Developed Oldowan remains problematical. Only Olduvai hominid 3 is directly associated with a Developed Oldowan assemblage, and this is represented by only a very large molar tooth (almost certainly deciduous) and a canine. The molar is generally regarded as belonging to a robust australopithecine (cf. *A. boisei*). This, however, brings one no nearer to unravelling the authorship of the Developed Oldowan industries, as *A. boisei* seems to have been present *throughout* the times of Beds I and II, along with one or more other hominids at each level in time.

If the claim is correct that *H. habilis* is the maker of the Oldowan culture, this adds another trait to the list of features serving to distinguish *H. habilis* from *A. africanus,* as Leakey, Tobias and Napier suggested some years ago (1964).

Sterkfontein

In December 1966, excavations were renewed by Tobias and Hughes (1969) at Sterkfontein, Transvaal, and have continued uninterruptedly ever since. A number of new hominid fossils have come to light (Tobias 1973c), including a cranium (StW/Hom 12/13/17) and a mandible (StW/Hom 14), both with teeth, found *in situ* at approximately the same level as the famous skull of "Mrs. Ples" (Sts 5). There are also a maxilla with teeth, some 15 isolated teeth, and 4 articulated lumbar vertebrae. Most of these remains can be assigned provisionally to *A. africanus.* A fragment of jaw with teeth from the West Pit (formerly called "Extension Site") may belong to *Homo* sp., as is true of some of the dental remains previously recovered from that uppermost part of the cave deposit (Robinson 1962; Tobias 1965b).

Detailed studies have been in progress for some time on the Sterkfontein hominids. For comparison with *A. boisei,* Tobias (1967) recorded many new data on the Sterkfontein *A. africanus.* Wallace, J. A., (1972, 1973a, b, c) and Sperber (1973) have made new detailed studies on the dentition—the former emphasizing wear patterns and other functional aspects of the masticatory apparatus; the latter concentrating on morphology (including odontometry, cf. Wolpoff 1971a,b) of the premolars and molars.

The cranial capacity of the Sterkfontein hominids has been subject to restudy and reanalysis (Bilsborough 1973, 1974; Holloway 1970, 1972b; Lestrel and Read 1973; Robinson 1966), and thus it is now clear that earlier estimates of 2 out of the 4 available Sterkfontein capacities were too high. Thus, the value formerly cited for Sts 71, 480–520 cc., (Broom *et al.* 1950) has been recomputed as 428 cc. (Holloway 1970) and that for Sts 19/58 has "dropped" from 530 cc. (Broom and Robinson 1948) to 436 cc. The values for Sts 60 (428 cc.) and Sts 5 (480 cc.) have remained largely unchanged following these new studies of Holloway. Thus, estimates of the mean for the 4 Sterkfontein capacities have been lowered from 486 cc. to 443 cc. A similar reassessment of the Makapansgat MLD 37/38 capacity has "lowered" it from 480 cc. (Dart 1962) to 435 cc. (Holloway 1970). For the total sample of 5 nearly contemporary *A. africanus* adult specimens from Sterkfontein and Makapansgat, the mean capacity estimate has been decreased from 485 cc. to 441 cc. If Holloway's (1970) revised adult estimate for Taung (440 instead of 540 cc.) is included in this sample of gracile australopithecine capacities, the estimated mean for 6 crania generally assigned to *A. africanus* "drops" from 494 cc. (Tobias 1971a) to 442 cc. (Holloway 1970) or 441 cc. with a standard deviation of 19.6 cc. (Tobias, in press c). With 5 degrees of freedom, the 95% confidence limits of the population are 390.8–491.6 cc.

This lowered estimate for *A. africanus* throws into strong relief the estimated mean value for *H. habilis* of 640.5 cc. ($n = 4$). The difference is highly significant as Pilbeam (1969), Campbell (1972), Holloway (1973) and Tobias (1971a, in press c) have pointed out.

Aguirre (1970, 1972) has reexamined the mandibles of some South African early hominids. He suspects that more than one hominid is represented in the Sterkfontein sample (excluding the late specimens from the West Pit), namely: *A. robustus* as well as *A. africanus,* but he is not as convinced of this for Sterkfontein as he is for Makapansgat (see below).

Rosen and McKern (1971) have been reexamining the values in some fossil hominid crania for Le Gros Clark's (1952) three cranial indices. They suggest that the supraorbital height index and a new supraorbital upper facial height index, which they have devised so effectively, distinguish *A. africanus* (represented in their study by Sterkfontein 5) from the robust australopithecines (represented by Robinson's 1963 reconstruction of Swartkrans 48 and by Olduvai hominid 5) as to justify the generic separation of the two groups of australopithecines, a view Robinson has long held. Since there are no precise lists of properties whose possession would qualify the taxon or taxa in question for membership of the category genus (cf. Rowell 1970) and since, according to Mayr (1963), the generic name emphasizes not a greater degree of difference but rather the belonging together of the species included in the genus, it would perhaps be difficult, on such a concept of the genus, to support the inference of generic distinctness based upon the indices studied by Rosen and McKern (1971). (For a further discussion on the relationship between the gracile and robust early hominids, see below.)

The postcranial bones from Sterkfontein have had a good deal of attention lately. In a recent restudy of the capitate bone from Sterkfontein, Lewis (1973) has shown that this wrist bone, far from being essentially human in appearance, "conserves, with but little progressive modification, important biomechanical characteristics still found in *Pan.*" Oxnard (1968) believes that the suggestion currently in the literature—that the Sterkfontein scapular fragment is relatively less specialized than the corresponding region of the gibbon and the chimpanzee (Oxnard 1967)—may not be justified. The femur has been the object of a number of studies by Heiple and Lovejoy (1970, 1971, 1972), by Preuschoft (1971) and by Day (1973). Heiple and Lovejoy have been led to reiterate the distinctly hominid position of *A. africanus* and, too, where comparable parts are available of *A. robustus* and *A. africanus,* how closely they conform to the same total morphological pattern. Likewise, the recent studies of the pelvis by Zihlman (1967, 1970; Zihlman and Hunter 1972) and by Lovejoy *et al.* (1973) have served to emphasize the close resemblance between the South African species, *A. robustus* and *A. africanus.* On the other hand, some studies have stressed the distinctive features of the australopithecine femur as compared with that of *Homo* (Day 1973; Robinson 1972; Walker, A. C., 1973). Robinson's new work on *Early Hominid Posture and Locomotion* (1972), including detailed descriptions of the locomotor apparatus of the South African early hominids, leads him to infer—in contradistinction to Lovejoy, Heiple and Zihlman—that the robust australopithecine ("Paranthropus") was characterized by a partly bipedal and partly quadrupedal-climber adaptation, whereas the gracile australopithecine ("Homo africanus") was distinguished by a "fully human type of locomotor adaptation." The essence of this view appears to be well summarized by Robinson, Freedman and Sigmon as follows (for *Paranthropus* read *A. robustus* and for *Australopithecus* read *A. africanus,* to bring the terminology into line with that used throughout the rest of the present chapter):

> The available early hominid postcranial material suggests that both *Paranthropus* and *Australopithecus* had a re-orientated occiput, spinal column curved in the human manner, widened sacrum, and short and wide ilia expanded backward in the human manner and orientated essentially as in Man. Both appear to have had the human type of lateral balance control mechanism, orientation of the ischium and arrangement of gluteus maximum. *Paran-*

> *thropus*, however, appears to have had an ischium of pongid-like relative length, somewhat short lower limbs and a relatively mobile foot, indicating that, although the basic adaptation for bipedality was already present, the propulsive mechanism was still largely power-oriented. *Australopithecus*, on the other hand, appears to have had a relatively extremely short ischium, a fully elongated lower limb and a remarkably man-like foot; it thus had a speed-oriented propulsive mechanism. (1972:369)

Lovejoy, Heiple and Burstein (1973), on the other hand, have been led to conclude that "There is no morphological or biomechanical feature by which the two presently recognized allomorphs of *Australopithecus* differ significantly with respect to their locomotor adaptations. Rather, both clearly conform to the same total morphological pattern and more significantly, total biomechanical pattern." (Lovejoy *et al.*:778). Functionally, they conclude that the australopithecine pattern is "as fully commensurate with erect striding as is that of modern man," a view with which Wolpoff (1973) concurs [". . . the available evidence . . . indicating that australopithecines were striding hominids, fully adapted to the upright posture and striding gait characteristic of modern man" (Wolpoff 1973:382)]. Perhaps it is relevant here to mention that Day (1973) has recently drawn attention to the imperfect preservation of several of the South African postcranial parts, including the Sts 14 pelvis and femur from Sterkfontein and Sk 50 pelvic fragment from Swartkrans. Wolpoff has noted a probable variation in the technique of measuring the bicondylar angle by Robinson (1972) and by Lovejoy and Heiple (1970): though Wolpoff finds reason to conclude that "there is no reason not to accept the results of the Lovejoy and Heiple reconstruction" (Wolpoff 1973:379). It is doubtful whether this difference alone is sufficient to account for the two diametrically opposed views based upon morphological and biomechanical studies of the same specimens (or casts of them).

Leutenegger (1972) has computed the size of the head of the newborn *A. africanus* in relation to the size of the pelvic inlet of Sts 14 and has demonstrated that, even on a maximum estimate of head diameter at birth, the head would readily have passed through the pelvic inlet of Sts 14. From this he is led to suggest that "in the early stages of hominid evolution selective pressures for enlarging the pelvic canal to ensure parturition may have been minor or even absent; selective forces toward highly efficient construction of the pelvis for bipedalism could have been stronger . . ." (p. 569).

It is interesting to note that Lovejoy *et al.* (1973) have lighted upon exactly the same feature, as has Leutenegger, to explain such differences of biomechanical significance as they find between the pelvi-femoral morphology of *Australopithecus* and of modern man. These differences, they suggest, are related to the combination of a fully bipedal striding gait with different degrees of encephalization, rather than to differences in the gait pattern itself. Their intricate argument shows that the need to enlarge the pelvic ring of the female could be achieved, without loss of efficiency of stride, by just such morphological changes as they find between *Australopithecus* and *Homo sapiens,* e.g., enlargement of the femoral head and increase in the iliac pillar and general robusticity of the ilium.

Another series of studies has lately attempted to assess the stature of the australopithecines (Burns 1971; Helmuth 1968; Lovejoy and Heiple 1970, McHenry 1974a, Wolpoff 1973). Individuals attributed to both robust and gracile *Australopithecus* seem, in general, to have been short to medium in stature, shorter on the average than most populations of modern man. Wolpoff's (1973) estimate for the South African gracile australopithecines gives a mean of 138 cm., McHenry's (1974a) a mean of 145 cm. ($n = 4$ in both studies). The robust forms seem to be somewhat taller with a mean of 152.7 cm. for 3 South African forms and of 163 cm. for 7 East African early hominids (McHenry 1974a).

Researches and discussions, some published and some not, have shed much light on the cave stratigraphy and extent (Butzer 1971b), pers. comm.; Edmund Gill, pers. comm.; Hughes and Tobias, unpublished; Partridge and A. B. A. Brink, pers. comm.; Tobias, in press b,d). These newer observations point to (a) deposition of the breccia over a considerable period of time, running to hundreds of thousands of years; (b) the existence of an earliest bone-bearing breccia well below the main *Australopithecus*-bearing cave deposit; (c) a much greater east-west extent of the cave deposit than was formerly suspected.

Attempts at dating by radioisotopes, paleomagnetism reversals and the racemization of isoleucine (Bada 1972) have so far not been successful, nor have fission-tracking of calcite crystals in the marrow cavities of fossil bones (Macdougall and Price 1974) and amino-acid analysis of bone (Carmichael *et al.* 1974). Two other lines of investigation have provided clear-cut pointers. Partridge's (1973) geomorphological approach has provided him with an estimate of the time period at which the Sterkfontein cave *first* opened to the surface: the date is 3.3 m.y. Although this method has been criticized (De Swardt 1974), Partridge (1974) has rebutted those criticisms. Meantime, new faunal comparisons have permitted H. B. S. Cooke (1970) to suggest provisionally that the Sterkfontein breccias may be about twice as old as has hitherto been thought, possibly "in the vicinity of 2.5 to 3.0 million years old."

This tentative view is supported by V. Maglio (1973, pers. comm.) who has kindly permitted me to state that his inference is based on a comparison of fossil suids from Sterkfontein and Makapansgat, and of *Elephas recki* from the latter site (according to Wright and Skaryd 1972), with their well-dated counterparts from East Rudolf and the Omo (Shungura Formation). Although the faunal evidence is far from adequate and further studies are needed, we may at this stage note Maglio's tentative statement that both Makapansgat and Sterkfontein appear to correlate best with the East African faunal succession of about the middle Shungura and lowest Koobi Fora Formations. In terms of absolute chronology, Maglio believes that 2.5 m.y. is a fair estimate for the age of these faunas. W. Maier (1973) quotes a recent personal communication from Cooke to the effect that he now considers 2.5 m.y. an appropriate date for the australopithecine breccia from Sterkfontein and 3.0 m.y. for that of Makapansgat.

Thus "faunal dating" points to 2.5 m.y. or 2.5–3.0 m.y. for Sterkfontein, while Partridge's earliest date for the opening of the Sterkfontein cave is 3.3 m.y. (Partridge 1973). It is at present not clear if this is a real discrepancy, a geomorphological methodological difficulty, or a faunal sampling problem. If it is a real discrepancy, it would suggest that the early stages in the accumulation of deposit in the Sterkfontein cave, when the recently exposed lowest breccia was forming, may have been far longer than we had imagined, since the identified fauna, which has been correlated with East African lineages, comes from much higher in the Sterkfontein deposit (Tobias, in press b,d).

The Sterkfontein artifacts come from the uppermost parts of the deposit, just beneath the roof, in the area of the West Pit. Those recovered earlier and studied by Mason (1962) have been restudied by Mary D. Leakey (1970, 1971a) who is inclined to relate them to Developed Oldowan B assemblages of Olduvai Bed II. Her classification of the Sterkfontein material differs from that of Mason, mainly in that most of the specimens which he regards as cores she would class as either choppers or polyhedrons. These earlier studies were made on limited samples. The excavations of 1967–1973 have yielded many new artifacts from the area of the West Pit: some of these have been examined by G. Ll. Isaac and all are to be studied by R. J. Mason. The presence, in only the uppermost part of the breccia, of stone implements with fragmentary hominid remains, including teeth which are small by the standards of the main Sterkfontein assemblage of *A. africanus,* suggests that further excavation in the latest strata may confirm the presence of a toolmaking *Homo* sp., such as *H. habilis,* as a late arrival in the apparently lengthy sequence represented in the Sterkfontein breccias, as suggested by Tobias (1965a).

It is a striking fact that foreign stone manuports and artifacts occur in only the uppermost part of the Sterkfontein deposit; but not the somewhat lower breccia which has yielded the australopithecine fossils. The two breccias with and without stone implements correspond to those called by Robinson (1962) the Middle and Lower Breccias, respectively, although these terms no longer seem to be appropriate in the light of the newer knowledge of the stratigraphic sequence now available. The older of these two breccias contains the wealth of australopithecine remains attributed to the species, *A. africanus,* but no implements; while the more recent breccia contains fragments of a hominid identified by Robinson as *A. africanus* (Robinson and Mason 1957; Robinson 1962), but possibly belonging instead to a more advanced hominid with smaller teeth (cf. *Homo habilis*) (Tobias 1965a).

A most important new development in the investigation of Sterkfontein is the discovery by Vrba (1974) that the bovids of Sterkfontein fall into three distinct groups, A, B, and C. Only the phase A bovids are associated with the Sterkfontein Faunal Span including the australopithecines. The phase B bovids are associated with the probably more advanced hominid and the stone tools, and they resemble most closely the bovids of the more recent Swartkrans Faunal Span. Thus, the evidence of the hominids, the bovids and the cultural remains points to an appreciable time lapse separating two levels within the Sterkfontein breccias. Both of these levels lie within the uppermost 12 meters of the deposit, which has a total depth of at least 30 meters and possibly even more. It has thus become clear that the Sterkfontein breccias have entombed an appreciable slice of time, to be measured in terms of hundreds of thousands of years. In the light of this development, the difference between Partridge's proposed date of cave opening (3.3 m.y.) and Cooke's and Maglio's faunal dating for the *Australopithecus*-rich breccia in the upper one-third to two-fifths of the cave depth (2.5 m.y.) becomes quite intelligible.

Makapansgat

Detailed studies of the form, measurements and function of the Makapansgat hominid teeth have been carried out by Sperber (1973), Tobias (1967), J. A. Wallace (1972) and Wolpoff 1971a,b, 1973). In addition, the crania and mandibles have been reexamined by a number of workers. The Makapansgat hominids are usually classified in *A. africanus,* along with the Sterkfontein earlier hominids and the type specimen from Taung. Yet in 1967 Tobias drew attention to some "robust" features in some of the crania, jaws and teeth from Makapansgat. He went so far as to state:

> In these respects, the Makapansgat specimens seem to show a somewhat nearer approach to *A. robustus* than do the Sterkfontein specimens. This reduces the distinctness of the lineages and renders it less likely that they represented two clades, the members of which should be regarded as generically distinct from each other (1967:244).

Since then, Aguirre (1970) has studied the early hominid mandibles from South Africa. He has identified a constellation of morphological features characterizing the mandibles of *A. robustus* from Swartkrans. These features, he believes, are clearly shown by MLD 2 of Makapansgat which he therefore regards as a young male *A. robustus.* To Aguirre it appears indubitable that there is more than one species of hominid at Makapansgat—a thought earlier raised as a tentative suggestion by Tobias (1968c, 1969).

Perhaps another way of looking at the somewhat intermediate features shown by the Makapansgat fossil hominids is that they resemble a population closer to the point of speciation between *A. africanus* and *A. robustus.* In such a population, anatomical polymorphisms could have co-existed, foreshadowing the speciation of the *A. robustus-A. boisei* lineage from the basic *A.* cf. *africanus* ancestor (Tobias 1973d).

In another study on mandibles from Makapansgat, Swartkrans, Olduvai and Peninj, Tobias (1971b) showed that the *relative space width* of hominid mandibles is—without a consistent definition—of no value in deciding taxonomic status (according to Robinson 1953b, 1965, 1966). There is no hard and fast dividing line between australopithecines and hominines with respect to the disto-mesial pattern of variation of these relative widths. Robinson had attempted to use this feature to support his claim that Olduvai hominid 7 (the type specimen of *H. habilis)* was a member of *Australopithecus* and not a member of the genus *Homo.* Unfortunately, three different definitions of the position of measurement were offered by Robinson in 1953, 1965 and 1966. The diagnosis of any specimen varied according to where the measurements were taken. Hence, Tobias (1971b) showed that this criterion did not and could not disqualify Olduvai hominid 7 from a place in the genus *Homo.*

The endocranial capacity of MLD 37/38 had been estimated by Dart (1962) as being approximately the same as that of Sterkfontein 5, that is, about 480 cc. Holloway (1970, 1972a) has made a "quite provisional" (pers. comm.) recalculation of the capacity of this specimen and obtained a somewhat lower value (435 cc.). He has used the MLD 1 parieto-occipital part to compute the bone thickness of MLD 37/38 in various regions. This may impart a source of error into his recomputation as MLD 1, with its converging temporal crests leading, in all probability, to an anteriorly placed sagittal crest, that may well have belonged to a young adult *male,* whereas MLD 37/38, with no comparable development of temporal crests, could have belonged to a young adult *female.*

MLD 1 with its probable sagittal crest is an interesting specimen. Although its median sagittal contour is similar to that of Sterkfontein 5 (Robinson 1954), its transverse dimensions are appreciably greater (unpublished original data). Hence it may be suggested that its capacity was probably greater than that of Sterkfontein 5 (with 485 cc.). This may well prove a feature relating another of the Makapansgat hominids to *A. robustus* with its capacity of 530 cc., based on a single specimen from Swartkrans (Holloway 1972a), and supported by 530 cc. for Olduvai hominid 5, *A. boisei* (Tobias 1963). It contrasts sharply with the bigger-brained early hominids of Olduvai *(H. habilis)* in which a larger cranial capacity is accompanied by temporal lines placed much more widely apart on the calvaria.

Sperber (1973) has supplemented his odontoscopic and odontometric study by a radiological investigation. For the first time, practically every single maxilla and mandible of the South African early hominids has been X-rayed. The resulting skiagrams are throwing new light on the size range of pulp cavities, the extent of the secondary dentine response to marked attrition, and details of root structure and number.

The search for dateable materials at Makapansgat has been no less rigorous than at Sterkfontein. Thus far, positive results have been yielded by Partridge's (1973) geomorphological assessment—3.7 m.y. as the earliest date at which the cave opened to the surface—and Cooke's (1970) and Maglio's (1973) faunal comparisons with East African suids and *Elephas recki.* As mentioned under Sterkfontein, the faunal picture matches the East Rudolf and Omo faunas of 2.5–3.0 m.y. or, on Cooke's latest view, 3.0 m.y. (pers. comm. in Maier 1973). Non-calibrated faunal comparisons among the South African sites had already led Wells (1969, 1971) to infer that, while both Makapansgat and Sterkfontein belong to the Sterkfontein Faunal Span, some of the differences between the faunas of the two sites might plausibly suggest that Makapansgat is older than Sterkfontein.

The cercopithecoids of Makapansgat have been the object of new studies by Maier, Freedman, and Eck. Maier (1970) has described some 20 further specimens, over and above the 70 fossils, belonging to 5 species, previously recorded from the Makapansgat Limeworks. A beautifully complete male skull of *Cercopithecoides williamsi* has enabled him to amplify the available descriptions of this species, to confirm its membership of the Colobidae (so far the only colobid monkey described from the South African early hominid deposits) and to confirm the invalidity of *C. molletti.* The other species represented are *Parapapio jonesi, P. broomi, P. whitei* and *Simopithecus darti.* Maier has recently described a complete skull of *S. darti* from the Makapansgat Limeworks and found it to show "an interesting mixture of primitive and advanced features." This "would seem to indicate a great age for this site." (1972:403). Freedman (1957, 1970, Freedman and Stenhouse 1972) has shown that the three *Parapapio* species occur at Makapansgat in the same proportions as at Sterkfontein: *P. broomi* $> 50\%$; *P. whitei* least common.

In the light of Wells's inference, it is noteworthy that Partridge's geomorphological analysis also suggests that the Makapansgat cave was opened to the surface earlier than the Sterkfontein cave, the two dates being 3.7 and 3.3 m.y., respectively. Again, there is an apparent discrepancy between the proposed date of cave opening (3.7 m.y.) and the "faunal date" (3.0 m.y.). As at Sterkfontein, errors, systematic or otherwise, may have entered into one or both dates. On the other hand, though the evidence from Makapansgat is at this stage less clear-cut, there are indications of a lengthy cycle of deposition after cave opening, but antedating the suid and elephantid remains which have been correlated with the calibrated lineages in East Africa. There is an appreciable depth of breccias with intercalated travertines below the level from which were derived the pig and elephant fossils (Tobias 1973a). Furthermore, the occurrence of differences in the faunal content of various strata within the Makapansgat deposit has been suggested by Dr. J. Kitching (quoted by Maier 1970). In other words, at Makapansgat, too, Partridge's proposed date of cave opening is probably compatible with Cooke's and Maglio's faunal date.

Butzer's (1971b) analysis of the Makapansgar cave fill indicated that it differed from those in the Krugersdorp area, inasmuch as it contains appreciable deposits constituted for the most part of insoluble cave residues and precipitates, as well as typical *limons rouges.* The other sites (Sterkfontein, Swartkrans and Kromdraai) are apparently comprised largely of a sediment matrix typical of *limons rouges,* similar to those developed in limestones of the Mediterranean region.

On the archaeological side, there is little newly published on the osteodontokeratic objects of Dart. Foreign stone objects, including artifacts, have been reported to be present in all levels of the deposit (Maguire 1965, Partridge 1965), but it has so far not proved possible to assign them to a specific cultural horizon which could be equated with either the Sterkfontein or the Swartkrans industries, or with any part of the cultural sequence established in East Africa (Leakey, M. D., 1970, footnote to p. 1224).

Swartkrans

C. K. Brain's (1967, 1970, 1973) excavation at Swartkrans has continued to yield new hominid specimens. Apart from the excellent endocast, the vertebrae and other specimens reported earlier, the most recent finds include an isolated adult left 5th metacarpal, which has been studied by Day and Scheuer (1973). Detailed morphological studies have been carried out on the very rich assemblage of hominids from Swartkrans, as well as on the other South African early hominids—by J. A. Wallace (1972) and Sperber (1973) on the teeth, Holloway (1972a) on the new endocast, Wolpoff, Day, Wood and Tobias on various features. Aguirre's (1970) study on the mandibles of the South African early hominids revealed that, out of 20 morphological traits investigated, the Swartkrans *A. robustus* jaws were characterized by a constellation of 14 of these traits. These same features occurred also in the mandible of Peninj (Lake Natron), as well as in some of the earlier discovered jaws from Omo and East Rudolf, and in MLD 2 from Makapansgat.

The superbly preserved endocranial cast, SK 1585, found by Brain (1970) on January 17, 1966, has a cranial capacity of 530 cc. (Holloway 1972a). This was the first secure evidence in support of the long held, though unsubstantiated, view that *A. robustus* from South Africa had a somewhat larger cranial capacity than *A. africanus* ($\bar{x}$ = 441 cc.); previously, it had been shown that the hyper-robust *A. boisei* of East Africa has a capacity of 530 cc. (Tobias 1963, 1967). Since then, Holloway (1973) has published two more estimates of cranial capacity in East African hyper-robust hominids, 510 ± 10 cc. for KNM-ER-406 and 500 cc. for KNM-ER-732, both from East Rudolf. These 4 values for South and East African representatives of the superspecies *A. robustus/A. boisei* (Tobias 1973b) give a sample mean of 519 cc., a standard deviation of 12.8 cc. and 95% confidence limits of the population of 478–560 cc. (Tobias, in press c). The available robust australopithecines thus have a larger mean capacity than *A. africanus* ($t = 2.69$, $p < 0.05 > 0.02$) (Holloway 1973).

On the morphology of the SK 1585 endocast, Holloway (1972a), like Tobias (1967) on the Olduvai hominid 5 endocast, recognizes "more modernness" in the shape and disposition of the cerebellum of the robust than of the gracile australopithecines. He also finds that in overall size and shape, and in the expansion of the parietal cortex, the robust endocasts reflect more modern features of the brain than do the endocasts of the gracile forms. In the parietal, occipital and temporal lobes, in particular, the robust australopithecine endocasts reflect a brain which is "not that of a pongid" but which "has been reorganized to a hominid pattern" (Holloway 1972a: 185, 173).

The composite cranium from Swartkrans assembled by R. J. Clarke in July 1969 has been examined in more detail (Clarke and Howell 1972, Clarke *et al.* 1970). These studies seem to be justified in attributing the specimen to a species of the genus *Homo* (according to Wolpoff 1971c). The new evidence is of such importance that Clarke is now devoting a detailed and comprehensive study to the specimen in comparison with other fossils assigned to early South and East African members of the genus *Homo*.

The postcranial bones of Swartkrans have received much further attention. The height estimates of Burns (1971) have suggested that SK 82 and SK 97 were both short individuals. The height of SK 82 was predicted from the femoral head diameter as 151 cm. (4 ft. 11.5 in.), the range of probable heights being 146 to 156 cm.; and that of SK 97 was estimated as 157 cm. (5 ft. 1.5 in.), with a range of probable heights of 152 to 161 cm. McHenry's most recent estimates, based on as many upper and lower limb bones as possible (1972, 1974a), confirm the generally medium-short stature of *A. robustus.* He infers that it is clear that the robust forms of South African australopithecines were only slightly taller than the gracile forms, though the robust form does appear to have been "a good deal heavier," to judge by the size of the teeth and skulls. Robinson (1970) has inferred, from his study of two new vertebrae from Swartkrans, that the females of the robust species and males of *A. africanus* were probably similar in robustness as well as in stature.

The most recent descriptions of the pelvic bones have led Zihlman (1970) and Lovejoy *et al.* (1973) to conclude that both *A. africanus* and *A. robustus* show essentially the same functional anatomical complex. While Zihlman tends to regard this complex as being *sui generis,* and not quite the same as that of *Homo,* the latter two workers recognize but little difference between the complex in *Australopithecus (sensu lato)* and the pelvifemoral complex associated with the bipedal, striding gait of modern man. Robinson (1972 and with Steudel 1973), on the other hand, holds firmly to the view that *A. robustus* had "a relatively long, pongid-like ischium, which appears to have been part of a power-oriented propulsive mechanism." *A. africanus,* on the contrary, he believes, "had an ischium proportionately shorter even than that of *H. sapiens* and had an elongated femur and thus a fully human speed-oriented propulsive mechanism (Robinson 1972). The anatomy of the forms thus differed in a manner indicating considerable adaptive difference . . ." (Robinson and Steudel 1973, p. 521).

A discriminant analysis of the Swartkrans left first metacarpal, SK 84, has suggested to Rightmire (1972) that this fossil bone may be functionally similar to that of the chimpanzee. On the other hand, the left fifth metacarpal, SK 14147, studied by Day and Scheuer (1973) led them to infer that it was hominid; though, having no securely identified fifth metacarpals of early hominids with which to compare it, they suspended judgement on the question of whether it belonged to *Australopithecus* or *Homo.*

On the archaeological and ethological side, Mary D. Leakey's (1970) study of the Swartkrans implements convinced her of resemblances between them and the Developed Oldowan B assemblages from Bed II, Olduvai. However, the Swartkrans collection is still too limited for quantitative analysis. Brain (1972) has continued his search for and careful study of evidences of early hominid behavioral patterns. In his latest work, he has critically reviewed the evidence for interpersonal violence. Following Roper's (1969) comprehensive review of all the published claims for inter-australopithecine violence, especially those of Dart (1949, 1953, 1957), Brain has reexamined all the original fossil hominids from the five South African australopithecine sites. He has stated that, "In most instances invalid conclusions have been drawn because ante-mortem damage to specimens has not been isolated conclusively from post-fossilization effects" (1972:379). He concludes that the question of the incidence of interpersonal violence in this group must for the time being remain an open one.

The reconstruction of the cave sequence at Swartkrans is continuing to receive Brain's attention, and a major study of it is presently under way (Brain 1973). Meantime, Butzer (1971b) has analogized the Swartkrans deposit to the *limons rouges* developed in limestone of the Mediterranean region. He does not accept the inferences drawn by Brain (1958, 1967) that Swartkrans was characterized by moist conditions and Sterkfontein and Makapansgat by dry conditions. The Swartkrans fauna is regarded as defining a Swartkrans Faunal Span, younger than the Sterkfontein Faunal Span, but older than the fauna of Kromdraai and the Cornelia Faunal Span (from Cornelia, Orange Free State) (Wells 1969). Hendey (1973a) has identified 17 species of Carnivora, "not all of them contemporaries of the early hominids." In a reassessment of the known carnivores from Swartkrans and a study of new material, Hendey (1974) has inferred that the carnivore assemblage is clearly heterogeneous in age. The main Swartkrans fauna, he states, may be divisible into at least two chronometric age categories. Vrba is making a detailed study of the bovid remains from Swartkrans and other South African sites; already she has added a number of new species to the available lists of mammals from the early hominid cave sites (1971, 1973). Freedman's (1970) checklist of cercopithecoids shows that of the genus *Parapapio,* only *P. jonesi* is represented at Swartkrans, along with *Papio robinsoni, Simopithecus danieli, Dinopithecus ingens* and the colobid *Cercopithecoides williamsi.*

Cooke (1970) has suggested that, on faunal comparisons, Swartkrans is at least as old as Bed I, Olduvai (1.7–1.8 m.y.), or about 2.0 m.y. (pers. comm. in Maier 1973); while Petter (1973) concludes that the Swartkrans carnivores were contemporaries of those of Bed II, Olduvai. Partridge's (1973) geomorphological study has set a date of 2.6 m.y. as the earliest period for the opening of the Swartkrans cave. Again there is a discrepancy of about half-a-million years between the two estimates, but both estimates confirm what faunal comparisons have indicated—that Swartkrans is younger than Makapansgat and Sterkfontein. That the difference between the two tentative dates may find an explanation similar to that which seems to apply at Makapansgat and Sterkfontein is suggested by the great vertical depth of the deposit ("almost as great as its lateral extent," Brain 1973). Furthermore, Brain has pointed out that "the deposit consists of several stratigraphic units, spanning a considerable period of time" and his study "has suggested that the fossil hominids may be restricted to quite a small part of this time sequence" (1973:5).

Kromdraai

The hominid teeth from Kromdraai have been studied recently by Sperber (1973), Tobias (1967) and J. A. Wallace (1972). Oxnard (1972) has interpolated the talus into a matrix of extant species. He concludes that on generalized distances the Kromdraai talus, like that from Olduvai (Day and Wood 1968, Lewis 1972), is "completely dissimilar from both African ape and modern human tali." On the other hand, McHenry's (1974b) study of early hominid humeri has shown that the Kromdraai humerus, like that of Kanapoi, closely resembles the humeri of modern man. McHenry (1974a) has estimated a stature of 154.1 cm. (5 ft.) for the individual represented by the Kromdraai humerus (TM 1517)—short, like the other australopithecines.

The Kromdraai fauna generally cited are those from the Faunal Site, not the Hominid Site. Freedman and Brain (1972) have recently demonstrated that the cercopithecoids from the two Kromdraai sites are different and point to different ages for the two deposits. The Hominid Site contains *Papio robinsoni, P. angusticeps* and the colobid *Ceropithecoides williamsi.* The Faunal Site includes the two species of *Papio* (though in different proportions), *Gorgopithecus major,* but not *C. williamsi.* From this and other evidence it is clear that the former practice of using the Faunal Site species as a guide to the faunal dating of the Hominid Site at Kromdraai will have to be abandoned. However, both Kromdraai sites are clearly younger than Sterkfontein and Makapansgat, and the Kromdraai fauna is still classified as belonging to the Swartkrans Faunal Span. Hendey (1973b) has described the scanty remains of Carnivora from Kromdraai B (the Hominid Site), those from the Kromdraai Faunal Site having been described earlier by R. F. Ewer (Bishop and Clark 1967). Hendey finds that the carnivores give little indication of whether or not the Kromdraai faunas were contemporaneous.

Taung

Recent geomorphological studies carried out independently and with different approaches by Butzer and by Partridge have suggested a much younger age for Taung than had been thought previously. Already in 1967 and again in 1969, Wells drew attention to the uncertainty of the dating of Taung in relation to the other australopithecine sites. Long regarded as the oldest, or among the oldest of the South African sites, Taung has a fauna which—at least in respect to the species most closely connected with the type specimen of *A. africanus*—do not warrant the view that the Taung child is the earliest South African australopithecine. The hominid-associated fauna, Wells believed, could just as readily be equated with Swartkrans or even with Kromdraai. He added, "Some of the animals recorded from Taung may however belong to parts of the deposit appreciably older than the *Australopithecus* breccia" (Wells 1969).

The point had been raised at the Wenner-Gren Foundation Symposium in 1965, when Wells raised the question, "What, if any, is the justification for considering the Taung fauna as belonging to the Sterkfontein rather than the Swartkrans stage?" R. F. Ewer replied, "This is slight and is based on the smaller forms which *seem* to indicate closer resemblances to Makapansgat and Sterkfontein than to Swartkrans. However, the designation was very tentative and statistically the numbers present are not significant" (Bishop and Clark 1967:105). Wells (1967) added that the short-faced baboon, *Papio wellsi,* from Taung seemed to be close to one from Swartkrans. Freedman's (1970) checklist shows the significant absence from Taung of *Parapapio broomi*—which is the commonest cercopithecoid at Sterkfontein and Makapansgat! There are two species of *Papio* at Taung, another two at Kromdraai and one at Swartkrans, but none at Makapansgat or Sterkfontein. It would certainly seem that the cercopithecoid fauna support the notion that Taung is younger than the Sterkfontein Faunal Span.

Partridge's (1973) geomorphological estimate indicates that the Taung fissure did not open until < 0.9 m.y. Butzer's (1974) separate estimate also points to a very young age for Taung, approximately the same as that arrived at by Partridge! If there is no systematic error in these estimates, we are forced to the astonishing conclusion that, far from being the oldest hominid site in Southern Africa, Taung is much younger and, according to Partridge, is probably the youngest! It would bid fair to being the site of the most recent survival of an australopithecine anywhere in Africa. (Peninj, Chesowanja and the youngest Omo hominids are a little *over* a million years, while the dating of Kromdraai Hominid Site is still not clear.)

The implications of this relatively recent date for the place of the Taung hominid in phylogeny will be considered below.

TAXONOMY: SOME GENERAL CONSIDERATIONS

The accumulation of new specimens and the closer study of them and their variability, as well as comparisons with living hominoids, have not led automatically to the resolution of all of the taxonomic problems associated with the fossil hominids. The position has not been made easier by the manifestly wide differences of opinion and approach that exist among taxonomists. S. M. Walters, introducing the general discussion following the Symposium on "Phenetic and Phylogenetic Classification" organized by the Systematics Association (1964), confessed himself "somewhat confused by the conflicting usages and definitions of 'natural,' 'phenetic' and 'phylogenetic,'" in relation to taxonomy. In the same symposium, R. E. Blackwelder (1964) questioned certain basic "acceptances" (as the term was used and explained by the philosopher of biology, J. H. Woodger): these include the acceptances (1) "that taxonomy in the past was exclusively 'morphological' and that the New Systematics differs in being 'biological'"; (2) "that there can be a direct basis of classification in phylogeny, or that the aim of classification is to reflect phylogeny"; and (3) "that only a phylogenetic classification is a natural one." It is worth remembering that taxonomists differ among one another, so that such assumptions, frequently implied and glibly quoted by classifiers of hominids, are not without a considerable body of critics.

Even Mayr's (1969:92) pragmatic definition of a genus ["a taxonomic category containing a single species, or a monophyletic group of species, which is separated from other taxa of the same rank (other genera) by a decided gap"] has not escaped criticism. Rowell (1970) has drawn attention to (a) the practical problems of establishing whether the group of taxa of specific rank are monophyletic (the application of this difficulty to the Hominidae is obvious); and (b) the fact that Mayr's definition applies to *any* supra-specific category, not just to a genus. As a definition, it is open ended and the same definition, *mutatis mutandis,* applies equally well to the phylum—and to intermediate higher categories such as the family (Rowell 1970:265). Another problem is the frequent failure to distinguish between clades and grades, as defined by Huxley (1959). It is indeed only five years since T. W. Amsden, introducing a symposium on "The Genus: a Basic Concept in Paleontology" at the North American Paleontological Convention, said ". . . the question 'What is a genus?' is as elusive today as in the time of Linnaeus" (Amsden 1970)—and Linnaeus, as Ernst Mayr (1963) has reminded us, pronounced this guiding principle: "It is the genus that gives the characters, and not the characters that make the genus."

Two clear statements on the difficulties of recognizing genera are those of Simpson (1963) and Mayr (1963). As many facile assumptions are made by palaeoanthropologists—such as the "fact" or Woodgerian "acceptance" that ecological or adaptive differences necessarily connote generic distinctness—it is worth citing both passages in full here:

> The category genus is necessarily more arbitrary and less precise in definition than the species. A genus is a group of species believed to be more closely related among themselves than to any species placed in other genera. Pertinent morphological evidence is provided when a species differs less from

> another in the same genus than from any in another genus. When in fact only one species of a genus is known, that criterion is not available, and judgment may be based on differences comparable to those between accepted genera in the same general zoological group. There is no absolute criterion for the degree of difference to be called generic, and it is particularly here that experience and common sense are required . . . It must be kept in mind that a genus is a *different* category from a species and that it is in principle a *group* of species (Simpson 1963:8).
>
> There is no non-arbitrary yardstick available for the genus as reproductive isolation is for the species. The genus is normally a collective category, consisting of a group of species believed to be more closely related to each other than they are to other species. Yet, every large genus includes several groups of species that are more closely related to each other than to species of other species groups within the same genus. For instance, in the genus *Drosophila* the species belonging to the *virilis* group are more closely related to each other than to those belonging to the *repleta* group, yet both are included in *Drosophila*. They are not separated in different genera because the species groups have not yet reached the degree of evolutionary divergence usually associated with generic rank. As Simpson (1961) has pointed out, the genus usually has also a definite biological significance, indicating or signifying occupation of a somewhat different adaptive niche. Again, this is not an ironclad criterion because even every species occupies a somewhat different niche, and sometimes different genera may occupy the same adaptive zone. . . .
>
> It is particularly important to emphasize again and again that the function of the generic and the specific names in the scientific binomen are different. The specific name stresses the singularity of the species and its unique distinctness. The generic name emphasizes not a greater degree of difference but rather the belonging-together of the species included in the genus. (Mayr 1963:340–341)

These views have been cited here not with any misguided or presumptuous hope that this analysis might resolve the vexed question of the varying *approaches* to hominid taxonomy in the literature. Rather, the intention is to show frank recognition that the problems of classifying hominids are universal ones. They are not peculiar to that strange breed of persons, the paleoanthropologists—as some of one's deprecating or self-abnegating colleagues are all too apt to think! These are real difficulties inherent in all taxonomy: and the solution has to be sought with patience, understanding and, hopefully, wisdom.

HOMINID TAXONOMY: SOME PARTICULAR PROBLEMS

The question of the taxonomic and phylogenetic relationship between the robust and gracile australopithecines still remains a point of high contention (Campbell 1972; Pilbeam 1972a; Robinson 1967, 1972; Tobias

1968b, 1973b). Most serious students of hominid paleontology find that all of the hominid specimens known from the Upper Pliocene onwards may be satisfactorily classified in two genera, most commonly regarded as *Australopithecus* and *Homo.* Tobias (1973b), in a review of the problem, has suggested that the morphological, temporal and phylogenetic relationships between the two or three groups of australopithecines justify the following systematic grouping:

One genus *Australopithecus* comprising: one superspecies consisting of *A. robustus* and *A. boisei;*

one polytypic species, *A. africanus.*

Among other workers who concur in regarding the gracile and robust australopithecines as separate species of a single genus, *Australopithecus,* or who recognize only the two hominid genera, *Australopithecus* and *Homo* from the late Pliocene onwards, are Campbell (1972), von Koenigswald (1973), Mayr with some qualifications (1963), Pilbeam (1972a,b), Simons (1972), Simons and Ettel (1970) and Simpson (1963). Campbell's (1972) view is a slight modification of the others in that he prefers to include the South African robust australopithecines from Swartkrans and Kromdraai in *A. africanus,* while recognizing the specific distinctness of the East African robust australopithecines as *A. boisei.*

Robinson (1972) also recognizes only two genera, *Homo* (in which he includes the gracile australopithecines) and *Paranthropus* (comprising the robust australopithecines of all geographical zones). His main reason for separating generically the robust and gracile australopithecines seems to be that "The two types appear to have occupied quite different adaptive zones, and for this reason to lump them in a single genus does not seem to be a taxonomically sound practice" (Robinson 1972:3). Against this view, there is the claim by Tobias (1967), underlined by Campbell (1972:45), that "There is considerable evidence that the differences between *A. africanus* and *A. robustus* emphasized by Robinson (1963) and Napier (1964) have been exaggerated." Mayr (1963), referring to the dental and cranial traits distinguishing the gracile from the robust australopithecines, states, "These differences are no greater than among species in other groups of mammals. . . . The two australopithecines *(africanus* and *robustus)* seem to represent the same 'grade' as far as brain evolution is concerned, but the differences in their dental equipment and facial muscles indicate that they may have occupied different food niches . . ." (1963:341–342). It is a moot point whether differences of this order are sufficient to justify generic separation between the gracile and robust australopithecines.

As if to confirm Mayr's statement about brain evolution, Leutenegger (1973) has recently analyzed brain–body weight relationships in australopithecines and concluded that the gracile and robust australopithecines were equally advanced in the degree of encephalization. In more detail,

Leutenegger finds that the *constant of cephalization* (CC), as proposed by Hemmer (1971), with an exponent of allometry $a = 0.23$, yields an almost identical degree of encephalization for gracile and robust australopithecines. On the other hand, the *index of progression* (IP), as proposed by Bauchot and Stephan (1966, 1969) and Stephan (1972), with an exponent of allometry $a = 0.63$, shows the gracile early hominids "clearly more progressive than robust australopithecines." Since the IP values grossly "overestimate the degree of encephalization in forms with small body size" and "since we are dealing with closely related forms, which can be much better fitted by regression equations based on a low exponent of allometry than by equations based on a high exponent," the CC values should yield more reasonable results (Stephan 1972:13).

Kinzey (1972), in his analysis of brain–body relationships, supports the view that gracile australopithecines show a higher degree of encephalization than robust ones. Stephan (1972) is said by Leutenegger (1973) to support the latter view, but a careful reading of Stephan's chapter suggests that this is not a correct interpretation. The matter is reflected only in Stephan's Figure 6-4, in which the encephalization index of Taung (and of no other gracile australopithecines) is shown as slightly higher than that of Swartkrans "I" and the latter, in turn, as appreciably higher than the index of "Zinjanthropus." However, this part of Figure 6-4 is based upon Thenius (1969), whose figures, in turn, were apparently culled from Jerison (1963) and Tobias (1967). This was before Holloway (1970) had shown that estimates of the Taung capacity made by earlier workers were too high and that its adult value on his estimate was only 440 cc. Jerison (1973), in his magnum opus on brain evolution, appears to demonstrate that in brain–body comparisons, the robust australopithecines lie somewhat behind *A. africanus,* whereas in his "extra neurons" analysis, the robust forms are in advance of the gracile ones! (1973: Table 16.2, Figures 16.8 and 16.10, pp. 390–398). These conflicting views find a possible reconciliation in Holloway's (1972a) analysis which Leutenegger (1973:14) interprets as suggesting that gracile and robust australopithecines were on a similar evolutionary level of brain development, the former being more advanced in some respects and the latter more advanced in other respects.

Mayr has reminded us that "The generic name emphasizes *not a greater degree of difference but rather the belonging-together of the species included in the genus.*" (1963:340–341, italics mine). In his proposed definitions of the genus *Australopithecus* and of the species *A. africanus, A. robustus* and *A. boisei,* Tobias (1967:234–235), based on a close study of all the South and East African early hominid fossils available at that time, as well as most of the Indonesian ones, attempted to show just how closely the different forms of australopithecine resembled one another and how compellingly, therefore, they *belonged together* in a single genus.

The specimens discovered since then do not seem to me to have overthrown that analysis of *Australopithecus,* though in two respects they may have created the need for modifications. First, the accumulation of new specimens, and especially postcranial bones, may by now have placed us in a position in which the definitions offered in 1967 can be amplified. Clearly, however, this modification must wait upon some consensus on the interpretation of the pelvi-femoral anatomy and locomotor function of the gracile and robust australopithecines. Secondly, further study of the South and East African robust australopithecine fossils led me to suggest in 1971 that, since the robust and hyper-robust australopithecines are allopatrically distributed, *A. boisei* being confined to East Africa and *A. robustus* to South Africa, their phylogenetic and taxonomic relationship would be adequately and appropriately expressed if they were regarded as forming a superspecies. This taxonomic device has been defined by Mayr (1931, 1963); Mayr, Linsley and Usinger (1953) and by Simpson (1961). Allopatry seems crucial to the concept, though Cain (1963:70) would argue that *slight* geographical overlap be permitted (see discussion of the superspecies and of its applicability to the robust and hyper-robust australopithecines in Tobias 1973b:69–71).

In the final analysis, as Campbell pointed out on the question of the appropriate generic name for hominids dated in the time zone between 2.0 and 1.3 m.y., "only an international committee could remove controversy . . . by assessing the majority of informed opinion" (1972:46).

TENTATIVE SYNTHESIS AND CONCLUSIONS

We may now marshall the evidence for the existence of various kinds or taxa of hominids at various time levels and attempt to construct a model of the later stages of hominid phylogenesis. In doing so, we shall accept the provisional estimates for Makapansgat and Sterkfontein (2.5–3.0 m.y.), Swartkrans (2.0 m.y.), Kromdraai (late in the Swartkrans Faunal Span, ± 1.5–2.0 m.y.), and Taung (c. 0.8 m.y.), though appreciating fully that these estimates are tentative and may well need later amendment. We shall accept for purposes of our model the date of c. 2.9 m.y. for the large-brained cranium of KNM-ER-1470 of East Rudolf and the provisional assignment of this specimen as an early member of *Homo.*

On this basis, Figure 1 gives the distribution of site samples of hominids in time. The hominoid molar of Ngorora, *Ramapithecus* of India, *Kenyapithecus* (? = *Ramapithecus*) of Kenya, *Graecopithecus* of Greece and *Gigantopithecus* of China and Bilaspur are omitted from this discussion, as are the early hominids of Indonesia.

This chart permits one to see which hominids were living contemporaneously at each time level in the Plio-Pleistocene. Thus, at about 5

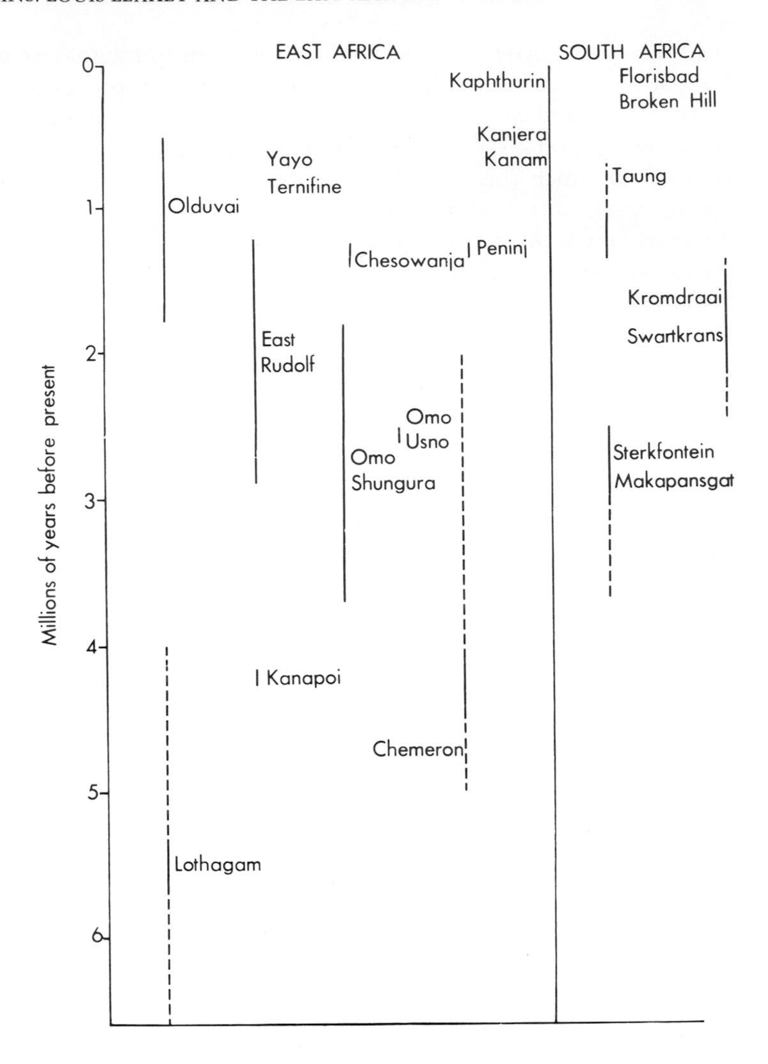

FIGURE 1
Chronological distribution of early hominids in the Plio-Pleistocene of Africa. For each East African site (data from Maglio 1973), the full duration in time of the deposit(s) at that site is indicated or suggested. A vertical solid line implies that hominid remains have been found throughout the sequence of deposits; an interrupted line with a solid segment indicates that only at the level of the solid segment have hominid fossils been found, though the deposit(s) at that site, as a whole, cover a greater range of time. For the South African sites, the solid parts of the lines represent the tentative dates of the hominid-bearing deposits, based mainly upon faunal comparisons by H. B. S. Cooke, V. J. Maglio and L. H. Wells; the interrupted lines indicate the dates estimated by T. C. Partridge for the opening of the caves. (In an earlier version of this chart, in the *Annual Review of Anthropology,* Chesowanja was erroneously shown as younger than 1 m.y.; there are also differences in the ranges of dating shown for the South African sites.)

m.y. the available sample comprises a single mandible from Lothagam—it resembles that of *A. africanus.* At just over 4 m.y. we have the humeral fragment of Kanapoi *(A.* aff. *africanus)* and perhaps the temporal bone of Chemeron, probably *Australopithecus* (Patterson 1966; Patterson *et al.* 1970; Patterson and Howells 1967), though later analysis indicates that the Chemeron temporal is not so old (Howell 1972a).

At 3.5 to 3.0 m.y. we have only some isolated teeth from the lower part of the Shungura Formation: both a large-toothed australopithecine and a small-toothed early hominid *(A.* aff. *africanus* or *Homo* aff. *habilis)* seem to be present.

From 3.0 to 2.5 m.y. the series comprises *A. africanus* at Sterkfontein and Makapansgat; the superspecies, *A. robustus/boisei,* and a gracile, bigger-brained hominid, probably *Homo* sp. or *Homo* cf. *habilis,* from Omo (both Usno and Shungura) and East Rudolf. This combination of *A. robustus/boisei* and *Homo* aff. *habilis* persists from 2.5 to 2.0 m.y.

From 2.0 to 1.5 m.y., we have in East Africa *A. boisei* and *H. habilis,* as represented at Olduvai, East Rudolf and Omo. In South Africa, there are *A. robustus* from Swartkrans and Kromdraai and *H.* aff. *habilis* from Swartkrans (the former "Telanthropus") and probably also from Upper Sterkfontein.

From 1.5 to 1.0 m.y., the form of *Homo* represented gives way from *H. habilis* to *H. erectus* (at Olduvai), while the very large-toothed *Australopithecus (A.* aff. *boisei)* lingers on at Olduvai, Peninj, Chesowanja, East Rudolf and Omo/Upper Shungura.

Below 1.0 m.y., *Homo erectus* is represented at Olduvai, Ternifine (in Algeria), probably Yayo (in Chad). If Partridge's and Butzer's claim that Taung is only c. 0.8 m.y. is correct, it would appear to have existed about 1.75 m.y. later than the youngest of the other hominids attributed to *A. africanus.* Its near contemporaries are *Homo* (most probably *H. erectus*) and perhaps the last of the *A. robustus/boisei* lineage. Under these circumstances, it would appear, on the face of it, to be highly unlikely that the Taung child represents the gracile lineage, as known from the Makapansgat and Sterkfontein hominid assemblages. The possibility of isolated relict populations of gracile australopithecines having survived in the southerly cul-de-sac of the African continent for over 1.5 m.y. after they had disappeared or given rise to the earliest *Homo* species elsewhere cannot, of course, be excluded at this stage of our knowledge. On the other hand, since its small brain size and dental characters would exclude it from *H. erectus* we must seriously consider whether it may be a late-surviving robust australopithecine (Tobias 1973a; 1974a,b; in press b). Despite the lapse of almost 50 years since its discovery, it is an amazing fact that the Taung skull has never yet been fully analyzed and described. Such a study is now urgently necessary in the light of new discoveries and newer dating evidence. The taxonomic implications of the transfer of the

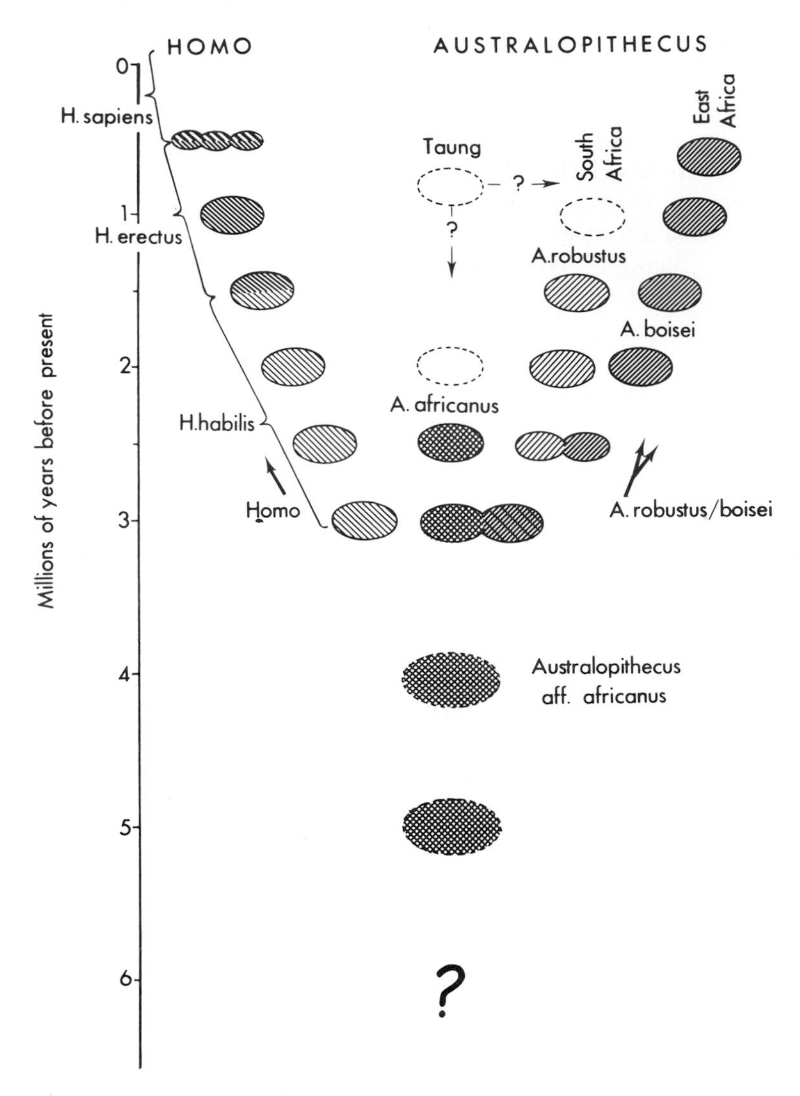

FIGURE 2
Hominid populations in Africa at various time levels in the Plio-Pleistocene. The oval figures at each time level represent the systematically identified, synchronic hominid populations known from the fossil record of Africa at that time. At any one time level, the horizontal distance between a pair of ovals is roughly proportional to the morphological and taxonomic distance between the populations or taxa represented by the ovals. The diagram indicates the anomalous position of the Taung skull, if the younger dating proposed by Partridge and by Butzer is accepted, its only near contemporaries being *H. erectus* and the late-surviving robust australopithecines. The queries (?) convey uncertainty as to whether the affinities of the Taung child lie with the gracile australopithecines, as traditionally believed and taught, or with the robust early hominids, as now seems to be at least a *prima facie* possibility.

Taung specimen to the robust australopithecine lineage are discussed elsewhere (Olson 1974, Tobias 1974c). This analysis of the chronological dispersal of the African early hominids permits one now to draw a diagram based on a succession of cross sections of the contemporaneous hominid populations (Figure 2).

In Figure 2, the flat ovals represent cross sections of the putative populations at each time level. The horizontal distance between any two such ovals at the same time level is approximately proportional to the morphological, taxonomic and phylogenetic distance between the two populations represented. Finally, in Figure 3, these cross-sectional population ovals are joined to provide a phylogenetic tree.

This reconstruction of the hominid family tree is amenable to the following interpretation:

(a) The ancestral population was a gracile form of hominid increasingly resembling *Australopithecus africanus.*

(b) Some time before 3 m.y., some East African populations diverged from the *A. africanus* lineage and, emphasizing cerebral enlargement and "complexification" as well as increasing cultural dependence, entered upon the very special and peculiar lineage of *Homo.* There followed a sequence of three time-consecutive species, *H. habilis, H. erectus* and *H. sapiens.*

(c) The South African (and perhaps some East African) populations of gracile australopithecines persisted for some time as such, even after the emergence of *Homo* in other parts of Africa. Just how long they persisted we do not know—possibly for another 2 m.y. (if Taung is correctly placed upon this lineage), or the gracile australopithecines as such may have died out soon after they had spawned the *Homo* lineage.

(d) Some time before 3 m.y. (on present evidence), some populations of the ancestral australopithecine began to differentiate in another direction, emphasizing cheek-tooth expansion, bodily enlargement with a moderate degree of concomitant brain enlargement, but with no major cultural-implemental component. Thus was produced the "robust" lineage characterized by modest morphological and probably ecological and ethological specialization in South Africa *(A. robustus)* and extreme degrees of such specialization in East Africa *(A. boisei).* The lineage of the *A. robustus/boisei* superspecies was apparently sufficiently distinct in behavior and ecological preferences from the line of *Homo* as to have permitted the two lineages to co-exist in Africa for something like 2 m.y. or even more.

The main difference between this reconstruction and that put forward in 1965 (Tobias 1965b) is that the recent discovery in the Rudolf Basin of early hominids attributable to *Homo* has forced the division between *Homo* and *Australopithecus* back a million years earlier than was hitherto held necessary.

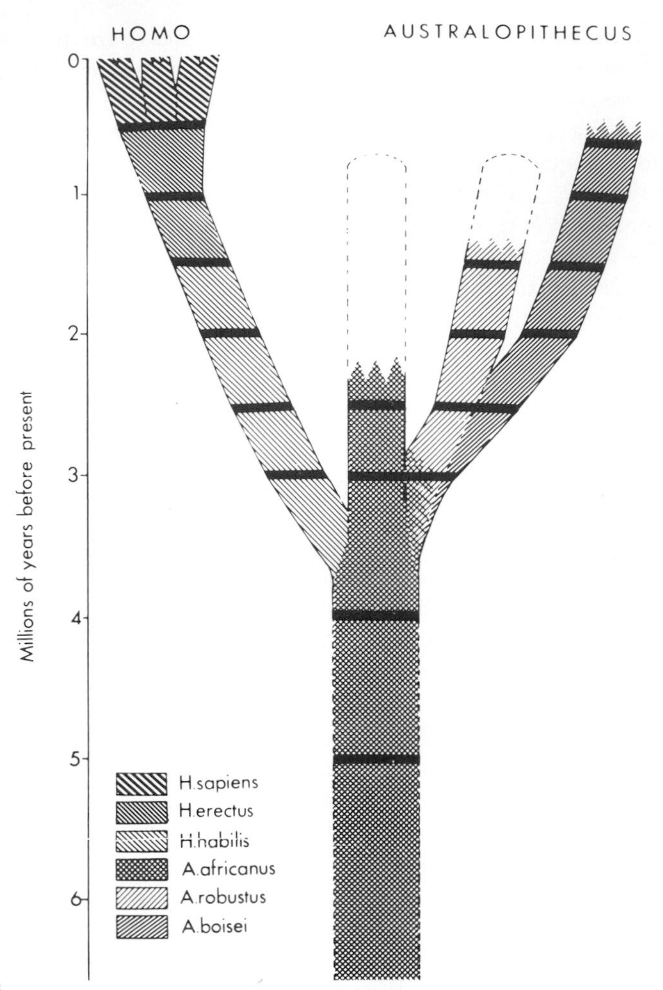

FIGURE 3
Provisional phylogenetic tree of the Hominidae during the Upper Pliocene and the Pleistocene. The unbranched trunk of the tree gives an almost certainly over-simplified picture of hominid evolution during the Upper Pliocene, owing to the paucity of hominid fossils dated to this period. The branching of *Homo* from *Australopithecus* is shown as an early event (before 3.0 m.y. B.P.), in order to accommodate the East Rudolf specimen, KNM-ER-1470; this cranium has been tentatively identified as possibly the earliest fossil attributable to *Homo,* its date being greater than 2.6 m.y. B.P. and perhaps as old as 2.9 m.y. B.P. (Leakey, R. E. F., 1973b). The diagram suggests that either *A. africanus* or *A. robustus* (or both) possibly survived to a very recent period (<0.9 m.y. B.P.); this enables the Taung skull to be accommodated in one or other lineage, if provisionally one accepts its proposed new young date of 0.7 to 0.8 m.y. B.P. The diagram makes clear the necessity for postulating a lengthy survival of the gracile lineage, if the Taung skull is still to be regarded as a member of this lineage; on the other hand, if the Taung skull is transferred to the robust lineage, it is not necessary to postulate a lengthy survival of either the South or East African robust lineages, in order to accommodate it.

It is clear, too, that there was a major element of cladistic evolution in the Hominidae from 3–4 m.y. onwards. The apparently purely phyletic pattern shown earlier than those dates might well be an artifact stemming from the paucity of hominid remains between 4 m.y. and 12–14 m.y.

On this interpretation, the evolving hominid lineages between 5 and 2 m.y. were African phenomena; thereafter, *Homo* appeared in Asia at a stage when *H. habilis* was changing into *H. erectus* (cf. Tobias and von Koenigswald 1964). It is not yet clear whether *Australopithecus* accompanied *Homo* in Indonesia—as Robinson (1953a, 1955) holds—or whether the robust forms there were within the range of variation of *H. erectus*—as Le Gros Clark (1964) believed.

Acknowledgments

I thank the Wenner-Gren Foundation for Anthropological Research, the University of the Witwatersrand and especially its Bernard Price Institute for Palaeontological Research, as well as the Council for Scientific and Industrial Research, Pretoria, for financial assistance, which made possible many of the studies cited above. I am indebted to Dr. C. Jolly, Dr. C. K. Brain, Mr. A. B. A. Brink, Dr. K. W. Butzer, Mr. R. J. Clarke, Dr. H. B. S. Cooke, Professor R. A. Dart, Professor M. H. Day, Mr. A. R. Hughes, the Leakey family, Dr. C. O. Lovejoy, Dr. H. M. McHenry, Dr. V. Maglio, Dr. A. E. Mann, Dr. Margaret Marker, Dr. T. C. Partridge, Professor G. Sperber, Dr. A. C. Walker, Dr. J. A. Wallace, Dr. M. H. Wolpoff, Mrs. Kay Copley, Miss Noreen Gruskin, Mr. P. Faugust, Miss Jeanne Walker; and Mrs. E. Hibbett.

References Cited

Aguirre, E.

1970 Identificacion de "Paranthropus" en Makapansgat. Cronica del XI Congreso Nacional de Arqueologia, Merida 1969. pp. 98–124.

1972 Africa y el origen de la humanidad. Documentacion Africana, Madrid, 15:1–36.

Amsden, T. W.

1970 The genus: a basic concept in palaeontology. Proc. N. Am. Paleont. Convention, Sept., 1969.

Aarmbourg, C.

1969 La nouvelle expédition scientifique de l'Omo. Riv. Sci. Preistoriche 24:3–13.

Arambourg, C., J. Chavaillon and Y. Coppens
1969 Résultats de la nouvelle mission de l'Omo (2e campagne 1968). C. R. Acad. Sci., Paris, 268-D:759–762.

Archibald, J. D., C. O. Lovejoy and K. G. Heiple
1972 Implications of relative robusticity in the Olduvai metatarsus. Am. J. Phys. Anthrop. 37:93–95.

Bada, J. L.
1972 The dating of fossil bones using the racemization of isoleucine. Earth Planet. Sci. Lett. 15:223–231.

Bauchot, R., and H. Stephan
1966 Données nouvelles sur l'encéphalisation des insectivores et des prosimiens. Mammalia 30:160–196.

1969 Encéphalisation et niveau évolutif chez les simiens. Mammalia 33:235–275.

Behrensmeyer, A. K.
1970 Preliminary geological interpretation of a new hominid site in the Lake Rudolf Basin. Nature 226:225–226.

Bilsborough, A.
1973 A multivariate study of evolutionary change in the hominid cranial vault and some evolution rates. J. Hum. Evol. 2:387–403.

1974 Some aspects of the evolution of the human brain. Symp. Inst. Biol. 21:195–213.

Bishop, W. W.
1971 The late Cenozoic history of East Africa in relation to hominoid evolution. *In* The Late Cenozoic Glacial Ages. K. K. Turekian, ed. New Haven and London: Yale University Press. pp. 493–527.

1972 Stratigraphic succession 'versus' calibration in East Africa. *In* Calibration of Hominoid Evolution. W. W. Bishop and J. A. Miller, eds. Edinburgh: Scottish Academic Press. pp. 219–246.

Bishop, W. W., and J. D. Clark, eds.
1967 Background to Evolution in Africa. Chicago: Chicago University Press.

Blackwelder, R. E.
1964 Phyletic and phenetic 'versus' omnispective classification. *In* Phenetic and Phylogenetic Classification. V. H. Heywood and J. McNeill, eds. London: The Systematics Association. pp. 17–28.

Bonnefille, R., J. Chavaillon and Y. Coppens
1970 Résultats de la nouvelle mission de l'Omo (3e campagne 1969). C. R. Acad. Sci. 270:924–927.

Brain, C. K.
1958 The Transvaal Ape-Man-bearing cave deposits. Transvaal Museum Mem. 11:1–125.

1967 The Transvaal Museum's fossil project at Swartkrans. S. Afr. J. Sci. 63:368–384.

1970 New finds at the Swartkrans australopithecine site. Nature 225:1112–1119.

1972 An attempt to reconstruct the behaviour of australopithecines: the evidence for interpersonal violence. Zool. Afr. 7:379–401.

1973 Seven years' hard labour at Swartkrans. Transvaal Mus. Bull. 14:5–6.

Brock, A. and G. L. Isaac
1974 Paleomagnetic stratigraphy and chronology of hominid-bearing sediments east of Lake Rudolf, Kenya. Nature 247:344–348.

Broom, R. and J. T. Robinson
1948 Size of the brain in the ape-man, *Plesianthropus*. Nature 161:438

Broom, R., J. T. Robinson and G. W. H. Schepers
1950 Sterkfontein ape-man *Plesianthropus*. Transvaal Museum Mem. 4:1–117.

Brown, F. H.
1969 Observations on the stratigraphy and radiometric age of the 'Omo beds,' lower Omo basin, southern Ethiopia. Quaternaria 11:7–14.

1972 Radiometric dating of sedimentary formations in the lower Omo valley, southern Ethiopia. *In* Calibration of Hominoid Evolution. W. W. Bishop and J. A. Miller, eds. Edinburgh: Scottish Academic Press. pp. 273–287.

Burns, P. E.
1971 New determination of australopithecine height. Nature 232:350.

Butzer, K. W.
1971a The lower Omo basin: geology, fauna and Hominids of Plio-Pleistocene formations. Naturwissenschaften 58:7–16.

1971b Another look at the australopithecine cave breccias of the Transvaal. Am. Anthropol. 73:1197–1201.

1974 Paleo-ecology of South African australopithecines: Taung revisited. Current Anthropology.

Butzer, K. W., and D. L. Thurber
1969 Some late Cenozoic sedimentary formations of the Lower Omo Basin. Nature 222:1132–1137.

Cain, A. J.
1963 Animal Species and their Evolution. London: Hutchinson.

Campbell, B. G.
1972 Conceptual progress in physical anthropology: fossil man. Ann. Rev. Anthropol. 1:27–54.

Carmichael, D. J., P. V. Tobias and C. M. Dodd
1974 A partial biochemical characterisation of fossilized bone from Makapansgat, Swartkrans and Queen Charlotte Islands. Comp. Biochem. Physiol.

Carney, J., A. Hill, J. A. Miller and A. Walker
1971 Late australopithecine from Baringo district, Kenya. Nature 230:509–514.

Chavaillon, J.
1970 Découverte d'un niveau Oldowayen dans la basse vallée de l'Omo (Ethiopie). C. R. Séances Soc. Préhist. Franç. 1:7–11.

Clark, Lė Gros, W. E.
1952 A note on certain cranial indices of the Sterkfontein skull no. 5. Am. J. Phys. Anthropol. 10:119–121.

1964 The Fossil Evidence for Human Evolution. Chicago: University of Chicago Press. (2nd ed.)

Clarke, R. J., and F. C. Howell
1972 Affinities of the Swartkrans 847 hominid cranium. Am. J. Phys. Anthrop. 37:319–336.

Clarke, R. J., F. C. Howell and C. K. Brain
1970 More evidence of an advanced hominid at Swartkrans. Nature 225:1219–1222.

Cooke, H. B. S.
1970 Notes from members: Canada: Dalhousie University, Halifax. Soc. Vert. Paleontol. Bull. 90:2.

Cooke, H. B. S., and R. F. Ewer
1972 Fossil Suidae from Kanapoi and Lothagam, Northwestern Kenya. Bull. Mus. Comp. Zool. 143:149–296.

Cooke, H. B. S., and V. J. Maglio
1972 Plio-Pleistocene stratigraphy in East Africa in relation to proboscidean and suid evolution. *In* Calibration of Hominoid Evolution. W. W. Bishop and J. A. Miller, eds. Edinburgh: Scottish Academic Press. pp. 303–329.

Coppens, Y.
1970-1971 Localisation dans le temps and dans l'espace des restes d'hominidés des formations Plio-Pléistocènes de l'Omo (Ethiopie). C. R. Acad. Sci., Paris 271:1968–1971, 2286–2289; 272:36–39.

1973a Les restes d'Hominidés des séries inférieures et moyennes des formations plio-villafranchiennes de l'Omo en Ethiopie (récoltes 1970, 1971 et 1972). C. R. Acad. Sci., Paris 276:1823–1826.

1973b Les restes d'Hominidés des séries supérieures des formations plio-villafranchiennes de l'Omo en Ethiopie. C. R. Acad. Sci., Paris 276:1981–1984.

Cox, A.
1969 Geomagnetic reversals. Science 163:237–245.

1972 Geomagnetic reversals—their frequency, their origin and some problems of correlation. *In* Calibration of Hominoid Evolution. W. W. Bishop and J. A. Miller, eds. Edinburgh: Scottish Academic Press. pp. 93–105.

Cox, A., R. R. Doell and G. B. Dalyrmple
1968 Radiometric time-scale for geomagnetic reversals. P. J. Geol. Soc. London 124:53–66.

Curtis, G. H., and R. L. Hay
1972 Further geological studies and potassium-argon dating at Olduvai Gorge and Ngorongoro Crater. *In* Calibration of Hominoid Evolution. W. W. Bishop and J. A. Miller, eds. Edinburgh: Scottish Academic Press. pp. 289–301.

Dart, R. A.
1949 The predatory implemental technique of *Australopithecus.* Am. J. Phys. Anthrop. 7:1–38.

1953 The predatory transition from ape to man. Internat. Anthrop. and Linguistic Review 1:201–218.

1957 The Osteodontokeratic Culture of *Australopithecus prometheus.* Transvaal Museum Mem. 10:1–105.

1962 The Makapansgat pink breccia australopithecine skull. Am. J. Phys. Anthrop. 20:119–126.

Day, M. H.
1967 Olduvai hominid 10, a multivariate analysis. Nature 215:323–324.

1969 A robust australopithecine fragment from Olduvai Gorge, Tanzania (Hominid 20). Nature 221:230–233.

1971 Postcranial remains of *Homo erectus* from Bed IV, Olduvai Gorge, Tanzania. Nature 232:383–387.

1973 Locomotor features of the lower limb in hominids. Symp. Zool. Soc. Lond. No. 33:29–51.

Day, M. H., and R. E. F. Leakey
1973 New evidence of the genus *Homo* from East Rudolf, Kenya. I. Am. J. Phys. Anthrop. 39:341–354.

Day, M. H., and J. L. Scheuer
1973 SKW 14147: a new hominid metacarpal from Swartkrans. J. Hum. Evol. 2:429–438.

Day, M. H., and B. A. Wood
1968 Functional affinities of the Olduvai hominid 8 talus. Man 3:440–455.

De Swardt, A. M. J.
1974 Geomorphological dating of cave openings in South Africa. Nature 250:683.

Evernden, J. F., and G. H. Curtis
1965 The potassium argon dating of late Cenozoic rocks in East Africa and Italy. Current Anthropology 6:343–385.

Fitch, F. J.
1972 Selection of suitable material for dating and the assessment of geological error in potassium-argon age determination. *In* Calibration of Hominoid Evolution. W. W. Bishop and J. A. Miller, eds. Edinburgh: Scottish Academic Press. pp. 77–91.

Fleischer, R. L., and H. R. Hart
1972 Fission track dating: techniques and problems. *In* Calibration of Hominoid Evolution. W. W. Bishop and J. A. Miller, eds. Edinburgh: Scottish Academic Press. pp. 135–170.

Freedman, L.
1957 The fossil Cercopithecoidea of South Africa. Ann. Transvaal Museum 23:121–262.

1970 A new check list of fossil Cercopithecoidea of South Africa. Paleontol. Afr. 13:109–110.

Freedman, L., and C. K. Brain
1972 Fossil cercopithecoid remains from the Kromdraai australopithecine site (Mammalia: Primates). Ann. Transvaal Museum 28:1–16.

Freedman, L., and N. S. Stenhouse
1972 The Parapapio species of Sterkfontein, Transvaal, South Africa. Paleontol. Afr. 14:93–111.

Grommé, C. S., and R. L. Hay
1967 Geomagnetic polarity epochs; new data from Olduvai Gorge, Tanzania. Earth Planet. Sci. Lett. 2:111–115.

Hay, R. L.
1971 Geologic Background of Beds I and II: stratigraphic summary. *In* Olduvai Gorge, Volume III, Excavations in Beds I and II, 1960–1963. M. D. Leakey. Cambridge: The University Press. pp. 9–18.

Heinzelin, J. de
1971 Observations sur la formation de Shungura (Vallée de l'Omo, Ethiopie). C. R. Acad. Sci., Paris 272:2409–2411.

Heinzelin, J. de, and F. H. Brown
1969 Some early Pleistocene deposits of the lower Omo Valley: the Usno formation. Quaternaria 11:29–46.

Heiple, K. J., and C. O. Lovejoy
1971 The distal femoral anatomy of *Australopithecus*. Am. J. Phys. Anthropol. 35:75–84.

Helmuth, H.
1968 Körperhöhe Gliedmassenproportionen der Australopithecinen. Z. Morphol. Anthropol. 60:147–155.

Hemmer, H.
1971 Beitrag zur Erfassung der progressiven Cephalisation bei Primaten. Proc. 3rd Congr. Primatol., Zurich. 1:99–107.

Hendey, Q. B.
1973a The Transvaal Museum's collection of fossil carnivores. Transvaal Museum Bull. 14:7–8.

1973b Carnivore remains from the Kromdraai australopithecine site. Annals Transvaal Museum 28:100–112.

1974 New fossil carnivores from the Swartkrans australopithecine site (Mammalia: Carnivora). Annals Transvaal Museum 29:27–48.

Holloway, R. L.
1970 Australopithecine endocast (Taung specimen 1924): a new volume determination. Science 168:966–968.

1972a New australopithecine endocast, SK 1585, from Swartkrans, South Africa. Am. J. Phys. Anthrop. 37:173–185.

1972b Australopithecine endocasts, brain evolution in the Hominoidea, and a model of hominid evolution. *In* The Functional and Evolutionary Biology of Primates. R. Tuttle, ed. Chicago/New York: Aldine-Atherton. pp. 185–203.

1973 Endocranial volumes of early African hominids and the role of the brain in human mosaic evolution. J. Hum. Evol. 2:449–459.

Howell, F. C.
1968 Omo Research Expedition. Nature 219:567–572.

1969 Remains of Hominidae from Pliocene/Pleistocene formations in the lower Omo Basin, Ethiopia. Nature 223:1234–1239.

1972a Pliocene/Pleistocene Hominidae in eastern Africa: absolute and relative ages. *In* Calibration of Hominoid Evolution. W. W. Bishop and J. A. Miller, eds. Edinburgh: Scottish Academic Press. pp. 331–368.

1972b Recent advances in human evolutionary studies. *In* Perspectives on Human Evolution. S. L. Washburn and P. C. Dolhinow, eds. New York: Rinehart and Winston. pp. 51–128.

Howell, F. C., and Y. Coppens
1973 Deciduous teeth of Hominidae from the Pliocene/Pleistocene of the lower Omo basin, Ethiopia. J. Hum. Evol. 2:461–472.
1974 Inventory of remains of Hominidae from Pliocene/Pleistocene formations of the lower Omo basin, Ethiopia (1967–1972). Am. J. Phys. Anthropol. 40:1–16.

Hurford, A. J.
1974 Fission track dating of a vitric tuff from East Rudolf, North Kenya. Nature 249:236–237.

Huxley, J.
1959 Clades and grades. *In* Function and Taxonomic Importance. A. J. Cain, ed. London: The Systematics Association. pp. 21–22.

Isaac, G. L.
1972 Chronology and the tempo of cultural change during the Pleistocene. *In* Calibration of Hominoid Evolution. W. W. Bishop and J. A. Miller, eds. Edinburgh: Scottish Academic Press. pp. 381–430.

Isaac, G. L., R. E. F. Leakey and A. K. Behrensmeyer
1971 Archaeological traces of early hominid activities, east of Lake Rudolf, Kenya. Science 173:1129–1134.

Jerison, H. J.
1963 Interpreting the evolution of the brain. Hum. Biol. 35:263–291.
1973 Evolution of the Brain and Intelligence. New York and London: Academic Press.

Kinzey, W. G.
1972 Allometric transposition of the brain-body size relationships in hominid evolution. 41st Annual Meeting, Am. Assoc. Phys. Anthropol. Lawrence, Kansas.

Koenigswald, G. H. R. von
1973 *Australopithecus, Meganthropus* and *Ramapithecus*. J. Hum. Evol. 2:487–491.

Leakey, L. S. B.
1973 Was *Homo erectus* responsible for the hand-axe culture? J. Hum. Evol. 2:493–498.

Leakey, L. S. B., P. V. Tobias and J. R. Napier
1964 A new species of the genus *Homo* from the Olduvai Gorge. Nature 202:7–9.

Leakey, M. D.
1967 Preliminary survey of the cultural material from Beds I and II, Olduvai Gorge, Tanzania. *In* Background to Evolution in Africa. W. W. Bishop and J. D. Clark, eds. Chicago: University of Chicago Press. pp. 417–446.
1970 Stone artefacts from Swartkrans. Nature 225:1222–1225.
1971a Olduvai Gorge, Volume III. Excavations in Beds I and II, 1960–1963. Cambridge: The University Press.
1971b Discovery of postcranial remains of *Homo erectus* and associated artefacts in Bed IV at Olduvai Gorge, Tanzania. Nature 232:380–383.

Leakey, M. D., R. J. Clarke L. S. B. Leakey
1971 New hominid skull from Bed I, Olduvai Gorge, Tanzania. Nature 232:308–312.

Leakey, R. E. F.
1970 Fauna and artefacts from a new Plio/Pleistocene locality near Lake Rudolf in Kenya. Nature 226:223–224.

1971 Further evidence of Lower Pleistocene hominids from East Rudolf, North Kenya. Nature 231:241–245.

1972a Further evidence of Lower Pleistocene hominids from East Rudolf, North Kenya, 1971. Nature 237:264–269.

1972b New fossil evidence for the evolution of man. Social Biology 19:99–114.

1973a Further evidence of Lower Pleistocene hominids from East Rudolf, North Kenya, 1972. Nature 242:170–173.

1973b Evidence for an advanced Plio-Pleistocene hominid from East Rudolf, Kenya. Nature 242:447–450.

1973c Australopithecines and hominines: a summary on the evidence from the early Pleistocene of eastern Africa. Symp. Zool. Soc. London 33:53–69.

Leakey, R. E. F., J. M. Mungai and A. C. Walker
1971 New australopithecines from East Rudolf, Kenya. Am. J. Phys. Anthropol. 35:175–186.

1972 New australopithecines from East Rudolf, Kenya (II). Am. J. Phys. Anthropol. 36:235–251.

Leakey, R. E. F., and A. C. Walker
1973 New australopithecines from East Rudolf, Kenya (III). Am. J. Phys. Anthropol. 39:205–222.

Leakey, R. E. F., and B. A. Wood
1973 New evidence of the genus *Homo* from East Rudolf, Kenya (II). Am. J. Phys. Anthropol. 39:355–368.

Lestrel, P. E., and D. W. Read
1973 Hominid cranial capacity versus time: a regression approach. J. Hum. Evol. 2:405–411.

Leutenegger, W.
1972 Newborn size of pelvic dimensions of *Australopithecus*. Nature 240: 568–569.

1973 Encephalization in australopithecines: a new estimate. Folia. Primat. 19:9–17.

Lewis, O. J.
1972 The evolution of the hallucial tarsometatarsal joint in the Anthropoidea. Am. J. Phys. Anthropol. 37:13–34.

1973 The hominid os capitatum, with special reference to the fossil bones from Sterkfontein and the Olduvai Gorge. J. Hum. Evol. 2:1–11.

Lovejoy, C. O., and K. G. Heiple
1970 A reconstruction of the femur of *Australopithecus africanus*. Am. J. Phys. Anthropol. 33:33–40.

1972 The proximal femoral anatomy of *Australopithecus*. Nature 235:175–176.

Lovejoy, C. O., K. G. Heiple and A. H. Burstein
1973 The gait of *Australopithecus*. Am. J. Phys. Anthropol. 38:757–779.

MacDougall, D., and P. B. Price
1974 Attempt to date early South African hominids using fission tracks in calcite. Science 185:943–944.

McHenry, H. M.
1972 Postcranial skeleton of early Pleistocene hominids. Ph.D. thesis. Harvard University, Cambridge, Mass.

1974 How large were the australopithecines? Am. J. Phys. Anthropol. 40:329–340.

McHenry, H. M., and R. Corruccini
1975 Distal humerus in hominid evolution. Folia Primatologia 23:274–277.

McKinley, K. R.
1971 Survivorship in gracile and robust australopithecines: a demographic comparison and a proposed birth model. Am. J. Phys. Anthropol. 34:417–426.

Maglio, V. J.
1971 Vertebrate faunas from the Kubi Algi, Koobi Fora and Ileret areas, East Rudolf, Kenya. Nature 231:248–249.

1972 Vertebrate fauna and chronology of hominid-bearing sediments east of Lake Rudolf, Kenya. Nature 239:379–385.

1973 Origin and evolution of the Elephantidae. Trans. Am. Phil. Soc., n.s. 63:1–149.

Maguire, B.
1965 Foreign pebble pounding artefacts in the breccias and the overlying vegetation at Makapansgat Limeworks. S. Afr. Archaeol. Bull. 20:117–130.

Maier, W.
1970 New fossil Cercopithecoidea from the lower Pleistocene cave deposits of the Makapansgat Limeworks, South Africa. Paleontol. Afr. 13:69–107.

1972 The first complete skull of *Simopithecus darti* from Makapansgat, South Africa, and its systematic position. J. Hum. Evol. 1:395–405.

1973 Paläoökologie und zeitliche Einordnung der süd-afrikanischen Australopithecinen. Z. Morph. Anthrop. 65:70–105.

Mann, A. E.
1968 The Paleodemography of Australopithecus. Ph.D. thesis, University of California, Berkeley.

1973 Australopithecine age at death. Transvaal Museum Bull. no. 14:11.

In press Australopithecine demographic patterns. *In* African Hominidae of the Plio-Pleistocene. C. Jolly, ed.

Mason, R. J.
1962 Australopithecines and artefacts at Sterkfontein, Part II. The Sterkfontein stone artefacts and their maker. S. Afr. Archaeol. Bull. 17:109–125.

Mayr, E.
1931 Notes on *Halcyon chloris* and some of its sub-species. Amer. Mus. Novitates No. 469:1–10.

1963 The taxonomic evaluation of fossil hominids. *In* Classification and Human Evolution. S. L. Washburn, ed. Chicago: Aldine. pp. 332–346.

1969 Principles of Systematic Zoology. New York: McGraw-Hill.

Mayr, E., E. G. Linsley and R. L. Usinger
1953 Methods and Principles of Systematic Zoology. New York: McGraw-Hill.

Miller, J. A.
1972 Dating Pliocene and Pleistocene strata using the potassium-argon and argon-40/argon-39 methods. *In* Calibration of Hominoid Evolution. W. W. Bishop and J. A. Miller, eds. Edinburgh: Scottish Academic Press. pp. 63–76.

Napier, J.
1964 The evolution of bipedal walking in the hominids. Arch. Biol. Liège (suppl.) 75:673–708.

Napier, J. R., and M. H. Day
1964 Hominid fossils from Bed I, Olduvai Gorge, Tanzania. Nature 202:969–970.

Olson, T. R.
1974 Taxonomic implications of the transfer of the Taung skull to the robust australopithecine lineage. Nature 252:85.

Oxnard, C. E.
1967 The functional morphology of the primate shoulder as revealed by comparative anatomical, osteometric and discriminant function techniques. Am. J. Phys. Anthropol. 26:219–240.

1968 A note on the fragmentary Sterkfontein scapula. Am. J. Phys. Anthropol. 28:213–217.

1972 Some African fossil foot bones: a note on the interpolation of fossils into a matrix of extant species. Am. J. Phys. Anthropol. 37:3–12.

Partridge, T. C.
1965 A statistical analysis of the Limeworks lithic assemblage. S. Afr. Archaeol. Bull. 20:112–116.

1973 Geomorphological dating of cave openings at Makapansgat, Sterkfontein, Swartkrans and Taung. Nature 246:75–79.

1974 Geomorphological dating of cave openings in South Africa, reply to A. M. J. De Swardt. Nature 250:683–684.

Patterson, B.
1966 A new locality for early Pleistocene fossils in northwestern Kenya. Nature 212:577–578.

Patterson, B., A. K. Behrensmeyer and W. D. Sill
1970 Geology and fauna of a new Pliocene locality in northwestern Kenya. Nature 226:918–921.

Patterson, B., and W. W. Howells
1967 Hominid humeral fragment from early Pleistocene of northwestern Kenya. Science 156:64–66.

Petter, G.
1973 Carnivores Pléistocènes du ravin d'Olduvai (Tanzanie). Fossil Vert. Afr. 3:43–100.

Pilbeam, D. R.
1969 Early Hominidae and cranial capacity. Nature 224:386.

1972a The Ascent of Man: an Introduction to Human Evolution. MacMillan Series in Physical Anthropology. New York: MacMillan.

1972b Evolutionary changes in hominoid dentition through geological time. *In* Calibration of Hominoid Evolution. W. W. Bishop and J. A. Miller, eds. Edinburgh: Scottish Academic Press. pp. 369–380.

Preuschoft, H.
1971 Body posture and mode of locomotion in early Pleistocene hominids. Folia Primatol. 14:209–240.

Rightmire, G. P.
1972 Multivariate analysis of an early hominid metacarpal from Swartkrans. Science 176:159–161.

Robinson, J. T.
1953a *Meganthropus, Australopithecus* and hominids. Am. J. Phys. Anthropol. 11:1–38.

1953b *Telanthropus* and its phylogenetic significance. Am. J. Phys. Anthropol. 11:445–501.

1954 The genera and species of the Australopithecinae. Am. J. Phys. Anthropol. 12:181–200.

1955 Further remarks on the relationship between *Meganthropus* and australopithecines. Am. J. Phys. Anthropol. 13:429–446.

1962 Australopithecines and artefacts at Sterkfontein: Part I. Sterkfontein stratigraphy and the significance of the extension site. S. Afr. Archaeol. Bull. 17:87–107.

1963 Adaptive radiation in the australopithecines and the origin of man. *In* African Ecology and Human Evolution. F. C. Howell and F. C. Bourlière, eds. Chicago: Aldine. pp. 385–416.

1965 *Homo 'habilis'* and the australopithecines. Nature 205:121–124.

1966 Comment on "The distinctiveness of Homo habilis." Nature 209:957–960.

1967 Variation and the taxonomy of the early hominids. *In* Evolutionary Biology. T. Dobzhansky, M. K. Hecht and W. C. Steere, eds. New York: Appleton-Century-Crofts. pp. 69–100.

1970 Two new early hominid vertebrae from Swartkrans. Nature 225:1217–1219.

1972 Early Hominid Posture and Locomotion. Chicago: University of Chicago Press.

Robinson, J. T., L. Freedman and B. A. Sigmon
1972 Some aspects of pongid and hominid bipedality. J. Hum. Evol. 1:361–369.

Robinson, J. T., and R. J. Mason
1957 Occurrence of stone artefacts with *Australopithecus* at Sterkfontein. Nature 180:521–524.

Robinson, J. T., and K. Steudel
1973 Multivariate discriminant analysis of dental data bearing on early hominid affinities. J. Hum. Evol. 2:509–527.

Roper, M. K.
1969 A survey of the evidence for intrahuman killing in the Pleistocene. Current Anthropology 10:427–459.

Rosen, S. I., and T. W. McKern
1971 Several cranial indices and their relevance to fossil man. Am. J. Phys. Anthropol. 35:69–73.

Rowell, A. J.
1970 The contribution of numerical taxonomy to the genus concept. Proc. N. Am. Paleont. Convention, September 1969. Lawrence, Kansas. Allen Press. pp. 264–293.

Simons, E. L.
1972 Primate Evolution. New York, MacMillan.

Simons, E. L., and P. C. Ettel
1970 Gigantopithecus. Scientific American 222:77–85.

Simpson, G. G.
1961 Principles of Animal Taxonomy. New York: Columbia University Press.

1963 The meaning of taxonomic statements. *In* Classification and Human Evolution. S. L. Washburn, ed. Chicago: Aldine. pp. 1–31.

Sperber, G. H.
1973 The morphology of the cheek teeth of early South African hominids. Ph.D. thesis, Department of Anatomy, University of the Witwatersrand, Johannesburg.

Stephan, H.
1972 Evolution of primate brains: a comparative anatomical investigation. *In* The Functional and Evolutionary Biology of Primates. R. Tuttle, ed. Chicago: Aldine and Atherton. pp. 155–174.

Thenius, E.
1969 Stammesgeschichte der Säugetiere (einschliesslich der Hominiden). Handb. d. Zool., Band 8 (Mammalia) 47:1–368.

Tobias, P. V.
1963 Cranial capacity of *Zinjanthropus* and other australopithecines. Nature 197:743–746.

1965a *Australopithecus, Homo habilis,* tool-using and tool-making. S. Afr. Archaeol. Bull. 20:167–192.

1965b Early man in East Africa. Science 149:22–33.

1966 The distinctiveness of *Homo habilis.* Nature 129:953–957.

1967 Olduvai Gorge, Volume II. The cranium and maxillary dentition of *Australopithecus (Zinjanthropus) boisei.* Cambridge: The University Press.

1968a The age of death among the australopithecines. The Anthropologist, special volume 23–28.

1968b The taxonomy and phylogeny of the australopithecines. *In* Taxonomy and Phylogeny of Old World Primates with References to the Origin of Man. Proc. Round Table at Inst. Anthropol. Centre Primatol., Univ. Turin, Italy. Turin: Rosenberg and Sellier. pp. 277–318.

1968c New African evidence on human evolution. Wenner-Gren Found. Supper Conf., New York City, April 1968.

1969 Commentary on new discoveries and interpretations of early African fossil hominids. Yearbook of Physical Anthropology 1967. S. Genoves, ed. pp. 24–30.

1971a The Brain in Hominid Evolution. New York/London: Columbia Univ. Press.

1971b Does the form of the inner contour of the mandible distinguish between *Australopithecus* and *Homo? In* Perspectives in Paleoanthropology: D. Sen,

Festschrift Volume. A. K. Gosh, ed. Calcutta: Firma K. L. Mukhopadhyay. pp. 9–17.

1972a Progress and problems in the study of early man in sub-Saharan Africa. *In* The Functional and Evolutionary Biology of Primates. R. Tuttle, ed. Chicago/New York: Aldine and Atherton. pp. 63–93.

1972b "Dished faces," brain size and early hominids. Nature 239:468–469.

1973a Implications of the new age estimates of the early South African hominids. Nature 246:79–83.

1973b Darwin's prediction and the African emergence of the genus Homo. *In* L'Origine dell'Uomo. Atti del Colloquio Internazionale, Roma, Ott. 1971. 182:63–85.

1973c A new chapter in the history of Sterkfontein early homonid site. J. S. Afr. Biol. Soc. 14:30–44.

1973d New developments in hominid paleontology in South and East Africa. Ann. Rev. Anthrop. 2:311–334.

1974a New African evidence on the dating and the phylogeny of the Plio-Pleistocene Hominidae. Paper delivered to IX INQUA Congress, Christchurch, December, 1973. Trans. Roy. Soc. New Zealand.

1974b The Taung skull revisited. Nat. Hist. Mag. (N.Y.) 83:38–43.

1974c Taxonomic implications of the transfer of the Taung skull to the robust australopithecine lineage. (Reply to Dr. T. R. Olson). Nature 252:85–86.

In press a Aspects of pathology and death among early hominids. The Leech (Johannesburg).

In press b The South African australopithecines in time and hominid phylogeny, with especial reference to the dating and affinities of the Taung skull. *In* African Hominidae of the Plio-Pleistocene. C. Jolly, ed.

In press c Brain Evolution in the Hominoidea. IX Internat. Cong. Anthrop. Ethnol. Sci., Chicago 1973.

In press d Recent Studies on Sterkfontein and Makapansgat and their bearing on hominid phylogeny in Africa. S. Afr. Archaeol. Bull.

Tobias, P. V., and A. R. Hughes

1969 The new Witwatersrand University excavation at Sterkfontein. S. Afr. Archaeol. Bull. 24:158–169.

Tobias, P. V., and G. H. R. von Koenigswald

1964 Comparison between the Olduvai hominines and those of Java and some implications for hominid phylogeny. Nature 204:515–518.

Vondra, C. F., B. E. Bowen, G. D. Johnson and A. K. Behrensmeyer

1971 Preliminary stratigraphical studies of the East Rudolf Basin, Kenya. Nature 231:245–248.

Vrba, E. S.

1971 A new fossil alcelaphine (Artiodactyla: Bovidae) from Swartkrans. Ann. Transvaal Mus. 27:59–82.

1973 Some recent developments in research on the Bovidae of the australopithecine caves. Transvaal Museum Bull. No. 14:8–9.

1974 Chronological and ecological implications of the fossil Bovidae at the Sterkfontein australopithecine site. Nature 250:19–23.

Walker, A. C.
1973 New *Australopithecus* femora from East Rudolf, Kenya. J. Hum. Evol. 2:545–555.

Wallace, J. A.
1972 The Dentition of the South African Early Hominids: a study of Form and Function. Ph.D. Thesis, Department of Anatomy, University of the Witwatersrand, Johannesburg.

1973a Molar occlusion in the ape-man *(Australopithecus).* Am. J. Orthodont. 63:606–609.

1973b Tooth chipping in the australopithecines. Nature 244:117–118.

1973c Dietary adaptations of *Australopithecus* and early *Homo.* IX Internat. Cong. Anthropol. Ethnol. Sci., Chicago 1973.

Walters, S. M.
1964 Introduction to General Discussion. *In* Phenetic and Phylogenetic Classification. V. J. Heywood and J. McNeill, eds. London: The Systematics Assoc. pp. 157–159.

Wells, L. H.
1967 Antelopes in the Pleistocene of Southern Africa. *In* Background to Evolution in Africa. W. W. Bishop and J. D. Clark, eds. Chicago: University of Chicago Press. pp. 99–107.

1969 Faunal subdivision of the Quaternary in southern Africa. S. Afr. Archaeol. Bull. 24:93–95.

1971 Africa and the ancestry of man. S. Afr. J. Sci. 67:276–283.

Wolpoff, M. H.
1971a Metric Trends in Hominid Dental Evolution. Case Western Reserve Univ. Stud. Anthropol. 2:1–244.

1971b A functional measure of tooth size. Southwestern J. Anthropol. 27:279–286.

1971c Is the new composite cranium from Swartkrans a small robust australopithecine? Nature 230:398–401.

1973 Posterior tooth size, body size and diet in South African gracile australopithecines. Am. J. Phys. Anthropol. 39:375–393.

Wright, G. A., and S. Skaryd
1972 Do fossil elephants date the South African australopithecines? Nature 237:291.

Zihlman, A. L.
1967 Human locomotion. A reappraisal of the functional and anatomical evidence. Ph.D. Thesis. University of California, Berkeley.

1970 The Questions of locomotor differences in *Australopithecus.* 3rd Internat. Congr. Primatol., Zurich.

Zihlman, A. L., and W. S. Hunter
1972 A biomechanical interpretation of the pelvis of *Australopithecus.* Folia Primatol. 18:1–19.

PART FOUR

Archaeological Reconstruction of Early Hominid Behavior in East Africa

Introduction to Part Four

Archaeological research in East Africa effectively began in 1926 with the expedition that Louis Leakey organized and led immediately after graduating from Cambridge University (see Clark this volume, Tobias this volume). The first area to be studied intensively was the sector of the Rift Valley around Lakes Naivasha and Nakuru in Kenya where Leakey and his colleagues made a pioneer study of a long and important late Quaternary geological and cultural sequence. However in 1971, Louis Leakey, in the company of Dr. Hans Reck, made a visit to Olduvai Gorge in northern Tanzania. Olduvai was already known as a rich source of Pleistocene fossils, but in that momentous visit Leakey found clear evidence of very early stone tools. This discovery strengthened Louis Leakey's intuition that East Africa would be an important source of archaeological and paleontological evidence bearing on human origins. From then until his death 41 years later in 1972, Louis restlessly strove to find new evidence by exploration and to promote the study of all promising localities.

This oil painting by Jay H. Matternes reconstructs a scene from early hominid life in East Africa, made from information supplied by the East Rudolf Research Project regarding site FxJj3. Reproduced by kind permission of Jay H. Matternes and Aubrey Buxton of Survival Anglia Ltd., who hold copyright.

An important part of Leakey's tremendous contribution to the development of archaeology in Eastern Africa was his recruitment in 1935 of Mary Nicol to come and work with him. Shortly afterwards they were married and continued to work together as a team. Division of labor within families is an important hominid characteristic—and Louis and Mary practiced it. By temperament Louis was a scientific explorer, while Mary is a meticulous and patient excavator. As time went on Mary increasingly took responsibility for the conduct of archaeological excavations and the detailed interpretation of findings. Together they tried out the possibilities of exposing by large-scale excavation the stratified remains of ancient camp sites. This was a novel procedure in the study of the Lower Paleolithic, and they proved that under the excellent conditions of preservation prevailing in Eastern Africa sedimentary basins, occupation floors could often be uncovered. At Olorgesailie, Mary Leakey pioneered these techniques with Louis' support, and she continues to apply and improve them at Olduvai Gorge. The research movement spread and such important sites as Kalambo Falls, Isimila, Latamne, Torralba came to be excavated by methods inspired in part by the Leakeys' work in East Africa.

Olduvai Gorge was the first find spot for definite evidence of very early man in East Africa. Now, after 40 years, it continues to yield a rich and detailed record of fossil hominids, archaeological traces and paleoenvironmental conditions. In Part Two of this volume Richard Hay provides a summary of aspects of the geological work that is still proceeding and of its implications in understanding the conditions under which early hominids lived. In this section, we present an annotated reprint by Mary Leakey of a summary of the Olduvai archaeology which has been taken from Mary Leakey's major monograph on excavations in Beds I and II. This monograph is by far the most important source of primary evidence about the culture and activities of lower Pleistocene hominids, and all students with particular interests in the development of culture should consult it directly; however, the concluding chapter reproduced in this tribute volume will serve to introduce some of the issues involved in the research begun by Louis and Mary and which is being continued by Mary.

While Olduvai remains pivotal for archaeological reconstruction of early hominid behavior, other sites in East Africa have in recent years also begun to yield evidence. In southern Ethiopia, the Shungura Formation of the Omo Group has been explored by the international expeditions led by F. Clark Howell and Yves Coppens (see Howell this volume). Here H. V. Merrick and J. Chavaillon have found numerous archaeological sites, particularly in Member F which is believed to be about 2 million years old—that is, slightly older than Olduvai Bed I. Harry Merrick's article is a brief summary of his findings.

As a consequence of exploration by Louis and Mary's son Richard, other important early sites have come to light in the East Rudolf area (see R. Leakey and G. Isaac, this volume). Archaeological sites of two different ages have been discovered. The older set has been dated by potassium-argon measurements at 2½ million years and this dating receives strong support from paleomagnetic stratigraphy. On a balance this estimate seems to be the best available, but there have been questions raised regarding its compatability with certain proposed paleontological correlations. There is also continuing debate over difficulties with the K-Ar measurements. Research is currently in progress to try to resolve these uncertainties. If the dates are confirmed, then these sites are the oldest definite archaeological traces of hominid behavior. Aspects of the information emerging from archaeological research at East Rudolf are covered in the general review paper in this section by Glynn Isaac.

Olduvai, the Omo, and East Rudolf are at present the most extensively investigated and best-dated early archaeological localities in East Africa, but it is clear that many more remain to be studied. In 1932 Louis Leakey made promising finds in the Kanam area of western Kenya, which are now being followed up by an expedition from Yale University. In Ethiopia Jean Chavaillon has been studying sites at Melka Kontouré, which range from Middle Pleistocene back to Lower Pleistocene ages, though geophysical dates are not yet available (Chavaillon 1967, 1970). In the Afar and Danakil depression, where vast sedimentary formations of Pliocene-Pleistocene age are being discovered and explored, there is already promise of more important early hominid and artifact finds. In addition, recent work at the site where the Chesowanja cranium was found suggests that there is a good stratified archaeological sequence there spanning from Oldowan or Developed Oldowan to Acheulian and later (Bishop, pers. comm.; J. W. K. Harris pers. comm.). More entirely new discoveries can confidently be predicted as the exploration of the Rift system proceeds.

All these discoveries in East Africa make it very clear that renewed searches for comparable traces in the early Pleistocene deposits of Asia are very important. The essay by J. Desmond Clark with which this volume opened helps to show how the archaeological evidence from various parts of Africa, including East Africa, can be fitted together and integrated with other kinds of paleoanthropological data.

The final item in this section is a general review of the archaeological evidence for the activities of early African hominids written by one of the editors, Glynn Isaac, as a tribute to Louis Leakey. Isaac was recruited into East African archaeology by Louis Leakey in 1961 and has worked since then in close association with Louis, Mary, and Richard. The review offers an outline of a pattern in the evidence and a personal assessment of the light that archaeology is shedding on human origins.

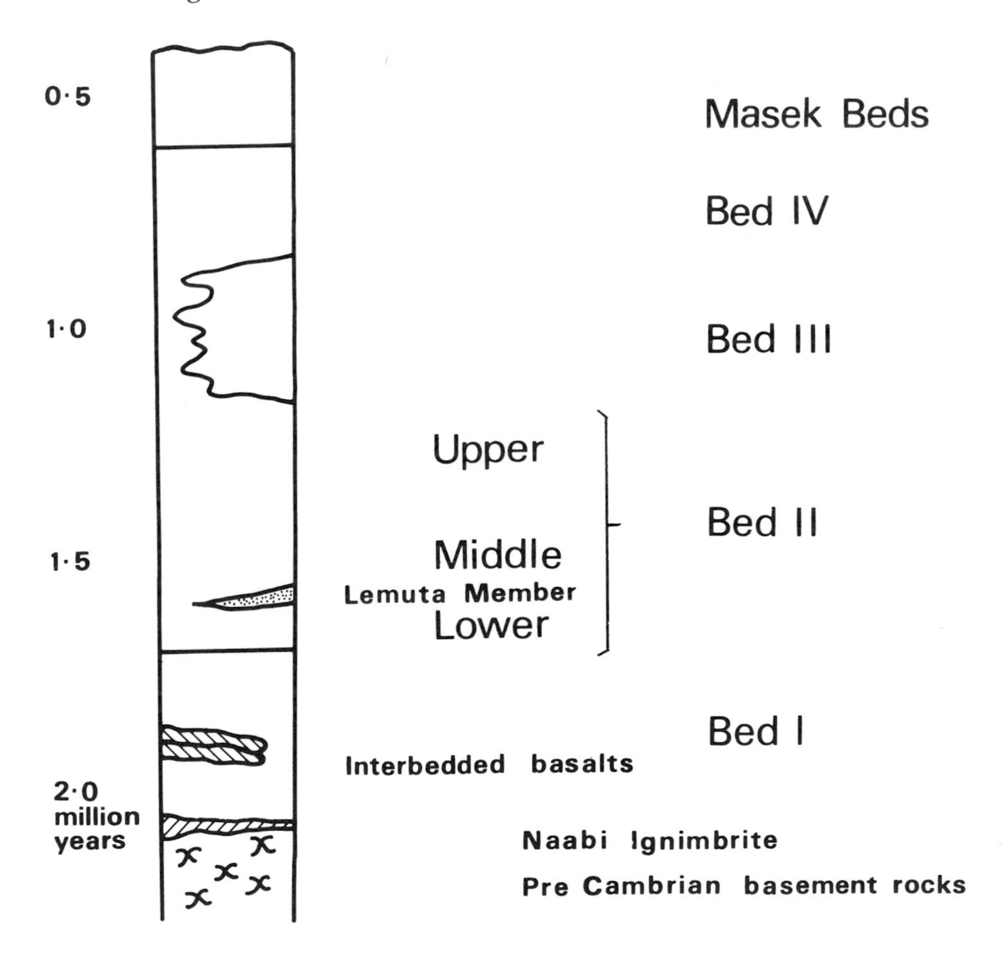

Readers will find agreement among the archaeology papers on some points and varying degrees of divergence in interpretation regarding other matters. This is because archaeology of early man in East Africa has become the subject of keen debate in recent years. There has been particularly lively discussion over the question of whether diversity among the stone artifact assemblages is most likely due to differences between stable cultures that co-existed over long time spans or due to differences in activities carried out at different times by members of a single cultural system. We hope that the archaeology papers included in this volume will give readers some sense of the fruitful ferment that has been occasioned by the digging up of the oldest known sets of traces of human behavior.

In the note that precedes the reprint from Mary Leakey's volume on Olduvai Gorge, she refers to revisions made in the chronology of the sequence at Olduvai. Dr. R. L. Hay has kindly contributed data that have allowed the editors to draw up the accompanying diagram showing the stratigraphy and chronology of Olduvai, as currently understood by him.

References Cited

Chavaillon, J.

1967 La Préhistoire Ethiopienne á Melka Kontouré. Archéologia, Nov.–Dec. pp. 56–63.

1970 Découverte d'un Niveau Oldowayen dans la Basse Vallée de l'Omo (Ethiopie). Bulletin de la Société Préhistorique Française 67:7–11.

Mary Leakey and Louis Leakey show visiting scientists the fossil skeleton of a deinotherium that Mary has exposed by excavation in Bed II, Olduvai Gorge. Among the bones, they found stone artifacts which were evidently used by early hominids as butchery tools. The visitors include Raymond Dart, Camille Arambourg, and J. Desmond Clark. © Des Bartlett, reproduced by his kind permission, through the help of the National Geographic Society.

Mary D. Leakey:

A Summary and Discussion of the Archaeological Evidence from Bed I and Bed II, Olduvai Gorge, Tanzania

*Olduvai Gorge has provided the longest record of cultural development yet obtained from any locality anywhere in the world. The lower levels in Beds I give clear evidence of the establishment by 1.8 million years ago of such fundamental human behavior patterns as toolmaking, hunting, and food sharing. In Bed II the record is more intricate and more than one cultural system may be represented. In the concluding chapter of her monograph on the archaeology of Olduvai Beds I and II, Dr. Mary Leakey offers an interpretative summary of her views on the significance of the unique set of documents that she has excavated and studied. This summary is reproduced here with a note that helps to bring it up to date.**

Since this summary of results from excavations in Beds I and II was written, work has been carried out in Beds III, IV, and the Masek Beds. The most important aspects of these excavations are (1) revision of the estimated age of Bed IV, based on palaeomagnetic data, and (2) the continuation of the Developed Oldowan into upper Bed IV.

When series of samples with reversed polarity were obtained from Bed III and others with normal polarity from Bed IV it was assumed that the

---

*Reprinted from M. D. Leakey 1971 with permission of the author and Cambridge University Press.

Brunhes-Matuyama boundary probably coincided with the Beds III–IV interface. Further sampling by Professor Allan Cox of the Department of Geophysics, Stanford University, however, has shown that the Matuyama reversed epoch continued into lower Bed IV, giving this part of the sequence an age of not less than 700,000 years, instead of 500,000 years as previously estimated.

It is now possible to subdivide Bed IV into upper and lower units, separated by a tuff known as IVB. The industries from all the sites excavated in lower Bed IV during the last season proved to be Acheulean with a proportionately high percentage of large bifacial tools, usually made in large flakes, with a greater or lesser degree of trimming. In upper Bed IV, however, a number of sites on the south side of the gorge, in an area hitherto unexplored, yielded Acheulean assemblages and also an industry that stands close to the Developed Oldowan B from upper Bed II. There are the same small, crude, step-flaked bifacial tools associated with some choppers and a variety of light-duty tools. An additional feature of this later stage of the Developed Oldowan is the presence of large numbers of pitted anvils which also occur in Acheulean sites at approximately the same horizon.

Further consideration of the stone industries of middle and upper Bed II has led me to doubt whether the allocation to the Developed Oldowan of the industries from MNK, FC West, and the lower floor at TK is, in fact, correct. They were attributed to Developed Oldowan B on the basis of the overall tool-kit and the proportionately small numbers of bifacial tools. These are certainly scarce and their percentage in the tool aggregate is similar to that in the Developed Oldowan, but in technology and size they resemble more closely the Acheulean bifaces.

THE OCCUPATION SITES

The sites excavated in Beds I and II can be subdivided into four groups, as follows:

(a) Living floors, in which the occupation debris is found on a palaeosol or old land surface with a vertical distribution of only a few inches (0.3 ft.).

(b) Butchering or kill sites, where artifacts are associated with the skeleton of a large mammal or with a group of smaller mammals.

(c) Sites with diffused material, where artifacts and faunal remains are found throughout a considerable thickness of clay or fine-grained tuff.

(d) River or stream channel sites, where occupation debris has become incorporated in the filling of a former river or stream channel.

Sites where artifacts are sparsely scattered on a former surface or palaeosol or old land surface with a vertical distribution of only a few inches (0.3 ft.).

floors, namely DK Level 3, FLK NN Levels 1 and 3, FLK the *"Zinjanthropus"* Level, HWK East Level 1, EF–HR, the Floor at FC West, the Annexe site at SHK and the Upper and Lower Floors at TK. Two kill sites were found at FLK North, in the upper part of Bed I and the lower part of Bed II respectively. Levels of clay or fine-grained tuffs with diffused artifacts and faunal remains were uncovered at DK Levels 1 and 2, FLK NN Level 2, FLK Levels 7 and 10 to 21, FLK North Levels 1–5, HWK East Level 2, the MNK Skull and Main Sites, the reworked tuff at FC West, the tuff above the channel at SHK and the Upper and Intermediate tuffs at TK. Stream channel occurrence were found at SHK, TK and BK.

Debris on most of the living floors and at the butchering sites is unweathered and shows little evidence of orientation by running water. Material from the clays and fine-grained tuffs is also generally unweathered, but some of the artifacts and faunal remains from the river channels have been abraded. This is particularly evident in the material from the channel at SHK and from the coarser parts of the channel filling at BK.

The geological and faunal evidence now available[1] indicate that the occupation sites in Beds I and II were situated close to water, either by the lake shore or by rivers and streams. In Bed I sites DK and FLK NN were almost certainly near the margin of the lake. This is indicated by the presence of many crocodile and fish remains, bones of aquatic birds, reed casts and fossil rhizomes resembling those of papyrus. The occurrence of flamingo bones at FLK NN and at DK, at a level beneath Tuff I^B suggests that the lake water was alkaline in these areas and contained the microorganisms on which flamingos subsist. [See also the contribution of R. L. Hay to this volume.]

At TK the Lower Occupation Floor was adjacent to a stream channel in which the filling was entirely fine-grained, consisting of clays and silts. This probably represents a small stagnant stream similar perhaps to the larger clay-filled channel at BK which lies close to the main channel and where many remains of *Pelorovis* and other large mammals were found.

Re-worked concentrations of occupation debris occurred in stream-channel deposits at SHK and BK, but are unknown in Bed I. At BK large quantities of artifacts and faunal debris became incorporated in the filling of a river channel as much as 8 ft. deep and of considerable width. It seems likely that at this site and at SHK the living sites were situated on the banks of the rivers and that the debris was swept into the channels by flood waters or by a shifting of the channel. Only a small percentage of the

[1]Studies of the pollen content of Olduvai sediments are currently being undertaken by Dr. R. Bonnefille.

specimens is rolled and it is evident that most of the material came to rest before it had travelled far enough to become abraded.

A re-worked tuff and sandy conglomerate which reaches a thickness of 2 ft. in certain areas and which lies 10–15 ft. above the base of Bed II contains an exceptionally high concentration of artifacts over a wide area. It extends northwest from KK as far as FLK NN, a distance of 0.6 mile. The deposit is characterized by the presence of chert artifacts in addition to those of lava and quartz. This bed appears to represent a gravel sheet laid down during the period of erosion that followed the onset of faulting in Bed II.

The usual mode of occurrence of Oldowan and Developed Oldowan occupation sites on palaeosols or in clays or fine-grained tuffs is in contrast to the conditions at many Acheulean sites, where the material is often found in sands as, for example, at Olorgesailie, Isimila, Kariandusi, etc., and even the Acheulean sites in Bed IV*a* at Olduvai itself. (The artifacts at the early Acheulean site of EF–HR in Bed II, however, were found on a clay surface.) This presumably indicates a somewhat different environment for the Oldowan and Developed Oldowan and the Acheulean living sites, but the implications are difficult to interpret. It has been suggested that the habitation sites in sandy hollows at Olorgesailie might represent camping places in dry river beds, where it would be possible to dig for water (Isaac 1966). This explanation is possible, but the danger of the camps being swept away by sudden floods is an objection to this theory. By analogy with the Olduvai sites, it seems more likely that the Acheulean sites may have been situated on the banks of rivers or streams and that the debris was swept into the channels, as appears to have been the case at SHK and BK.

There is, unfortunately, no means either of assessing the length of time during which any camp site was occupied or of estimating the number of resident hominids. Some light may be thrown on this problem when the study of Bushman sites now being undertaken by De Vore, Lee and Yellen has been completed, since the Bushmen are undoubtedly one of the nearest modern parallels to the early hunter-gatherers of Africa.

On the basis of bushman economy it seems likely that the groups of early hominids were never very large, but comprised a sufficient number of active males to form hunting bands and to protect the females and young in case of attack. Hunting and fishing were unquestionably practiced in view of the remains found on the living floors, but it is probable that scavenging from predator kills was also a method of obtaining meat. Judging by the habits of present-day hunter-gatherers it seems likely that the major part of the diet consisted of plant foods, with the addition of small animals and reptiles, snails, grubs and insects.

The evidence of the small concentrations of crushed microfaunal bones from FLK North is not conclusive, but it is quite possible that they represent the residue of hominid faeces. If this is the case, then small mammals, lizards, chameleons and small birds were eaten—apparently whole—by the contemporary hominids. Somewhat surprisingly, fragments of snail shell have so far only been found at one locality, the MNK Skull Site, where snails almost certainly formed an article of diet.

The density of remains on the Oldowan and Developed Oldowan living floors varies greatly (see Table 1). The highest concentration was at FC West, where remains on the living floor occurred at a density of 27.6 objects per cu.ft. The remains on the Upper and Lower Occupation Floors at TK were also densely concentrated, with 21.9 and 17.4 objects per cu.ft. respectively. The lowest figure for a living floor was at Level 1, FLK NN where there were only 0.4 objects per cu.ft.

TABLE 1
Density of finds at the principal occupation sites in Beds I and II

SITE	DEPTH OF DEPOSIT (FT.)	DENSITY OF FINDS (PER CU.FT.)
BK	5.0 (av.)	1.6
TK		
Upper Floor	0.3	21.9
Lower Floor	0.3	17.4
FC West		
Tuff	2.4	2.3
Floor	0.3	27.6
MNK Main Site	4.5	1.5
EF–HR	0.3	3.9
MNK Skull Site,		
Level of H. 13	2.0	1.3
HWK East		
Levels 3–5	5.25	5.8
Level 2	6.75	0.4
Level 1 (Floor)	0.3	2.4
FLK North		
Levels 1–2	1.75	2.4
Level 3	0.5	2.8
Level 4	0.9	1.4
Level 5	1.5	1.4
Level 6	1.75	1.1
FLK *"Zinjanthropus"*		
Floor	0.3	6.5
FLK NN		
Level 1 (Floor)	0.3	0.4
Level 2	0.8	0.3
Level 3 (Floor)	0.3	3.3
DK		
Level 1	1.75	0.3
Level 2	2.25	0.7
Level 3 (Floor)	0.3	1.7

There is a marked rise in the number of artifacts in proportion to faunal remains in Middle and Upper Bed II where at all recorded sites they exceed 50.0 percent of the total remains, in contrast to Bed I, where the highest figure is 40.6 percent on the "*Zinjanthropus*" floor at FLK (see Table 2). This may be due in part to greater facility in toolmaking and greater accessibility of raw material or to the fact that remains of large mammals such as hippopotamus, *Libytherium,* giraffe and rhinoceros are more common at sites in Middle and Upper Bed II than in Lower Bed II or Bed I. It is evident that any group of hominids would require fewer such animals in order to subsist than would be the case if they relied mainly on antelopes and smaller game. The accumulation of bones at the living sites would thus be much less, particularly if the meat were generally removed from the carcass where the animal was killed, as is the

TABLE 2
Proportions of faunal remains, artifacts and manuports at living sites in Beds I and II (excluding micro-vertebrate and avian remains)

SITE	FAUNAL REMAINS		ARTIFACTS		MANUPORTS	
	NO.	%	NO.	%	NO.	%
BK	2,957	29.1	6,801	66.8	419	4.1
TK						
Upper Floor	230	4.2	5,180	93.3	139	2.5
Lower Floor	147	6.3	2,153	92.8	21	0.9
Channel	43	2.9	1,436	96.7	6	0.4
FC West						
Tuff	254	24.6	673	65.0	107	10.3
Floor	127	8.1	1,184	75.8	251	16.1
MNK Main Site	1,723	24.5	4,399	62.5	916	13.0
EF–HR	34	6.2	522	93.8	—	—
MNK Skull Site, Level of H. 13	378	33.3	689	60.7	67	6.0
HWK E						
Levels 3–5	269	7.8	1,989	57.8	1,184	34.4
Level 2	631	65.4	313	32.4	21	2.2
Level 1	425	57.3	154	20.7	163	22.0
FLK North						
Levels 1, 2	3,294	69.9	1,205	25.6	210	4.5
Level 3	1,254	85.5	171	11.8	39	2.7
Level 4	929	91.7	67	6.6	17	1.7
Level 5	2,210	92.5	151	6.3	29	1.2
Level 6	614	82.5	123	16.5	7	1.0
FLK						
Level 13	187	91.9	11	8.1	—	—
Level 15	259	93.2	9	6.8	—	—
"*Zinjanthropus*" level	3,510	57.8	2,470	40.6	96	1.6
FLK NN						
Level 1	275	89.0	16	5.2	18	5.8
Level 2	481	100.0	—	—	—	—
Level 3	2,158	96.8	48	2.2	24	1.0
DK all levels	9,984	89.3	1,198	10.7	—	—

usual custom today. Equids are also more common in Upper Bed II than in Lower Bed II or Bed I. This may be partly due to selective hunting by the hominid population, but the increase is more likely to have been caused by the change in ecology that followed the earth movements in Bed II. The Bed I lake was largely drained and areas that had previously been under water or marshes became open grassland.

Examination of the distribution of cultural and faunal remains on the occupation sites, shows that only in the case of living floors on palaeosols can any patterns of possible significance be detected. In so far as the reworked tuffs are concerned, where repeated accumulations of debris occur throughout a considerable thickness of deposit, the position is not clear, although the most likely explanation appears to be that these sites were reoccupied on successive occasions.

The pattern of distribution on the "*Zinjanthropus*" living floor at FLK is undoubtedly the most complete that was uncovered. The oblong central area of densely concentrated and very fragmentary remains was surrounded by a relatively barren zone, particularly to the south and east, beyond which debris again became more plentiful and consisted of more complete material than in the central area. It has been suggested in discussion that the central living area may have been enclosed, at least to the south and east, by a thorn fence or windbreak, which would correspond with the barren zone, while objects found on the outside might have been thrown out over the fence by the occupants of the camp.

Two adjacent circular areas containing concentrated artefacts and faunal remains were also uncovered in the uppermost occupation level at FLK North. At FC West, in Middle Bed II, a living floor occurred on a clay palaeosol similar to those in Bed I. Here, again, the artifacts and faunal remains were concentrated within two roughly circular areas, lying close together and reminiscent of the pattern uncovered at FLK North. At TK, however, although the debris on both the Upper and Lower Occupation Floors occurred on palaeosols and was confined to limited areas, there was no discernible pattern in the distribution. It is perhaps significant that the faunal remains and a proportion of the artifacts from this site are weathered, indicating that they were exposed on the surface longer than usual before being buried and so were possibly displaced from their original positions.

It is probable that the stone circle at DK formed the base of a rough windbreak or simple shelter. The two factors that are most suggestive of an artificial structure are the small heaps of piled-up stones that form part of the circle and the fact that occupation debris did not appear in comparable density within the circle and in the surrounding area.

The stone circle and the distribution patterns of the debris on the three living floors described above, where little or no disturbance appears to have taken place after the camps were abandoned, suggest that some form of crude shelter was probably constructed at Oldowan and Developed Oldowan living sites. This may well have been no more than a protective thorn fence, but the existence of some factor affecting the horizontal diffusion of debris on the living floors is indicated.

Some information concerning the methods by which animals were hunted and killed has come to light at certain sites. The discovery of skeletons of an elephant, a *Deinotherium* and remains of *Pelorovis oldowayensis* embedded in clay and associated with numbers of stone tools suggests that these animals may have been deliberately driven into swamps by the early hominids. The possibility that they became engulfed accidentally cannot be ruled out entirely, but the repeated discovery of large animals that appear to have died under identical conditions is suggestive of hominid activity. Remains of two herds of antelopes, similarly embedded in clay, have also been found: one in Bed II at SHK and the second in Bed IV*a* at the Fifth Fault.

The depressed fractures on the three frontlets of *Parmularius altidens* from FLK North can be taken to indicate that the animals were killed by means of a blow, delivered at close quarters, since the fractures are accurately placed above the orbits, on the most vulnerable part of the skull.

Although there is no direct evidence that spheroids were used as bolas, no alternative explanation has yet been put forward to account for the numbers of these tools and for the fact that many have been carefully and accurately shaped. If they were intended to be used merely as missiles, with little chance of recovery, it seems unlikely that so much time and care would have been spent on their manufacture. (Possible methods by which spheroids were used are discussed below.)

The raw materials employed for making artifacts have been discussed by R. L. Hay (1971) and need only be briefly referred to here.[2] Except in the lower part of Bed II, when chert became available for a time, the artifacts are made almost exclusively from fine-grained lavas, quartz and to a lesser extent from quartzite (see Table 3). In Bed I, with the exception of the *"Zinjanthropus"* level at FLK I, 80–94 percent of all the heavy-duty tools such as choppers, polyhedrons, discoids, spheroids, etc., are made from lava, while in the Middle and Upper Bed II quartz and quartzite are the most common material. The increase in the use of these materials is first evident in Level 2 at HWK East, in Lower Bed II, although there are two sites in the lower part of Middle Bed II where lava predominates, namely the MNK Skull Site and the early Acheulean site of EF–HR.

Quartz and quartzite (and chert when available) are the most usual materials for the light-duty tools, light-duty utilized flakes and *débitage.*

---

[2]Fieldwork carried out in conjunction with R. L. Hay during 1970 has resulted in the identification of the sources of the eight principal types of raw materials used for toolmaking in Beds I–IV. Some varieties were transported as far as 8–12 miles to the living sites.

DK, however, is an exception, since lava is the predominant material for all the artifacts with the exception of the light-duty utilized flakes. This is also the only known site where the amount of lava *débitage* is sufficient to suggest that the heavy-duty tools may have been made on the spot. At other sites where the majority of heavy-duty tools is of lava the scarcity of lava *débitage* indicates that the tools were made elsewhere, presumably at the sources of the raw materials.

The proportionate occurrences of the various tool types in Oldowan, Developed Oldowan and early Acheulean assemblages in Beds I and II are shown in the histograms of Figure 1. It will be seen that the predominant tools of the Oldowan are choppers, among which it is possible to distinguish several distinct types. There is also a variety of other heavy- and light-duty tools.

Five types of choppers have been described: side, end, two-edged, pointed and chisel-edged. Unifacial examples are rare and the most common form is a bifacially flaked side chopper. The maximum development of choppers both in diversity of types and proportionate abundance within an assemblage, is to be found in Upper Bed I and the Lower Bed II, where they amount to 78.7 percent of the tools. In Developed Oldowan A and B assemblages from Middle and Upper Bed II they do not exceed 35 percent and fewer types are represented. At the early Acheulean site of EF–HR, also in Middle Bed II, they amount to under 16 percent of the tools. Very little evolutionary change can be observed in the choppers of Beds I and II except in Upper Bed II, particularly at BK, where many of the choppers are made on blocks of quartz. In these specimens the cutting edges are often evenly curved and extend round a greater part of the circumference than is usual in choppers made on cobbles. The butts are generally formed by a flat vertical face, blunted on the upper and lower edges.

The tools termed "proto-bifaces" are always rare and are restricted in time span from Upper Bed I to the Sandy Conglomerate, the lowest horizon of Middle Bed II. They do not conform to any particular pattern or technique of manufacture but appear to represent attempts to achieve a rudimentary handaxe by whatever means was possible.

It has been shown that no true bifaces occur in Bed I or in Lower Bed II. In fact, none has been found below Tuff II^{B}. They occur first in the upper part of Middle Bed II, in both early Acheulean and Developed Oldowan B industries.

The bifaces from Bed II are not consistent in type and exhibit such a degree of individual variation that it is impossible to classify them satisfactorily on the basis of typology. In so far as is possible, the series from each site has ben divided into broad groups. The descriptions will not be repeated, but the occurrence of a few cleavers in Bed II, at both Acheulean and Developed Oldowan sites, may be noted, since these tools were previously believed to occur only in Bed IV*a*.

TABLE 3
Showing the incidence (as percentages) of the various types of raw materials in the five artifact groups from excavated sites in Beds I and II

	A Choppers, "proto-bifaces," polyhedrons, discoids, spheroids, sub-spheroids, heavy-duty scrapers			B Light-duty scrapers, burins, awls, laterally trimmed flakes, sundry small tools		
	Lava	Q/zite	Other	Lava	Q/zite	Other
BK	17.5	78.1	4.4 G, P	0.7	98.8	0.5 P, C
TK						
Upper Floor	18.9	80.0	1.1 G	—	100.0	—
Lower Floor	13.2	84.2	2.6 G	—	100.0	—
SHK, all levels	19.0	80.0	1.0 G, P, WT	0.9	99.1	—
FC West						
Tuff	16.9	77.5	5.6 G, P	—	100.0	—
Floor	29.9	67.8	2.3 G	—	100.0	—
MNK Main Site	20.3	77.0	2.6 G, P	4.3	95.7	—
EF–HR	72.7	26.1	1.1 P	66.7	33.3	—
MNK Skull Site	73.2	24.4	2.4 G	50.0	50.0	—
FLK North, Sandy Conglomerate	21.8	74.7	3.5 C	—	—	100.0 C
HWK East						
Level 4	27.6	59.0	13.1 C, M	—	7.4	92.6 C
Level 3	35.3	58.0	6.7 C, G	—	18.2	81.8 C
Level 2	31.3	67.2	1.5 C	14.3	28.6	57.1 C
Level 1	80.0	20.0	—	100.0	—	—
FLK North						
Levels 1/2	82.1	17.2	0.7 G	6.7	93.3	—
Level 3	85.7	14.3	—	—	—	—
Level 5	83.3	16.7	—	—	—	—
FLK, "*Zinj*" Level	47.4	50.0	2.6 G	—	100.0	—
DK all Levels	94.3	5.7	—	64.5	35.5	—

(C = Chert, G = Gneiss, P = Pegmatite, O = Obsidian, M = Migmatite, WT = Welded Tuff.)

Polyhedrons are never a common tool type, but they are more plentiful in the Oldowan than in later assemblages and are best represented at DK, on the "*Zinjanthropus*" floor at FLK and in Levels 3 and 4 at FLK North. There is virtually no difference in the percentage of polyhedrons at other Oldowan or Developed Oldowan sites, or at the lower Acheulean site of EF–HR, but they become noticeably scarce at Developed Oldowan B sites in Upper Bed II.

Spheroids and subspheroids must be regarded as tools of considerable significance in the various assemblages under review. They occur at all sites and at all levels with the exception of FLK, although in Bed I they are relatively scarce and a total of only twenty-four specimens was recovered from the four excavated sites. The first appearance of symmetrical stone balls, smoothed over the entire surface, is at FLK North in Upper Bed I. They are always rare, however, and the spheroids are generally fa-

C			D			E		
Anvils, hammerstones, utilised cobbles, nodules, and blocks			Light-duty utilized flakes, etc.			*Débitage*		
Lava	Q/zite	Other	Lava	Q/zite	Other	Lava	Q/zite	Other
65.1	29.4	5.5 G, WT	0.2	99.6	0.2 G	4.2	95.6	0.2 G, P
55.6	44.4	—	0.8	99.2	—	2.6	97.4	—
27.3	72.7	—	—	100.0	—	0.1	99.9	—
7.7	92.3	—	2.4	96.7	0.8 C	17.5	82.4	0.1 G
93.8	6.2	—	—	100.0	—	2.6	97.4	—
86.4	12.5	1.1 P	2.6	97.4	—	5.9	94.0	0.1 C
81.1	17.2	1.7 G	2.2	97.8	—	2.3	96.6	0.6 G
75.0	25.0	—	60.0	30.0	10.0 C	63.5	35.5	1.0 C
84.7	12.5	2.8 G	—	100.0	—	9.7	90.3	—
92.6	7.4	—	—	22.2	77.8 C	8.2	57.4	34.4 C
90.0	7.5	2.5 C	1.2	50.0	48.8 C	0.2	71.6	26.4 C
91.4	7.5	1.1 G	4.3	40.0	55.7 C	3.3	69.6	27.1 C
96.3	3.7	—	5.9	70.6	23.5 C	6.1	86.7	7.2 C
94.0	4.5	1.5 G	—	100.0	—	45.5	54.5	—
74.0	26.0	—	5.9	94.1	—	17.4	82.6	—
76.3	23.7	—	—	100.0	—	15.0	85.0	—
75.9	24.1	—	—	—	—	22.8	77.2	—
82.3	17.7	—	—	100.0	—	3.4	96.5	—
92.6	7.4	—	47.1	52.9	—	64.0	35.8	0.1 C

ceted, although projecting ridges are battered and partly reduced. Both spheroids and the less symmetrical specimens that have been termed subspheroids become markedly more numerous in Middle and Upper Bed II at all Developed Oldowan B sites, where they are never less than 20 percent of the tools and where the range in size is also more extensive than in the earlier assemblages.

In Figure 2 the range in weight has been plotted for the spheroids and subspheroids from three sites. They consist of seventeen specimens from HWK East, Levels 3, 4 and 5 (in the lower part of Middle Bed II), 159 from the Main Site at MNK (in the upper part of Middle Bed II) and 199 from BK (in Upper Bed II). It will be seen that there is a greater number of specimens weighing less than 4 oz. at BK than at either of the earlier sites, where the most common weight is between 4 oz. and 1 lb. The earliest series from HWK East, Levels 3–5, is also noticeably more limited

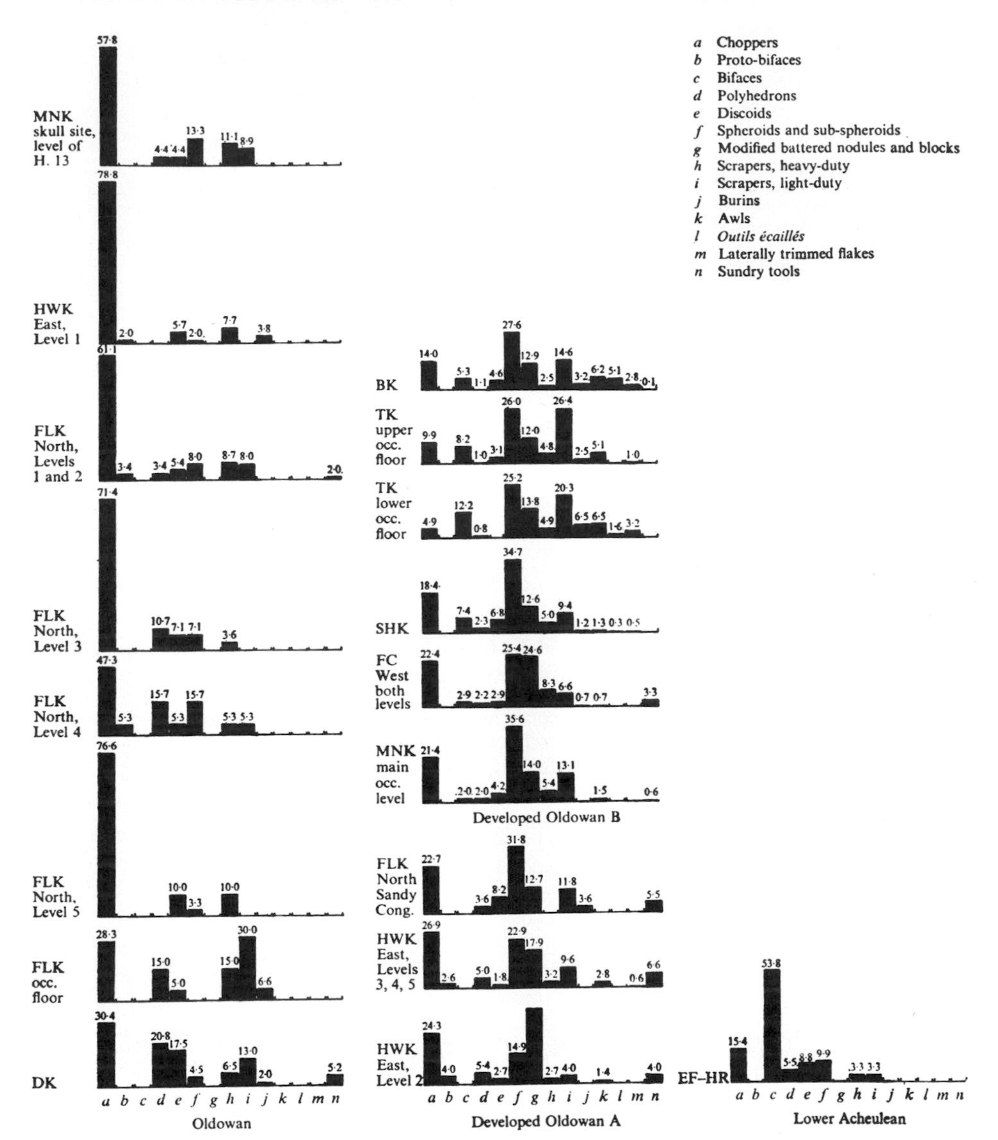

FIGURE 1
Histograms of tool percentages in Oldowan, Developed Oldowan A and B and early Acheulean assemblages in Beds I and II.

in range. No specimens from this site exceed 3 lb. in weight, whereas in both the later series a few examples weigh considerably more. Although any interpretation of the use to which these tools were put must be largely speculative, they appear more likely to have served as some form of missile than any other purpose. Their use as bolas stones has been strongly supported by L. S. B. Leakey and may well be correct. Experiments by R. E. Leakey in this connection have shown that a pair of spheroids, each

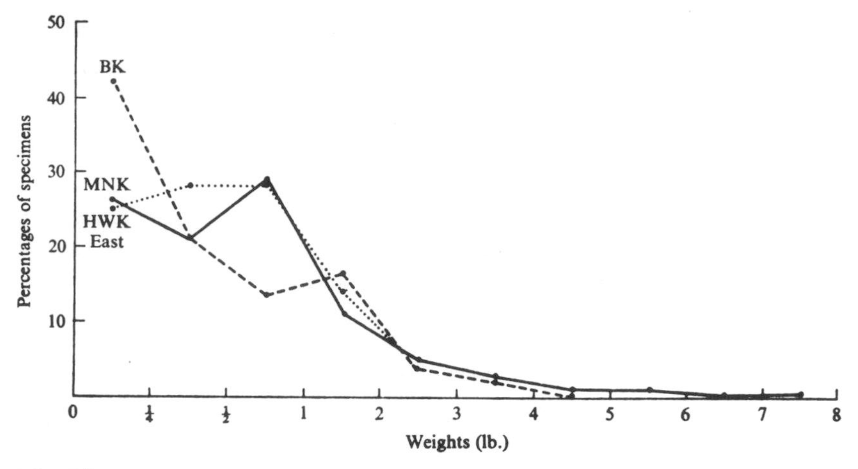

FIGURE 2
Graph to show the weights of spheroids and subspheroids from three sites: HWK East Levels 3, 4, 5; MNK Main Occupation Site; and BK.

attached to a piece of cord about a yard long, entangle more effectively round upright posts than does a group of three; a pair is also easier to aim accurately.

The fact that spheroids are rare in Bed I but occur in relatively large numbers in Bed II probably reflects an important change in the tool requirements, which may possibly have been the introduction of missiles as hunting weapons.

The artifacts that have been termed "modified battered nodules and blocks" are also numerous at Developed Oldowan sites. Unlike the utilized nodules and blocks of the Oldowan, they show some degree of flaking and shaping in addition to extensive battering. The amorphous appearance of the whole group at first suggests reject material, but this does not seem to be consistent with the evidence of utilization.

The scrapers from the Oldowan, the Developed Oldowan and the early Acheulean fall into two groups, namely heavy-duty and light-duty, with virtually no borderline cases. Certain heavy-duty scrapers would probably be termed "push planes" by some workers, but since they are morphologically identical to specimens in the light-duty group, except for size, the term "scraper" has been preferred. Heavy-duty scrapers are usually less numerous than light-duty specimens. Side scrapers are the most common form in both groups. A number of different types of light-duty scrapers are well represented at all Developed Oldowan B sites in Upper Bed II. They are also numerous on the "*Zinjanthropus*" living floor, where they constitute the highest percentage of any of the six tool categories represented, even exceeding choppers. Although burins are not common in any assemblage, they occur at the majority of Oldowan and Developed Oldowan sites and also at the early Acheulean site of EF–HR.

They are most numerous at TK and include a variety of angle burins that would not be out of place in an Upper Palaeolithic industry. The small retouched points that have been termed "awls" have not been found at any Oldowan site. They occur first in Lower Bed II, at HWK East, Level 2, where they are made exclusively of chert. They are rare in Middle Bed II but become comparatively common in Upper Bed II at TK and BK and exceed the number of burins at these two sites. A few *outils écaillés* are known from SHK, but they only occur in any quantity in Upper Bed II, at BK and on the Lower Occupation Floor at TK. They are made exclusively from quartz and quartzite and both single- and double-ended examples are present.

There remains only one further group of light-duty tools to be mentioned, namely laterally trimmed flakes. These occur for the first time in Level 2 at HWK East, in common with the awls. They are always relatively scarce but are more common at SHK, TK and BK than at any other sites.

The various forms of utilized material require no comment with the exception of the light-duty flakes. These are common at nearly all sites throughout Beds I and II and are made almost exclusively from quartz and quartzite, except for the levels where chert was available. There is no indication that these flakes were made to accord with any particular pattern, but three types of utilization are present which appear to be related to the form of the edge, suggesting the possibility of some specialization in use. On rounded edges the chipping is usually seen on one face only, as in scrapers. Similarly, on notched edges, the chipping within the notches is also usually on one face. On relatively straight edges, however, it is generally present on both faces. There are also a number of flakes and other fragments with haphazard chipping.

Evidence of utilization is usually present on the working edges of choppers and other heavy-duty tools. It varies from light chipping to extensive battering which has completely blunted the edges. Some specimens also exhibit small, deeply indented crushed notches, which appear to have been caused by a grinding action against a hard substance.

The greater part of the *débitage* from the occupation sites in Beds I and II consists of broken flakes and chips. Complete flakes are relatively scarce and only the ten levels listed opposite have yielded a sufficient number for analysis to be of any significance. End-struck flakes are invariably more numerous than side-struck, particularly in the nine Oldowan and Developed Oldowan assemblages where, on the average, they amount to 68 percent. Side-struck flakes are proportionately more common at the early Acheulean site of EF–HR, where only 54 percent are end-struck.

Although the dimensions and the angles of the striking platforms of the *débitage* flakes have been recorded, the variations that exist appear to be largely dependent on the nature of the raw materials employed and are probably of little significance. It can be stated, however, that the average

SITES	TOTAL NO. OF COMPLETE FLAKES	END-STRUCK FLAKES (%)
BK (1963)	652	72.0
TK		
Upper Floor	183	62.0
Lower Floor	84	68.1
SHK	449	72.0
FC West, Floor	99	71.0
MNK Main site	226	77.2
EF–HR	147	54.0
HWK East, Level 3	88	73.0
FLK North, Levels 1, 2	178	62.6
FLK, the "*Zinjanthropus*" level	258	68.5
DK	242	67.0

figure for the angles of the striking platforms for the lava flakes tends to be higher in the early Acheulean and Developed Oldowan series than in the Oldowan. The average size for both the lava and quartz side-struck flakes is also greater in the early Acheulean, but a few unusually large end-struck lava flakes from the Main site at MNK exceed those from EF–HR. The average size of the quartz series, however, is less. Attention has been drawn to the quartz flakes from Oldowan and Developed Oldowan assemblages in which the striking platforms are reduced to a minimum and consist merely of a point of impact. These contrast very strongly with many of the early Acheulean lava flakes in which the striking platforms—and even the bulbs of percussion—extend across almost the whole width of the flakes. A few similar specimens occur in Developed Oldowan B assemblages. They appear to be typical roughing-out flakes derived from the manufacture of bifaces.

It seems likely that a proportion, at least, of the sharp flakes among the *débitage* may have served as small tools, possibly for cutting meat or other soft substances that would leave no evidence of use on the edges. That they are not merely waste material is indicated by the occurrence of large numbers of such flakes, nearly always made of quartz and quartzite, at sites where the majority of heavy-duty tools is of lava and where the number of quartz and quartzite core tools is insufficient to account for the number of flakes.

An interesting present-day example of unretouched flakes used as cutting tools has recently been recorded in South-West Africa and may be mentioned briefly. An expedition from the State Museum, Windhoek, discovered two stone-using groups of the Ova Tjimba people who not only make choppers for breaking open bones and for other heavy work, but also employ simple flakes, un-retouched and un-hafted, for cutting and skinning (McCalman and Grobbelaar 1965).

The question as to whether certain specimens should be described as cores is largely subjective; but if the term "core" is intended to imply a simple nucleus which was put to no further use after flakes had been struck from it, then, in the writer's opinion, it is inapplicable to any of the material from Beds I and II, with the possible exception of a few specimens from SHK and BK. It is true, however, that the majority of choppers, polyhedrons, discoids and bifaces, as well as many of the heavy-duty scrapers, are in fact made on "cores," but this is tacitly implied in the descriptions of any tools that are made on cobbles or blocks of raw material, as distinct from flakes.

In concluding this review of the lithic material from Oldowan and Developed Oldowan Sites the grooved and pecked phonolite cobble found in Upper Bed I at FLK North must be mentioned. This stone has unquestionably been artificially shaped, but it seems unlikely that it could have served as a tool or for any practical purpose. It is conceivable that a parallel exists in the quartzite cobble found at Makapansgat (Dart 1959) in which natural weathering has simulated the carving of two sets of hominid—or more strictly primate—features on parts of the surface. The resemblance to primate faces is immediately obvious in this specimen, although it is entirely natural, whereas in the case of the Olduvai stone a great deal of imagination is required in order to see any pattern or significance in the form. With oblique lighting, however, there is a suggestion of an elongate, baboon-like muzzle with faint indications of a mouth and nostrils. By what is probably no more than a coincidence, the pecked groove on the Olduvai stone is reproduced on the Makapansgat specimen by a similar but natural groove and in both specimens the positions of the grooves correspond to what would be the base of the hair line if an anthropomorphic interpretation is considered. This is open to question, but nevertheless the occurrence of such stones at hominid sites in such remote periods is of considerable interest.

Before considering industries that may be analogous to the Developed Oldowan and early Acheulean of Bed II, the grounds for differentiating these two industrial complexes must be stated. The Developed Oldowan appears to represent an uninterrupted local continuation from the Oldowan, in which the same tool forms persist, with the addition of some new elements and an increase in others that were rare in the Oldowan. These are notably bifaces, various small flake tools, spheroids and subspheroids. There was thus an overall increase in the tool kit. This is clearly demonstrated by the fact that, while the average number of tool types for seven Oldowan levels in Bed I and the base of Bed II amounts to only six, the average figure has risen to over ten for nine Developed Oldowan levels in Bed II. The enlargement of the tool kit is already evident in the industries of the lower part of Middle Bed II (Developed Oldowan A), although this horizon lacks bifaces. Up to and including this level there is no suggestion whatever of duality in the known lithic assemblages which are remarkably constant in character, in spite of minor differences in the proportionate representation of certain tool types.

In Middle Bed II, however, the first unquestionable bifaces appear, including a few cleavers as well as handaxes. At some sites they occur in sufficiently large numbers (over 50 percent of the tools) for the term "Acheulean" to be applied without hesitation to the industry. (The term "early or lower Acheulean" as used here denotes those stages that were previously termed "Chellean" or "Abbevillian.") At other sites bifaces occur in small numbers (an average of 6 percent of the tools) in assemblages that are otherwise wholly Developed Oldowan in character.

No bifacial tools that could be considered intermediate between the "proto-bifaces" of Upper Bed I and the true bifaces of Middle Bed II have come to light. This is not necessarily significant, since it is doubtful if the evidence for such a transition would be preserved. The possibility cannot be overlooked, however, that the Acheulean was intrusive to the area, as indeed *Homo erectus* appears to have been.

While there will be general agreement in applying the term "Acheulean" to those industries containing 50 percent or more bifaces, it may be argued that the term should also be applied to the contemporary assemblages where a low percentage of bifaces is found in an industry otherwise characteristic of the Developed Oldowan. This is perhaps a matter of opinion, but it does not appear reasonable to the writer to consider bifaces or any other single tool as an all-important index fossil to the exclusion of the predominant elements of an industry. For the present, the proposal put forward by M. R. Kleindienst (1962) that there should be not less than 40–60 percent of bifaces if an industry is to be classed as Acheulean seems acceptable, although a more specific diagnosis would be of advantage, but this will only become possible when detailed analysis of a substantial number of assemblages from sealed horizons has been carried out.

In addition to the overall dissimilarity of the early Acheulean and the Developed Oldowan B industrial complexes, there is a marked difference in the nature of the bifaces. This is not so much in typology, since there is too great a degree of individual variation for valid typological classification, but in size, morphology and method of manufacture. The contrast in size, breadth/length and thickness/length ratios is shown in the accompanying graphs (Figures 3–5). In Figure 3 the mean diameters of bifaces from EF–HR and from an Acheulean assemblage in Bed IV*a* (TK, Fish Gully) are shown, together with those of the bifaces from BK and SHK—the only two Developed Oldowan sites to yield a sufficient number of specimens for statistical purposes.[3] It will be seen that in both

[3]The series from Bed IV is predominantly of quartz and that from EF–HR predominantly of lava. Both materials are almost equally represented at BK and SHK.

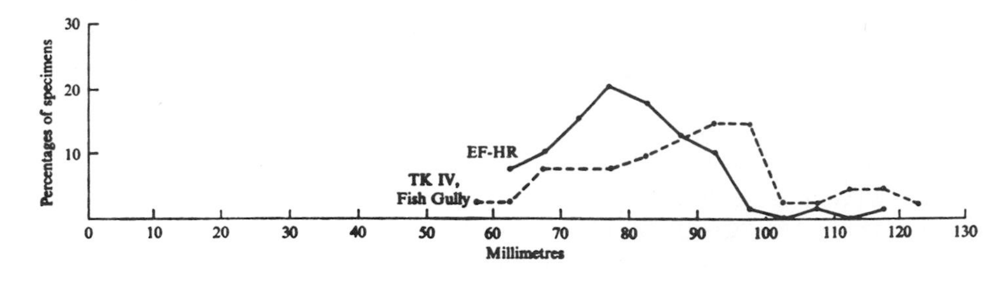

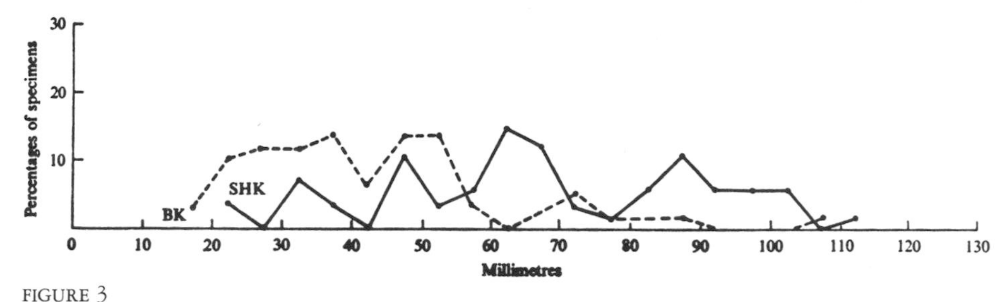

FIGURE 3
Graphs showing the mean diameters of Acheulean bifaces from EF-HR and TK IV, Fish Gully, and of Developed Oldowan bifaces from SHK and BK.

the Acheulean groups the curves are regular and follow a similar pattern, although the Bed IV series tends to be slightly larger than that from Bed II. In the Developed Oldowan groups, however, the curves are entirely haphazard and suggest that the size of the tools depended largely on whatever piece of raw material happened to be available. About half the specimens, however, are smaller than any in the Acheulean series.

The breadth-length ratios for the Bed IV Acheulean bifaces follow a simple curve, the maximum number of specimens having a ratio of between 40 and 50 percent. The series is more restricted in range than the Early Acheulean, which includes a number of specimens that are wider.

The thickness/length ratios of the Bed IV bifaces show an entirely even curve in the lower range, corresponding closely to that of the early Acheulean, although the latter includes a number of thicker specimens. In the Developed Oldowan the curve for SHK is remarkably even, but the ratio for the maximum number of specimens lies between 40 and 55 percent, a higher average than for either of the Acheulean groups.

There is also a considerable difference in the numbers of trimming flake scars on the bifaces from the early Acheulean and Developed Oldowan industries. Omitting quartz specimens, in which it is virtually impossible to count the number of flake scars accurately, the average for the early Acheulean series if 9.7, for the SHK series 18, and for the BK series 16.5.

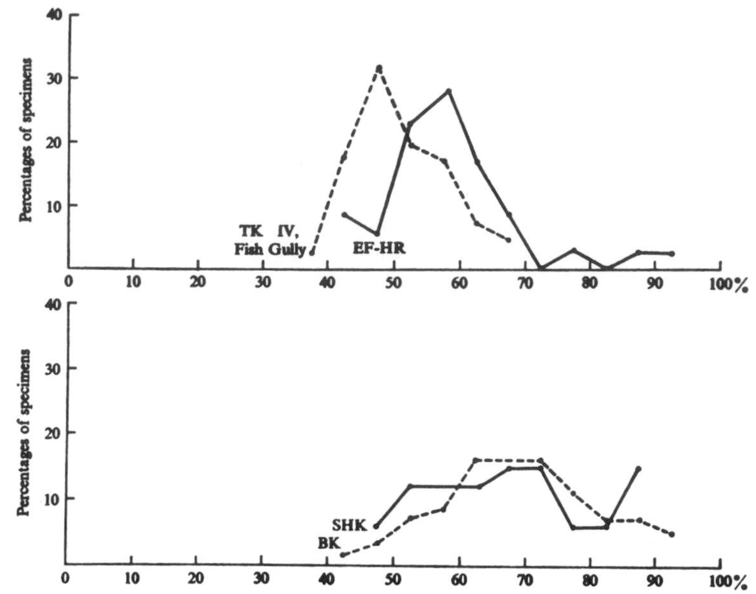

FIGURE 4
Breadth/length ratios of Acheulean bifaces from EF-HR and TK IV, Fish Gully, and of Developed Oldowan bifaces from SHK and BK.

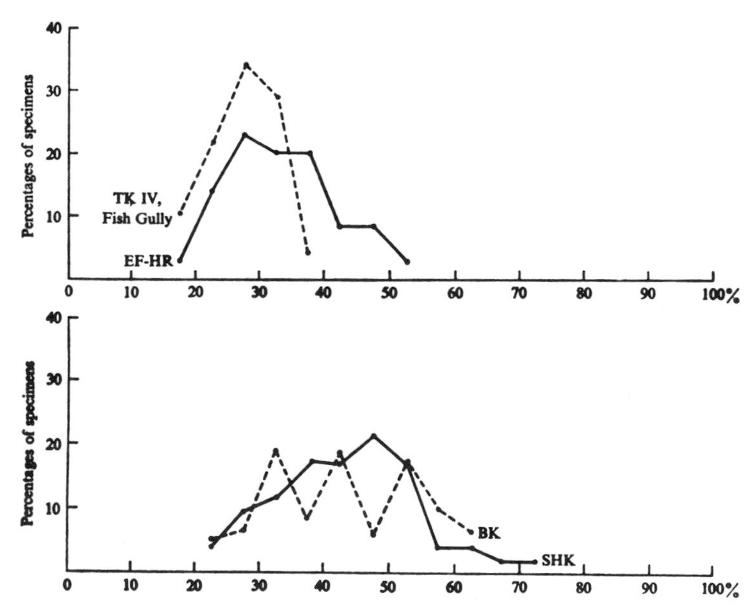

FIGURE 5
Thickness/length ratios of Acheulean bifaces from EF-HR and TK IV, Fish Gully, and of Developed Oldowan bifaces from SHK and BK.

Although the number of bifaces recovered from other Developed Oldowan sites is too small to permit statistical analysis, the few specimens from the Main site at MNK and from the Lower Floor at TK conform more closely in size and morphology to the early Acheulean than to Developed Oldowan series from BK, SHK and the Upper Occupation Floor at TK. This is an anomaly which cannot be explained for the present, although there are several possible interpretations.

It has been suggested that the industries termed here Developed Oldowan might in fact belong within the Acheulean and that their dissimilarity from the classic Acheulean might be due to different environment or ecology. An alternative interpretation put forward is that they might reflect a form of activity different from that at sites where bifaces are the predominant tools. While both hypotheses are possible, the evidence at Olduvai does not lend support to either since Acheulean and Developed Oldowan assemblages have been found under what appear to be identical environmental conditions. Furthermore, although the period of habitation certainly varied at each site, the usual criteria for living sites are always present i.e. tools, *débitage,* utilized lithic material and also a variety of broken animal bones.

Consideration of all the factors, therefore, would seem to indicate that the two dissimilar industrial complexes of Middle and Upper Bed II should be interpreted, for the present, as representing two distinct cultural traditions, perhaps made by two different groups of hominids. The position will almost certainly be clarified by further field work, but the existence of two contemporary cultural streams does not seem in any way impossible, particularly when it is known that *Homo erectus, Homo habilis* and a robust australopithecine all existed at Olduvai during Bed II times. Two of these, at least, must supposedly have been toolmakers and it would be surprising if their industries proved to be identical. If *Homo habilis* is accepted as the maker of the Oldowan—and in view of the cumulative evidence, this seems to be an inescapable conclusion—then it would be reasonable to assume that he was also responsible for the Developed Oldowan. Although both *Homo habilis* and the Oldowan had long existed in the area, there is as yet little to suggest the existence of antecedents either for the Acheulean or for *Homo erectus.* It is tempting therefore to link the two, but since the calvaria from LLK was not associated with any industry, this interpretation remains hypothetical for the present.

When analogies for the industries of Beds I and II at Olduvai are sought elsewhere in Africa the lack of total assemblages that can be used for comparative purposes proves a serious handicap. For the Oldowan itself, as represented in Bed I and the base of Bed II, it is doubtful whether any comparable assemblages exist. Choppers, "pebble tools" or *"galets aménagés"* are widely known and some are certainly from early contexts.

But since it has been amply demonstrated at Olduvai that these tools form only one element of the Oldowan, although admittedly the most obvious and the most abundant, it is of little value to consider them alone. This is especially true when it is realized that they are a tool form which has persisted from the Villafranchian until the present day. (McCalman and Grobbelaar 1965, and M. D. Leakey 1966).

P. Biberson has, in fact, separated the Moroccan "Pebble culture" into three stages: a lower, middle and upper (Biberson 1961). Only about 100 specimens are known for the earliest phase and these are said to consist largely of unifacial choppers. Many of the tools attributed to this stage have been included on the grounds that they are the most heavily rolled specimens in assemblages from raised beaches, which also contain moderately rolled and fresh groups. The site of Tardigueter-Rahla, however, appears to be promising from the point of view of obtaining more material under good stratigraphic conditions. It has yielded a stage of the "Pebble Culture' considered to be Middle Villafranchian in age (Biberson 1967).

The succeeding stage, for which a great deal more material is known, is dated as Upper Villafranchian. There is a greater variety in the forms of the choppers and many are bifacial. Polyhedrons also occur. The specimens illustrated from this stage undoubtedly resemble some from Olduvai, but unfortunately no tools other than choppers and polyhedrons are described. Several sites in the neighbourhood of Casablanca are said to be exceedingly rich in artifacts of this stage and would certainly also repay fuller investigation.

In the last stage of the Moroccan "Pebble Culture," recovered from the basal *"poudingue"* at the Sidi Abderrahman Extension quarry, choppers continue to occur, but a number of the figured specimens resemble crude bifaces. They approach true bifaces more closely than do the Oldowan tools that have been termed "proto-bifaces." Biberson questions whether this stage might not be more correctly included in the beginning of the Acheulean. This, in fact, appears to be the case. (Biberson 1961).

Undoubtedly the closest analogies to the Developed Oldowan B industrial complex from Olduvai is to be found in the collections of artifacts from Swartkrans and from the West Pit, Sterkfontein, both of which I have been able to examine personally during recent visits to South Africa through the kindness of Dr. C. K. Brain and Mr. Alun Hughes. Since work is now in progress at these sites, the collections will undoubtedly be augmented within the near future, but even in the somewhat limited series now available, the resemblances are very striking, both in the tool types represented and in their proportionate occurrence. (M. D. Leakey 1970a).

My classification of the Sterkfontein material differs in certain respects from the description published by R. Mason (1962a), mainly in that most of the specimens which he regarded as cores I would class either as choppers or polyhedrons. These, in my view, are the two most common tool types. Mason also considers that the scarcity of bifaces and lack of flakes indicates that the assemblage is incomplete. The bifaces, however, occur in almost the same proportions as to those in the Developed Oldowan. As regards the lack of flakes, it is now generally known that flakes derived from the manufacture of heavy-duty tools are lacking or very scarce at most early living sites, where the tools appear to have been introduced in a finished or half-finished state.

An aspect of the lithic material from the australopithecine caves that has received little attention so far is the number of small sharp-edged chert fragments which occur in the breccia alongside the orthodox tools and the faunal remains. Many of these have been chipped on the edge and it is possible that they are the equivalent of the small quartz flakes from Oldowan and Developed Oldowan sites which appear to represent small cutting tools. Unfortunately, the type of chert found in the dolomite formation of the Transvaal does not flake with a conchoidal fracture and there appears to be no sure means of determining whether these fragments were artificially fractured. When the position of material in the undisturbed breccia comes to be recorded, the distribution of the chert in relation to the heavy-duty tools may prove informative.

Although the lithic material from the australopithecine caves stands closest to the Developed Oldowan at present, the site at 'Ubeidiya in Israel may well provide even closer parallels when it is more fully explored. This site was similarly situated near the shore of a fluctuating lake, so that a number of successive occupation levels were buried rapidly enough for the debris to be preserved in place, as was the case at many of the Olduvai sites. The artifacts from 'Ubeidiya can be divided into two main groups. The earliest was named by the late Professor M. Stekelis the "Israel Variant of the Oldowan" (Stekelis 1966). It contains choppers of several types, polyhedrons, spheroids and cuboids, but few bifaces. The second group contains a high proportion of bifaces as well as some choppers and a few polyhedrons. Although this site was first claimed to be of Villafranchian age, the faunal material as well as the nature of the stone industries indicate that it is more likely to be early Middle Pleistocene and therefore approximately contemporary with the lower part of Middle Bed II at Olduvai.

Yet a third site that may provide a parallel with the Oldowan or Developed Oldowan is Melka Kontouré in Ethiopia. This site lies in the Awash valley, approximately 50 km. from Addis Ababa, and was found by G. Dekker in 1963. It consists of a series of sedimentary deposits within a former lake basin (Chavaillon and Taieb 1968). The deposits reach a known thickness of 40 m. and are predominantly lacustrine, consisting of tuffs, clays, and sands. Three distinct levels containing occupation debris are known. The industry from the highest level appears to be an Acheulean of farily advanced facies, but the lowest level has so far yielded only choppers, polyhedrons, spheroids, etc. (Chavaillon 1967). Excavations now being carried out will certainly yield a larger series of specimens than is available at present and make it possible to determine more precisely the nature of this industry. The dating also remains to be established satisfactorily.

The Acheulean assemblages from two sites at Peninj, on the western side of Lake Natron, bear a strong resemblance to the early Acheulean from Olduvai, as has been pointed out by G. L. Isaac (1965). The Peninj series, however, contains a higher proportion of cleavers and none of the specimens appears to be so highly finished as a few from EF–HR. Both at Peninj and at Olduvai the tools have been shaped by means of a minimum of well-directed flakes and the average number of 9.7 flake scars per specimen is identical in both series. The Peninj sites are within 50 miles of Olduvai and on the evidence of the geology and the faunal remains they appear to be approximately contemporary with Middle Bed II.

The industries from two sites in South Africa, Kliplaatdrif and Three Rivers, both of which are in the Transvaal, do not seem unlike the Acheulean of Bed II, Olduvai, judged on the basis of the published description (Mason 1962b). Comparison of both these assemblages is made difficult by the fact that the method of classification is not the same as that used for the Olduvai material. For example, specimens that would have been classed as choppers by the writer are listed as cores, in spite of evident utilization. The bifaces appear to be very similar to the East African series, with minimal, bold flaking; it is interesting to note that Mason comments that they have rarely had more than ten flakes removed, as in the series from Olduvai and Peninj.

There are also resemblances with the early Acheulean of North Africa, particularly in some of the bifaces from the Sidi Abderrahman quarry near Casablanca. Although the whole collection from this site was originally lumped together and described as "Clacto-Abbevillian" by Neuville and Ruhlmann (1941), Biberson considers that it is probably a mixed series containing material of different ages. He has divided it into three groups, on the basis of *état physique.* In the most heavily rolled series, choppers and faceted spheroids are said to predominate, but in both the slightly rolled and fresh series bifaces are the most common tool type. They include a series of trihedral specimens which appear very similar to those from the East African early Acheulean sites, characterized by a triangular cross-section near the tip, although the general form of the tools is variable. This resemblance, however, may well be coincidental, since the Moroccan specimens are made on large flakes struck from boulders as was the case at EF–HR, Olduvai. It is also noticeable that all the specimens illustrated for this stage exhibit the minimal bold flaking that appears to be characteristic of the early Acheulean bifaces from East and South Africa that have been referred to here.

THE MODIFIED BONES

The series of utilized and flaked mammalian bones that has been described in chapter VIII (Leakey 1971) indicates that bones were sometimes used as tools and that occasionally parts of massive bones were employed as a substitute for stone in toolmaking. The presence of relatively large flake scars and the splitting of the articular ends of *Libytherium* limb bones and others of comparable size are the factors which indicate most clearly that these bones have been modified by human agency. Certain types of flaking can undoubtedly be caused by carnivores who are also capable of splitting fair-sized limb bones, including the extremities. But even those of present-day giraffe are seldom split right through and these do not compare in robustness with the bones of *Libytherium*. It would also be impossible for animals to detach flakes 10 cm. long, or more, in the course of chewing; percussion would be required. Smaller flakes are, however, frequently removed by gnawing and a series of such flakes often simulates a "utilized" edge. Other forms of apparent utilization brought about by natural means can also be deceptive. Brain (1967) has shown that partial weathering in a dry climate and subsequent movement in sand, such as can be caused by the trampling of animals, can simulate artificial polishing and abrasion.

The occurrence of flaked fragments of bone has been recognized at the Acheulean site of Ambrona in Spain, a butchering site where the remains are principally of elephants (Biberson 1964). But there are no records of any unquestionable specimens from the East African Acheulean sites of Olorgesailie, Kariandusi and Isimila, although examples are known in the Acheulean of Bed IV*a* at Olduvai itself. These include a large biface made from a fragment of elephant bone.

As regards the "Bone and Antler Industry" from Chou Kou Tien (Breuil 1939), the illustrated specimens, with few exceptions, appear to represent food debris rather than bones that have been shaped or utilized. But without examination of the original specimens it is impossible to form a satisfactory estimation.

Further detailed analysis of human and carnivore food debris, such as that carried out by Brain on the Hottentot goat remains, should provide data by which it will be possible to determine whether bones from occupation sites have been artificially shaped as tools, utilized, or are merely debris. For the present it is evident that great caution is needed before accepting as artifacts bones with apparent hominid modification. In general, the presence of flaking appears to be the surest criterion, but certain forms of abrasion, such as that on the bone point from the West Pit, Sterkfontein and on the metapodials from Makapansgat, must also be considered artificial.

TOOLMAKING

In conclusion, the question of toolmaking by the early hominids must be discussed briefly. Before doing so, it should be stated that, in the writer's opinion, evidence for the manufacture of tools by means of using one tool as an instrument to make another is one of the most important criteria in deciding whether any particular taxon had reached the status of man. This stage of toolmaking, which marks the beginning of formal tool manufacture, should not be confused with simple modification of objects by means of the hands and teeth, such as is practiced by other higher primates. It is true, however, that man's ancestors must necessarily have passed through this preliminary phase, once they had progressed beyond the random use of sticks or stones. If evidence of toolmaking is not counted as a decisive factor for the human status it is difficult to see what alternative can be used for determining at what point it had been reached. Evolutionary changes must have been so gradual that it will never be possible for the threshold to be recognized on the evidence of the fossil bones alone. This would be true even if a far more complete evolutionary sequence of material were available for study: with the scanty and often incomplete material that has survived it is clearly out of the question. An arbitrary definition based on cranial capacity is also of doubtful value, since the significance of cranial capacity is closely linked with stature or body size, of which we have little precise information in respect of early hominids. However, as more and more material becomes available for study it should eventually become possible to recognize the morphological characters that existed when toolmaking came into general practice.

The toolmaking ability of *Homo habilis* has now been generally accepted. His remains have been found directly associated with Oldowan tools at no less than five sites in Bed I and a sixth site in the lower part of Middle Bed II (H. 13). *Australopithecus boisei* was undoubtedly also present at Olduvai at the same time, but on the evidence now available it seems unlikely that he was responsible for more than tool using, or possibly for simple modification of objects without employing another instrument for the purpose.

The small, probably female skull, H. 13, is the most recent known occurrence of *H. habilis* at Olduvai. So far, no taxonomically identifiable remains have been found in the upper part of Middle Bed II with the Developed Oldowan, but in Upper Bed II both *Homo erectus* (H. 9) and *Australopithecus* cf. *boisei* (H. 3) are present. If *H. erectus* is accepted as the probable maker of the Acheulean, as is now generally thought to be the case, his remains may also be expected to come to light in the upper part of Middle Bed II where the Acheulean first occurs. No identifiable hominid remains other than the robust australopithecine teeth of H. 3 have yet been found directly associated with the Developed Oldowan industry, so that at present we have no indication as to who the maker may have been.

The hominid remains so far recovered from Bed IV consist of an incomplete, small cranium (H. 12), two particularly thick vault fragments (H. 2), the right half of a mandible (H. 22), and an associated femur shaft and innominate bone found on an Acheulean living site during 1970 (H. 28). All these specimens appear to belong to *Homo* cf. *erectus* and, so far as is known, are from the upper part of Bed IV*a*. A small, slender femur and part of a tibia shaft (H. 34) that do not appear to be morphologically compatible with the remains provisionally attributed to *Homo erectus* were discovered in 1962 at a lower stratigraphic level within Beds III–IV*a*.

Recent discoveries of hominid fossils on the eastern side of Lake Rudolf in north Kenya and in the Omo basin, southern Ethiopia, have established the existence of two distinct taxa at a period considerably earlier than Bed I, Olduvai. These consist of a robust australopithecine that stands close to *A. boisei* and *A. robustus* although the degree of robustness is even more pronounced than in the later species. The second taxon is more lightly built. It seems likely to be linked with the artifacts found in the Koobi Fora area of East Rudolf, in a tuff dated at approximately 2.6 million years (M. D. Leakey 1970b) and may, perhaps, prove to be allied to *Homo habilis*.

The position in South Africa concerning the early toolmakers of the Transvaal caves has remained obscure for many years, in spite of the wealth of hominid fossils and the discovery of tools at both Sterkfontein and Swartkrans that bear such close resemblances to those of the Developed Oldowan from Middle Bed II (M. D. Leakey 1970a). The question of whether *Homo erectus* or *Australopithecus* was responsible for these tools has been discussed at length by various authors (Robinson 1962; Mason 1962a; Tobias 1965, Oakley 1968; *et al.*). Arguments in favor of one or other taxon (or both) being toolmakers continued to be largely hypothetical until 1969 when several cranial fragments from Swartkrans, including part of a palate that had formerly been ascribed to *Telanthropus capensis* and later to *Homo erectus* were fitted together by R. J. Clarke (Clarke and Howell 1970). The assembled pieces leave little doubt that the specimen belonged to the genus *homo* although not necessarily to *Homo erectus;* it may, in fact, possibly represent the South African counterpart of the East African *Homo habilis*.

The affinities of the robust australopithecine from South Africa to *A. boisei* and even to the more massive specimens from East Rudolf and the

Omo basin are unquestionable. All clearly belong to the same lineage with only minor variations. But the question of whether or not *Australopithecus africanus* is represented in East Africa is by no means clear. It has been suggested by some authorities that *Homo habilis* should be regarded as the East African counterpart of *A. africanus:* that he was an australopithecine who had reached a more advanced stage of development than *A. africanus* and had consequently become capable of organized toolmaking. Recent discoveries in East Africa, particularly at East Rudolf, do not support this view. They suggest, further, that the validity of *A. africanus* as a taxon distinct from *A. robustus* may be doubtful. The possibility that *A. robustus* and *A. africanus* represent the male and female of a single species deserves serious consideration. The nearly complete cranium from Bed I, Olduvai discovered in 1968, in which the facial region is preserved as well as the greater part of the vault and the base of the skull, indicate that *Homo habilis* differed significantly from *A. africanus* in a number of features generally regarded as critical in assessing the status of fossil hominids.

The wealth of hominid material that has been obtained in East Africa during the last decade—ranging in time from the Plio/Pleistocene through the Lower and Middle Pleistocene—should provide answers to many of the problems related to human evolution and the early toolmakers, once it has been fully studied.

References Cited

Biberson, P.

1961 Le Paléolithique inférieur du Maroc atlantique. Rabat: Pub. Services des Antiquités du Maroc, Fasc. 17.

1964 Notes sur deux stations Acheuléens de chasseurs d'éléphants de la vieille Castille. Monografias Inst. de Preh. y Arch., Vol. VI.

1967 Some aspects of the Lower Paleolithic of north—west Africa. *In* Background to Evolution in Africa. W. W. Bishop and J. D. Clark, eds. Chicago: Chicago University Press. pp. 447–475.

Brain, C. K.

1967 Bone weathering and the problem of bone pseudo-tools. South Afr. J. Sci. 63 (3):97.

Breuil, H.

1939 Bone and antler industry of the Choukoutien Sinanthropus site. Pal. Sin., New Series, D, No. 6.

Chavaillon, H.

1967 La préhistoire éthiopienne à Melka Knotouré. Archeologia 19:56–63.

Chavaillon, J. and M. Taieb.
1968 Stratigraphie du Quarternaire de Melka Kontouré (vallée de l'Aouache, Ethiopie): premiers résultats. C. R. Acad. Sci. 266:1210–1212.

Clarke, R. J. and F. C. Howell.
1970 More evidence of an advanced hominid at Swartkrans. Nature 225:1219–1222.

Dart, R. A.
1959 How human were the South African man-apes? South African Panorama, November. pp. 18–20.

Hay, R. L.
1971 Geologic Background of Beds I and II. *In* Olduvai Gorge, Vol. III Excavations in Beds I and II. 1960–1963. Cambridge: The University Press. pp. 9–18.

Isaac, G. Ll.
1965 The stratigraphy of the Peninj beds and the provenance of the Natron australopithecine mandible. Quarternaria 7:101–130.

1966 New evidence from Olorgesaillie relating to the character of Acheulean occupation sites. *In* Actas del V Congreso Panafricano de Prehistoria y de estudio del Cuaternario. L. D. Cuscoy, ed. Vol. VI, II. Pub. Mus. Arg. Santa Cruz de Tenerife, pp. 135–145.

Kleindienst, M. R.
1962 Components of the East African Acheulean assemblage: an analytical approach. *In* Actes du IVe Congrès Panafricain de l'Etude du Quarternaire. G. Mortelmans and J. Nenquin, eds. Vol. III. Tervuren. pp. 81–111.

Leakey, M. D.
1966 Primitive artefacts from the Kanapoi Valley. Nature 212:579–581.

1970a Stone artefacts from Swartkrans. Nature 225:1222–1225.

1970b Early artefacts from the Koobi Fora Area. Nature 226:228–230.

1971 Olduvai Gorge. Volume III. Excavations in Beds I and II. 1960–1963. Cambridge: The University Press.

MacCalman, H. R. and B. J. Grobbelaar.
1965 Preliminary report of two stone-working Ova Tjimba groups in the northern Kaokoveld of S.W. Africa. Cimbebasia 13.

Mason, R. J.
1962a Australopithecines and artefacts at Sterkfontein, Part II. S. Afr. Arch. Bull. 17:109–125.

1962b The Prehistory of the Transvaal. Johannesburg: Witwatersrand University Press.

Neuville, R. and A. Ruhlman.
1941 La place du paléolithique ancien dans le Quarternaire marocain. Inst. des Hautes-Etudes marocaines, Vol. VIII. Casablanca.

Oakley, K. P.
1968 The earliest tool-makers. Sond. aus Evolution und Hominisation. Stuttgart: Gustav Fischer Verlag.

Robinson, J. T.
1962 Australopithecines and artefacts at Sterkfontein. S. Afr. Arch. Bull. 17:87–107.

Stekelis, M.
1966 Archaeological Excavations at Ubeidiya. 1960–1963. Jerusalem: The Israel Academy of Sciences and Humanities.

Tobias, P. V.
1965 *Australopithecus, Homo habilis,* tool-using and tool-making. S. Afr. Arch. Bull. 20:167–192.

Erosion of Member F of the Shungura Formation has exposed two-million-year-old archaeological sites. Excavations by Harry Merrick at site FtJi1 have led to the recovery of one of the earliest known samples of hominid craftsmanship.

Harry V. Merrick:

Recent Archaeological Research in the Plio-Pleistocene Deposits of the Lower Omo Southwestern Ethiopia

What would a stone-age hominid do if he needed a knife and yet there were almost no stones around bigger than a pigeon's egg? Harry Merrick's researches on the archaeology of the Shungura Formation show that the hominids there solved the problem mainly by smashing the little pebbles that were available to obtain small sharp flakes and splinters of quartz. Patches of smashed quartz form telltale markers at some places where hominids were active two million years ago.

> At the north end of Lake Rudolph is the valley of the Omo River. Here there are extensive fossil beds which yield a fauna comparable to that from Oldoway, but so far no stone implements have been found in these beds. I am, however, confident that somewhere in the Omo region Stone Age material comparable to that from Oldoway will one day be found. (Leakey, L. S. B., 1936:130).

Louis Leakey's confidence founded on his uncanny intuition and perception of East Africa's prehistoric past was eventually rewarded. Over three decades later members of the International Omo Research Expedition, of which he was a co-organizer, discovered stone artifacts in the fossiliferous Lower Pleistocene deposits of the lower Omo River valley.

In spite of the early recognition of the potential importance of the Omo Valley region in the study of early man, intensive paleoanthropological research in the area only began in 1967 (Howell 1968). Since then the several contingents of the International Omo Research Expedition have conducted an annual program of extensive paleoanthropological, geological and paleontological research in the lower Omo Valley (see Howell this volume). This multi-disciplinary research has contributed greatly to our understanding of Lower Pleistocene hominids, their physical environment, and to the chronology of the Lower Pleistocene in eastern Africa. The archaeological investigations conducted as a part of this research program are just beginning to yield results. It has only been within the last four years that a number of undisturbed *in situ* archaeological sites have been found. These sites have proven to be broadly comparable in age to the earliest sites known from Olduvai Gorge, Tanzania (Leakey, M. D., 1971) and to the earliest sites known from the area east of Lake Rudolf in Kenya (Isaac *et al.* 1971). Thus far only a very small number of sites in the lower Omo Valley has been closely investigated, but the evidence which is accumulating adds another dimension to our knowledge of the variability of toolmaking/tool-using patterns of Lower Pleistocene hominids. This paper is intended to summarize recent archaeological research in the lower Omo Valley and to consider its findings in the broader context of archaeological research on the Lower Pleistocene in East Africa.

THE LOWER OMO RIVER VALLEY SEDIMENTS

The valley of the lower Omo River has been the scene of active fluvial, deltaic, and lacustrine sedimentation throughout much of later Cenozoic time. The Omo River, which is the main drainage for the southern Ethiopian highlands, flows southward from the Ethiopian highlands and empties into the northern end of Lake Rudolf. This drainage pattern was well established by late Pliocene times some 4.0 million years ago. Since then the river has carried vast quantities of sediments from the highlands into the Lake Rudolf basin. Much of this sediment has accumulated in a thick series of deposits in the lower Omo Valley, which forms the northernmost extension of the Lake Rudolf basin. Relatively recent tectonic activity and subsequent erosion has now re-exposed large areas of these deposits, particularly on the west side of the Omo River. Two major groups of fluvial and lacustrine deposits with markedly different ages have been recognized in the lower Omo Valley. The older set of deposits, of primary interest here, are the Plio-Pleistocene deposits collectively known as the "Omo Beds" which were deposited during the time range of roughly four to one million years ago. The second set of deposits, the Kibish Formation, is much younger and is of Upper Pleistocene and Holocene age. In many areas these deposits directly overlie the "Omo Beds." Kibish Formation deposits are very widespread and form the present-day surface mantle of sediments over much of the lower Omo Valley.

The fossiliferous Plio-Pleistocene "Omo Beds" have been divided into three geological formations—the Mursi, Usno, and Shungura Formations. Each is exposed in a different area of the valley (Figure 1). To date the most intensive research has been undertaken in the Shungura Formation area, with somewhat less in the Usno Formation area. The Mursi Formation area, the least well known, is the oldest of the three formations and dates to approximately 4.0 million years. The Usno Formation appears to be equivalent in age to the lower part of the Shungura Formation and the two together may span the time range between 3.2 and 0.9 million years. The Mursi, Usno, and the lower portion of the Shungura Formation are composed chiefly of fluvial, deltaic, and lacustrine sediments. All three formations contain rich terrestrial and aquatic vertebrate fossil assemblages.

Hominid fossils have not yet been found in the oldest of these deposits, the Mursi Formation. However, they have been recovered from the upper part of the Usno Formation and from most members of the Shungura Formation. These fossil remains attest to the presence of at least two varieties of hominids in the lower Omo Valley between 3.0 and 1.4 million years ago (F. C. Howell, pers. comm.). The archaeological traces of the possible toolmaking/tool-using capabilities and activities of these early hominids are exceedingly scarce. Thus far, despite moderately intensive surface survey of the Shungura and Usno Formations, archaeological sites and occurrences of undoubted stone artifacts are known only from the middle and upper portion of the Shungura Formation.

THE SHUNGURA FORMATION

Geology

The Shungura Formation is composed of a sequence of sediments over 800 meters in thickness. These sediments have a surface area of exposure of about 200 sq. km. in the southern end of the lower Omo Valley. The lower Omo Valley is part of the Eastern Rift Valley system, and as a result the Shungura Formation deposits have been subjected to extensive faulting since their deposition. Faulting along numerous north–south trending fault lines and erosion has created a staircase-like "badlands" topography in the Shungura Formation exposures (Figure 2). This faulting and erosion has provided good exposure of the entire stratigraphic sequence, however the degree of exposure of different portions of the sequence is

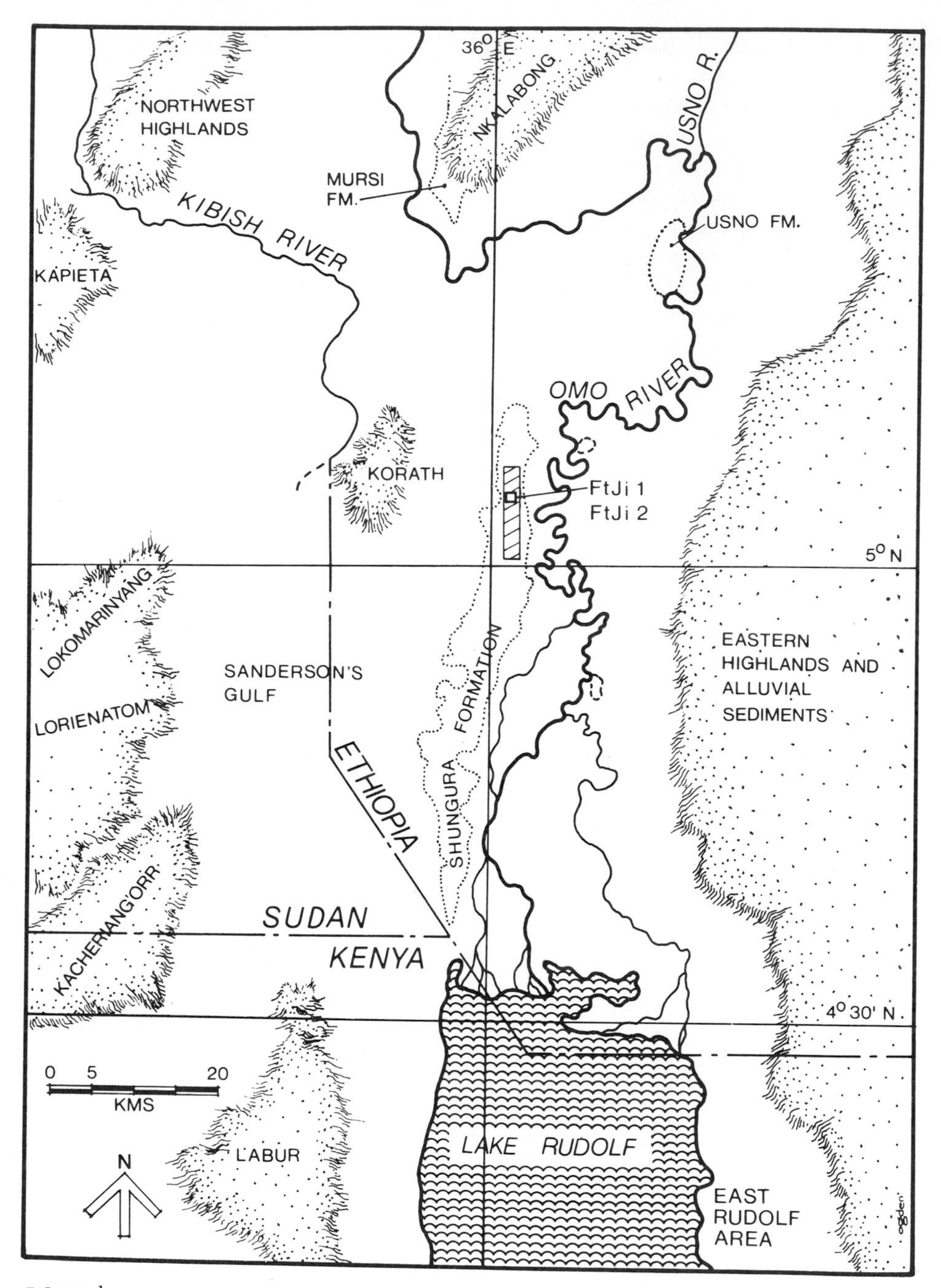

FIGURE 1
Generalized map of the lower Omo Valley indicating the locations of the Plio-Pleistocene exposures. The area containing the Member F sites in the northern Shungura Formation is indicated by hatching and highlands are stippled.

FIGURE 2
Typical exposures and topography in the Shungura Formation area.

highly variable. In general the lower portions (Basal Member through Member B) are not well exposed, while the middle portions (Member C through Member H) are better exposed. The uppermost part of the sequence, Members J, K, and L, are fairly well exposed, but only in the southernmost exposures northwest of Kalam.

In the northern portion of the Shungura Formation where the most intensive research has been conducted, the formation has been divided into 10 members on the basis of nine widespread volcanic tuffs, designated Tuffs A through J. This sequence of sediments is nearly 700 m. thick. In the southern exposures additional sediments above Tuff J are present, including several tuffaceous horizons. Detailed mapping and microstratigraphy are not yet available for this area, but the additional sediments are at least 120 m. thick and include tuffaceous horizons provisionally designated Tuffs K and L.

Each member of the Shungura Formation is composed of a major volcanic tuff and the overlying series of sediments (gravels, sands, silts and clays) up to the base of the next major tuff. The major tuffs are excellent time and stratigraphic markers for correlation as they represent recurrent episodes during which different distinctive volcanic ashes almost instantaneously covered the land surfaces and clogged the drainage channels throughout the lower Omo Valley. In addition the presence of six minor, but relatively widespread, tuffs within the sequence offers the potential for finer stratigraphic correlations within the individual members where they occur.

Dating

The widespread volcanic tuff horizons have proven to be as important for dating the Shungura Formation as they are for microstratigraphic correlations within it. Potassium-argon age determinations made on nine of these tuffs have provided a detailed chronology indicating the probable ages of the various members of the Shungura Formation. These K/Ar age determinations indicate that the base of the Shungura Formation (Tuff B) may be older than 3.75 m.y. and the top (Tuff L) younger than 1.3 m.y. (Table 1). Initial studies of the paleomagnetic reversal record preserved in the Shungura Formation sediments and comparison with other paleomagnetic reversal chronologies generally supports the dating suggested by the K/Ar age determinations. However, these studies indicate that the base of the formation is probably about one half m.y. younger than the K/Ar age determination for Tuff B indicates. They also indicate a probable age of about 0.9 m.y. for the uppermost Member L sediments (F. Brown and R. Shuey, pers. comm.).

Depositional Environments

The various microenvironments of the lower Omo Valley during Lower Pleistocene times are partially reflected by the different types of deposition (riverine, floodplain, lake margin, etc.) in the Shungura Formation sediments. These different microenvironments are undoubtedly a major factor in the types and distributions of flora and fauna, including the early hominids. It is therefore important to bear in mind that the presence of archaeological occurrences and/or their stratigraphic positions may reflect changes in microenvironmental conditions as much as changes in hominid behavioral patterns.

The depositional environments represented in the Shungura Formation sediments changed through time. The sediments of the lower half of the Shungura Formation (Basal Member through lower Member G) are primarily fluvial deposits. Detailed geological mapping has divided each of these lower members into a number of subunits which reflect some type of recurrent sedimentation cycle. Each cycle is composed of a set of gravels and sands at the base, overlain by silts and clays and often capped by a paleosol. These cyclic subunits have been interpreted as the deposits of a river channel system which slowly migrated back and forth across its

TABLE 1
Dating, archaeological occurrences and localities yielding hominid fossils in the Shungura Formation.

TUFFS	MEMBERS	AVERAGE K/AR AGE (M.Y.)	TESTED ARCHAEOLOGICAL OCCURRENCES	NUMBER OF LOCALITIES YIELDING HOMINID FOSSILS PER MEMBER
L	L	1.34	—	—
	KNW II	1.43		
K	K		—	3
J	J		—	0
	I_2	1.84		
H	H		—	2
G	G	1.93	Locality 7	21
			Omo 123	
			FtJi2	
			FtJi1, Omo 57	11
F	F	2.04	FtJi5	
E	E		—	10
	E_i	2.12		
D	D	2.35	—	7
C	C		—	18
	B_{10}	2.95	—	3
B	B	3.75 (3.0)*		
A	A		—	0
Basal Member			—	0

**Best estimate on the basis of preliminary paleomagnetic data.*

floodplain. Most of the sediments of the Basal Member through Member D appear to have been deposited by major river channels. The later sediments of Members E, F, and lower G appear to have been deposited in smaller meandering stream channels, similar to those found in river delta situations. The uppermost members of the Shungura Formation (middle Member G and above) were deposited primarily under prodeltaic and lacustrine conditions. Temporary land surfaces are present in these sediments, but they are rare.

Archaeological Occurrences in the Shungura Formation

Archaeological research in the Shungura Formation exposures began in 1967, the first year of the International Omo Research Expedition. Unhappily the results of this continuing investigation initially accrued very slowly because the identification of artifactual occurrences of Lower Pleistocene age in the Shungura Formation exposures is complicated by two factors. First, archaeological sites of this age are relatively rare and do not have high artifact densities. Secondly, the initial recognition of sites by surface survey in the Shungura Formation exposures is extremely difficult. In many areas of the Shungura Formation exposures small remnants of the Holocene Kibish Formation sediments overlie the Shungura Formation sediments. Often Kibish Formation sediments contain simple and seemingly unrefined artifacts which, if eroded from Kibish Formation sediments and found lying on exposures of Shungura Formation sediments, can be readily mistaken for simple unstandardized artifacts of Lower Pleistocene age. Therefore artifacts found on the surface of Shungura Formation sediments always remain somewhat suspect. To be absolutely certain an artifact is of Lower Pleistocene age it is necessary to recover it *in situ* by excavation in the Shungura Formation deposits. Since 1970, a program of intensive surface survey and excavations by both the United States and French contingents of the International Omo Research Expedition have increasingly located archaeological occurrences of Lower Pleistocene age *(in situ)* in the Shungura Formation deposits.

The lower portion of the Shungura Formation from the Basal Member through Member E, spanning roughly the 3.2 to 2.0 m.y. time range, has still to yield stone artifacts and occupational surfaces *in situ* in completely unambiguous situations. Since 1967 a number of quartz and lava artifacts have been found either singly or in low-density scatters on the surface of eroded outcrops of Members B, C, D, and E (Howell 1972; Chavaillon 1975). Many of these surface artifacts appear to have been originally associated with stream channel fillings. A smaller number may have been originally associated with land surfaces, including some Member D artifacts which may have been associated with temporary land surfaces on the top of Tuff D. While many of these surface finds, including the quartz chopper from Member E exposures reported by Chavaillon (1970), have a high probability of being derived from Shungura Formation sediments, many others may have been derived from the more recent Kibish Formation sediments. Unfortunately, excavation at the location of many of the surface finds generally has not produced any further artifacts *in situ,* although Chavaillon (Coppens *et al.* 1973) recently reported the finding of a single quartz flake in a stream channel deposit in Member E.

Members F and G, dated by potassium-argon to between about 1.8 and 2.0 million years, contain the oldest stone artifacts which are undoubtedly *in situ* (Merrick *et al.* 1973; Coppens *et al.* 1973). These members also contain the oldest traces of probable "living floors" or occupation horizons that have been located in the Shungura Formation. Since 1970 text excavations have located six occurrences with artifacts *in situ* in Members F and G. Five of these have been in Member F and one in Member G. Two of the excavations in Member F have tested what appear to be little disturbed primary occupation horizons (FtJi2 and Omo 123). The remaining three excavations in Member F (FtJi1, FtJi5, and Omo 57) and the Member G excavation (Locality 7) have tested secondarily derived concentrations of artifacts associated with small stream channel deposits.

The sediments above Member G have not yet yielded artifacts *in situ,* although artifacts of crude and unstandardized appearance are commonly found on the eroded surfaces of these upper members. The probability that any of these surface artifacts originated from the Shungura Formation deposits is low for several reasons. First, the upper members are largely lacustrine with few temporary land surfaces. This greatly reduces the likelihood that artifacts would have been originally incorporated in the sediments. Secondly, artifact-bearing Kibish Formation deposits often directly overlie these upper members and could easily be the source of many of the artifacts. A third slightly more remote source for some of the surface artifacts is the modern-day pastoral inhabitants of the area, who may be to a small extent redistributing stone artifacts across the landscape. Their abandoned camps have been observed to occasionally contain stone choppers and anvils which were apparently collected in the local vicinity and used to break long bones (J. de Heinzelin, pers. comm.).

Member F and Its Archaeological Occurrences

The earliest definitely *in situ* archaeological occurrences in the Shungura Formation are in Member F. The five excavated sites and the numerous surface occurrences which have a high probability of having originated from Member F sediments are located within an area approximately 14 km. long by 5 km. wide in the northern portion of the Shungura Formation exposures (Figure 1). The sediments of Member F in which these occurrences are found are of fluvial and floodplain origin. Member F has a measured thickness of some 35 m. Tuff F and its channel deposited facies, Tuff F′, at the base of the member usually vary between 4.0 and 4.5 m., but in places may attain a thickness of 7 m. Overlying Tuff F are 4 to 5 cyclic units of fluvial and floodplain sedimentation. Each cyclic unit is usually composed of a graded sequence of coarse sands, passing upward through medium and fine sands to silts and clays. Occasional weakly developed palaeosols, indicating temporary land surfaces, occur on the tops of some of the silt and clay subunits within the individual cyclic units. The lowest submember in Member F, F-1, contains a widespread minor tuff, Tuff T, which is always found filling small channels. Figure 3 illustrates the nature and variation of the sedimentary sequence in Member F and indicates the relative stratigraphic positions of three of the excavated sites in Member F.

A fairly precise estimation of the age of the Member F occurrences is possible. The Member F sediments have been neatly bracketed by potassium-argon age determinations for Tuffs F and G. The average of

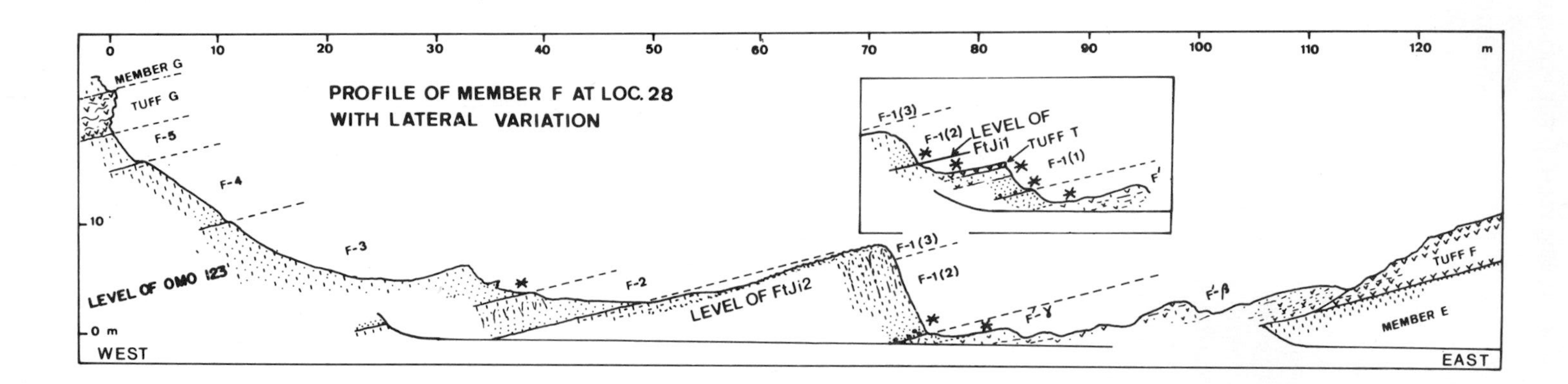

FIGURE 3
Profile of Member F at locality 28, near the FtJi2 occurrence. The relative positions of the FtJi1 occurrence (inset, profile) and the Omo 123 occurrence are also indicated.

two K/Ar age determinations for Tuff F gave an age of 2.04 ± .10 m.y. A single K/Ar age determination for Tuff G gave an age of 1.93 ± 10 m.y. (Brown and Lajoie 1971).

Four of the five sites excavated in Member F, (FtJi1, FtJi2, FtJi5 and Omo 57) occur low in the Member F sequence in submember F-1. Many of the untested surface occurrences on Member F exposures also appear to originate from the F-1 sediments. The fifth excavated occurrence, Omo 123, occurs somewhat higher in the sequence in submember F-3. Both the excavated sites and the untested surface scatters are associated with a variety of depositional microenvironments. The two possible occupation horizon occurrences (FtJi2 and Omo 123) are found in fine silt and clay deposits and both of the sites appear to have been in floodplain or marginal backswamp environments. Two other excavated occurrences (FtJi1 and Omo 57) and several of the surface scatters (FtJi3 and FtJi4) are associated with small sand- and gravel-filled meandering stream channels. Although it is not certain, this depositional situation is suggestive of an environment along small, vegetation-lined streams, if analogy with the modern Omo River deltaic environment is valid. Artifacts have also been excavated from small mudflow and slope wash type deposits which are associated with small-scale gully erosion (FtJi5). Presently similar steep gully erosion and unsorted deposits occur in some areas along the margins of the Omo River and also in the arid "Omo Beds."

Member F Sites and Settings

Brief sketches of the three sites excavated by the United States contingent of the International Omo Research Expedition will serve to indicate the nature and variation of the known Member F occurrences.

The FtJi1 Occurrence. The FtJi1 occurrence is situated on the steep slope of the western side of a heavily eroded spur of F-1 sediments in Locality 204 (Figure 4). A small excavation of 18 sq. m. yielded a total of 367 artifacts and probable artifacts. A number of mammalian fossils, including elephant, hippotamus and several species of bovids were also recovered. An additional 270 artifacts were collected from the surface of the adjacent erosion exposures and these almost certainly were derived from the FtJi1 occurrence. The excavated artifacts were recovered *in situ* from a small lens of sands and gravels, approximately 30 cm. thick, filling the base of a small meandering channel directly above Tuff T. This channel is situated at the base of the F-1(2) unit (see Figure 3, inset profile) and is stratigraphically 6 to 8 m. below the FtJi2 occurrence. It is of note in this regard that the Omo 57 occurrence (Coppens *et al.* 1973) is in an identical stratigraphic position and depositional context, but some 7.5 km. to the south of the FtJi1 occurrence.

FIGURE 4
Exposures in Locality 204, and the site of the FtJi1 occurrence.

The artifacts appear to be in a secondarily derived context; they were found in the gravel lenses of the channel fill and their physical condition varies from fresh to heavily abraded. The fossils also appear to be in a derived context, for they vary in condition from unrolled to heavily rolled. This suggests both artifacts and fossils were concentrated in and deposited as part of the bed load of the stream. As numerous channel fills in Member F contain abundant mammalian fossils but lack artifacts, the presence of both artifacts and fossils in this channel may be fortuitous. Any postulated association between the artifacts and fossils due to hominid activity before their incorporation into the channel deposits is problematical.

Stratigraphic studies are inconclusive as to the original surface(s) from which the artifacts were derived. They may have been derived from pri-

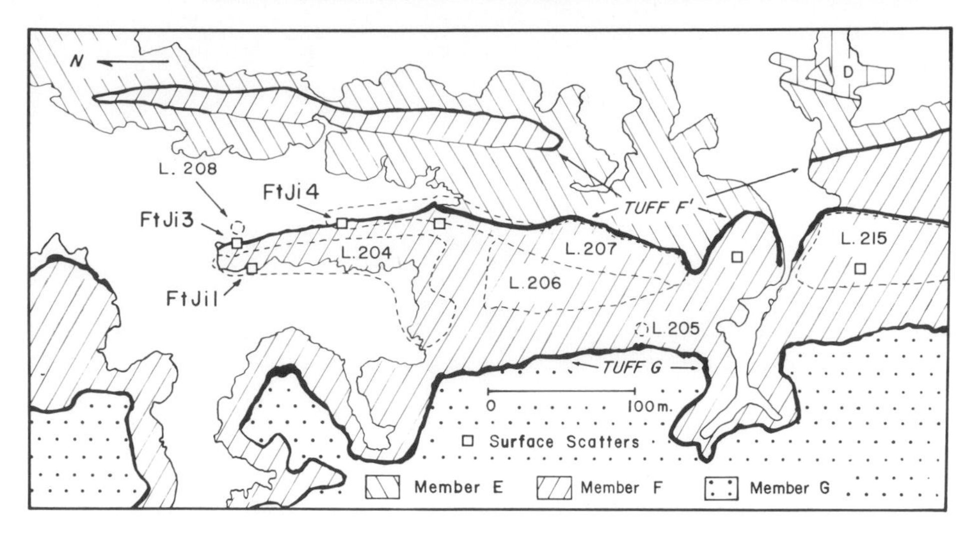

FIGURE 5
Location of the FtJi1, FtJi3 and FtJi4 occurrences and surface artifact scatters in the locality 204–208 and 215 area. Locality boundaries are shown with dashed lines. Reprinted from H. V. Merrick, J. de Heinzelin, P. Haesaerts, & F. C. Howell, in *Nature,* 242, 1973.

mary contexts on the channel banks and flood basins adjacent to the small channel, or they may have been discarded on the sandy substratum of the dry stream channel and subsequently reworked into the bed load of the stream.

In the vicinity of locality 204 and the FtJi1 occurrence are a number of surface scatters of artifacts which are almost certainly derived from the F-1 sediments. Figure 5 shows the distribution of surface artifacts in the Loc. 204–208 and 215 area. There are six small (± 25 m.^{2}) patches of densely concentrated artifacts in the 500 × 100 m. area included in these localities. The surface scatters are composed almost entirely of small quartz artifacts. The probable source of the artifacts in two of these scatters (FtJi3 and FtJi4) has been confirmed by small test trenches in the Member F sediments, but the occurrences have not yet been excavated. The positions of the remaining untested surface scatters at the bases and on the steep slopes of freshly eroded exposures makes it almost certain that they were also recently derived from Member F sediments.

The FtJi2 Occurrence. The FtJi2 site is presently being dissected by the erosion of a small spur of fluviatile sediments in fossil locality 396 (Figure 6). Thus far a small excavation of 22 sq. m. has recovered 224 artifacts *in situ.* Another 131 artifacts have been recovered from the surface of the adjacent erosion exposures. The stratigraphic sequence at this locality is summarized in Figure 7.

FIGURE 6
Exposures in Locality 396 and the FtJi2 occurrence.

The FtJi2 occurrence was initially interpreted as a little-disturbed single occupation horizon (Merrick *et al.* 1973). However, the results of the 1973 field season indicate the occurrence may be comprised of two areally overlapping scatters of artifacts which are separated vertically by about 15 cm. of sediments. Tentatively, the occurrence is now being interpreted as the result of two short-term occupations. The occurrence appears to be composed of two low-density concentrations of small quartz artifacts scattered in two horizontally overlapping lenses. Each lens of artifacts appears to be about 15 cm. thick and separated vertically by 10 to 15 cm. However, there appears to be some vertical overlap between the two lenses which makes the positive identification of the occurrences as the result of two distinct occupation horizons less certain. The artifact lenses are located in the lowest 50 cm. of a 2.5 m.-thick bed of fine silt and clay which forms the lower part of the F-1(3) unit of Member F. The artifacts, almost entirely small fragments of shattered quartz lumps and pebbles, are the only stone fragments in this silt and clay layer. The

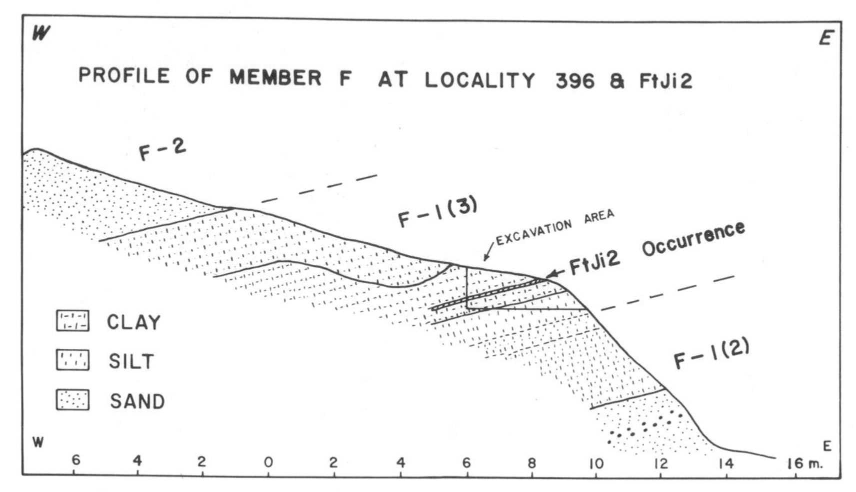

FIGURE 7
Profile of Member F at locality 396 and the FtJi2 occurrence. Reprinted from H. V. Merrick, J. de Heinzelin, P. Haesaerts, & F. C. Howell, in *Nature,* 242, 1973.

artifact-bearing horizon has not yielded bone in this locality, but fossil pollen is preserved (R. Bonnefille, pers. comm.). The excavation has not yet determined the areal extent of the occurrence. The surface scatter of artifacts along the adjacent erosion exposure suggests a maximum north–south extension of about 20 m., but the east–west extent of the occurrence cannot yet be determined.

It is not yet possible to reconstruct the detailed environmental setting of this occurrence. Stratigraphic studies suggest that initially the artifacts were scattered on temporary land surfaces of silty clay in a slowly agrading backswamp or in a marginal flood basin situation. Subsequently continued deposition under backswamp conditions buried the site. As a result of both secondary development of a calcium carbonate concretionary horizon and the location of the site in clays which are subject to marked swelling and shrinkage with wetting and drying, all traces of the original temporary land surfaces have been obliterated. Undoubtedly this has slightly disturbed the scatters of artifacts and accounts in part for their vertical dispersion.

The FtJi5 Occurrence. The FtJi5 occurrence is situated on the eastern face of a heavily eroded spur of F-1 sediments just west of locality 4 (Figure 8). During the 1973 field season a small excavation of 8 sq. m. was conducted to locate the source of the dense surface scatter of artifacts in the locality. A total of 24 artifacts were recovered *in situ* during the excavation. Another 78 artifacts were collected from the surface of the adjacent exposures.

FIGURE 8
The FtJi5 excavation west of Locality 4.

The FtJi5 occurrence is a secondarily derived concentration of artifacts in the base of a stream channel. The artifacts were recovered from a thin layer (20 cm.) of unsorted gravel, sand, rolled balls of clay and bits of reworked tuff which filled the very base of a broad channel in the lower part of the F-1(1) unit. Tuff T closely overlies this basal channel filling. The position of the occurrence below Tuff T makes it the earliest occurrence in Member F. Although the artifacts are in a derived context, their condition is usually fresh, suggesting they have not been transported a great distance. Fragmentary rolled bone and teeth fragments were also recovered from the channel fill, but as at the FtJi1 occurrence, the significance of the association of artifacts and bones is problematic.

Artifact Assemblages from Member F

The Member F excavated and surface artifact assemblages collected by both the United States and French contingents of the International Omo Research Expedition form a relatively homogenous group. The majority of artifacts recovered from both excavation and surface scatters in Member F are small angular fragments of shattered pebbles and lumps of milky white vein quartz. Artifacts of lava, chert and chalcedony are relatively rare, comprising less than 5% of the total artifacts. The physical size of the artifacts is small—the mean maximum length of the artifacts in each of the excavated assemblages is less than 20 mm. Large core tools, such as choppers, discoids, and polyhedrons and large manuports are lacking in all of the excavated assemblages and in all of the collected surface occurrences. Small "tools," possible cores, and other pieces exhibiting either secondary modification or traces of edge damage (possible utilization?) are rare. The artifact assemblages from the FtJi1, FtJi2 and FtJi5 occurrences are summarized in Table 2.

The quartz artifacts which predominate at these occurrences are largely fragments of shattered quartz lumps and pebbles. Whole flakes and flake fragments (pieces with a part of the bulb of percussion remaining) are rare. Pieces exhibiting secondary retouch are absent, although among the FtJi2 *in situ* artifacts where minimal transport in very fine grained sediments was involved during deposition, there are several pieces which exhibit minor traces of edge damage. A selection of the excavated quartz artifacts from FtJi2 is illustrated in Figure 9. In the absence of standardized tool forms and the lack of apparent systematic techniques of manufacture, these quartz artifacts can probably be interpreted best as the debris from the intensive smashing of small quartz lumps and pebbles.

GENERAL DISCUSSION AND COMPARISONS

The first definite traces of the "cultural" behavior of the early hominids who inhabited the lower Omo Valley appear, in archaeological perspective, suddenly, relatively numerous and geographically moderately widespread in the lower portion of Member F at about two million years ago. The suddenness of this appearance of relatively numerous stone artifacts presents an intriguing problem; particularly since fossil remains of at least two varieties of early hominids are well represented—but without traces of undoubted stone artifacts—much earlier in the Shungura Formation sequence and in the Usno Formation. The suddenness of the appearance of stone artifacts may involve the interaction of several factors. In part it may be due to the change in depositional characteristics of the Shungura Formation sediments after Member D times. The change from deposition in

TABLE 2
Artifacts recovered from the FtJi1, FtJi2 and FtJi5 Occurrences.

	FtJi1				FtJi2			FtJi5	
	EXCAVATED			SURFACE	EXCAVATED		SURFACE	EXCAVATED	SURFACE
ARTIFACT CATEGORY	quartz	lava	chert	quartz only	quartz	lava	quartz only	quartz only	
Debitage									
Flakes	16	—	1	19	3	—	6	1	6
Flake fragments	7	1	—	30	4	—	14	—	1
Angular fragments	344	1	6	221	215	1	110	23	67
Cores (?)	1	—	—	—	—	—	—	—	3
Total artifacts	368	2	7	270	222	1	130	24	77
Manuports	—	—	—	—	7	2	—	—	—
Mean maximum length (mm) of artifacts	16.4		13.8	24.4	13.7	11	19.7	16.2	25.2
Range of maximum length (mm) of artifacts	6—49	19—55	9—19	9—64	5—35		8—40	6—40	10—65

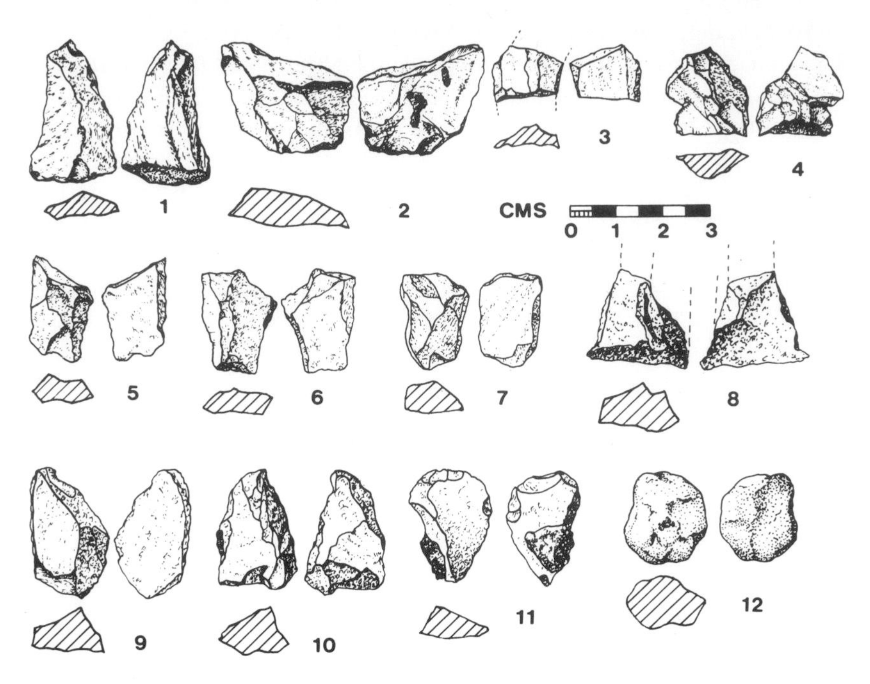

FIGURE 9
In situ quartz artifacts from the FtJi2 occurrence. Flake fragments (1–2), angular fragments (3–11), and small pebble (manuport) (12).

major river channel situations in the lower Shungura Formation to deposition in smaller meandering stream channel situations in Members E, F, and lower G, might increase the possibility of sites being preserved *in situ* or only being minimally disturbed. A second factor may be microenvironmental change. Such change, suggested by the changes in the depositional regime and also indicated by preliminary pollen and microfaunal studies (Bonnefille 1975, Jaegaer and Wesselman 1975) may have made the area more suitable to hominid occupation than before. Additionally (but attempting to avoid sounding too much like a strict environmental determinist) the microenvironments represented by the Member E, F, and G sediments may have been those which favored hominid activities including stone tool use. At present there is no evidence from the Omo archaeological occurrences to suggest which one, or if both, of the early hominid varieties present in the area produced the artifacts.

The Member F occurrences and artifacts present a number of interesting comparisons with other Lower Pleistocene occurrences in eastern Africa. The artifact densities for the primary context occurrences, FtJi2, and Omo 123, are variable. The FtJi2 density of 10.2 artifacts per sq.m. is probably much lower than that of Omo 123 (J. Chavaillon, pers. comm.). The artifact density at FtJi2 is most comparable to the low densities reported for the early East Rudolf sites (Isaac *et al.* 1971). The artifact density of the Omo 123 occurrence may be most similar to the higher densities reported from many of the Lower Pleistocene occurrences at Olduvai Gorge (Leakey, M. D., 1971). To date no bone has been found associated with the FtJi2 occurrence and only very rare fragmentary bone has been found associated with the Omo 123 occurrence (J. Chavaillon, pers. comm.). Although poor conditions for bone preservation at these particular Member F occurrences may be involved, this contrasts markedly with both the Olduvai and East Rudolf occurrences where bone in moderate quantities is often associated with artifacts on the occupation surfaces.

By comparison to both the early Olduvai Gorge and East Rudolf occurrences, the most notable features of the Omo occurrences are the absence of large tools, such as choppers, the preponderance of quartz as a raw material (although some Olduvai Gorge occurrences have relatively high percentages of quartz artifacts) and the generally smaller size of the artifacts. The lack of any standardized or even readily definable forms of "tools" in the Omo assemblages also offers an interesting contrast. Although part of these differences may relate to the sampling of different activity facies, at least part may be explained by the nature and proximity of the available raw material. The Member F sediments are almost totally lacking in suitable raw material for tool manufacture. The principal sources of raw material, both quartz and lava, during Member F times may have been 20 to 30 km. distant. The nearest source of any raw material would probably have been in small stream channels several kilometers to the east of the sites. These channels, which drained the highlands forming the eastern margin of the valley, would have provided only small quartz lumps and pebbles as raw material. This material definitely does not flake as regularly as lava or other materials might and this may account for the size and simplicity of the Member F artifacts.

Acknowledgments

Much of the research reported here, and the author's participation in it, has been supported by grants-in-aid of research to F. C. Howell from the National Science Foundation. The continued support and encouragement of the Government of Ethiopia is gratefully acknowledged as is the continued cooperation of the Kenyan Government. Many of the ideas and interpretations presented here have been the outgrowths of discussions with F. C. Howell, F. H. Brown, G. Ll. Isaac, and J. de Heinzelin; their continued help and encouragement has been appreciated. And in particular, I am indebted to Joan Self Merrick who assisted with all stages of this research, from excavation to artifact analysis; her aid has been immeasurable.

References Cited

Bonnefille, R.

1975 Palynological evidence for an important change in the vegetation of the Omo basin between 2.5 and 2 million years. *In* Earliest Man and Environments in the Lake Rudolf Basin: Stratigraphy, Paleoecology and Evolution. Y. Coppens, F. C. Howell, G. L. Isaac and R. E. F. Leakey, eds. Chicago: University of Chicago Press.

Brown, F. H., and K. R. Lajoie

1971 Radiometric Age Determinations on Pliocene/Pleistocene Formations in the Lower Omo Basin, Ethiopia. Science 229:483–485.

Chavaillon, J.

1970 Dècouverte d'un niveau oldowayen dans la basse vallée de l'Omo (Ethiopie). Bull. Soc. Préhist. Fr. 67(1):7–11.

1975 Evidence for the technical practices of early Pleistocene hominids. *In* Earliest Man and Environments in the Lake Rudolf Basin: Stratigraphy, Paleoecology and Evolution. Y. Coppens, F. C. Howell, G. L. Isaac and R. E. F. Leakey, eds. Chicago: University of Chicago Press.

Coppens, Y., J. Chavaillon and M. Beden

1973 Résultats de la nouvelle mission de l'Omo (campagne 1972). Découverte de restes d'Hominidés et d'une industrie sur éclats. C. R. Acad. Sc. Paris. t. 276, Série D, 161–164.

Howell, F. C.

1968 Omo Research Expedition. Nature 219:567–572.

1972 Pliocene/Pleistocene Hominidae in Eastern Africa: absolute and relative ages. *In* Calibration of Hominoid Evolution. W. W. Bishop and J. A. Miller, eds. Edinburgh: Scottish Academic Press. pp. 331–368.

Isaac, G. Ll., R. E. F. Leakey and A. K. Behrensmeyer

1971 Archaeological traces of Early Hominid Activities, East of Lake Rudolf, Kenya. Science 173:1129–1134.

Jaeger, J. J., and H. B. Wesselman

1975 Fossil remains of micromammals from the Omo group deposits. *In* Earliest Man and Environments in the Lake Rudolf Basin: Stratigraphy, Paleoecology and Evolution. Y. Coppens, F. C. Howell, G. L. Isaac and R. E. F. Leakey, eds. Chicago: University of Chicago Press.

Leakey, L. S. B.

1936 Stone Age Africa: An Outline of Prehistory in Africa. London: Oxford University Press.

Leakey, M. D.

1971 Olduvai Gorge. Vol. 3. Excavations in Beds I and II, 1960–1963. Cambridge: The University Press.

Merrick, H. V., J. de Heinzelin, P. Haesaerts and F. C. Howell

1973 Archaeological Occurrences of Early Pleistocene Age from the Shungura Formation, Lower Omo Valley, Ethiopia. Nature 242:572–575.

In 1961 Louis Leakey recruited Glynn Isaac as warden of the Olorgesailie Prehistoric Site in East Africa. Here he cleans artifacts on an occupation floor at that site. (Photo by Barbara Isaac.)

Glynn Ll. Isaac:

The Activities of Early African Hominids:

A Review of Archaeological Evidence from the Time Span Two and a Half to One Million Years Ago*

If we really want to know how the early hominids made their living, what can we do to find out? The essay by Glynn Isaac gives an imaginary reconstruction of a scene in early stone-age life, and then goes on to describe the main classes of archaeological evidence that throw light on dietary patterns, land use, technology, and novel social patterns that contributed to adaptation.

If an observer could be transported back through time and climb a tree in the area where the Koobi Fora Formation was accumulating—what would he see?

As the upper branches are reached, the climber would find himself in a ribbon of woodland winding out through open areas. A kilometer or so away to the west would be seen the swampy shores of the lake, teeming

---

*This paper was prepared as a tribute to Louis Leakey; a preprint of it was circulated in connection with a symposium held in New York in January, 1974. Essentially the same paper with a modified concluding section is being published also in the proceedings of that conference (C. Jolly, editor).

with birds, basking crocodiles, and *Euthecodons.* Here and there are schools of hippos. Looking east, in the distance some ten or twelve kilometers away lie low, rolling hills covered with savanna vegetation. From the hills, fingers of trees and bush extend fanwise out into the deltaic plains. These would include groves of large *Acacia, Celtis,* and *Ficus* trees along the watercourses, fringed by shrubs and bushes. Troops of colobus move in the tree tops, while lower down are some mangabees. Scattered through the bush, the observer might see small groups of waterbuck, impala, and kudu, while out in the open areas beyond, would be herds of alcelaphine antelope and some gazelle *(Megalotragus* and *Antidorcas).* Among the undergrowth little groups of *Mesochoerus* pigs rootle, munching herbiage.

Peering down through the branches of the tree, the climber would see below the clean sandy bed of a watercourse, dry here, but with a tidemark of grass and twigs caught in the fringing bushes and showing the passage of seasonal floods. Some distance away down the channel is a small residual pool.

Looking out beyond the bushes can be seen large open floodplains, covered with grasses and rushes, partly dry at those seasons of the year when the lake is low and when the river is not in spate. Far across the plains, a group of four or five men approach; although they are too far off for the perception of detail, the observer feels confident that they are men because they are striding along, fully upright, and in their hands they carry staves.

To continue the reconstruction in a more purely imaginative vein: as the men approach, the observer becomes aware of other primates below him. A group of creatures has been reclining on the sand in the shade of a tree while some youngsters play around them. As the men approach, these creatures rise and it becomes apparent that they too are bipedal. They seem to be female, and they whoop excitedly as some of the young run out to meet the arriving party, which can now be seen to consist mainly of males. The two groups come together in the shade of the tree, and there is excited calling, gesturing and greeting contacts. Now the observer can see them better, perhaps he begins to wonder about calling them men; they are upright and formed like men, but they are rather small, and when in groups they do not seem to engage in articulate speech. There are a wealth of vocal and gestural signals in their interaction, but no sustained sequential sound patterns.

The object being carried is the carcass of an impala and the group congregates around this in high excitement; there is some pushing and shoving and flashes of temper and threat. Then one of the largest males takes two objects from a heap at the foot of the tree. There are sharp clacking sounds as he squats down and bangs these together repeatedly. The other creatures scramble round picking up the small sharp chips that have been detached from the stones. When there is a small scatter of flakes on the ground at his feet, the stone worker drops the two chunks, sorts through the fragments and selects two or three pieces. Turning back to the carcass, this leading male starts to make incisions. First the belly is slit open and the entrails pulled out; the guts are set on one side, but there is excited squabbling over the liver, lungs, and kidneys; these are torn apart, some individuals grab pieces and run to the periphery of the group. Then the creatures return to the carcass; one male severs skin, muscle and sinew so as to disengage them from the trunk, while some others pull at limbs. Each adult male finishes up with a segment of the carcass and withdraws to a corner of the clearing, with one or two females and juveniles congregating around him. They sit chewing and cutting at the meat, with morsels changing hands at intervals. Two adolescent males sit at the periphery with a part of the intestines. They squeeze out the dung and chew at the entrails. One of the males gets up, stretches his arms, scratches under his arm pits and then sits down. He leans against the tree, gives a loud belch and pats his belly . . . . *End of scenario.*

Jane Lancaster (1967) pointed out that in dealing with early hominids we ought to make an imaginative leap and realize that we are dealing with an adaptive system that no longer exists. The system presumably had some structural features held in common with that of man, some in common with that of the apes, and very probably it also had unique components and configurations. In hindsight we can see that the system was continuously evolving and in that sense it never reached equilibrium—but in terms of those who lived it, it was presumably more or less constant over hundreds of generations, and it must have been not just promising, but downright effective, otherwise we could not be here to indulge our curiosity over its nature.

The scenario, fanciful though it is, does serve to bring into sharp focus how little we know about the aspects of early hominids that really matter. For instance, what if we included in the glimpse an encounter between two hominid bands or between different hominid species? How would those have looked? We have direct evidence for toolmaking at a place on a tree-lined stream channel where bones also accumulated, but the rest of the behavior pattern sketched here is an extrapolation. Food sharing, division of labor, paribonding with economic involvements are envisaged in this reconstruction, but all we can in fact say is that at some stage these behaviors did become critical ingredients of human adaptation.

In contrast with our ignorance of hominid behavior, we are steadily learning more about its context and concomitants. Thus as geological, palaeontological and palynological research progresses we can speak with more and more confidence about the setting of early hominid life and death, and the description of the scene for the scenario may not be too far from the truth. Similarly, as hominid fossil finds continue to mount in numbers, we are getting better and more complete information on the appearance and physical capabilities of early hominids. However, human evolution is characterized more by changes in behavior and social organization than by changes in anatomy, and what one would really like to know is how these came about—what were the formative stages like? What influence if any do they have on the character of the end product?

Table 1 (see p. 505) sets out in an oversimplified and elementary fashion some of the contrasts that I know between men and their closest living relatives. This crude list helps me to symbolize my stance which is that the evolutionary transformation with which we are concerned was a complex one, involving intricately interrelated anatomical, physiological, economic, technological and cultural components.

As an archaeologist, I am primarily concerned with the physical traces of activities, but these will remain as rather dull stones and bones unless they are considered as parts of broader patterns of habit, behavior and ecology.

The following components of the evidence are now widely recognized, and form convenient headings for the review:

1. *Artifacts and equipment.*
2. *Food refuse* (usually bone).
3. *Spatial configuration* within sites.
4. Relationships of sites and finds to local and regional *ecology* and *geography*.

Although our knowledge of the archaeology of early man is currently expanding in a most encouraging way, it should be borne in mind that it still depends on a mere handful of significant excavated sites:

Olduvai Gorge Beds I and II (Leakey, M. D., 1971).
Peninj Formation (Isaac 1967).
Shungura Formation (Coppens, Chavaillon and Beden 1973; Chavaillon 1975; Merrick *et al.* 1973).
Koobi Fora Formation (Leakey, M. D., 1970a; Isaac *et al.* 1971; Isaac *et al.* 1975).

The review will focus on these East African sites with Olduvai occupying a pivotal position, both because of the qualities of the evidence available there and because Mary Leakey has compiled and published more complete data for Olduvai than is available for any other site (see Leakey, M. D., this volume). For East Rudolf, I have drawn on our own field and laboratory data, of which a summary has recently been published (Isaac

1975). Comments on the Shungura evidence depend both on the published reports and on comprehension gained during field exchanges between the East Rudolf and the Omo teams.

The application of K-Ar dating and paleomagnetic stratigraphy to these East African sedimentary formations has placed them on a chronometric time scale. We can be sure that all these documents relate to the time span between about 2½ and 1½ million years. Within this span, some time placements are debatable, Olduvai Bed I is securely dated to 1.83 ± .13 m.y. (Curtis and Hay 1972), but there is current discussion about whether the KBS Industry at East Rudolf is really as old as 2½ million years, or whether it might be closer to 2 million years.[1] However, the chronological uncertainties within East Africa are on a much smaller scale than those affecting estimates of the age relations of paleoanthropological sites in most other regions. For instance estimated ages for Makapan and Sterkfontein range from more than 3 million years to perhaps less than 2 million years. These uncertainties notwithstanding, the following important sites outside East Africa are presumed to be at least in part coeval with the East African material under review:

North Africa: Ain Hanech (Arambourg 1950; Balout 1955; Howell 1960)
Sidi Abderahman and related sites (Biberson 1961)

Ethiopia: Melka Kontouré (Chavaillon 1967, 1970)

South Africa: Makapansgat (Dart 1957a,b; Tobias 1967)
Sterkfontein (Robinson and Mason 1962; Tobias and Hughes 1969; (Leakey, M. D., 1970b)
Swartkrans (Brain 1970; Leakey, M. D., 1970b)

Some of the Javanese paleoanthropological sequence may also overlap this time range, but because of chronological uncertainties and because archaeological investigation *(sensu stricto)* has not been carried out, this sequence has not been considered further. All archaeological sites in the temperate zone of Eurasia now appear to be younger than 700,000 years. (See Butzer and Isaac, in press; Isaac and Curtis 1974)

What follows is necessarily a rather cursory survey of the information now available under each of the four headings listed above. By way of

[1]Subsequent to the writing of this paper two significant changes have occurred: (1) New dates on the KBS tuff indicate an age of 1.6–1.8 m.y., so that the KBS industry is broadly coeval with the Oldowan at Olduvai (Curtis pers. comm.) and (2) Jean Chavaillon has found artifacts in member C at the Omo with an age between 2.4 and 2.7 m.y. (Announcement by Chavaillon at a geological society meeting London, Feb. 1975).

conclusion I offer some speculations on overall interpretation—one archaeologist's "model" of the evolution of hominid behavior. I have reviewed elsewhere the archaeology of the subsequent "Middle Pleistocene" period (Isaac, 1975).

STONE ARTIFACTS

Stones which show clear signs of having been fractured by hominids occur in Plio-Pleistocene deposits in several parts of Africa. These occur both as isolated stray pieces and in concentrated patches on old ground surfaces. For obvious reasons it is the material from the concentrations that is most often studied. At the best investigated and most important localities we can be certain that an unusual agency, presumably hominids, was concentrating stone and causing its fracture in a systematic way. This process generates two distinctive classes of objects: lumps of stone from which flakes have bee removed, and comparatively thin slivers of stone fractured from larger blocks. Clearly the two series are in part complementary, though assemblages are not always balanced with regard to numbers of each or of raw material composition. The first series includes choppers, discoids, polyhedrons, and some heavy-duty scrapers, etc., which can perhaps be conveniently grouped under the collective term "core tools." The second class includes flakes, flake fragments, and angular fragments, etc. Where no further modification is evident, the term "debitage" is often used. Items showing chipping and/or battering attributable to damage in use are termed "utilized"; items showing modification by the removal of small trimming flakes are designated as tools and are termed "scrapers," "awls," "burins," etc. as appropriate. The structure and definition of this terminology is clearly set out in Dr. Mary Leakey's monograph (1971:3–8) and certainly does not require detailed repetition here. For the benefit of those not extensively initiated into the mysteries of stone artifact terminology, let me stress that in spite of the literal, non-technical meaning of the words used, the distinctions involved are *morphological;* they do not depend on recurrent judgements about *function.* Thus the distinction between a tool and a non-tool depends largely on whether the piece is judged to show purposive design for use, not on whether the piece was actually used as a tool. For this reason such items as choppers, polyhedrons, and scrapers which are systematically shaped by flake removals are classified as tools, while raw flakes are relegated to debitage, in spite of the fact that many of us believe that these were very important as implements. The term "artifact" remains neutral and covers anything shaped by the hand of man, whether by design or as a by-product; it can be contrasted with the category of "manuports" which are items believed to have been introduced but not shaped by man. Figure 1 summarizes the frequency distributions of percentage incidence of these

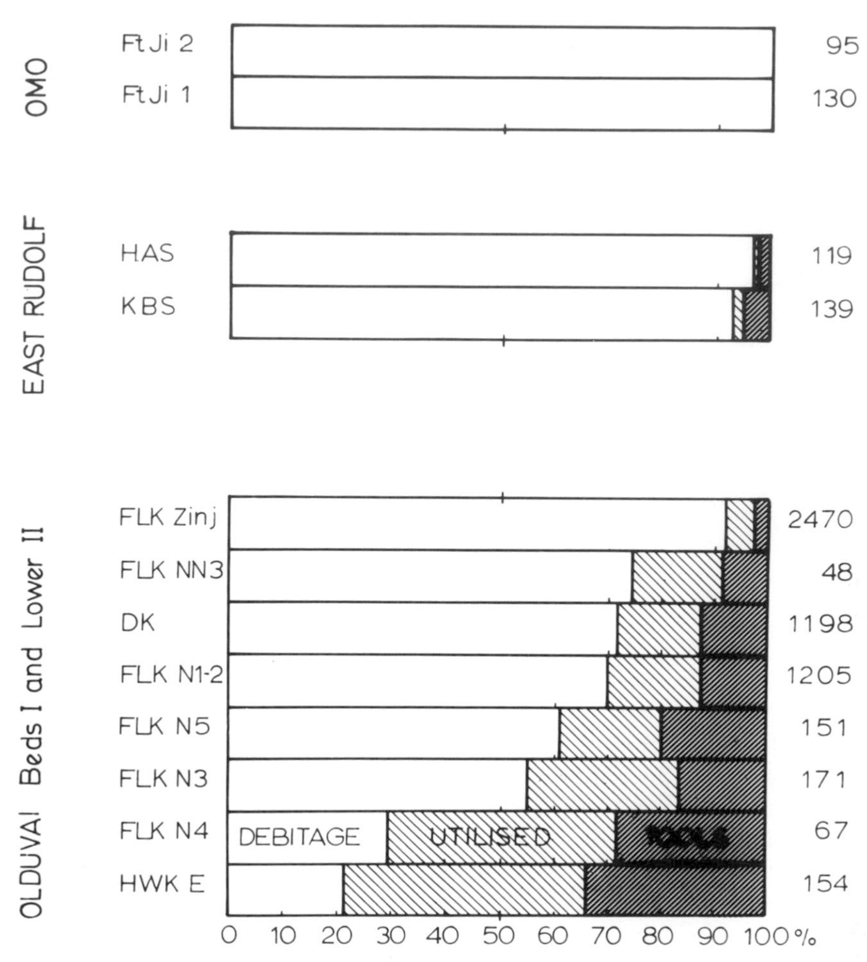

FIGURE 1
Percentages of "debitage," utilized, and tools in early stone industries from East Africa.

major categories in Plio-Pleistocene assemblages. The question may now be asked—what can these simple manufactures teach us about human evolution? What useful question can we pose with some hope of getting valid answers?

It seems to me that artifact assemblages contain potential information on the following broad topics:

1. The *levels of craft competence* of the makers and the *intensity of involvement* with or *dependence* on equipment (material culture).

2. *Function* and *adaptation,* the need for particular forms of artifacts in economic and social life.

3. *Tradition,* that is, culturally determined craft practices and design norms. This information can in turn be used to attempt assessments of culture-historic interrelations and distinctions.

4. The extent of *transportation of some materials.*

Until recently most of the endeavors of palaeolithic archaeology were concerned with topic number three and to a lesser extent with topic number one. We know distressingly little about the function of stone artifacts, partly because of lack of emphasis on this in enquiries and partly because clear evidence is difficult to obtain.

Before trying to unravel these strands of information from the tangle of an assemblage, it is necessary to ask what determines the form of a stone artifact? As a gross simplification one can enumerate four primary factors:

1. The physics of conchoidal fracture.

2. The size, form, and detailed mechanical properties of available stone.

3. The needs of the makers for certain kinds of working edges.

4. Prevailing cultural norms with regard to artifact form (= "fashion," "style," "tradition").

When comparatively simple artifact forms are compared, the possibility has to be borne in mind that some similarities and some differences may have been occasioned by purely mechanical contingencies; these should not be interpreted as being indicative of important cultural connections or distinctions. An example of this phenomenon may be the marked differences between the Shungura Formation assemblages made on small quartz pebbles and other Plio-Pleistocene assemblages made from lava cobbles and blocks. We simply cannot tell whether the Shungura craftsmen were more or less advanced, and also we cannot tell what cultural norms they held in common with Olduvai and Koobi Fora craftsmen.

Levels of Competence and Intensity of Involvement with Implements

The earliest known archaeological sites are marked as documents of hominid behavior by the presence of clear-cut artifacts. These relicts signal that a major behavior shift in the direction of the human condition had already occurred. Without the artifacts it is very difficult to identify traces of hominid behavior of any kind, as can be seen from the continuing uncertainty about the osteodontokeratic culture and the Makapan evidence for the predatory behavior of "*A. africanus.*"

One question that arises is as follows—was the process of initial involvement in stone artifact production a long, slow, gradual development, or have thresholds been crossed, that is to say, innovations which rapidly brought about a certain technological situation which may then have persisted for long periods with little change?

We do not yet have enough evidence to answer this question, but it seems to me that the actual trajectory of development probably lay somewhere between even, steady progression and a stepwise pattern of change. These are simply opposite extremes in possible conceptualization of the origins of stone tool manufacture. It does seem likely to me that the empirical discovery of the effects of conchoidal fracture was a threshold. This discovery immediately creates two families of artifact form—sharpened lumps of stone (core tools) and sharp slivers and fragments such as flakes. As I have indicated, these are in fact the fundamental components of the earliest known assemblages.[2]

Clearly some hominids, by the time of Olduvai Bed I, had firmly established the basic techniques of achieving conchoidal fracture (Leakey, M. D., 1971). If, as seems likely, the Koobi Fora Formation, Lower Member sites are substantially earlier, then this position had been achieved by about 2½ million years ago. (Leakey, M. D., 1970a; Isaac *et al.* 1971). There seems no way of judging how much further back the record of these practices will be traced. We have to get out into the field and search.

Louis Leakey found a stone in the excavation at the 14-million-year-old Miocene site of Fort Ternan, which seemed to him to have been damaged by use (Leakey, L. S. B., 1968). He also concluded that the variety of lava of which the stone is composed is extraneous to the deposits. In default of detailed geological studies it is hard to evaluate the possibility that the stone is a manuport. Unfortunately also the signs of use, while suggestive, are not unambiguous. For the time being, tool-using activities by *Ramapithecus wickeri* seem to me to be possible but unproven. (See Andrews and Walker this volume for more information on Fort Ternan.)[3]

One possible view of the three earliest excavated artifact assemblages (Olduvai I, Shungura F, and Koobi Fora Lower Member) is that they each represent the intersection of elementary control over stone fracture

---

[2]A partial exception may be the Shungura Formation Industry (Merrick *et al.* 1973; Coppens *et al.* 1973). In this instance it seems likely that the artifacts were generated by smashing small pebbles. Under these circumstances the core tool series is very poorly represented.

[3]I am omitting from discussion the so-called Kafuan artifacts of East Africa since there is now serious doubt as to whether they are really recognizable products of hominid activities (Bishop 1959).

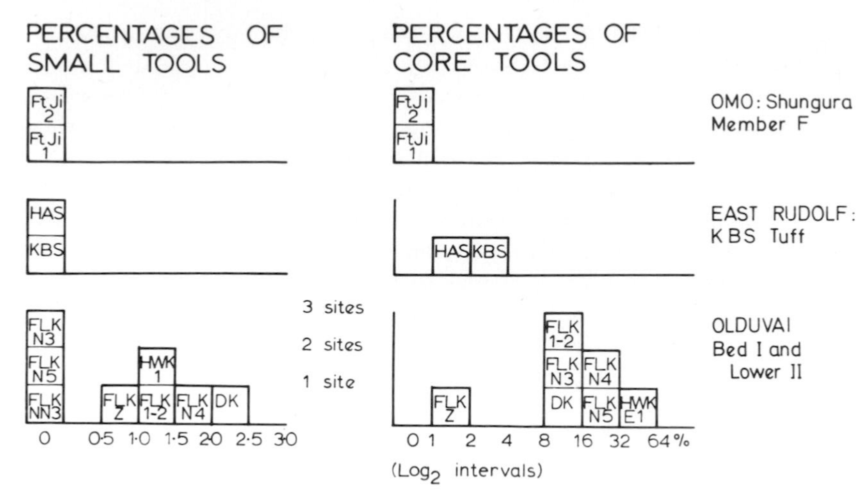

FIGURE 2 .
Frequency distributions of observed percentages of tool categories.

with the form in which stone was available in the particular area. Under this view, the specific design-tradition element in each of these would be small. This is a possibility that I think should be considered; I do not regard it as in any way established. Elimination or validation will require the development of fresh procedures of analysis since it cannot be ascertained from typological analysis.

Can we detect any progressive changes in the known Plio-Pleistocene record of technology, that is, in the period before one million B.P.? The crucial evidence in this connection comes from Olduvai Gorge where Dr. Mary Leakey has shown very clearly that there were important changes:

> There was an overall increase in the tool kit . . . the average number of tool types for seven Oldowan levels in Bed I and the base of Bed II amounts to only six; the figure has risen to over ten for nine Developed Oldowan levels in Bed II. (Leakey, M. D., 1971:269)

In the Koobi Fora Lower Member and the Shungura samples which are currently believed to be slightly older than the better known Olduvai series, even fewer tool type categories *sensu stricto* are represented (see Figure 2). This could be an illusion created by inadequate sampling, but it is suggestive.

Some time between one and one-and-a-half million years ago, handaxes and cleavers began to be made. In some cases these were made by the application of bifacial retouch to large flakes which were also a technical novelty. Where these forms preponderate the whole aspect of the assemblages appears suddenly quite different from all the previous assemblages and from some contemporary ones. Dr. Mary Leakey summed up the situation saying of Middle Bed II:

> Up to and including this level there is no suggestion whatever of duality in the known lithic assemblages which are remarkably constant in character. . . . In Middle Bed II, however, the first unquestionable bifaces appear, . . . At some sites they occur in sufficiently large numbers (over 50 percent of the tools) for the term "Acheulian" to be applied without hesitation to the industry. . . . At other sites bifaces occur in small numbers (an average of 6 percent of the tools) in assemblages that are otherwise wholly Developed Oldowan in character. (Leakey, M. D., 1971:269)

At East Rudolf a similar situation is being observed—the contrast between the KBS industry (Lower Member ~ 2½ m.y) and the Karari industry (Upper Member ~ 1½ m.y.) in the Koobi Fora Formation differs in its specifics, but it also involves a great increase in the number of types and a rise in assemblage diversity. Bifaces also seem to appear in the later Karari industry, but so far have never been found dominating an assemblage. Mary Leakey's data also show a marked rise in the maximum density of artifacts (Figure 3 based on Leakey, M. D., 1971:260). Our

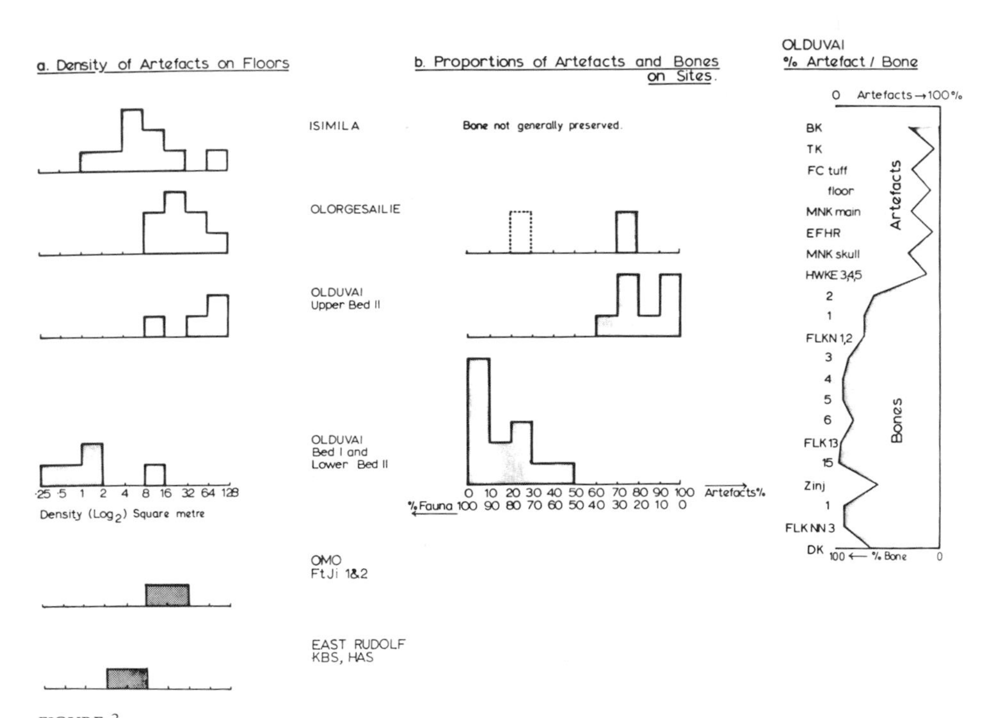

FIGURE 3
Diagrams representing changes in the density of artifacts and bone on occupation floors.

data at East Rudolf show a similar contrast between the early and the late sites with a rise of at least an order of magnitude. It is hard to know what may be the behavioral significance of this rise in density. Factors to bear in mind include the following:

(a) Stone toolmaking had become a more facile and habitual behavior.

(b) Sites were more intensively occupied (longer duration and/or more recurrent visits).

(c) Development of better bags/baskets for carrying raw material to site.

In summary the East African evidence does document progressive developments. Two arbitrary divisions in a continuum can therefore perhaps be recognized. First, an early set of industries is known which has less typological diversity and perhaps lower occurrence densities. This includes the classic Oldowan of Bed I and Lower II, the KBS industry, and presumably the Shungura industry, although this is different and distinctive. This series is believed to span a time interval of ~ 2.6 to ~ 1.6 million years. We do not know whether the phase began with the rapid establishment of stone knapping following the "discovery" of conchoidal fracture, or whether there is a long antecedent sequence of slowly developing stone craft. We will have to find this out by patient searches.

A later set of assemblages can also be recognized, which are still of Lower Pleistocene age (i.e., 1.5–0.7 m.y.). These involve the addition of a variety of new types to the total artifact repertoire, a propensity for assemblages to vary in composition from one site to the next and higher artifact densities. The Lower Acheulian and the Developed Oldowan Industries of Middle and Upper Bed II, Olduvai, fall into this set as does the Karari Industry of East Rudolf. How do the South African and North African occurrences compare with this apparent sequence? Having examined, by courtesy of C. K. Brain and P. V. Tobias, parts of the collections from Swartkrans and from Sterkfontein Extension site, I would venture the opinion that both of these belong morphologically with the later set of East African Plio-Pleistocene assemblages. This is in accord with M. D. Leakey's findings (1971) and with earlier reports by Robinson and Mason (1962). I do *not* wish to use the morphological matching to suggest a relative date for these assemblages, since we are all trying to get away from this kind of circular procedure.

Unfortunately we have no complete assemblages from the Plio-Pleistocene of North Africa which were recovered by excavation and which include a fair sample of both core tools and flakes, etc. In this region, as in East Africa, there seem to be both pre- and post-biface industries, but there is no means of knowing whether the change occurred at a similar time. The abundance of spheroids at Ain Hanech, coupled with the presence of bifaces in an immediately overlying layer (Balout 1955), is suggestive of similarities with the Developed Oldowan.

Function

We have as yet no direct evidence of the functions of early stone tools, but there is strong circumstantial evidence for associating them with some activities which we can surmise would have been facilitated by their use. At both Olduvai and Koobi Fora, the early artifacts form scatters coincident with patches of broken-up mammal bones which would seem to constitute evidence for the cutting up and consumption of animals ranging in size from rodents to pachyderms. It is hard to conceive of a large carcass being dismembered by smallish primates without cutting tools, so that tools may fairly be regarded as an integral part of an adaptive complex which also involved hunting and food sharing. In this connection it should be pointed out that there has been a growing awareness of the tendency for butchery to involve small, comparatively simple tools (Clark and Haynes 1969). To be useful, a sharp or pointed piece need only be large enough to be held between index finger and thumb with the point or edge projecting—that is to say, sharp pieces from 10–15 mm. in diameter are potentially useful. This realization received dramatic reinforcement during fieldwork at Lake Rudolf. In our presence, a young Shangilla pastoralist found himself without a metal knife while confronted by a fresh lion-killed antelope carcass. Spontaneously he tapped off a very small flake from a lava cobble and proceeded to slit the skin covering a cannon bone, which he then peeled and cracked open to obtain marrow (Figure 4). Clearly the minute flakes used would have been

FIGURE 4
A young boy of the Shangilla, or Dasenitch, tribe uses a small stone flake to slit the skin of an antelope carcass.

equally effective for cutting articular sinew. From this and related observations (e.g. Gould 1968) it emerges that the small sharp flakes and flake fragments, which by convention we label as waste or debitage, may well have had crucial adaptive importance. Perhaps it is significant that this is, in fact, the preponderant class of material in all the early stages.

It seems unlikely that any of the early stone artifacts were actually fashioned as a weapon. However, if improvements were made to the suitability of sticks and branches for use as spears or clubs, then presumably the sharp edges of the flakes would have served to notch and whittle, while the stouter, jagged chopper edges may have been effective for hacking. It is not inconceivable that the acquisition of bark trays, such as those found at Kalambo Falls would have been facilitated by the use of flakes and core tools.

In the regrettable absence of more direct evidence, the matter can also be viewed in another way—the known stone tool kit of Plio-Pleistocene sites would be quite adequate to perform all of the essential tasks of tropical, nonagricultural peoples, namely to butcher carcasses, to make wooden spears and to make containers. These are integral parts of human adaptation.

Tradition and Culture History

Distinctive features of stone artifact assemblages can be attributed to differences in the traditions or cultures of the hominids that made them. Clearly before this is done it is desirable to distinguish features which may have been induced largely by *differences in raw material,* and differences which may reflect *varied activities* by the same people at different times and places. The distinctiveness of the Shungura industries *vis-à-vis* Olduvai and Koobi Fora may be an example of differences induced by contrasting raw materials, which therefore cannot be interpreted as necessarily indicative of other cultural or developmental stage differences. There are differences also between the Olduvai Developed Oldowan and the Sterkfontein industry that could as well be due to the influence of material as to cultural difference *per se.*

The dichotomy between Acheulian and Developed Oldowan industries in Bed II at Olduvai cannot be explained in terms of raw material. A lively and as yet unresolved debate continues about the relative merits of several possible models which might help account for the observed pattern (see Isaac 1972). Distinct cultural groups perhaps stabilized by coincident species differences is one possibility (Leakey, M. D., 1971:269–273). Others have suggested differentiation of activities (e.g. Clark 1970:85–86).

Transport

All of the important Plio-Pleistocene archaeological sites involve the transport of some materials prior to their being discarded at the sites. Full details of the systems have not been worked out, but in most cases the distances involved need not be more than a few kilometers. The quantities at sites range from a few kilograms to perhaps a hundred or so. These are guesses, since total weights have been worked out for very few sites. Since we do not know any of the critical variables such as how many hominids for how long, we cannot yet work out the minimum number of hominid trips that might have been needed—with and without containers to augment the amount carried at one time.

A very early "factory site" was excavated at Olduvai by Marie-Lou Harms working in conjunction with Dr. M. D. Leakey. D. Stiles carried out laboratory analysis and has reported on this material (Stiles *et al.* 1974).

BONE REFUSE AND DIET

Many of the early sites at Olduvai and the two excavated sites in the Lower Member of the Koobi Fora Formation consist of coincident patches of stone artifacts and scatters of broken-up bones (Figure 5). The conclusion seems inescapable that the same hominids that made the artifacts concentrated the bone. It also seems virtually certain that the bone was the residue discarded after the consumption of meat. These sites provide us with positive evidence that at least some toolmaking Plio-Pleistocene hominids were partially carnivorous; but what we do not know at these or other sites is to what degree vegetable and gathered foods were also important. I have argued elsewhere that the evidence is at least consistent with the notion that tropical hominids have always had a two-pronged opportunistic subsistence strategy involving both hunting and foraging as reported for !Kung, Ba Twa, Hadza, Australian aborigines, etc. (Isaac 1971; Lee and DeVore 1968). I continue to regard this as highly likely.

The excavated archaeological sites of the Shungura Formation lack significant bone concentrations, as does the site of FxJj 10 at Koobi Fora; but these cases cannot be used as proof of meat-free diets, since they are just as likely to be due to poor preservation of bone.

Mary Leakey's report on the Olduvai sites is, as usual, the most important source of evidence (Leakey, M. D., 1971:ch.IX). It emerges that the early hominids were eclectic in their tastes, and bone remains range from mice to elephants. At most sites medium-sized antelopes predominate in the remains, as they do in the fauna itself when it is on the hoof.

FIGURE 5
Part of an antelope pelvic bone is uncovered at the site of KBS, in East Rudolf, where excavation has revealed an area in which discarded stone artifacts coincide with a patch of broken-up bones. This evidence along with comparable finds in Bed I at Olduvai suggests that by the early Pleistocene some hominids were not only eating meat, but were carrying it back to home bases to share.

The notion that these sites involved mainly small species and young animals has not been sustained.

There are also a small number of Plio-Pleistocene sites where the partly broken-up carcass of a single large animal has been found with artifacts scattered around. These are indistinguishable from sites termed butchery sites by the archaeologists. In the KBS Tuff (East Rudolf) a hippo carcass is represented at the site of HAS, while at Olduvai FLK N an elephant and a deinothere are present at different levels. At this time it is hard to tell the extent to which the meat involved in these cases was acquired by hunting as opposed to scavenging. Schaller, Kruuk, and others have shown that there is no fixed distinction between hunters and scavengers among large carnivores. (See Schaller and Lowther 1969). On this and other grounds I am inclined to think that the hominids were hunters, although they may also have pirated and scavenged other animals' kills.

Recent field studies of nonhuman primates have shown that some other species, notably chimpanzees and baboons, are predatory and carnivorous. However, meat appears as an occasional, opportunistic extra in diet rather than a regular component, and the prey species are all much smaller than the primate predator. The archaeological evidence from East Africa strongly suggests that by 2–2½ million years ago, some hominids, at least, were eating meat regularly and included prey as large or larger than themselves. I would join Washburn and others in regarding this as a highly significant adaptive shift (e.g. Washburn and Lancaster 1968; Campbell 1966; Pilbeam 1972, etc.)

Dr. Mary Leakey has shown that some changes occur in the character of bone refuse between Bed I and Upper Bed II (1971:248–256). Notable points of contrast include: a rise in the ratio of artifacts to bone fragments (Figure 3); an increase in the degree to which the bone is broken up and rendered unidentifiable; a change in species composition, with very large animals being very slightly better represented in the later sites. In particular, the extreme predominance of bovids drops from 70–80% in Bed I, to 40–50% in Upper Bed II, while equids and hippos become more conspicuous components; the first known instances of successful decimation of animal herds occurs in Upper Bed II (*Antidorcas recki* at SHK and *Pelorovis* at BK).

We do not yet have sufficient data to judge whether the rather slight and subtle changes observed at Olduvai reflect widespread progressive shifts or local adjustments. There is bone associated with the Karari industry at East Rudolf, but studies are not yet far enough advanced for comment on its implications.

I find myself unable to assess the significance of comparative data from other regions. The Transvaal cave sites yield hominids in association with an abundance of other animal bones, but in no case does the evidence appear compelling that the hominids were the principal bone concentrating agency. Brain (1970) has provided convincing arguments that at Swartkrans, leopards were involved in the emplacement of at least some of the hominid bones. Investigations are currently in progress at Makapansgat which should help to clarify the mode of accumulation involved there (Tobias pers. comm.).

In Morocco, the very early sites are largely devoid of significant bone accumulations. Bone was associated with the artifacts at Ain Hanech and included *Anacus, Elephas,* rhinoceros, *Stylohipparion, Equus, Hippopotamus, Omochoerus, Libytherium, Giraffa, Bubalus, Bos, Gazella, Oryx, Numidocapra,* etc. (Balout 1955:163). Unfortunately it is not clear to what extent these constitute food refuse associated with the artifacts.

SPATIAL CONFIGURATIONS AND ECOLOGICAL OPPORTUNITIES

I have recently reviewed aspects of this evidence and will offer only a very brief commentary here (Isaac 1972). Traces of the life and death of early hominids are not evenly scattered over the palaeolandscape. Once we have distinguished between effects due to life patterns and effects due purely to preservation differences, we can attempt to learn about ecology from the observed configurations. The first point to stress is that we have an extremely biased sample from the Plio-Pleistocene in East Africa. As yet no site is known which can inform us regarding hominid activities on the hills and uplands; all our data concerns the alluvial plains and shoreline marshes of sedimentary basins.

Within the sedimentary basins the archaeological material is not evenly distributed; there is commonly a widespread, low-density scatter of artifacts, and then there are concentrated patches of discarded artifacts. Eventually we will get around to systematic studies of the dispersed material and may learn much from them. However for the moment, one is obliged to restrict comments to the meaning and location of the concentrated patches. These are indistinguishable from vestiges that in later periods are known as camp sites, and they seem to indicate that the movements of Plio-Pleistocene hominids were organized around a home base. This is a distinctive feature of human behavior *vis-à-vis* other hominoids and has its closest parallels in social carnivores, birds, and social insects. (See concluding section.) The sizes of the early sites seem to be very similar to the sizes of later Pleistocene sites, namely 5 to 20 meters in diameter.

Comparative studies suggest that while sites occupy very diverse settings, the artifact concentrations tend to be associated with stream channel situations. Even sites that are on a floodplain soil substratum are often immediately adjacent to channels (e.g. Koobi Fora FxJi 11 and 20, Omo

FtJj 2, etc.) The association appears to become more pronounced after about 1–1½ million years ago. There has been some discussion of the reason for this site location pattern (Isaac 1966, 1972; Leakey, M. D., 1971:259). I tend to think that one important factor has been the propensity of trees and bushes to be distributed as ribbons fingering out along channels into the edaphic grasslands of the basins. These gallery strips would have provided shade, cover, fruiting trees, and bushes—and something to climb when predators came. It is almost a metaphysical point, but there is some interest in the fact that hominids, while colonizing the savanna, may have preferred to keep their home bases in strips of woodland that extended out into more open country, rather like tourists transporting themselves into Hilton hotels in alien lands! In this connection it is perhaps significant that the only horizon at the Omo where artifact concentrations have been found in large numbers, is Member F (Coppens, Chavaillon and Beden 1973; Merrick *et al.* 1973). The horizons are associated with indications that the normal axial hydrographic pattern of the proto-Omo river may have broken down in favor of a braided, anastomosing network of smaller stream courses (J. de Heinzelin pers. comm.).

In summary, the palaeoenvironmental reconstructions of the basins that preserve Plio-Pleistocene archaeological traces suggest a mosaic of diverse habitats—beaches, reed beds, swamps, edaphic grassland, savanna, riverine woodland and bush, some gallery forest. All we know for certain is that at least two species of hominids were active at least from time to time among this varied stock of opportunities. We do not know whether the presence of either or both was seasonal. By stressing some habitats rather than others one could reinforce arguments in favor of seed eating, rhizome eating or hunting—but I strongly suggest that it was the *diversity of resources* that may have been attractive. Hominid adaptation as we know it from later times has flexibility and opportunism as its hallmark. Very likely this began early in human evolution.

GEOGRAPHY AND CHRONOLOGY

The best-dated Plio-Pleistocene traces of hominids come from the savanna[4] country of Eastern Africa, where local conditions of preservation were optimal, and where perhaps also conditions of life were particularly favorable. The most prolific source of fossils are still the various Transvaal cave sites, which are also in savanna country *(sensu lato).* Although not well dated they certainly fall within the range 4 to 1 million years. Other occurrences within Africa that may be of this age range include the Chad

[4]The term *savanna* is used to cover vegetation that is neither desert nor woodland forest.

fossil and the Maghreb artifacts (Ain Hanech, the *Civilisation du galet amenagé,* and the *Acheuléen ancien,* etc. of Biberson 1961). All of these are in areas that are environmentally intermediate—that is, not forest or desert.

Distribution maps of African prehistoric remains commonly show a plethora of "Pebble culture" (= Oldowan) occurrences including sites in Central Africa, West Africa and all over the Sahara. Most of the plots represent undated, typologically designated surface collections. Given the improbability of the survival of recognizable artifacts for one or more million years, without their being encased in a sedimentary formation, most of these records must be discounted. The absence of confirmed and dated finds of this age in the forested zones of the Congo basin and of West Africa cannot be interpreted since suitable deposits have not been located and searched. However within East Africa there are indicators that the optimal zone of early hominid occupation lay more in the drier east than in the moister west. Extensive prospecting on partially fossiliferous exposures of great thickness from Plio-Pleistocene deposits in the Western Rift has not revealed hominids or stratified artifacts (e.g. Bishop 1971). It seems inconceivable that these could have been as well represented as in the Eastern Rift. The only possible exception is the locality of Kanyatsi where de Heinzelin (1960) found artifacts that may be of Plio-Pleistocene age.

Outside Africa there are very few artifactual or fossil traces of Plio-Pleistocene hominid presence that are more than 0.7 million years old (i.e., the age of the Matuyama-Brunhes palaeomagnetic boundary and a convenient base of the "Middle Pleistocene"). These include Ubeidiya in Israel which is probably ~ 0.6–0.8 million years old (Horowitz *et al.* 1973; Bar Yosef pers. comm.). The Djetis and Trinil faunal zones of Java which possibly span from 1½ to ½ million years ago (Jacob 1972; Isaac 1972; Pilbeam, in press) and perhaps the Grotte du Vallonet near Nice.

For what it is worth then, available archaeological data is consistent with a tropical distribution of Plio-Pleistocene hominids, perhaps centered in Africa and extending into Southeast Asia. Within Africa the hominids may well have been predominantly savanna forms.

ARCHAEOLOGY AND MODELS OF HUMAN EVOLUTION

It has become fashionable for all kinds of research workers to present models intended to elucidate the process by which the early hominids became differentiated. To mention but a few: Lancaster (1967) wrote from the stance of primate studies; Fox (1967) tackled the questions from the viewpoint of a social anthropologist; Reynolds (1966, 1968) used chimpanzee studies as a basis for suggestions; and Jolly (1970) offered a very specific seed-eating model inspired by patterns in comparative anatomy and by his interpretation of the development of masticatory mechanisms. More or less comprehensive models have been put forward by human biologists such as Washburn (1960, 1965, 1968a,b), Campbell (1966), and Pilbeam (1972). These latter are all in my view convincing in their broad outlines. All share the practice of treating anatomy, diet, tool involvement, and behavior as interrelated parts of an integrated system of adaptation. These versions draw in a general way on the main features of the archaeological record, but do not treat it very specifically. In effect, the only discipline which has not yet had a fling at articulating a hominid differentiation model are the archaeologists. I venture to offer the sketch of one, because archaeologists may have a vantage point that can lead to improved reconstructions. One of the most distinctive aspects of human evolution has been the development of self-recording behaviors. The great apes for all their intricacy and interest leave no archaeological record.

I find it easiest to approach the model building proposition by way of a comparison of human behavior with the behavior of man's closest living relatives, the African apes (Figure 6). Clearly the most relevant material for comparison on the human side are the living arrangements of peoples not dependent on agriculture. Among the apes, the chimpanzee is closest, for reason of ecology and also of phylogeny, on the evidence of biochemistry. (See Goodall in this volume.) The comparison of behaviors defines the challenge involved in seeking to make models; it is done without any intention to imply that human behavior is derived from chimp behavior. However it seems certain that behavior patterns that are prominent in man but absent or very weakly expressed in other primates must have been developed during the evolutionary divergence of the hominids. The chimp studies provide a means of preparing such a list as a basis for discussion. In Table 1 I have set out an oversimplified summary of features that I would judge to have particular relevence. I have distinguished items for which direct physical evidence can be sought, while those which have some parallels among social carnivores are marked with an asterisk. Let me turn now to the Plio-Pleistocene record and see what we know about the position of at least some hominids with regard to the points of contrast.

- Some hominids were fully upright bipeds (Napier 1967; Robinson 1972).

- Most hominids had slightly larger brain/body weight ratios than the apes, and some had reached a condition intermediate between apes and men: brain size 600–800 cc. with small to moderate body size.

- Teeth and the masticatory mechanisms had been rearranged, including loss of large canines.

- Infancy may have been even more prolonged than in the apes (Mann 1968).

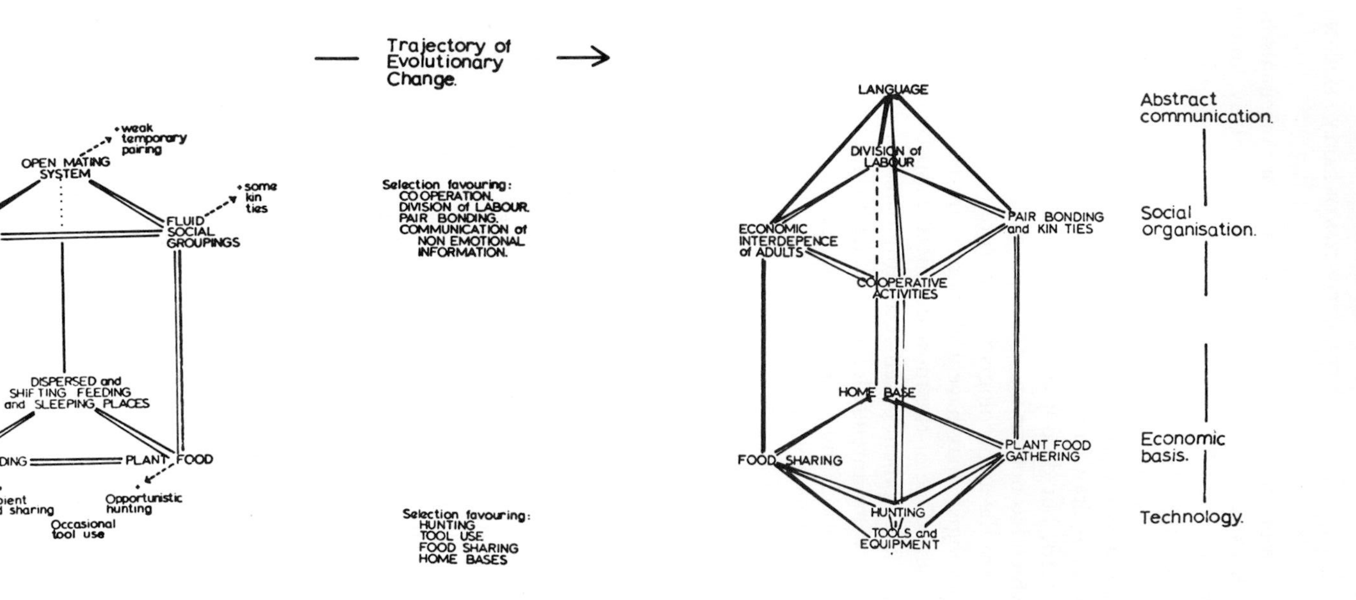

FIGURE 6
Diagrammatic representation of graphic "models" of hominoid behavior and of human behavior. Each is shown with a material, ecological and economic substructure and a "social" superstructure. The development of the human configuration has involved both reorganization of the central "framework" and the addition of new levels—equipment and language. Archaeological evidence suggests that some Plio-Pleistocene patterns had the human central configuration but they seem to have lacked elaborate development of the extensions.

- Some hominids were extensively involved in tool manufacture, but we do not know to what degree they were dependent on equipment.

- Significant quantities of meat were eaten and this derived from animals far larger than any hunted by other primates.

- Concentrated patches of discarded artifacts and food refuse strongly suggest home-base behavioral arrangements and food sharing to some degree.

- The known extent of ecological and geographic distribution *may* not have been much greater than that of several other primate species.

Fragmentary as the evidence is, it strongly suggests the existence, by the Plio-Pleistocene, of some hominids which had already diverged from the last common ancestor with an ape, to the point where a whole series of adaptive novelties had become firmly established. To find parallels for some of the traits, one has to turn not to other primates but to social carnivores, or even social insects. Some hominids had become free ranging, striding bipeds, using tools (and weapons?), hunting large animals, and carrying at least part of the meat back to camps for sharing. But this is not at all the same as saying that these hominids were men in the way we

TABLE 1
Changes which have occurred in the evolutionary development of mankind since divergence from the last common ancestor shared with any living ape

I ANATOMICAL AND PHYSIOLOGICAL CHANGES	
A. Visible in fossils.	B. Not readily visible in fossils.
Modification of hind limbs for full bipedal locomotion.	Changes in skin and skin glands.
Modification of hands and arms for more effective carrying and tool use.	Reduction in body hair.
Enlargement of the brain.	Continuous sexual receptivity of female.
Reduction and remodelling of the jaws.	Partial reorganization of the brain.
	Modification of the vocal tract.
II BEHAVIORAL CHANGES	
A. Directly detectable by archaeology.	B. Not directly detectable by archaeology.
Increasing dependance on manufactured *equipment* and *tools*.	*Language.*
*Great increase in *meat eating* (= hunting until recently).	Subtle controls on displays of emotion.
*Increasing interdependence through *food sharing.*	*Great increase in *cooperation* and *division of labor.*
Reorganization of behavior around a "camp" or *home base.*	Great increase in social bonding mechanisms: "marriage," kinship, reciprocation.
	Great increase in symbolism.

**Traits shared with social carnivores.*

understand the term today. With Jane Lancaster (1967) I am inclined to think that we are dealing with traces of an original adaptive system that was perfectly successful in its day and which has no living counterpart. One variant of it happens to have been transformed by further evolution into the human situation which thereby has features in common.

If the basics were established by 2 to 3 million years ago, what has happened since? Palaeoanthropology can fairly readily document the following changes:

- Further enlargement, and presumably continued reorganization, of the brain.
- Increased complexity of tools and equipment with regard to both the number of designs and the degree of precision in execution.
- The augmentation of art, ritual, and symbol systems.
- Great increases in ecological and geographic range.

It is also tempting to surmise that the development of human levels of capability in communication ("language") has been extensively post-Pliocene, though I am inclined to guess that the early hominids already had rudimentary sound communication signal systems beyond those of other primates. Of course the enlargement of the brain has in part been connected with language and association.

I suggest therefore that one can conveniently divide the continuum of human evolution into two phases. Probably the first involved shifts in the basic systems of locomotion and subsistence, plus two new ingredients—tools and food sharing. This led to a pattern of adaptation which in hindsight we see as protohuman, but which is probably better termed "early hominid," since it was probably a nonhuman system that was effective in its own right. As I envisage it, the early hominid adaptive pattern created a situation in which, during the Pleistocene, natural selection favored the exaggeration of certain qualities that facilitate making a living by being an opportunistic, food-sharing hunter and forager. These qualities include communication and information exchange, restraint, cunning, economic and social insight, plus ability to augment anatomy with diverse equipment.

Now these suggestions are speculative in that we have no direct information on the stages by which man's linguistic capabilities developed. However it seems reasonable to me to suggest that the main phase of selection for this capacity has been since the Plio-Pleistocene. I and others have argued that there is archaeological evidence of a critical threshold in design complexity that was crossed in the Upper Pleistocene, and which may have been related to an analogous threshold in linguistic efficacy. (Isaac 1972).

The existence of pair bonds between human males and females and the incorporation of family modules into social and economic structures have been universal characteristics of our species. Since there is no direct fossil evidence, we really do not know whether this arrangement in some degree was primitive to the basal hominoid stock, whether it was part of the first early hominid adaptive pattern, or whether it has been a Middle Pleistocene development. Since it could easily have been functionally interconnected with food sharing and division of labor I would surmise its development during the first adaptive shift.

The divergence of the hominid lineage(s) from the last common ancestor with one or more pongid lineages lies in the time range prior to that on which the attention of this essay is focused. We have no archaeological evidence and very little fossil evidence regarding either the steps by which the transformation proceded, nor do we know much about the environmental context of the population involved. If one follows some biochemists (e.g. Sarich and Wilson 1968) then one infers that the divergence occurred no more than 4 to 8 million years ago; while if one accepts *Ramapithecus (Kenyapithecus)* as an acestral hominid, then the point of divergence was at least 15 million years ago (Leakey, L. S. B., 1962; Simons and Pilbeam 1965). It seems to me that the Plio-Pleistocene pattern of skeletal and archaeological traces (3–1½ m.y. B.P.) represents a situation which could equally well have been derived by either a shortish period of relatively fast evolutionary change, or by a long period of slower, perhaps sporadic development. To decide finally between these rival hypotheses, we badly need more complete *Ramapithecus* fossils and an interpretable series from the time span between 10 million years and 4 or 5 million years ago. (See also Andrews and Walker this volume.)

In accounting for the origins of the first adaptive shift, it has been customary to invoke as causes, more or less drastic environmental changes such as a Pliocene drought (Ardrey 1961; and many others), or dramatic dietary specializations such as hunting or seed eating (Jolly 1970). However as research progresses, it becomes increasingly apparent that there has probably not been any great environmental trauma in the late Tertiary of Africa. The continent seems to have supported throughout this time a fluctuating mosaic of forest, woodland, savanna, grasslands and steppe. It is an exaggeration to think of a Pliocene decimation of continuous Miocene forests. The most dramatic faunal change in Africa appears to have occurred 4–5 million years ago and to relate more to intercontinental connections than to climatic change (Maglio, in press). Similarly the view expressed by C. Jolly (1970:19) that almost all grasslands and savannas are recent artifacts of man can be specifically disproved even from the fragmentary pollen records now available. (cf. Kendall 1969; Livingstone 1967, 1971; Bonnefille 1972; Van Zinderen Bakker and Coetzee 1972). It seems to me that deterministic environmental factors of these kinds are not only undocumented, they are unnecessary as explanations. (See also Bishop this volume.)

If one accepts that evolution is a consequence of mutation and natural selection and that it is a restless opportunistic process (cf. Simpson 1949), then it does not seem difficult to accept that the normal pressures of ecological competition could transform the versatile behavioral system of ancestral hominoids into the novel early hominid pattern. It is widely believed that the process involved extensive feedback among the several subsystems, thus: hunting facilitated food sharing since meat was more readily carried than any other common food stuff; missiles, weapons, and tools facilitated the killing and butchering of larger and larger animals; bipedalism facilitated weapon use, the carrying of food for sharing and long-range mobility; gathered vegetable foods remained as a staple and as an insurance policy against failure in the hunt, so that division of labor (and pair bonding?) gave stability to the system; bags and baskets facilitated food sharing and division of labor, etc. It can thus be argued that savanna *(sensu lato)* constituted a vacant ecological niche for an animal at the hominoid grade, capable of basing its subsistence on a combination of hunting and foraging. Opportunism is the hallmark of mankind as we see the species in ethnography and history, and I would suspect that it was important, perhaps crucial from start to finish in human evolution.

Clearly Mio-Pliocene patterns of hominoid life and the circumstances of evolutionary divergence of men and apes need to be identified and studied before we will *know* about the actual history of what happened. But before the data are gathered, I would be ready to bet that we will find a steady trajectory of change, guided by evolutionary processes involving opportunistic exploitation of diverse microenvironments, rather than dramatic causation through occupation of a restricted habitat and/or sudden adoption of a radically different diet, whether chosen or imposed. Archaeology does not as yet contribute anything substantial to the problem of species divergence within the hominidae. At Olduvai, Omo, East Rudolf, and Swartkrans there seem to be at least two species of hominid with overlapping habitats. These formations each contain artifacts but, as Mary Leakey (1967:442) has pointed out the attribution of the tools and camp sites to one or the other or both hominid species is a matter of intuition. Hindsight makes us feel sure that the ancestors of *Homo sapiens* would have been among those involved with tools, hunting, and food sharing—but can we be sure that these traits were not part of a primitive hominid adaptive pattern shared by both species? We will need much more data on artifacts and fossil distribution before anything beyond an intuitive answer can be offered. Clearly the general model for hominid divergence presented above can accommodate internal differentiation between lineages which placed uneven emphasis on the exploitation of the multitude of subsistence opportunities presented by savanna mosaics. In default of a detailed late Tertiary fossil record we do not yet know whether the divergence took place between ecologically differing sympatric populations, or was initially related to geographic isolation which later broke down, leading to a pattern of overlapping ranges such as that we seem to observe in the sedimentary basins of East Africa. Both hominid species were by our standards powerful, grinding chewers; but in one species this condition seems to have been more marked than in the other. Whether this contrast really relates to dietary differences, patterns of tool use, or to other factors remains to be determined.

EPILOGUE

This review, along with the contributions of Desmond Clark, Mary Leakey and Harry Merrick, shows clearly how the context of the archaeology of early man has been expanded in the past two decades. From being narrowly concerned with the morphology of stone artifacts, the enquiry has broadened to the extent where effective factual discussion of diet, land use, ecology, and even aspects of social organization is possible.

This research movement has had many contributors. Raymond Dart, although an anatomist, long ago refused to confine his attention to the wrinkles on teeth and the ridges on bones. He asked behavioral questions that have demanded reorientations in research (Dart 1957a,b). Sherwood Washburn has been another pioneer in showing the relevance of comparative behavior studies for our understanding of human evolution, and his formulations of the questions have provided fresh starting points for many archaeologists and human biologists (Washburn 1968a,b). As already mentioned, Desmond Clark has been a strong driving force in furthering an ecological orientation in African archaeology. However, it is to Louis and Mary Leakey that credit goes for finding the sites which preserve detailed factual evidence of what hominids were doing nearly 2 million years ago. Through patient work, these sites have been uncovered, and through imaginative interpretation, they have been brought to the attention of people all over the world.

Louis Leakey's genius for paleontological exploration together with his knowledge of natural history and his restless imagination helped to bring about a major extension of anthropological science. Now, after his death in 1972, the field is the poorer for the loss of his tireless energy in stirring things up by his new discoveries and his new ideas.

References Cited

Arambourg, C.
1950 Traces possibles d'une industrie primitive dans un niveau Villafranchien de l'Afrique du Nord. Bull. de la Soc. Préhist. Franç. 47:350.

Ardrey, R.
1961 African Genesis—a personal investigation into the animal origins and nature of man. London: Collins.

Balout, L.
1955 Préhistoire de l'Afrique du Nord. Paris: Arts et Métiers Graphiques.

Biberson, P.
1961 Le paléolithique inférieru du Maroc Atlantique. Rabat: Publ. du Service des Antiquités du Maroc, 17.

Bishop, W. W.
1959 Kafu, stratigraphy and Kafuan artefacts. So. Afr. Jn. of Science 55:117–121.

1971 Late Cenozoic history of East Africa in relation to hominoid evolution. *In* Late Cenozoic glacial ages. K. V. Turekian, ed. New Haven and London: Yale University Press. pp. 493–528.

Bonnefille, R.
1972 Association polliniques actuelles et Quaternaires en Ethiopie (Vallée de l'Awash et de l'Omo). No. d'enrég. au CNRS Ao 7229, Paris.

Brain, C. K.
1970 New finds at Swartkrans Australopithecine sites. Nature 225:1112–1118.

Butzer, K. W., and G. Ll. Isaac, eds.
In press After the Australopithecines. Proceedings of Burg Wartenstein symposium #58 'Stratigraphy and patterns of cultural change in the Middle Pleistocene.' The Hague: Mouton.

Campbell, B. G.
1966 Human evolution: an introduction to man's adaptations. Chicago: Aldine.

Chavaillon, J.
1967 La préhistoire éthiopienne à Melka Kontouré. Archéologia Nov.–Dec. 56–63.

1970 Découverte d'un oldowayen dans la basse vallée de L'Omo (Ethiopie). Bull. de la Soc. Préhist. Franç. 67:7–11.

1975 Evidence for the technical practices of early Pleistocene hominids: Shungura Formation, Lower Valley of the Omo, Ethiopia. *In* Earliest man and environments in the Lake Rudolf Basin. Y. Coppens, F. C. Howell, G. Ll. Isaac and R. E. F. Leakey, eds. Chicago: University of Chicago Press.

Clark, J. D.
1970 The Prehistory of Africa. London: Thames and Hudson.

Clark, J. D., and C. V. Haynes
1969 An elephant butchery site at Mwanganda's Village karonga, Malawi, and its relevance for palaeolithic archaeology. World Archaeology 1:390–411.

Coppens, Y., J. Chavaillon and M. Beden
1973 Résultats de la nouvelle mission de l'Omo (campagne 1972). Découverte de restes d'hominidés et d'une industrie sur éclats. Comptes rendus de l'Académie des Sciences, Paris (Série D) 276:161–164.

Curtis, G. H., and R. L. Hay
1972 Further geological studies and potassium-argon dating at Olduvai Gorge and Ngorongoro Crater. *In* Calibration of hominoid evolution. W. W. Bishop and J. A. Miller, eds. Edinburgh: Scottish Academic Press. pp. 289–301.

Dart, R. A.
1957a The osteodontokeratic culture of *Australopithecus Prometheus*. Memoir of the Transvaal Museum.

1957b The Makapan australopithecine osteodontokeratic culture. *In* Proceedings of the Third Panafrican Congress on Prehistory, Livingstone, 1955. J. D. Clark and S. Cole, eds. London: Chatto and Windus. pp. 161–171.

Fox, R.
1967 In the beginning: aspects of hominid behavioural evolution. Man 2:415–433.

Gould, R. A.
1968 Chipping stone in the outback. Natural History 77(2):42–49.

Heinzelin, J. de
1960 Le paléolithique aux abords d'Ishango. Exploration du Parc National Albert. Fasc. 6:10–11. L'Institut Belge pour l'Encouragement de la Recherche Scientifique Outre-Mer.

Horowitz, A., G. Siedner and O. Bar-Yosef
1973 Radiometric dating of the 'Ubeidiya Formation, Jordan Valley, Israel. Nature 242:186–187.

Howell, F. C.
1960 European and northwest African Pleistocene hominids. Current Anthropology 1:195–232.

Isaac, G. Ll.
1966 New evidence from Olorgesailie relating to the character of Acheulian occupation sites. *In* Actas del V Congreso Panafricano de Prehistoria y de Estudio del Cuaternario, Vol. 2. L. D. Cuscoy, ed. Publicaciones del Museo Arqueologico Santa Cruz de Tenerife, 6. pp. 135–145.

1967 The stratigraphy of the Peninj Group—early Middle Pleistocene formations west of Lake Natron, Tanzania. *In* Background to Human Evolution. W. W. Bishop and J. D. Clark, eds. Chicago: University of Chicago Press. pp. 229–257.

1971 The diet of early man: aspects of archaeological evidence from lower and middle Pleistocene sites in Africa. World Archaeology 2:278–298.

1972 Comparative studies of Pleistocene site locations in East Africa. *In* Man, Settlement and Urbanism. P. J. Ucko and G. W. Dimbleby, eds. London: Duckworth. pp. 165–176.

In press Middle Pleistocene stratigraphy and cultural patterns in East Africa. *In* After the Australopithecines. K. W. Butzer and G. Ll. Isaac, eds. The Hague: Mouton.

Isaac, G. Ll., R. E. F. Leakey and A. K. Behrensmeyer

1971 Archaeological traces of early hominid activities east of Lake Rudolf, Kenya. Science 173:1129–1134.

Isaac, G. Ll. and G. Curtis

1974 The age of early Acheulian industries: from the Peninj Group, Tanzania. Nature 249:624–627.

Isaac, G. Ll., J. W. K. Harris and D. Crader

1975 Archaeological evidence from the Koobi Fora Formation. *In* Earliest Man and Environments in the Lake Rudolf Basin: Stratigraphy, Paleoecology and Evolution. Y. Coppens, F. C. Howell, G. Ll. Isaac and R. E. F. Leakey, eds. Chicago: University of Chicago Press.

Jacob, T.

1972 The absolute date of the Djetis Beds at Modjokerto. Antiquity 182:148.

Jolly, C.

1970 The seed-eaters: a new model of hominid differentiation based on a baboon analogy. Man 5:5–26.

In press African Hominidae of the Plio-Pleistocene. London: Duckworths.

Kendall, R. L.

1969 An ecological history of the Lake Victoria Basin. Ecological Monographs 39:121–176.

Lancaster, J. B.

1967 The evolution of tool-using behaviour. American Anthropologist 70:56–66.

Leakey, L. S. B.

1962 A new Lower Pliocene fossil from Kenya. Annals and Magazine of Natural History 13:689–696.

1968 Bone smashing by Late Miocene hominidae. Nature 218:528–530.

Leakey, M. D.

1967 Preliminary summary of the cultural material from Beds I and II Olduvai Gorge, Tanzania. *In* Background to Evolution in Africa. W. W. Bishop and J. D. Clark, eds. Chicago: Chicago University Press. pp. 417–442.

1970a Early artefacts from the Koobi Fora area. Nature 226:228–230.

1970b Stone artefacts from Swartkrans. Nature 225:1222–1225.

1971 Olduvai Gorge. Volume III. Excavations in Beds I and II, 1960–1963. Cambridge: The University Press.

Lee, R. B., and I. DeVore, eds.

1968 Man the Hunter. Chicago: Aldine.

Livingstone, D. A.

1967 Postglacial vegetation of the Ruwenzori Mountains in Equatorial Africa. Ecological Monographs 37:25–52.

1971 Speculations on the climatic history of mankind. American Scientist 59:332–337.

Maglio, V. M.
In press Pleistocene faunal evolution in Africa and Eurasia. *In* After the Australopithecines. K. W. Butzer and G. Ll. Isaac, eds. The Hague: Mouton.

Mann, A. E.
1968 The paleodemography of *Australopithecus*. Unpublished doctoral dissertation, University of California, Berkeley.

Merrick, H. V., J. de Heinzelin, P. Haesaerts and F. C. Howell
1973 Archaeological sites of early Pleistocene age from the Shungura Formation, Lower Omo Valley, Ethiopia. Nature 242:572–575.

Napier, J. R.
1967 The antiquity of human walking. Scientific American 216:56–66.

Pilbeam, D. R.
1972 The ascent of man. New York and London: MacMillan.

In press Trends in Middle Pleistocene hominid evolution. *In* After the Australopithecines. K. W. Butzer and G. Ll. Isaac, eds. The Hague: Mouton.

Reynolds, V.
1966 Open groups and hominid evolution. Man 1:441–452.

1968 Kinship and the family in monkeys, apes and man. Man 3:209–223.

Robinson, J. T.
1972 Early Hominid posture and locomotion. Chicago: Chicago University Press.

Robinson, J. T., and R. J. Mason
1962 Australopithecines and artefacts at Swartkrans. So. Afr. Archaeol. Bulletin 17:87–125.

Sarich, V. M., and A. C. Wilson
1967 Immunological time scale for hominid evolution. Science 158:1200–1203.

Schaller, G. B., and G. R. Lowther
1969 The relevance of carnivore behaviour to the study of early hominids. Southwestern Journal of Anthropology 25:307–341.

Simons, E. L., and D. R. Pilbeam
1965 Preliminary revision of the Dryopithecine (Pongidae, Anthropoidea). Folia Primatologia 3:81–152.

Simpson, G. G.
1949 The meaning of evolution. New Haven: Yale University Press.

Stiles, D. N., R. L. Hay and J. R. O'Neil
1974 The MNK chert factory site, Olduvai Gorge, Tanzania. World Archaeology 5:285–308.

Tobias, P. V.
1967 Cultural hominisation among the earliest African Pleistocene hominids. Proceedings of the Prehist. Soc. 13:367–376.

Tobias, P. V., and A. R. Hughes

1969 The new Witwatersrand University excavation at Sterkfontein. So. Afr. Archaeol. Bulletin 24:158–169.

Van Zinderen Bakker, E. M., and J. A. Coetzee

1972 Reappraisal of late Quaternary climatic evidence from tropical Africa. Palaeoecology of Africa 7:151–181.

Washburn, S. L.

1960 Tools and human evolution. Scientific American 203:3.

1965 An ape's eye-view of human evolution. *In* The origin of man. P. L. DeVore, ed. New York: Wenner-Gren Foundation. Private circulation. pp. 89–96.

1968a The study of human evolution. Condon Lectures, Oregon State System of Higher Education, Eugene, Oregon.

1968b Behaviour and the origin of man. Proceedings of the Royal Anthropological Institute of Great Britain and Ireland for 1967. pp. 21–27.

Washburn, S. L. and C. S. Lancaster

1968 The evolution of hunting. *In* Man the Hunter. R. B. Lee and I. DeVore, eds. Chicago: Aldine. pp. 293–303.

PART FIVE

The Life Work of Louis Leakey

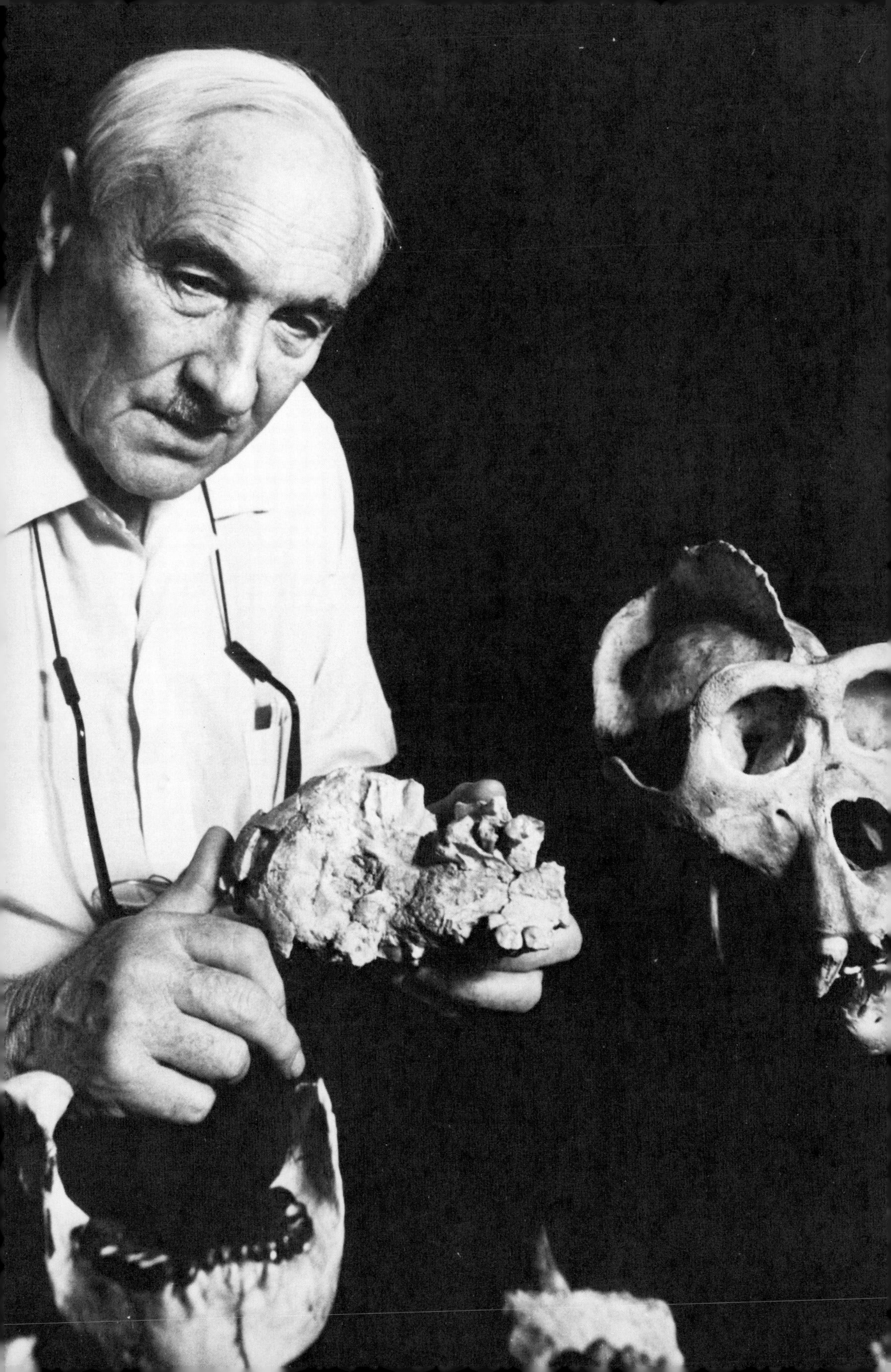

Introduction to Part Five

In this volume we have chosen to pay tribute to Louis Leakey by bringing together a varied series of papers and essays that convey something of the scope and effectiveness of the research that has grown out of Leakey's pioneer explorations. However, it should be clear that Louis Leakey did much more than inspire other people to do research. He himself was possessed of phenomenal energy and of an almost uncanny instinct for where to look for things of interest and importance to science. In this final section of the book we offer additional documentation on Louis Leakey's personal contributions to the subject of human origins.

We are reproducing here a biographic essay, originally published in the Proceedings of the British Academy for 1974, which provides a summary of Louis Leakey's career and a commentary on some of his most important achievements. It is written by J. Desmond Clark, who has been a fellow pioneer with Louis Leakey in the exploration of Africa's prehistoric past. Clark worked in Central Africa during much of the time that Leakey was at work in East Africa. He was a frequent vis-

L. S. B. Leakey. Courtesy, the L. S. B. Leakey Foundation.

itor to the Museum in Nairobi and to sites such as Olduvai, so that his insight is drawn from firsthand acquaintance. In part, the material of this biographic sketch overlaps with that in Phillip Tobias' biography in the opening part of the book; however, the two portraits are penned by two different scientists who each had a distinct and independent perspective on the work and the personality of their friend and colleague. Readers can gain additional insight into Louis' character and ideas by comparing these two biographies.

The bibliography of Leakey's work bears further testimony to his restless energy and to his varied interests. This was compiled by Shirley Coryndon (now Shirley Savage) who for many years participated in the East African discoveries as a research assistant to Louis. She is now a leading authority on "hippopotamology" and continues to visit Africa for paleontological research. Louis Leakey did not keep records or files of his publications, so the task of compiling a complete and authoritative bibliography has been an arduous one, involving the interest and cooperation of many people. We present the outcome of this herculean task with the entries grouped according to conventional disciplines such as archaeology, paleontology, ethnography, natural history, etc.—but the diversity and originality of Leakey's contribution make satisfactory classification of many items difficult. Readers with an interest in Leakey's work can explore the material of this massive bibliography and even a casual perusal will serve to indicate the unusual scope and scale of the activities of this restless, visionary pioneer.

Louis Leakey demonstrates the feasibility of cutting up a fresh carcass with the kinds of stone tools recovered from two-million-year-old sites.

J. Desmond Clark:

Louis Seymour Bazett Leakey 1903–1972

Reprinted from the proceedings of the British Academy, Volume LIX (1973). London: Oxford University Press, Ely House, W.1 By permission of the British Academy.

INTRODUCTION

One of the most characteristic memories not a few of us have of Louis Leakey is, dressed in his mechanic's overalls, stretched full length on the floor of an excavation or on an eroded slope somewhere in East Africa, carefully and methodically working with a dental pick at some partially exposed fossil from which, from time to time, he would blow away the loosened deposit, finally exposing the complete specimen. Mary, his wife, would never be far away, engaged, most likely, in some similar occupation, while under an adjacent thornbush, panting from the heat or the exertions of a hunt, lay never less than three or four Dalmatians, the

Leakeys' constant companions in the field as in the city. Anyone who had an opportunity to see him at work in the field, to feel something of the triumph with which each new find was received, and to assess with him the significance of these finds, could hardly fail to absorb some of the excitement and infectious enthusiasm for his search for the origins of man and for the causes of our humanity. Besides a great breadth of scientific knowledge and understanding in both paleo-anthropology and the natural sciences, he possessed the rare ability to treat each enterprise as an adventure that his self-assurance told him would be certain to succeed.

He would walk into a meeting of colleagues or into a lecture room, equipped with yet more and unique discoveries and brimming over with enthusiasm, and his forthright personality would command and hold the attention of all those present. A second characteristic memory we have of Louis Leakey, therefore, is as an outstanding speaker, lecturing in Europe and America, often to audiences of several thousands of students, colleagues, and the general public. He had the ability to hold the interest of all who heard him and to give to some the help and encouragement that would make them outstanding research workers in their own right. For others, who had the necessary means, it was his conviction and his record of unparallelled successes that persuaded them to provide the financial backing for further search for additional definitive fossil evidence of the ancestry of man.

Louis Leakey is unique in this century for the discoveries he made towards understanding the biological and cultural evolution of man. It is easier to see why he made so great an impression on the scientific world than it is to understand why his work should have generated the overwhelming interest it has received from the layman throughout the world. He was not a great orator but he had the ability to catch the feeling of his audience and to react to it and what he had to say was said simply and convincingly so that all could understand. As one looked at the evidence in the slides with which he illustrated his lectures, each specimen was not just another fragmentary piece in the huge jigsaw puzzle of human evolution, but assumed a real and often exciting new meaning when explained by Louis. That his interpretations did not always receive the acceptance of his colleagues and were shown sometimes to be at fault is, perhaps, not unexpected when one takes into account his provocative personality. However, such criticism detracted nothing either from the discoveries themselves or from our admiration for the discoverer, and they stimulated both colleagues and laymen to probe more searchingly and meaningfully into the foundations for the established principles and beliefs on which understanding of human evolution was based and to turn to other sources of evidence, besides that of the actual fossils or the archaeology, for interpreting the behaviour of the earlier hominids.

His discoveries in the late 1950s and 60s came at a time when a large segment of the Western world, out of sympathy, because of the unrest caused by two world wars, with the established religious tenets of their society, was looking to some other explanation—turning to humanism and like concepts—as a means of understanding man's uniqueness in the animal kingdom. In such circumstances, the regularity with which the numerous new finds were made and announced attracted immediate attention and set off a new interest in palaeo-anthropological research. He had an unremitting faith that East Africa held the key to the ancestry of man and the unique conditions of fossil preservation and the primary-context archaeological sites in the East African Rift Valley and adjacent regions are proving him correct. This belief would have availed him little, however, had it not been for his equally strong perseverance and courage, often in the face of considerable practical difficulties, in returning again and again to the field as he did to Olduvai Gorge convinced that somewhere in that unique series of exposures and cultural sequence he would find the remains of the tool-makers themselves. Twenty-eight years after he first visited the Gorge, Louis's search was rewarded by Mary's discovery of *Zinjanthropus boisei.* Thereafter, not a year went by without the announcement of the discovery of further hominid fossils.

Louis Leakey was a man of tremendous energy, great conviction, and unusual intellectual ability. His was a sparkling personality that endeared him to the younger generation of palaeo-anthropologists. Always something of an iconoclast, he brought a critical mind to the problems of interpretation and his views were often highly influential in shaping new concepts. There were, however, other times, albeit few, when his critics might accuse him of using 'Nelson's telescope,' though one could never be quite sure whether it may not have been as much from perversity and perhaps a certain obstinacy as from any conviction of the correctness of his diagnosis of the evidence, as, for example, on the controversy over the 'early, early man' site at Calico Hills, California, that he took the stand he did.

Leakey was the most amazing and influential of the palaeo-anthropologists of this century and it was in no small part due to the circumstances of his early life in Africa that the success of his remarkable career can be ascribed.

BIOGRAPHY, 1903–1926

Louis Seymour Bazett Leakey was born at Kabete, Kenya, on 7 August 1903, the third of the four children of Harry and Mary Leakey. His parents were missionaries with the Church Missionary Society working among the Kikuyu in one of the more heavily populated parts of the Kenya highlands west of Nairobi and overlooking the Rift Valley where dwelt the Masai with whom the Kikuyu had only recently ceased to fight. The outbreak of the First World War prevented his going back to school in England and, except for two periods of furlough with his family there, the first sixteen years of his life were spent at Kabete. During this period he was taught by a governess and later by his father but he was free to associate with Kikuyu boys of his own age-group and, apart from his mother and sisters, they were his constant companions. It was from them that he first learned to know Africa and, as he said, he was more a Kikuyu than a European. One Kikuyu chief called him 'the black man with a white face because he is more an African than a European and we regard him as one of ourselves.' He learned as one of them the language and customs of the Kikuyu and this inherent understanding of, and sympathy with, Africans enabled him to know them and the country of his adoption with a rare insight and understanding.

He also acquired a close knowledge of animal and bird behaviour that was always put to maximum use in later life and from his observations and experiments he was an inexhaustible source of knowledge on the subject. I recall in particular two such useful pieces of information with implications for the palaeo-anthropologist: one, that it is possible to stalk close enough to a Thompson's gazelle to capture it with a final dash if you do not show your arms, and the other that the apparently man-made 'pick marks' and hollows scooped in the bank at the side of the road up Ngorongoro caldera were done by elephants with their tusks digging out the mineral earth to eat. Later in his life his house at Langata was always well stocked with animals—genets, hyrax, vervets, and tropical fish for the breeding of which he won several awards. Much of what he learned about birds he obtained from the subsequent world authority, Arthur Loveridge, who had recently come out to Nairobi as the Curator of the East African Natural History Society's museum. It was then also, at the age of thirteen, that he acquired his interest in archaeology and made his first collection of obsidian artifacts round Kabete—almost the first to be found in Kenya. During the war years he was active also in helping with the recruitment of labour for the Volunteer Carrier Corps which had been formed, mostly with missionaries as officers, to carry supplies to the combatant troops.

In 1919, after the war, he returned to England and was sent to school at Weymouth College where he spent the next three years attempting to make up some of the deficiencies that an education in the African bush had left in the knowledge he would need for passing the entrance examination to Cambridge. His time at Weymouth was not a particularly happy one since his independence and comparative freedom in Kenya made it impossible for him to conform to the stereotype of an English public schoolboy, as was expected of him. What he lacked in knowledge of some required subjects, however, he more than made up for in his understanding of people and independence of thought and action.

The problems he faced in getting into Cambridge were not only connected with examinations but, in addition, he had almost no funds. Most boys in his position would have followed the advice of his headmaster to 'try to get into a bank,' but such were his determination and the impression he made at his interview at St. John's that he sat and was awarded a sizarship of £ 40 and went up to Cambridge in October 1922.

His decision to read for the Modern Languages Tripos and to offer French and Kikuyu begins one of those never-to-be-forgotten stories that have gathered around Louis Leakey. The consternation following the discovery of this unsuspected loophole in the University regulations governing the Tripos can only be guessed at and, indeed, these were subsequently amended, but he was allowed to offer Kikuyu. His second year at John's was interrupted by an accident on the rugger field. A rugger accident to his leg also resulted, in later life, in his having to undergo an operation to the femur which left him more or less lame until his death. In October 1923 he was concussed and the need to take a year away from any studying resulted in his being chosen to accompany W. F. Cutler on a British Museum expedition to collect upper Jurassic dinosaur fossils in the vicinity of Tendaguru Hill in southern Tanganyika. This year in the field proved to be immensely useful to him in his later research for it gave him experience of fossil collecting and preservation and a knowledge of vertebrate anatomy as well as insight into the general running of an expedition and the organization of a camp. While in Tanzania he had two narrow escapes from snakes and other animals and one or two unusual experiences which he explained in terms of his African upbringing. It was on this occasion that he was a witness to the activities of a poltergeist—the spirit of a man who had died away from his village and who returned to his hut to demand the necessary ritual burial by his relatives. On another occasion a clairvoyant described to him the landing of a motor-car at Lindi over fifty miles away—an incident that he considered most unlikely and, therefore, carefully verified because in 1923 there was only some half mile of road in Lindi.

He supported himself at Cambridge and paid for his extra coaching in France by lecturing and by such ingenious means as selling ebony walking-sticks bought in Tanganyika. It was at the end of his second year that there took place the now famous incident of his very nearly finding himself in the position of being his own examiner in Kikuyu and of teaching Kikuyu to the second examiner. He received a First Class in the Modern Languages Tripos as well as a scholarship and, in his third year, read Archaeology and Anthropology in which he similarly gained First Class Honours in May 1926. Here, under A. C. Haddon, he studied the ethnography of pre-literate peoples and was able to systematize his understanding of Kikuyu society, and under M. C. Burkitt his awakening interest in prehistory received the best training available in the country.

One of his interests was in experimental stone flaking and he paid regular visits to the flint knappers at Brandon to study the basic techniques. This knowledge he put to good effect thus becoming one of the very first prehistorians to study stone working by means of his own experiments. He remained an expert at flaking flint and was always willing to demonstrate the manufacture of a range of stone tools. One of the best attended of his demonstrations was at the meeting of the Third Pan-African Congress in Livingstone in 1955 when he went on to show how to use cleavers and stone knives to skin and dismember a carcass. On this occasion also the depth and extent of his observation in all matters of technology were well shown by his remark that if a stone implement becomes blunt in the course of using it for butchering, it is really the meat fibres sticking to the stone that make it blunt and that by sucking these off the knife can be quickly 'resharpened'! This was but one of many such keen observations that he used to help reach a more meaningful understanding of the scatter of artifacts and associated remains on an occupation site.

Inaccurate or incomplete factual information was often a means of stimulating him to carry out his own research in the interests of finding the truth or expanding understanding. One of his first such pieces of research was 'A new classification of the bow and arrow in Africa,' carried out while still an undergraduate and following visits to European museum collections. It was published in 1926 in the *Journal of the Royal Anthropological Institute* and still remains one of the main reference works on this subject.

EARLY EXPEDITIONS

Between 1926 and 1935 Leakey led four highly successful expeditions to East Africa after the first of which he was awarded a Research Fellowship at St. John's and submitted his Ph.D. on the results of the work. Initially research was concentrated in the closed lake basins of Naivasha-Nakuru where he and his associates were able to show the existence of a long cultural sequence that could be interpreted in relation to episodes of climatic fluctuation. In this he was influenced by the work of E. J. Wayland, Director of the Geological Survey in Uganda, by the meteorologist C. E. P. Brooks, and by Eric Nilsson who was mapping the raised lake levels in the East African lakes. In the Western Rift, Wayland had studied a sequence of several stages of lacustrine and fluviatile sediments that had often been subjected to tectonic movement and were found at varying heights above the exisiting levels of the lakes. To account for this succession of high and low lake levels he had developed the pluvial/interpluvial hypothesis, which was adopted and expanded by Leakey and his geologist, J. D. Solomon, and was later widely used by prehistorians throughout the continent. It was not until 1956 that the critical studies of the geologists H. B. S. Cooke and R. F. Flint showed the inadequacies of a chronology based on hypothetical climatic interpretations of geological phenomena in what tectonically is the most unstable part of the continent. The abandonment of the 'pluvial hypothesis' gave rise to a number of important studies of sedimentary rock units and volcanics that are the new basis for chronological studies and with which the cultural sequence can be more precisely correlated.

The work of these four early expeditions not only established the nature of the cultural sequence in some detail in Kenya and northern Tanzania, but it showed beyond all doubt the great antiquity of that sequence and the enormous potential for finding fossil animal and hominid material in association with the stone artifacts. His finds came, perhaps, as something of a shock to scholars in England and on the continent where the discoveries of French and other prehistorians since the middle of the previous century carried the implication that most other parts of the world were only peripheral to Europe so far as man's origins were concerned. The report on the work in the Kenya Rift (*The Stone Age Cultures of Kenya Colony*) published in 1931 was, for its time, a model of clarity and speediness in publishing. It has remained a standard reference work for Kenya on the Upper Pleistocene and later industries, in particular receiving confirmation and clarification very recently through further fieldwork by Dr. G. Ll. Isaac. Leakey was one of the first to initiate the practice of studies by teams of investigators which is the general rule today. With the geologist Solomon he established his cultural chronology in the Naivasha-Nakuru basin by the study of certain key sediment sections in the gorge of the Melawa river, at Enderit Drift, Little Gilgil river, and Long's Drift—now historic names in African perhistory—and, in particular, by his excavation at Gamble's Cave where the greater part of the later cultural sequence was preserved and where he obtained the first evidence, and the only good, stratified, developmental sequence of the Kenya Capsian Industry, as well as burials showing the physical characteristics of the makers. There has subsequently been much speculation over the origins of the Kenya Capsian and whether it is Pleistocene or post-Pleistocene in age. Leakey always maintained that its roots lay in the Upper Pleistocene and, although the sequence at Gamble's Cave itself appears to fall into the early Holocene, confirmation of the existence of a terminal Pleistocene stage is now forthcoming from Glynn Isaac's work in the Long's Drift locality.

In 1929 Leakey and some of the field party drove down from Kenya to Johannesburg for the joint meeting of the British and South African Associations for the Advancement of Science. Here he took an active part in the discussions that formulated the South African prehistoric terminology as well as giving an account of that established in Kenya through his own work. He took several distinguished British scientists back with him to Kenya and showed them the excavation and the finds at Gamble's Cave.

Wherever he went, he made friends with Africans and, if language was a barrier, as it was when he found himself at Abercorn (Mbala) in what is now Zambia, he was never at a loss as to how to establish the best relations by, for example, making string figures which he had learned from Haddon at Cambridge, or by conjuring tricks, with which he used later to regale the children of his friends and, on occasions, his colleagues besides, as I well remember once when some of us met at the house of 'Peter' van Riet Lowe in Johannesburg. He took evident enjoyment in these accomplishments and rumour has it that, early in his career, he was not far from being inducted as a member of the Magicians' Inner Circle.

On his trip to Johannesburg in 1929 he visited known sites and discovered new ones. In particular, he was able to spend some time at Broken Hill (Kabwe) where he studied what was left of the stratigraphy and made collections of fossil fauna from the dumps. These bones he sent back to the British Museum (Natural History) but the boxes were not opened until the mid 1950s when the material was found, packed in the grass of the Broken Hill *dambo*, together with his pencilled identifications and notes. It is pleasing to record that this material was subsequently published by Leakey himself when he carried out a reexamination of the whole faunal assemblage from Broken Hill. This confirmed him in his impressions that *Homo rhodesiensis* was considerably older than previous assessments had suggested and it is a matter for regret that he did not live to hear of the date of 120,000 years B.P. obtained recently by the amino-acid racemization method.

Leakey's pioneer work on the Rift Valley and Lake Victoria basin laid the foundations for the prehistoric cultural succession in East Africa and served as an important link between those in North and South Africa, showing the continent-wide extent of the Middle Pleistocene Acheulian Industrial Complex and the essentially local or regional nature of the post-Pleistocene industrial patterns. The tectonic troughs and mantles of volcanic dejecta of the East African region are probably unique in the world in respect to the preservation of fossils and Leakey was also able to show what type of man was responsible for most of the later industrial entities and to compare them to evidence from the rest of Africa. These findings he published in 1935 *(The Stone Age Races of Kenya),* having found time to develop, with a colleague, a new machine for drawing human palaeontological material, of which they had published the details in 1930.

When he was not in the field in Kenya, he was back at St. John's supervising undergraduates, lecturing, and writing. Glyn Daniel, who was one of those he supervised, describes him as 'enthusiastic, energetic, exciting, an erratic teacher—always inspiring. We didn't write weekly essays, we just listened to LSBL talk and watched him chipping flints. There was no furniture in his College rooms: we sat on packing cases, listened and sorted flints.' His continual insistence on getting to grips with the material is well emphasized here: his interpretations and inspiration came from a study of the artifacts themselves in their original context, not from second-hand sources in textbooks, and he had picked the one part of the world where the contexts of his finds from the most crucial time period, the Plio-Pleistocene, were largely undisturbed.

The experience gained during his early expeditions enabled Leakey to examine from a new perspective the classic European palaeolithic sequence and 'Adam's Ancestors,' published in 1934, was both refreshing and thought-provoking for the questions it raised and the answers it proposed. Two years later he was invited to give the Munro Lectures at Edinburgh University and these were published in *Stone Age Africa*—the first time that anyone had attempted to view as a whole the African prehistoric field. This was a milestone in its time and probably did more than any other book to focus the interest of prehistorians on the possibilities of research in Africa. It also shows the great breadth of his interests which ranged from one end of the cultural succession to the other and included, besides the artifact assemblages themselves, the fossil evidence of the hominids and of the changing faunas with which they were associated. His insistence on sound stratigraphical sequences confirmed by excavation and the importance of the associated animal fossils for building up a relative chronology may be said to have established the basis for subsequent archaeological research in the continent.

It was probably inevitable that such a critical and outspoken personality should raise some reciprocal and equally valid criticisms, particularly among older associates, and such was the case in regard to the inquiry into the age and associations of the two very significant hominid fossils from Kanam and Kanjera which he found in 1932. Had he had a geologist working with him, as he invariably did later, no problem would have arisen, but lack of funds and other circumstances prevented this and the results of the ensuing field inquiry ran contrary to Leakey's contention that the fossils were of Lower and Middle Pleistocene age respectively. Both have now been eclipsed by later finds, but his claim that the Kanjera skulls represented an early form of *Homo sapiens* has to a great extent been vindicated by the more recent finds from the Kibish Formation in the Lower Omo basin.

RETURN TO KENYA AND THE SECOND WORLD WAR

The circumstances of the investigation of these controversial finds, as well as personality clashes, were some of the reasons behind Leakey's decision to return to Kenya in 1937 where he made a two and a half years' study of the Kikuyu whose social and economic customs, ritual, and beliefs he already knew intimately, having been admitted as an elder of the second grade. This was an impressive and lengthy work and was nearing completion at the outbreak of the Second World War. He never had the opportunity later to return to it until after his retirement from the Directorship of the Centre for Prehistory and Palaeontology in Nairobi in January 1972, when he again took it up to revise for publication and had virtually completed it at the time of his death. It is gratifying to learn that it is being edited for posthumous publication by his sister, Mrs. Gladys Beecher.

Louis continued to work for the advancement of the Kikuyu and other African peoples of Kenya and was an adviser on Kikuyu customs, land tenure, and other matters to the Government of Kenya. His intimate knowledge was of particular value during the emergency at the time of the Mau Mau movement and he published two books (*Mau Mau and the Kikuyu,* 1952, and *Defeating Mau Mau,* 1954) that helped to clarify understanding of the situation and of the means to overcome it. He was asked to write the first of these by Methuens while he was on a brief visit to London at the height of the emergency and, consenting, he took a writing pad with him on the plane and by the time he arrived at Nairobi he had already completed the draft! Although he spoke little about it, there can be no doubt that he must have been under considerable strain and anxiety during the time of the Mau Mau troubles. He acted as interpreter during the trial of Jomo Kenyatta and it says much for both their personalities that they remained the best of friends. The title of his autobiography—*White African*—(1937) seems to show his identification with Africa and his love of Africans and it is in keeping that, following independence and the establishment of the Republic of Kenya in 1963, he became a citizen of that country.

During the Second World War Leakey was put in charge of the African section of Special Branch C.I.D. Headquarters, a post he held until the end of the war. Apart from hearing him lecture on his Olduvai finds in Cambridge in 1937, it was during the war in Nairobi in 1941 that I first met him when he and Mary found time to take me in their ancient Ford pick-up on a quick, very bumpy, but exhilarating and never-to-be-forgotten tour of the classic sites in the Naivasha–Nakuru Rift. He was most assiduous in keeping in close and regular touch with current opinions and rumours in the city and the surrounding country and he would suddenly stop his car and get out at some house or village, whereupon he would be joined by two or three of his Kikuyu friends or associates and, sitting together in the grass, they would spend the next ten minutes or so in animated conversation.

One of Louis's special interests was that of handwriting, a competence he had acquired while studying the authorship of emendations in French medieval manuscripts at Cambridge, and he was recognized as one of the leading handwriting experts.

Also during the war years Jack Trevor, then with Military Intelligence in Nairobi, was a constant visitor at the house and it was Louis who arranged for Trevor's field trip to look for chimpanzees in Tanzania, a forerunner of the major research programme later carried out so successfully with Louis's help by Jane van Lawick-Goodall.

It was during the war that Leakey was made a Trustee of the Coryndon Museum and, in his spare time, he entered fully into its administration. In 1945 he was appointed full-time Curator and so successful was his reorganization of the museum, that it became one of the foremost research centres in the whole of East Africa as well as one of the finest museums in the continent. The exhibition galleries were completely redesigned and he especially encouraged schools, learned societies, and visiting scholars to make use of its facilities. He moved into the Curator's house adjacent to the museum where, if one was successful enough to fight off the pack of Dalmations that greeted one at the gate, a friend or colleague was assured of a warmth of welcome for which he and Mary were renowned.

When Louis and Mary were not in the museum, where they would often work late into the night, they were in the field searching for new sites. Mary's excavations at Hyrax Hill near Nakuru and their joint undertaking at Njoro River Cave on the Mau Escarpment greatly extended understanding of the 'Neolithic' in the East African Rift. The physical population that made the 'Stone Bowl Cultures' and that he described, confirmed what Gamble's Cave had previously shown, that peoples exhibiting many physical characteristics of the 'Nilotes' were living there from early in post-Pleistocene times, if not before. The excavation of the burial cave at Njoro was, incidentally, the occasion for another seeming manifestation of a poltergeist.

It was during the war years also that Louis and Mary found the now famous Acheulian site of Olorgesailie where handaxes and cleavers, eroded from the lake beds, lay in extraordinary profusion over the surface. This was, I believe, the first site they excavated to expose the concentration of artifacts as a unit on a single horizon in such a way as to reveal the association between different kinds of artifacts, fauna, and features such as rubble concentrations or butchering places. This application of excavation techniques previously reserved for much more recent archaeological periods revolutionized the study of Palaeolithic culture and behaviour throughout the world. From the work at Olorgesailie it became apparent that the fine-grained lake and stream sediments preserved occupation sites of Acheulian man that had undergone minimal disturbance since the time they were abandoned. Excavation of similar sites followed, for example of Kariandusi, and other workers applied the same techniques and refined them so that today we possess an understanding of the specific features of an increasing number of earlier Pleistocene primary context sites and a closer appreciation of early hominid behaviour than was ever thought possible before.

THE PAN-AFRICAN CONGRESS

Shortly after the war, Leakey was able to realize his wish to bring together prehistorians working in all parts of the continent—and not only prehistorians but geologists, palaeontologists, and others concerned with the study of Quaternary problems—to exchange information, to establish personal relationships, to effect collaboration, and to call to the attention of the governments concerned the need actively to support and extend the study of Africa's past. It was, therefore, through Leakey's efforts and the support he was given by the Kenya Government and the City Council of Nairobi, that the First Pan-African Congress on Prehistory met in January 1947 in Nairobi. It was attended by most of the leaders in African Quaternary studies—men such as Robert Broom, Raymond Dart, Sir Wilfrid LeGros Clark, Camille Arambourg, Alex du Toit, Sidney Haughton, the Abbe Breuil, 'Peter' van Riet Lowe, John Goodwin, Armand Rulhman, Tony Arkell—and a number of younger men, more recently started in the field, who, like Raymond Mauny, Bernard Fagg, or Basil Cooke, were destined later themselves to make major contributions to Quaternary research and understanding. So successful was the Congress in laying the foundations for friendship, mutual assistance, and collaboration between workers from one end of the continent to the other, that it remains the major instrument for exchange and rapport in Quaternary studies in Africa today. Every four years these relationships are revitalized by a new meeting of the Congress.

Leakey himself took an active part in all but one of the six Congresses that followed, being President of the Third in Livingstone in 1955 and receiving a standing ovation in the plenary session for his discoveries and report on early hominid evolution at the Seventh in Addis Ababa in 1971. The detailed and critical examination that he always gave to the sites visited on the Congress excursions sometimes resulted in the need to modify the established interpretation as, for example, following his discovery of handaxes at Aïn Hanech!

THE SEARCH FOR EARLY MAN

As a colleague on some of his earlier expeditions and at the Coryndon Museum, Leakey had a young palaeontologist, Donald McInnes, who was responsible for four important contributions to the British Museum's 'Fossil Mammals of Africa' series. From close association with McInnes

Leakey gained much factual knowledge of how to work on fossil mammals and two of his earlier important contributions were, one, on the East African fossil pigs and, the other, on the giant baboon *(Simopithecus).* He also received much help from his friend Dr. Kenneth Oakley at the British Museum (Natural History) when he was there working on faunal collections and on dating his discoveries. Dr. Oakley records that the authorities deemed it incredible that Leakey should ask for permission to work at the museum from 8:30 in the morning instead of the statutory 10:00 in order to get through the immense amount of work involved in the very short time available! He was particularly concerned that too few people were working on the rich collections of fossil mammalian remains from the Miocene sediments on Rusinga Island and adjoining parts of Kavirondo in the Lake Victoria basin. Accordingly, he set out to revisit the island himself and with the financial support provided by Mr. Charles Boise he purshased a boat and he and Mary and their associates carried out several seasons' work at this and other Miocene sites. Especially important was the discovery in 1948 on Rusinga of a nearly complete skull of *Dryopithecus (Proconsul) africanus.* More than twenty million years old, it is the oldest skull of an ape yet known and may well be close to the ancestral form prior to the separation of the pongid and hominid lineages. Over six hundred fossils of Miocene hominoidea have now been recovered from East Africa, mostly by Leakey and his co-workers, as well as a wide spectrum of the associated flora and fauna right down to uniquely preserved insects—caterpillars, beetles, grasshoppers, and even butterflies.

It was during the late 1940s and 50s that Leakey began to turn his attention more to a search for and study of the fossil evidence for man, and less to the cultural evidence of his activities. Although he continued to work on the occupation sites, the responsibility for this side of their joint researches was now largely assumed by Mary while Louis was actively engaged in a study of the fossils themselves at the Olduvai Gorge and other sites. Over the years, Leakey had collected a team of African workers who were not only skilled excavators in the techniques that he and Mary had pioneered but were also very knowledgeable in the recognition of fossils and it is to their sharp eyes and perceptions that we owe some of the now famous hominid fossils.

Leakey owed not a little of his understanding of and thought on the origins of man to Sir Arthur Keith and Sir Grafton Elliot Smith with whom he had been associated in England, and he believed that the earliest hominid fossils would be found in the Tertiary. At Fort Ternan, in Kavirondo, where he had been encouraged to dig by the owner of the farm, his friend Dr. Fred Wicker, he discovered an East African form of *Ramapithecus, R. (Kenyapithecus) wickeri,* which not only confirmed the earlier Indian evidence that this was probably a hominid, but was dated to the late Miocene some twelve to fourteen million years ago. Moreover, a single battered piece of stone adjacent to the smashed bones was for him suggestive evidence that *Ramapithecus* may have been a tool-user.

Further search of other Kavirondo sites, including Rusinga, brought to light other fragments which he believed represented an earlier form of *Ramapithecus (Kenyapithecus)* dating to the Lower Miocene, some six to seven million years or more earlier. Although most palaeontologists would now see these remains as those of Dryopithecine ape, they serve to point up the difference of opinion existing between the palaeontologists favouring a long chronology for the emergence of the hominid line and the molecular biologists who inclined to a short chronology with the separation of ape and man lineages taking place some time between ten and four million years ago.

Another study he undertook in Kavirondo was at the request of the amateur archaeologist Archdeacon Owen. While Louis studied the later Pleistocene sequence in the red colluvial soils exposed by the many steep erosion gulleys in this region of forest/savanna mosaic, Mary worked on sites yielding a new kind of pottery and later gave us the first knowledge of early Iron Age 'dimple based wares.' Louis's sequence of cultural stages of what has later come to be called the East African Lupemban showed the close behavioural associations that existed between the populations of the Lake Victoria basin and those of Zaïre. The geologist, Jean Janmart, asked his help in unravelling the succession in the diamond-mining area of northeast Angola which he visited early in 1948 with Mary and produced a pioneer work that formed the basis for later studies of Pleistocene climatic and cultural changes in the Zaïre basin.

OLDUVAI GORGE, 1931–1972

The site with which the name of Louis Leakey will always be coupled is Olduvai Gorge in the eastern part of the Serengeti Plain in northern Tanzania, not far west of the Rift Valley. While still at Cambridge, he had discussed with their discoverer, Hans Reck, the fossils brought back from there by a German expedition immediately before the First World War, and he felt convinced that somewhere in the lacustrine and terrestrial sediments in the three-hundred-foot sections exposed in the Gorge, there waited to be found the tools and living places of early man. Reck accompanied him on his first visit to the Gorge in 1931 but warned him that he would be disappointed because there were no stone implements there. Leakey made a bet with him that he would find some within twenty-four hours of arriving. He won the bet by quickly finding tools of volcanic rock which he showed to Reck who told him that, as a student in Germany, he had been taught that early Stone Age implements were always made of flint!

Reck was invited to accompany the expedition, not only because of his prior geological work at the site in which, incidentally, he had discovered a human burial, but also because of Louis's recognition of the need to establish the stratigraphical sequence and to interpret this as a basis for understanding the behaviour and the stone tool equipment of the hominids. This insistence on expert geological investigation simultaneously with his archaeological excavations characterized all his field-work but he was not always able to put it into effect and, although he was knowledgeable in geological matters from his association with E. J. Wayland and K. P. Oakley, he realized that the only successful way to pursue large-scale programmes of investigation of this kind was through teamwork.

One of the early matters for investigation was whether the skeleton excavated by Reck was contemporary with the bed in which it had been found, or had been buried into it. Wayland from Uganda and others were called in and it was convincingly established that this was an in-burial of Kenya Capsian age. The cause and origin of the 'Red Bed' (Bed III) was a matter of some concern at this time and the bulky correspondence between Leakey, Wayland, and others and the conflicting views expressed show something of the importance attached to the ecological interpretation of the geological sequence.

The results of the earlier expeditions to the Gorge were collated in the 1951 publication, *Olduvai Gorge.* Although remains of earlier Pleistocene man still eluded him, the impressive Oldowan/Acheulian cultural sequence established by Leakey was, for a year or two, used as a yardstick against which to compare finds in other parts of the continent. This work had, however, been based on collections made mostly from the outcrops rather than on excavation and the way in which such assemblages can obscure the true state of affairs became clear only after Leakey's work at the Gorge from 1959 onwards.

It was on 17 July of that year that Mary made her famous discovery of the skull of *Australopithecus (Zinjanthropus) boisei* in Bed I, a discovery that was the reward of constant visits to the Gorge during the twenty-eight years since he had first set foot there. This discovery came about as a result of yet another of Louis's techniques, simple enough in itself but not previously adopted by palaeo-anthropologists, namely that of *crawling* over the exposures, minutely examining every inch of the ground for teeth and bone fragments of the hominids. The profusion with which well-preserved, large animal fossils occur at Olduvai and other East African sites and the relative ease with which they can be found and collected helped to obscure the presence there of the small fossil material, including hominid teeth. Once the 'small tooth comb' was applied the Gorge began to produce early Pleistocene hominid remains in relative abundance, to an extent that no other site had done before.

I well remember Louis's and Mary's arriving for the Leopoldville (Kinshasa) Pan-African Congress in August 1959. Louis was sitting in the bus holding on his knees a small square box and, by the expression on his face, it was obvious that he had something momentous to report. Some of us were later able to handle the specimen and discuss the importance of his find in some secrecy prior to the official announcement of the discovery of *'Zinjanthropus'* which extended the known range of the Australopithecines into East Africa and showed them to be contemporaneous with the makers of the Oldowan Industry, if not the makers themselves.

From then on began the intensive study of Olduvai Gorge that has made it the best-known early-man site in the world. With financial help from the National Geographic Society of Washington, Leakey was able to get together a brilliant team of specialists such as had never before collaborated at a single site. Not only is it true to say that Olduvai Gorge is unique in the continuity and length of the record (from 1.89 million to *c.* 200,000 years B.P.), the biological evolution (from *Australopithecus boisei, Homo habilis* to *H. erectus*) and the earlier cultural history (from Oldowan to Upper Acheulian), that it preserves, but it is the best understood and dated of any Lower or Middle Pleistocene site. The meticulous excavation of primary-context living sites has provided a remarkable understanding of the changing behavioural patterns; the studies of Leakey himself and his colleagues, P. V. Tobias, M. H. Day, J. Napier, and others have resulted in the development of the hominid phylogeny while R. L. Hay has provided an invaluable insight into the sequence of geological events marking the history of the Olduvai basin and thus thrown light on the palaeo-ecology of the changing preferences in the siting of the hominid camping places. The dating carried out by J. F. Evernden, G. H. Curtis, R. L. Hay, C. S. Gromme, and others by means of the K/Ar, palaeomagnetism, and other methods and the study of the many different faunal species undertaken by, amongst others, P. M. Butler, H. Greenwood, J. J. Jaeger, R. Lavocat, R. F. Ewer, G. Petter, and S. C. Savage, have shown more completely and reliably than before, the composition of the main Lower and Middle Pleistocene faunal zones represented in the Gorge.

Although hominid fossils of equal importance are now being found at the East Rudolf localities by his son Richard Leakey, and camping places and butchery sites dating from 2.6 million years ago are being investigated by his colleague and friend Glynn Isaac, Louis's faith in the potential of Olduvai is fully vindicated for there is no other known site which provides, for this crucial time period, such a long, continuous, and well-dated record in which culture and biological change can be documented with such completeness.

Louis and Mary, their son Jonathan, and their trained African assistants have, between them, given us some forty-eight hominid fossils from Olduvai. These fall into three distinct groups—the robust Australopithecine, a more gracile form more persistently associated with tool-making, which he named *Homo habilis,* and a later, larger-brained, robust form, *Homo erectus.* While some would see an evolutionary development of the hominid line from a gracile Australopithecine (tool-user), through a small-brained form of tool-maker *(H. habilis)* to the larger-brained *H. erectus,* Leakey himself considered it more probable that the *Homo* line had diverged from the Australopithecine some time in the late Pliocene or earliest Pleistocene and he believed it likely that a larger-brained form would one day be found in geological deposits of that age. Although he was not destined to find this proof himself, nevertheless, it gave him immense pleasure to study the KNM-ER-1470 skull from Koobi Fora at East Rudolph, discovered that summer by his son Richard, which he did just before flying to London in September 1972. This, he believed, fully substantiated his contention that a larger-brained hominid (estimated cranial capacity *c.* 800 cc), distinct from *Australopithecus,* was present in East Africa nearly three million years ago. His death in London on 1 October 1972 deprived palaeo-anthropology of its most lovable, quixotic, and brilliant scholar but he died in the knowledge that his lifetime work was being successfully continued by his wife Mary and his son Richard.

THE CENTRE FOR PREHISTORY AND PALAEONTOLOGY

In 1962 Leakey gave up the directorship of the museum and, with funds provided by the Wenner-Gren Foundation for Anthropological Research, New York, and the National Science Foundation, Washington, established the Centre for Prehistory and Palaeontology under the Trustees of the National Museums of Kenya; he remained its Director until his retirement early in 1972. This gave him the opportunity to devote all his energies to the work at Olduvai, Rusinga, Fort Ternan, and elsewhere as well as providing housing for the large number of fossils from these and other sites (Baringo, Omo, East Rudolph, Songhor, Peninj, etc.) and for the archaeological material that some of them have also yielded. In ten years the Centre became a focus for research in East Africa and a base for distinguished palaeo-anthropologists and prehistorians. The Lake Rudolph research teams are based here and the Centre, which houses the most complete collection of East African fossil hominids and fauna, is used by numerous distinguished specialists—Clark Howell, Phillip Tobias, Bryan Patterson, Alan Gentry, Yves Coppens, Basil Cooke, Dick Hooijer, Dick Hay, Glynn Isaac, and Vincent Maglio to mention only a few.

It is most gratifying to be able to record that, in place of the Centre and to continue and expand its aims and work, there is being established by the Trustees of the National Museums and his son Richard, the Administrative Director, the 'Louis Leakey Memorial Institute for African Prehistory,' a worthy tribute to Leakey's great and lasting contribution to palaeo-anthropology.

PRIMATE BEHAVIOUR STUDIES

Leakey's interest in trying to reach a clearer understanding of the behavioural pattern of early man led him to look at that of man's closest relatives, the chimpanzee and gorilla. Finding that little was known about the behaviour of free-ranging groups, he set about finding and training students to undertake such studies. In his choice of those to do the work, he was particularly successful, and the chimpanzee studies of Jane van Lawick-Goodall in the Gombe Stream Reserve, Tanzania, and Dian Fossey's studies of the mountain gorilla in Rwanda as well as his own establishment of the National Primate Research Centre at Tigoni outside Nairobi, have added considerably to knowledge of primate behaviour. Louis continued to take the liveliest interest in these primate projects as well as to advise on them and, at the time of his death, he was training two other students, one to work on the lowland gorilla and the other on orangs for he believed that the clue to understanding the behaviour of extinct animal species lay in studying that of their closest living relatives.

His interests were catholic, his knowledge staggering in its range and depth and his influence world wide. A truly international scientist, he had the great ability to inspire people—whether it was scientists from many different disciplines or a general audience several thousand strong. While as a leader of a carefully picked research team he was unsurpassed, it must be admitted that not all those with whom he came into contact found his enthusiasm and almost dogmatic interpretation attractive. He could be extremely intolerant of the views of others, if not even verging on the vindictive at times and, like most of us, he did not like to admit that he had been mistaken. Nor did he easily accept criticism though he minded rebukes and criticisms from the press less, on the whole, than he did those of his colleagues, possibly because he felt that their foundations were second hand and uninformed. When he was challenged by a contemporary with a personality as strong as his own, the continent was hardly large enough to hold them both. This may, perhaps, be why he was generally more successful with the women whom he encouraged and trained than he was with his male associates. For many, however, such shortcomings, if they manifested themselves, were accepted as only what might be expected from one with such a breadth and depth of knowledge and understanding of nature and human behaviour, past and present.

He was a fluent, stimulating, and entertaining speaker and his writings were equally exciting. He used to gain both publicity and financial support for his Centre, chiefly by means of the exacting lecture tours in Europe and America that he undertook throughout the 60s and up to his death. These tours were highly successful and the publicity he received made his name and that of Olduvai Gorge household words. There is, however, no doubt that, while they brought him the finance that enabled him to run the Centre, they were strenuous undertakings that must have helped to shorten his life. His enthusiasm was infectious, no matter with whom he made contact, whether through his persuasion of the Emperor of Ethiopia to initiate investigation by international teams of the Plio-Pleistocene sediments in the Ethiopian section of the Rift Valley or as adviser to the research team excavating what may be the oldest occupation site in Eurasia at 'Ubeidiya in the Jordan Rift; in his dealings with undergraduates or with the group of influential business men and their wives in southern California who started the L. S. B. Leakey Foundation for furthering the study of man and his origins.

Leakey's contribution to palaeo-anthropology received world-wide recognition and he was the recipient of many honours and awards.

He was elected a Fellow of the British Academy in 1958. In the same year an honorary D.Sc. degree was conferred upon him by the University of Oxford. He was given an honorary LLD by the University of California in 1963, a D.Sç. by the University of East Africa in 1965, and a further LLD by Guelph University in 1969.

He was awarded the Cuthbert Peek Prize by the Royal Geographical Society of London (1933); the Andrée Medal of the Swedish Geographical Society (1933); the Henry Stopes Memorial Medal of the Geologists' Association, London (1962); the Hubbard Medal of the National Geographic Society, Washington, D.C., jointly with Dr. Mary Leakey (1962); the Richard Hopper Day Memorial Medal of the Academy of Natural Sciences of Philadelphia (1964); the Medal of the Svenska Sallskapet fur Antropologi och Geographi (1963); the Royal Medal of the Royal Geographical Society of London (1964); the Viking Fund Medal of the Wenner-Gren Foundation for Anthropological Research, New York (1965); the Hailie Selassie I Award (1968); the Wellcome Medal of the Royal African Society (1968); the Science Medal, Academy for Biological Sciences, Italy (1968), and, jointly with Dr. Mary Leakey, the Prestwich Medal of the Geological Society of London (1969).

From 1929 to 1934 he was Fellow of St. John's College, Cambridge, and in 1934 he was Jane Ellen Harrison Memorial Lecturer there. He was a Leverhulme Research Fellow from 1933 to 1935; Munro Lecturer at the University of Edinburgh in 1936; the Herbert Spencer Lecturer at the University of Oxford (1960–61); Huxley Memorial Lecturer at Birmingham University (1961); Regents' Lecturer at the University of California (1963); Silliman Lecturer at Yale University (1963–4); George R. Miller

Professor at the University of Illinois at Urbana (1965); Andrew R. White Professor at Large of Cornell University (1968), and Honorary Professor of Anatomy and Histology at the University of Nairobi.

He was the author of some sixteen books, several of which have gone into second editions, and more than one hundred and seventy papers in scientific journals on a great range of subjects. He was also General Editor of the new *Fossil Vertebrates of Africa* series (Academic Press, London) and of *Adam or Ape* (Schenkman, New York).

In the autumn of 1972, while on a visit to London and preparing to fly to the United States for a further lecture tour he suffered a heart attack from which he never recovered. He died on 1 October and was buried by the side of his parents at Limuru overlooking the Rift Valley which was the scene of so much of his active life and momentous discoveries.

In 1928 he married Henrietta Wilfreda Avern by whom he had a son and a daughter. This marriage was dissolved and in 1936 he married Mary Douglas Nicol by whom he had three sons. Mary was his constant companion in the field and in the laboratory and her tireless encouragement, help, and consummate archaeological ability made this husband and wife team so remarkably successful. He remained close to all his children and took a lively interest in the grandchildren, one of whom was observed making string figures at the gathering after the Memorial Service, arranged by his friends Dr. and Mrs. Glyn Daniel, in St. John's College chapel on 20 October.

Louis spoke often of what he called 'Leakey's luck.' Lucky he certainly was, in that he was working in just about the one part of the world where the evidence of the emergence of man the tool-maker had been so uniquely preserved. That he concentrated his energies here was not simply due to his love of Africa but to his keen observation of its potential, while his success came more especially from his tireless energy and perseverance in the pursuit of the evidence which was both its own reward and the means of making known so many of the missing elements of the ancestry of man. It was he who provided much of the evidence that formed the basis for the long and detailed time-scale against which to measure early hominid biological and cultural evolution; it was his work that made possible the recognition of the dry savanna habitats they favoured and the remains of the hominids themselves in association with the concentrations of artifacts, food waste, and other behavioural factors at the oldest living sites of man in the world.

As Robert Ardrey so dramatically puts it in *African Genesis:*

> L. S. B. Leakey has made the discoveries that will tantalise the future. Dr. Leakey and his wife Mary have been the finders beyond equal and they have uncovered in East Africa enough significant remains of man's origin to keep a regiment of analysts busy for a generation.

And again—

> I have frequently wondered at the sight of Dr. Leakey, grey hair flying, charging like a Spanish bull down the corridors of his museum, as along the crimson gravel roads so like straight bloody cuts on the face of Kenya: gasping to keep up with him, I have wondered, 'Must he be in such a hurry?'

With the great zest for life that he had and so much material awaiting discovery, so many lines of evidence requiring investigation and so few trained to do it—small wonder he frequently gave the impression of being in a hurry: he was! but gathering up as he went along new friends and new collaborators to share with him the excitement of the hunt and the reward of the successful venture. It will be long before we shall again see his like in energy, enthusiasm, depth of knowledge and experience, dedication, courage, and a perseverance that amounted to genius. It would not be an exaggeration to say that, by his discoveries and his brilliant personality, he gained the admiration and captured the imagination of the world.

Acknowledgements

Thanks are most gratefully recorded to Dr. Mary Leakey, Dr. K. P. Oakley, Mrs. Sonia Cole, Professor Phillip Tobias, and Mrs. M. C. Burkitt for their valued help in compiling this record of an outstanding life and to Dr. Jeffrey Bada for permission to quote the new date for *Homo rhodesiensis*. With the limitation of space, the problem has been to decide what to leave out, so catholic were his interests; the length is, I feel, justified by the fullness of his life and the significance of his contribution.

Shirley C. Coryndon:

A Bibliography of the Written Works of Louis Seymour Bazett Leakey

It is typical of the extensive interests and constant enthusiasm characteristic of Louis Leakey that he never kept a record of his published works—by the time a book or article was published he was several steps ahead preparing future works. The extent of his interests is only partly exemplified in the variety of subjects on which he wrote. As all who knew him realized, he could switch his mind completely from one subject to another of completely different aspect in a moment and apply himself to a new idea with unequalled concentration and depth of knowledge. Many aspects of his knowledge were unfortunately kept in his head and never committed to paper, though he was always generous in passing on ideas and experiences to others.

Because of the lack of comprehensive records, the compilation of an absolutely complete bibliography has been fraught with problems. The first attempts were made by P. V. Tobias and S. C. Coryndon in the mid-1960s, with contributions recently from the East Africa Natural History Society, the L. S. B. Leakey Foundation and Miss A. Thurston, to all of whom many thanks are due. But this work could never have been undertaken without the concentrated efforts of Mrs. Pat Barrett of Nairobi, whose indefatigable investigations into the L. S. B. Leakey papers, co-ordination of all contributions and compilation of data made the publication of this bibliography possible. It is with pleasure that her major contribution to this publication is most gratefully acknowledged.

There are almost certainly some published articles which have been missed from this bibliography, and for any omissions and/or errors S. C. Coryndon is alone responsible. The majority of works have been checked, read, and a brief summary of content written. If no summary follows an entry, the publication has not yet been seen.

Sincere thanks are due to Dr. Mary Leakey for her warm hospitality and help, and for permitting access to their private library—a source of many entries otherwise difficult to obtain. Thanks are also due Elizabeth McCown and Alice Davis for many hours of devoted work in adjusting bibliographic format to that used in *Perspectives,* and for careful searches for some obscure items of information needed for final completion of the bibliography.

The bibliography is presented under seven subject matter divisions. Each of these in turn is divided into a list of books and monographs (Section A) and a list of papers and articles (Section B).

Two items of autobiography have been placed first of all. The remaining publications follow in this order:

- I. Archaeology and Prehistory (Arch.)
- II. Human Osteology, Human Evolution, Hominid and Hominoid Fossils (H. Pal.)
- III. Paleontology (non-hominoid) (Paleo.)
- IV. Geology and Dating (Geol.)
- V. Ethnography (Ethno.)
- VI. Natural History (Nat. Hist.)
- VII. Miscellaneous (Misc.)

Those articles attributed to two categories have the second category noted in the margin.

AUTOBIOGRAPHY

1937 White African. London: Hodder and Stoughton. New Edition: 1966. Cambridge, Ma.: Schenkman. 320 pp.

Autobiography of early years, including expeditions to Tendaguru, Oldoway, Zimbabwe, Kanam.

1974 By the Evidence. Memoirs 1932–1951. New York/London: Harcourt Brace Jovanovich. 278 pp., 12 pls., 1 map, index.

Continuation of autobiography.

I. ARCHAEOLOGY AND PREHISTORY

A. Books and Monographs on Archaeology

PALEO. GEOL. 1931 The Stone Age Cultures of Kenya Colony. Cambridge, England: Cambridge University Press. 283 pp., 31 pls., 47 figs., 2 maps. Reprinted in 1971 by F. Cass, London, as No. 2 in the series African Prehistory, with a new introductory note by the author.

Human remains and associated cultures described from Kenya Rift Valley in Nakuru and Naivasha Basins. Appendices: A: J. D. Solomon, Geology and implementiferous deposits in the Nakuru and Naivasha Basins and surrounding areas. B: C. E. P. Brooks, The correlation of Pluvial Periods in Africa with climatic changes in Europe. C: A. T. Hopwood, Preliminary report on the fossil Mammalia. D: M. Connolly, The Mollusca from deposits of Gamblian Pluvial date, Kenya Colony. E: J. W. Gregory, Prehistoric iron in British East Africa. F: H. C. Beck, Notes on beads from the Upper Kenya Aurignacian and the Gumban B cultures.

H.PAL. 1934 Adam's Ancestors. London: Methuen. 4th Edition rewritten and reset: 1953. Rewritten with addenda: 1960. 235 pp., 22 pls., 34 figs.

The search for man's ancestors, his environment, cultures and evolution.

H. PAL. 1936 Stone Age Africa: An Outline of Prehistory in Africa. London: Oxford University Press. 218 pp., 13 pls., 28 figs., 1 map.

Ten chapters representing ten Munro lectures delivered in Edinburgh, February 1936, giving general picture of African Stone Age with climate, geography, fauna, culture, art, anthropology and comparisons with Europe.

1949 Tentative Study of the Pleistocene Climatic Changes and Stone Age Culture Sequence in North-Eastern Angola. Lisbon: Diamang Museu do Dundo Publicações Culturais, No. 4. 82 pp., 30 pls., 7 figs.

Sangoan culture associated with dry conditions but not so dry as in other parts of Africa. Economic importance of prehistoric research.

PALEO. GEOL. 1951 Olduvai Gorge: A Report on the Evolution of the Hand-axe Culture in Beds I–IV. Cambridge, England: University Press. 163 pp., 38 pls., 62 figs., 2 maps.

Hand-axe cultures described, with correlation of faunal and climatic changes. Contributions by others: Chapter II: H. Reck, A preliminary survey of the tectonics and stratigraphy of Olduvai. Chapter III: A. T. Hopwood, The Olduvai fauna. Chapter IV: D. G. MacInnes, Geological and palaeontological evidence.

1952 . . . and S. Cole, Eds. Proceedings of the First Pan-African Congress on Prehistory, Nairobi. Oxford: Blackwell. H. PAL.

B. Papers and Articles on Archaeology

1927 . . . and Newsam, B. H. Preliminary report on excavations in Kenya Colony. The Stone Age In Kenya. Leeds: Reports of the British Association for the Advancement of Science 95:357–358.

1929 An Outline of the Stone Age in Kenya. South African Journal of Science 26:749–757. H. PAL. GEOL. PALEO.

Geology, climate, culture, human remains and fauna from Kenya in the Pleistocene and Post-Pleistocene.

1929 . . . and Solomon, J. D. East African Archaeology. Nature 124:9. 1 tbl.

Correlation of pluvials and glacials premature. Nomenclature for pluvials proposed, and tabulated with cultures in sequence.

1931 Notes on the Stone Age Cultures of East Africa. Journal of the East Africa Natural History Society 42–43:200–204. 1 tbl.

Notes to accompany exhibit in Coryndon Memorial Museum. Table of climatic change in the Pleistocene with changes in lake levels. Olduvai tools from Pre-Chellean to advanced Acheulean displayed, also tools from Nakuru of Aurignacian, Mousterian types.

1931 Prehistoric Man in Kenya. Nature 127:814–815. GEOL.

Description of Gamble's Cave and Bromhead's Site with explanation of geology and archaeological finds.

1934 The Sequence of Stone Age Cultures in East Africa. *In* Essays Presented to C. G. Seligman. London: Kegan, Paul, Trench, Trubner. 143–147. 1 tbl.

Culture stages from basic Oldowan upwards.

1934 The Oldoway Culture Sequence. Proceedings of the 1st International Congress of Prehistory and Protohistoric Science. London: Oxford University Press. 73 pp., 1 tbl. PALEO.

Very brief outline of culture and fauna throughout Oldoway succession from Bed I to present.

1935 East African ruined cities contrasted. The Illustrated London News 187:610–611. 15 figs.

Ruins described at Engaruka in Northern Tanganyika, and Gedi on the Kenya Coast. Engaruka about 300 years old, with dry stone walling, probably built by Wambulu people. Gedi dated about 600 A.D. of Persian or Arab origin; architecture suggests early English.

1936 Preliminary Report on Examination of the Engaruka Ruins. Dar es Salaam: Tanganyika Notes and Records 1:57–60.

Ruins and burials not more than 300 years old, probably of Mbulu people ousted by Masai. Burial cairns and circular stone houses. Middens with beads, bits of iron, copper wire and chain work, maybe Masai and not connected with ruins. Estimate of population about thirty to forty thousand.

1938 Excavations at Njoro, Kenya. Nature 142:319–320. 1 fig.
Neolithic. Cremation site with stone bowls, charred basketry and woodwork. Beads include amazonite and much opal from local mines. Trading likely.

1943 The Industries of the Gorgora Rock Shelter, Lake Tana. Journal of East Africa Natural History Society 17:199–203. 1 tbl.
Early Stillbay to Magosian followed by degenerate microlithic cultures.

1945 . . . and Owen, W. E. A Contribution to the Study of the Tumbian Culture in East Africa. Nairobi: Coryndon Memorial Museum Occasional Papers, No. 1. 59 pp., 23 figs., 1 map.
Kavirondo area. Evolution from Sangoan. No human remains.

PALEO. 1946 A Prehistorian's Paradise in Africa: Early Stone Age Sites at Olorgesaillie. The Illustrated London News 209:382–385. 1 pl., 8 figs.
Discoveries at Olorgesaillie.

1946 Early Man in Kenya. London: The Times. 4, 10, 1946.
Acheulean camp sites.

1947 Prehistory Notes. Nairobi: Nature East Africa 1:8.
Negotiations to open Tanganyika's prehistoric rock paintings for the public.

1947 Archaeological Note. Nairobi: Nature East Africa 2:11.
Plan to preserve and protect ruins of Gedi.

1947 Palaeolithic Nomenclature. Man 14:19–20.
Letter in reply to criticism by Van Riet Lowe and defending Menghin in naming cultures after the site where they are first clearly recognized.

1947 Archaeological Note. Nairobi: Nature East Africa 3:15.
New excavations at Olorgesaillie open to public. Resident curator appointed.

1948 Leakey, M. D., and Owen, W. E., and . . . . Dimple Based Pottery From Central Kavirondo, Kenya Colony. Nairobi: Coryndon Memorial Museum Occasional Papers, No. 2. 43 pp., 8 pls., 12 figs.
Pottery belongs to Iron Age. Unlike extant pottery. Climate wetter than today.

1950 South African Stone Age Terminology II. Nature 165:1026.
South African Magosian as a term. Cultural terms must not be confused with terms for techniques which are widespread in time and space.

1950 Terminology in Prehistory. Claremont, Cape: South African Archaeological Bulletin 5:20–22.
Recent research shows that earlier terminology is often archaic and misleading. Slovenliness in science makes for chaos.

1951 Preliminary Excavations of a Mesolithic Site at Abinger Common, Surrey. Research Papers of the Surrey Archaeological Society, 3. 44 pp., 5 pls., 12 figs.
Pit dwellings. Flint implements of microlithic assemblage possibly ancestral to Horsham culture.

1952 The archaeological aspect of the Tanganyika Paintings with a tentative note on sequence. Section 4. *In* Tanganyika Rock Paintings: A Guide and Record. Dar es Salaam: Tanganyika Notes and Records 1:15–19.

Thirteen distinct superimposed styles representing fauna and human figures.

1952 The Tumbian Culture in East Africa. *In* Proceedings of the First Pan-African Congress on Prehistory, Nairobi. L. S. B. Leakey and S. Cole, Eds. Oxford: Blackwell. 201–202.

Indecision of nomenclature for "African Tumbian." Possibly Lupemban = Upper Tumbian, Sangoan = Middle/Lower Tumbian as described by Leakey and Owen 1945.

1952 Capsian or Aurignacian?: Which Term Should be Used in Africa? *In* Proceedings of the First Pan-African Congress on Prehistory, Nairobi. L. S. B. Leakey and S. Cole, Eds. Oxford: Blackwell. 205–206.

Suggestions for nomenclature. Capsian of East Africa a blade and burin culture.

1952 Olorgesaillie Prehistoric Site. *In* Proceedings of the First Pan-African Congress on Prehistory, Nairobi. L. S. B. Leakey and S. Cole, Eds. Oxford: Blackwell. 209.

Acheulean culture on living floors and a "Hope Fountain" culture on land surface 6. Faunal remains very broken. Several bolas stones in groups of three. Site opened as "museum on the spot."

1952 *Review of* A. J. Arkell, The Old Stone Age in the Anglo-Egyptian Sudan. Man 222:152.

1953 Results of recent research in Kenya. *In* Acts of the 3rd International Congress of Prehistory and Protohistoric Science, Zurich. E. Vogt, Ed. 169–170.

Excavations at Olorgesaillie, Naivasha, Njoro, and Tanganyika. Pluvials explained.

1954 Olduvai Gorge. Scientific American 190:66–71. 6 figs.

GEOL.
PALEO.

Description of geology, palaeontology, and archaeology of Olduvai Gorge and its discovery.

1954 A new method of exhibiting prehistoric art. London: Museums Journal 54:11–12. 1 pl.

Construction of museum exhibit showing natural size reproductions of rock paintings from Tanganyika Territory on simulated rock face.

1954 Working Stone, Bone and Wood. *In* A History of Technology, Vol. 1. C. Singer, E. J. Holmyard and A. R. Hall, Eds. London: Oxford University Press. 128–143. 14 figs.

Techniques used by Palaeolithic man to make tools.

1954 Graphic and Plastic Arts. *In* A History of Technology, Vol. 1. C. Singer, E. J. Holmyard and A. R. Hall, Eds. London: Oxford University Press. 144–153. 10 figs.

Types and techniques of Palaeolithic art in Europe.

1955 Preliminary Notes on a Survey of Prehistoric Art in Tanganyika. *In* Congrès Panafricain de Préhistoire: Actes de la II^{eme}Session. Algiers. L. Balout, Ed. Paris: Arts et Metiers Graphiques. 723–724.

Four or five major art styles superimposed. Sixty-seven sites with paintings worth tracing. Age from Upper Palaeolithic to end Mesolithic, with poor Wilton culture at top, and at site Kisese 3 Stillbay below. Three sites open to the public under supervision.

PALEO. 1957 Preliminary Report on a Chellean I Living Site at BK II, Olduvai Gorge, Tanganyika Territory. *In* Proceedings of the 3rd Pan-African Congress on Prehistory. Livingstone. J. Desmond Clark and S. Cole, Eds. London: Chatto and Windus. 1955. 217–218.

Description of living floor at BK II with Chelles-Acheulean stage I artifacts and numerous pebble tools of evolved Oldowan, few rough hand-axes, few small worked flake tools, utilized flakes. Upper Bed II fauna.

PALEO. 1957 Techniques of Recording Prehistoric Art. *In* Proceedings of the Third Pan-African Congress on Prehistory, Livingstone. J. Desmond Clark and S. Cole, Eds. London: Chatto and Windus. 1955. 304–305.

Tracing of rock paintings in Tanganyika. Discussion of method, paper, ink and colour media, etc.

H. PAL.
PALEO. 1958 Recent Discoveries at Olduvai Gorge, Tanganyika. Nature 181:1099–1103. 8 figs.

Chellean I living site at BK II. Two hominid teeth. Extensive fauna. Chellean II living site at SHK II with tool of hippo tusk.

H. PAL. 1962 The Archaeology of East Africa. Reports on Natural Resources of East Africa. East African Literature Bureau. Nairobi. 17–21.

Summary of discoveries relating to fossil man and his cultures from Miocene to historic times.

1964 Gamble's Cave, Kenya: A Field Museum in a Cave. Studies in Speleology 1:22–25. 1 fig.

Two rock shelters above Lake Nakuru: Gamble's Cave I and II. II very rich in cultural material. Thirty feet deep occupation levels from Upper Kenya Capsian to Iron Age. "Museum on the spot" opened to show public results of excavation and interpretation of occupation levels.

1968 . . . and Leakey, M. D. Archaeological Excavations at Olduvai Gorge. Washington, D.C.: National Geographic Society Research Report. 1963 179–182.

Completion of work at DK I, HWK II E, and VEK II. Stratigraphical sequence of Bed II and correlation of different sites ascertained. Fences and shelters erected at certain key sites. Hominid remains include MNK II *Homo habilis.*

1968 Bone Smashing by Late Miocene Hominidae. Nature 218:528–530. 5 figs.

Bones found in Upper Miocene fossil beds of Fort Ternan, Kenya, show evidence of having been broken up by some kind of blunt instrument.

1968 . . . , Simpson, R. de E., and Clements, T. Archaeological Excavations in the Calico Mountains, California: Preliminary Report. Science 160:1022–1023.

Very early humanly made artifacts and flakes from alluvial fan dated on geological evidence of possible age 50–80,000 years. Proposal for symposium to discuss pros and cons.

PALEO. 1970 Archaeological Research at Olduvai Gorge, Tanzania and Fort Ternan, Kenya, 1961–1962. Washington, D.C.: National Geographic Society Research Report. 1961–1962. 131–133.

Excavation results at Olduvai: DK I, FLK NN I, FLK N I, HWK E, JK2 (III and base IV), Hominid skull VEK IV, etc. Geological studies initiated.

1971 Archaeological and Palaeontological Investigations in East Africa, 1965. Washington, D.C.: National Georgraphic Society Research Report. 1965. 153–154. — H. PAL. PALEO.

Excavations at MKI, Olduvai. Fencing of sites. Baringo (Kapthurin) hominid.

1972 Archaeological Investigation at Olduvai Gorge, Tanzania, 1960. Washington, D.C.: National Geographic Society Research Report. 1955–1960. 109–110. — H. PAL.

Two distinct hominids in Bed I. Age of deposit 1.7 million years with tools representing seven different forms.

1973 Was *Homo erectus* responsible for the hand-axe culture? Journal of Human Evolution 2:493–498.

II. HUMAN OSTEOLOGY, HUMAN EVOLUTION, HOMINID AND HOMINOID FOSSILS

A. Books and Monographs

1934 Adam's Ancestors. See Section IA above.

1935 The Stone Age Races of Kenya. London: Oxford University Press. 143 pp., 37 pls., 52 figs., 1 map.

Contains letters to Nature 1932; report on conference at Cambridge by Royal Anthropological Society 1953. Hominid remains from Kanam, Kanjera, Elementeita, Nakuru, etc. Comparison with discoveries in Europe, Asia and other parts of Africa. Appendices: A: J. W. P. Lawrence, A note on the pathology of the Kanam mandible. B: G. Eliot Smith, A note on the endocranial cast of Kanjera Man. C: F. Colyer, Notes on the odontological pathology of various specimens from East Africa. D: J. W. P. Lawerence, Notes on the pathology of a neolithic skeleton, and also some certain pathological bones from Bromhead's Site, Elementeita. E: L. S. B. Leakey, A note on a mandible from a cave at Kiambu. F: L. S. B. Leakey, List of the fragmentary remains from Kenya not described in the main part of this book. Second edition, 1970, with introduction by the author. Oosterhout, N. B., Netherlands. Anthropological Publications.

1950 Leakey, M. D., and . . . Excavations at the Njoro River Cave: Stone Age Cremated Burials in Kenya Colony. Oxford: Clarendon Press. 78 pp. 14 pls., 22 figs., 3 maps. — H. OST ARCH.

Archaeological remains by M. D. Leakey. The Crania by L. S. B. Leakey. New characters revealed in culture of Neolithic and Mesolithic burial methods. Industry: obsidian, pottery, stone bowls, beads, etc. People of Njoro Cave not negroid.

1961 The Progress and Evolution of Man in Africa. London: Oxford University Press. 30 pp., 2 pls., 2 figs.

Two lectures: A: "The Progress of Man in Africa." Africans and Europeans differ physically and psychologically. Africans should be given chance to develop wisdom, rather than copy European politicians. B: "Africa's Contribution to Human Evolution." Discusses *Zinjanthropus* and the South African australopithecines.

1969 . . . and Goodall, M. Unveiling Man's Origins. New York: Schenkman; London: Methuen. 185 pp.

History of major discoveries of fossil man, the interpretation of man's evolution and attitude of the world to them since Darwin. Latest finds in East Africa discussed and major gaps emphasized.

ARCH. 1970 Adam or Ape. New York: Schenkman. J. Prost and S. Prost, Eds.

General collection of previously published papers on discoveries relating to fossil hominids and their cultures.

B. Papers and Articles on Human Osteology and Human Evolution

1927 Stone Age Man in Kenya Colony. Nature 120:85–86. 2 figs.

Skulls from Elementeita and Nakuru not of types of people found in Africa today.

1928 The Oldoway Skull. Nature 121:499.

Skull part of a burial. Morphology similar to skulls from Elementeita.

ARCH. 1930 Exhibition of Some of the Skeletal Material Collected by the East African Archaeological Expedition and Some Associated Finds. Catalogue of the Museum of the Royal College of Surgeons. London. 13 pp.

Finds from excavations in Nakuru and Elementeita with table of climatic changes. Skulls, tools and stone bowls displayed.

PALEO. 1932 The Oldoway Human Skeleton. Nature 129:721–722.

Age of skeleton. Agreement that it was buried into Bed II. First reports of excavations at Kanam, Kendu, Rusinga.

1932 The Oldoway Human Skeleton. Nature 130:578.

Letter in reply to D. M. S. Watson and C. Foster Cooper. L. S. B. Leakey considers skeleton buried into Bed II before Bed III laid down. Mention of Kanam and Kanjera skulls as *Homo sapiens*.

1933 Die Menschenreste von Kanam und Kanjera, Kenya-Kolonie. Anthropologischer Anzeiger 10:238–243.

Kanam and Kanjera hominid finds and summary of Cambridge Conference.

1933 . . . , Reck, H., Boswell, P. G. H., Hopwood, A. T., and Solomon, J. D. The Oldoway Human Skeleton. Nature 131:397–398.

Burial of Oldoway skeleton into an older deposit. Probably Bed V age for skeleton.

ARCH. 1933 Notes on the Comparative Series of Skulls and Cultures in the Museum. Journal of the East Africa Natural History Society 47–48: 142–145.

Exhibit of the plaster casts of prehistoric skulls from many parts of the world. Cultures found in association with East African skulls explained.

PALEO. 1933 The Status of the Kanam Mandible and the Kanjera Skulls. Man 33:200–201.

Evidence for Lower Pleistocene age of Kanam.

1936 Fossil Human Remains from Kanam and Kanjera, Kenya Colony. Nature 138:643.

Answer to letter from P. Boswell explaining difficulty of pinpointing Kanam jaw site, and the muddle over photographs.

1936 A New Fossil Skull from Eyasi, East Africa. Discovery by a German Expedition. Nature 138:1082–1084. 3 figs.

Skull of Upper Pleistocene age found with Levalloisian artifacts. Possibly new genus? Demonstrates diversity of hominine genera and species still evident in Upper Pleistocene.

1942 The Naivasha Fossil Skull and Skeleton. Journal of the East Africa Natural History Society 16:169–177. 1 pl., 2 figs. — ARCH.

Homo sapiens male. Closing phase of Gamblian pluvial. Associated industry Upper Kenya Aurignacian phase 'C'.

1943 A Miocene Anthropoid Mandible from Rusinga, Kenya. Nature 152:319–320.

Proconsul nyanzae mandible shows morphological links toward human line. No simian shelf, premolars reduced, 3rd molar dryopithecine.

1944 Stone Age Man in East Africa. Bulletin of the Uganda Society 2:11.

Prehistory in context of geography, climatology, zoology and culture in attempt to learn of prehistoric ecology.

1945 Notes of the Skulls and Skeletal Material from Hyrax Hill. Part IV *In:* M. D. Leakey, Report on the Excavations of Hyrax Hill, Nakuru. Transactions of the Royal Society of South Africa 30:374–405. 17 figs. — ETHN.

Finds from Neolithic cemetery and from Iron Age burial pits.

1946 Report on a Visit to the Site of the Eyasi Skull, found by Dr. Kohl-Larsen; Part I. Journal of the East Africa Natural History Society 19:40–43. 1 map. — GEOL. ARCH.

Skull of Upper Pleistocene age. Maker of Levalloisian artifacts. Not *Homo sapiens* but neanderthaloid.

1946 Fossil Finds in Kenya. Ape or Primitive Man? Antiquity 20:202–204. — PALEO. ARCH.

Discovery of *Proconsul.* Description of Olorgesaillie and its discovery.

1946 Was Kenya the Centre of Human Evolution? The Illustrated London News 209:198–201. 14 figs., 1 tbl. — PALEO.

Proconsul discoveries and comparisons with modern man and apes.

1946 African Ancestries. Interpretation of Fossil Finds in Kenya. Ape or Primitive Man? London: The Times 23:8. 1 pl. — PALEO.

Proconsul finds from East African Lower Miocene. Photograph of *P. nyanzae* from Rusinga.

1947 Fossil Apes in Kenya. Finds near Lake Victoria. London: The Times 30:12.

1948 Skull of *Proconsul* from Rusinga Island. Nature 162:688. 1 pl. — PALEO.

First report and photograph of skull of *P. africanus.*

1948 Fossil and Sub-fossil Hominoidea in East Africa. Robert Broom Commemorative Volume. A Special Publication of the Royal Society of South Africa. 165–170. — PALEO.

Review and summary of recent work.

PALEO. 1948 Palaeontological Notes. Nature East Africa 8:14.

Discovery of skull of *Proconsul africanus.*

PALEO. 1950 Clark, W. Le Gros and . . . . Diagnoses of East African Hominoidea. Appendix *In:* Quarterly Journal of the Geological Society of London 105:2, 260–262.

Diagnoses recorded for *Proconsul africanus* Hopwood, *P. nyanzae* sp. nov., *P. major* sp. nov., *Sivapithecus africanus* sp. nov., *Limnopithecus legetet* Hopwood, *L. macinnesi* sp. nov.

PALEO. 1950 The Fossil Apes of Lake Victoria. Have Man's Ancestors Been Found? The Listener Sept. 21:143–148. Also *In:* South African Archaeological Bulletin 5:143–148.

Proconsul and *Limnopithecus* morphology, environment and food.

PALEO. 1950 The Fontéchavade Skulls from Charente, France. Archaeological News Letter 2:196.

Homo sapiens as a victim of Tayacian making Neanderthalers.

1950 The Age of *Homo sapiens.* Mankind 4:196–200.

1950 Early Man in Australia. Mankind 4:214.

PALEO. 1951 Clark, W. Le Gros and . . . . The Miocene Hominoidea of East Africa. Fossil Mammals of Africa: No. 1, British Museum (Natural History). 117 pp., 9 pls., 28 figs.

Description of Miocene fossils from East Africa with map of localities, age, occurrence, analysis of distribution of primates, stratigraphy of Rusinga Island. Diagnoses and description of *Proconsul africanus, P. nyanzae, P. major, Sivapithecus africanus, Limnopithecus legtet, L. macinnesi.* Description of limb bones. East African hominoids in relation to primate evolution.

PALEO. GEOL. ARCH. 1952 Age of the Eyasi Skull. *In:* Proceedings of the First Pan-African Congress on Prehistory. Nairobi. . . . and S. Cole, Eds. Oxford: Blackwell. 133–134.

Upper Pleistocene age confirmed from geology, palaeontology and stone age culture. Probably *Pithecanthropus* sp.

1953 Kenya and Tanganyika entries: *In:* Catalogue des Hommes Fossiles, E. V. Vallois and H. L. Movius, Eds. Compte Rendu du 19th Congress of International Geologists. Algiers, 1952, V, 282–286.

Human remains catalogued from: Bromhead's site; Elementeita; Eyasi; Gamble's Cave; Garusi; Kanam; Kanjera; Naivasha Railway Sites; Oldoway (Olduvai).

1955 The Environment of the Kenya Lower Miocene Apes. *In* Congrès Panafricain de Préhistoire: Actes de la IIeme Session. Algiers. L. Balout, Ed. Paris: Arts et Metiers graphiques 18:323–324.

Miocene apes not forest dwellers but savanna and gallery forest dwellers.

1958 A Giant Child Among the Giant Animals of Olduvai? A Huge Fossil Milk Molar Which Suggests That Chellean Man in Tanganyika May Have Been Gigantic. The Illustrated London News 232:1104–1105. 10 figs.

Comparison of large Olduvai teeth with modern *Homo* milk teeth, australopithecines and apes. Must have been large jawed. Found on Chellean I living site.

1958 Problems Relating to Fossil Man. Johannesburg. Leech 28:116–119.
Theories on evolution of man critically discussed.

1959 The Fossil Skull from Olduvai. British Medical Journal 5152:635.
Letter in reply to criticism of creating new genus for Olduvai australopithecine.

1959 A New Fossil Skull from Olduvai. Nature 184:491–493. 2 figs.
Discovery and description of *Zinjanthropus boisei* gen. et. sp. nov. from FLK I Olduvai. Age similar to Omo and Taungs. Tools on living floor possibly made by *Zinjanthropus*.

1959 The Newly-discovered Skull from Olduvai: First Photographs of the Complete Skull. The Illustrated London News 235:288–289. 10 figs.
Nutcracker Man skull *(Zinjanthropus boisei)*, photo of skull and site. Compared with gorilla and *Australopithecus*.

1959 The Origin of the Genus *Homo*. *In:* Evolution after Darwin. Sol Tax, Ed. University of Chicago Press 2:17–32. PALEO.
Darwin suggested in 1871 that Africa was probably the cradle of man's origin. History of hominid discoveries in Africa from Oligocene of Fayum through East African Miocene and Pleistocene of East and South Africa. Suggestions for future lines of research.

1959 The First Men: Recent Discoveries in East Africa. Antiquity 33:285–287.
Summary of work at Olduvai and finding of *Zinjanthropus* among tools.

1960 The Affinities of the New Olduvai Australopithecine. Nature 186:456–458.
Answer to letter from J. T. Robinson, Pretoria, point by point on morphology and taxonomic position of *Zinjanthropus boisei.*

1960 From the Taung Skull to "Nutcracker Man." Africa as the Cradle of Mankind and the Primates. Discoveries in the Last Thirty-five Years. The Illustrated London News 236:44. 2 figs.
Discoveries in Africa of *Proconsul, Australopithecus; Homo* from Kanjera. Help to prove Darwin was right in seeing Africa as the cradle of Man.

1960 . . . and Howell, F. C. The Newest Link in Human Evolution (The discovery by L. S. B. Leakey of *Zinjanthropus boisei.)* Current Anthropology 1:76. 1 pl.
Discovery of *Zinjanthropus* and resemblances with other hominids. Age earlier than South African australopithecines.

1960 Recent Discoveries at Olduvai Gorge. Nature 188:1050–1052. 4 figs.
Tibia and fibula of hominid from FLK I. Animal bones broken, high proportion juveniles. Tools Oldowan. FLK NN I, many fossils, few tools. Hominid remains include fragments skull, teeth, 2 clavicles, left foot, 6 finger bones, two ribs, bone tool. Start of work on FLK N I.

1960 Finding the World's Earliest Man. The National Geographic Magazine 118:420–435. 13 pls. ARCH. PALEO.
Olduvai and discovery of *Zinjanthropus*, artifacts and fossil mammals.

1961 L'Evolution des Primates Supérieures et de L'Homme. *In* Problèmes actuels de paléontologie (Evolution des vertébrés). Colloques Internationaux du Centre National de la Recherche Scientifique. Paris, 104:451–453.

Five families for Hominoidae: *Pongidae, Hylobatidae, Proconsulidae, Oreopithecidae, Hominidae. Propliopithecus* might be ancestral to first three families.

PALEO. ARCH.

1961 New Finds at Olduvai Gorge. Nature 189:649–650. 4 figs.

FLK NN I finds include hominid mandible, two parietals, all immature. Few tools. LLK II, brain case large hominid with Chellean III tools. FLK N I, rich fauna, evidence of greater skill of hominids in hunting than in lower levels. Tools intermediate between Oldowan and Chellean Stage I.

1961 New Links in the Chain of Human Evolution: Three Major New Discoveries from Olduvai Gorge, Tanganyika. The Illustrated London News 238:346–348. 7 figs.

Discovery of Chellean Man skull, *Homo habilis* foot, hand bones and parietals. Tibia and fibula from *Zinjanthropus* level. Bone tool from FLK NN I.

1961 The Juvenile Mandible from Olduvai. Nature 191:417–418. 1 fig.

Description of immature hominid mandible from FLK NN I. Premolars unlike australopithecines; possibly primitive ancestor of *Homo.* Possibly made tools at FLK I.

PALEO.

1961 Exploring 1,750,000 Years Into Man's Past. The National Geographic Magazine 120:564–589. 28 pls.

Discoveries of fossil hominids at Olduvai and details of excavations. People and fauna of the area.

1962 Man's African Origin. Annals of the New York Academy of Science 96:495–503.

African origin of higher primates including man, also probable point of origin of hand-axe culture. Need for further excavation emphasized. Discussions.

PALEO.

1962 A New Lower Pliocene Fossil Primate From Kenya. Annals and Magazine of Natural History, London. 13:689–696. 1 pl., 1 fig.

Fort Ternan discoveries include mastodont more evolved than *Trilophodon* of Lower Miocene, but not so advanced as *Anacus.* Small giraffid; antelopes, pig. Carnivora and Rodentia. Description of new hominid *Kenyapithecus wickeri* gen. et sp. nov. Is it allied to *Sivapithecus africanus?*

1962 Africa's Contribution to the Evolution of Man. South African Archaeological Bulletin 16:3–7.

Higher primates from Oligocene through to present *Homo.* Speed of evolution in Pleistocene.

ARCH.

1962 Remains of Man with Oldowan Culture at Olduvai. *In:* Actes du IVeme Congrès Panafrican de Prehistoire et de l'Etude du Quaternaire 1959. G. Mortelmans and J. Nenquin, Eds. Musée Royal de l'Afrique Centrale, Tervuren, Annales Sciences Humaines 40:361–364.

Zinjanthropus boisei named as an australopithecine. Found on living floor with Oldowan tools.

1962 The Olduvai Discoveries. Antiquity 36:119–121.

Reply to letter from Napier and Weiner on phylogeny of Olduvai hominids and human origins.

1963 Very Early East African Hominidae and Their Ecological Setting. *In:* African Ecology and Human Evolution. F. C. Howell and F. Boulière, Eds. Viking Fund Publications in Anthropology 36:448–457. GEOL. PALEO. ARCH.

Geology of Olduvai. Fossil fauna, climatic indications of Bed I, absolute age, living floors, cultural levels and hominid remains. Bed II hominid from LLK with Chellean III tools.

1963 Adventures in the Search for Man. The National Geographic Magazine 123:132–152. 26 pls. ARCH. PALEO.

Discoveries from Fort Ternan and Olduvai.

1963 East African Fossil Hominoidea and the Classification Within this Superfamily. *In:* Classification and Human Evolution. S. L. Washburn, Ed. Viking Fund Publications in Anthropology 37:32–49.

Classification at family and generic level. Stages of evolution.

1963 Early Man in East Africa. L. A. Bates, Ed. London: Teamwork Magazine.

1964 Prehistoric Man in the Tropical Environment. *In:* Pre-Industrial Man in the Tropical Environment. H. F. L. Elliot, Ed. Proceedings and Papers of the International Union for Conservation of Nature and Natural Resources, Ninth Technological Meeting, Nairobi, 1963. 4:24–29.

Importance of water to life in times past just as today. Necessity to institute water conservation.

1964 The Evolution of Man. Discovery 25:48–49. 1 fig.

Reply to letter from W. Le Gros Clark on taxonomy and morphology of *Homo habilis. Kenyapithecus* different from *Ramapithecus.*

1964 . . . and Leakey, M. D. Recent Discoveries of Fossil Hominids in Tanganyika: At Olduvai and Near Lake Natron. Nature: 202:5–7. 7 figs.

VEK II Hom. 12, skull; MNK II Hom. 13, vault, maxillae, and mandible; FLK II Maiko Gully, cranial vault and teeth. Lake Natron mandible of robust *Australopithecus.*

1964 The Evolution of Man. Discovery 25:68.

Must not include culture in defining taxa.

1964 . . . , Tobias, P. V., and Napier, J. R. A New Species of the Genus *Homo* from Olduvai Gorge. Nature 202:5–7.

Family Hominidae Genus *Homo.* Revised diagnosis given. Species *habilis* sp. nov. described. Upper Villafranchian and Lower Middle Pleistocene. Type mandible Hom. 7, from FLK NN I. Paratypes from MNK II, FLK NN I, FLK I, MK I. Probable maker of Oldowan tools and DK I stone circles. Relationships to *Tchadanthropus,* Kanam mandible, *Telanthropus,* Ubedeiyah and *Australopithecus.*

1965 Facts Instead of Dogmas on Man's Origin. Chicago Symposium on The Origin of Man. The Wenner-Gren Foundation for Anthropological Research, Inc. Symposium, (Mimeographed transcript distributed by Current Anthropology.) University of Chicago. 3–9. 1972 Republished in: Climbing Man's Family Tree. T. D. McCown and K. A. R. Kennedy, Eds. New Jersey, Prentice Hall. 386–399.

Polyphyletic origin for *Homo* based on African fossil hominids.

GEOL. 1965 The New Olduvai Hominid Discoveries and Their Geological Setting. London: Proceedings of the Geological Society 1617:104–109.

Zinjanthropus and *Homo habilis* discoveries at Olduvai. Which made the tools? Man's polyphyletic evolution.

1965 Man's Beginnings. The World Year Book 107–122. 14 figs.

Olduvai Gorge Discoveries. *Homo habilis, Zinjanthropus.* Evolution of man discussed.

1966 *Homo habilis, Homo erectus* and the Australopithecines. Nature 209:1279–1281. 6 figs., 1 tbl.

Morphological aspects of mandibles of australopithecines with V-shaped mandible and *Homo* with U-shaped mandible. Morphology of cranium; *Australopithecus* and *H. erectus* have maximum cranial width near base, others have greatest width higher on parietals. Stratigraphical relationships of Olduvai hominids.

1967 Development of Aggression as a Factor in Early Human and Pre-human Evolution. *In:* University of California; Los Angeles Forum in Medical Science. no. 7. Fifth Conference on Brain Function 1965. C. D. Clements and D. B. Lindsey, Eds. Brain Function 5:33.

Large carnivores dislike human flesh and will be aggressive only if human is aggressive. Protection for early man. When vegetable foods in short supply, hominids turned to flesh, but canine teeth not good enough. Incentive for breaking of bones with tools, then making of tools. Once tools are there hominids can be aggressive. Long discussion follows.

PALEO. 1967 An Early Miocene Member of the Hominidae. Nature 213:115–163. 6 figs. 1968 Reprinted *In:* Perspectives On Human Evolution 1:61–84. New York. Holt, Rinehart, and Winston.

Discoveries at Rusinga and Songhor, early Miocene, make possible the separation of the Hominidae from the Pongidae in Lower and Early Middle Miocene.

1967 Kenya and Tanzania. *In:* Catalogue of Fossil Hominids, Part I, Africa K. P. Oakley and B. G. Campbell, Eds. Bulletin of the British Museum (Natural History) 22–28, 106–118.

Catalogue of all known hominid specimens up to 1964.

PALEO. 1968 Lower Dentition of *Kenyapithecus africanus.* Nature 217:827–830.

Mandible of *Kenyapithecus africanus* from Early Miocene of Kathwanga, Rusinga, with dentition described. *Kenyapithecus* and *Proconsul* possibly derived from pre-Miocene *Proconsul*-like stock.

1968 Un Spécimen d'Hominidés du vieux Miocènes. Revue Anthropologique 35–49. 1 tbl.

Specimens of *Kenyapithecus africanus* from Rusinga and Songhor described.

PALEO. 1968 Upper Miocene Primates from Kenya. Nature 218:527–528.

Fort Ternan primates include *Kenyapithecus wickeri,* a hylobatid resembling *Limnopithecus,* teeth resembling *Oreopithecus,* teeth of *Proconsul* sp., part of a jaw suggesting *Dryopithecus* and some cercopithecid teeth.

PALEO. 1969 Ecology of North Indian *Ramapithecus.* Nature 223:1075–1076.

Reply to letter by Tatersall questioning validity of extrapolating ecology of *Ramapithecus* from fossil forms such as pigs and giraffes as forest forms. Could just as easily be plains and swamp.

1969 Archaeological and Palaeontological Investigations at Olduvai Gorge, Lake Natron, Tanzania and Fort Ternan, Kenya. National Geographic Research Report 1964, 119–121. ARCH. PALEO.

Discovery of *Zinjanthropus* mandible from Peninj; also living floor; geological mapping. Delineating fossil areas at Olduvai by thorn fencing to prevent damage by Masai cattle. Fort Ternan excavations extended to include many new discoveries.

1969 . . . , Simpson, R. de E., and Clements, T. Man in America: The Calico Mountains Excavations. *In:* 1970 Britannica Year Book of Science and the Future. Encyclopaedia Britannica, Inc. London: 65–79. 15 figs.

History of search for Stone Age Man in Calico Hills of America and postulation of migration routes. Criticism by C. Vance Haynes—considers flints flaked naturally, not by man.

1969 Fort Ternan Hominid. Nature 222:1202. PALEO.

Questioning Simon's 1968 evaluation of Fort Ternan hominid.

1970 The Evolution of Man. Abbotempo. Abbot Universal Ltd. 2–7.

Evolution of *Homo sapiens* from *Homo habilis* with numerous over-specialized side branches.

1970 'Newly' recognised mandible of *Ramapithecus.* Nature 225:199–200. 1 fig. PALEO.

Questioning of Siwalik mandible BMNH M/15423 as *Ramapithecus/Dryopithecus.*

1970 The Relationship of African Apes, Man and Old World Monkeys. Proceedings of the National Academy of Science. Washington. 67(2):746–748.

Questioning conclusions of Wilson and Sarich 1969 that human lineage diverged from African Apes four to five million years ago based on unsupported assumption that Hominoidea and Cercopithecoidea separated thirty million years ago. To the contrary, the Hominoidea were separated by the Oligocene.

1971 Basic Rectangle of the Mandible in the Hominoidea. Nature 231:60.

Reply to letter by Kinzey criticising the BRM index. Confirms *Kenyapithecus africanus* as primitive member of Hominidae.

1971 Leakey, M. D., Clark R. J., and . . . . New Hominid Skull from Bed I Olduvai Gorge, Tanzania. Nature 232:308–312. 3 figs., 1 tbl.

Resembles *Homo habilis* from Bed II but differs in some respects from Type from Bed I. ? sexual dimorphism.

1971 Review: The Emergence of Man by John Pfeiffer. Current Anthropology 12:380–381.

1972 Our African Ancestors. UNESCO Courier (August-September 1972) 25:24–29.

Popular account of fossil finds in Kenya and their significance for *Homo sapiens* today.

1972 The Origins of Oldest Man. *In:* Africa, Tradition and Change. Chapter 3. Early Man in Africa. E. J. Rich and I. Wallerstein, Eds. New York: Random House, 71–74. 2 figs.

School textbook. Summarizes rise of hominids in Africa from *Aegyptopithecus* to *Homo. H. sapiens* in the middle Pleistocene of Africa—earlier than in Europe.

1972 Pleistocene Man in America, 9–12, Problems of Calico, some background thoughts, 13–16, The task ahead, 75. *In:* Pleistocene Man at Calico. W. C. Schuiling, Ed. San Bernardino County Museum Report, International Conference on Calico Mountains Expedition. San Bernardino Museum Association Association. October 1970.

1972 *Homo sapiens* in the Middle Pleistocene and the evidence of *Homo sapiens'* evolution. *In:* The Origin of *Homo sapiens.* F. Bordes, Ed., Paris, U.N.E.S.C.O. (Ecology and Conservation 3), 25–29, 241, 297–299. With French summary.
Mid-Pleistocene *Homo* from Africa and Europe described and seen as widespread. Possible crossbreeding between *H. sapiens* and *H. erectus.*

III. PALEONTOLOGY (OTHER THAN HOMINID AND HOMINOID PALEONTOLOGY)

A. Books and Monographs on Palaeontology

1958 Some East African Pleistocene Suidae. Fossil Mammals of Africa, No. 14. British Museum (Natural History). 69 pp., 31 pls.
Key to genera of Pleistocene Suidae. Descriptions of: *Nyanzachoerus kanamensis* gen. et sp. nov., *Potamochoerus majus, P. koiropotamus, Mesochoerus paiceae, M. limnetes, M. olduvaiensis, M. grabhami, Metridiochoerus andrewsi, M. heseloni, M. pygmaeus* sp. nov., *Pronotochoerus jacksoni, P. nyanzae* sp. nov., *Notochoerus capensis, N. euilus, N. compactus* sp. nov., *N. hopwoodi* sp. nov., *N.* cf. *hopwoodi, Tapinochoerus meadowsi, T. minutus* sp. nov., *Hylochoerus antiquus* sp. nov., *H. meinertzhageni, Orthostonyx brachyops* gen. et sp. nov., *Afrochoerus nicoli, Phacochoerus a. altidens, P. a. robustus* ssp. nov., *P. africanus.*

GEOL. 1965 Olduvai Gorge 1951–1961. Vol. I. A Preliminary Report on the Geology and Fauna. Cambridge: Cambridge University Press. 109 pp., 97 pls., 3 maps.
With contributions by P. M. Butler, M. Greenwood, G. G. Simpson, R. Lavocat, R. F. Ewer, G. Petter, R. L. Hay, M. D. Leakey. Greatly enlarged faunal list. Several new species described. Suidae: *Promesochoerus mukiri,* gen. et sp. nov., *Ectopotamochoerus dubius,* gen. et sp. nov., *Potamochoerus intermedius* sp. nov. Giraffidae: *Giraffa jumae,* sp. nov. Bovidae: *Strepsiceros grandis,* sp. nov., *S. maryanus,* sp. nov., *Taurotragus arkelli,* sp. nov., *Gorgon olduvaiensis,* sp. nov., *Hippotragus gigas,* sp. nov., *Damaliscus antiquus,* sp. nov., *Parmularius rugosus,* sp. nov., *Alcelaphus howardi,* sp. nov., *Beatrgus antiquus,* sp. nov., *Xenocephalus robustus,* gen. et sp. nov.

1969 Ed., Fossil Vertebrates of Africa. Vol. I London: Academic Press.
New series on African fossil faunas. Contributions from P. M. Butler, D. A. Hooijer, G. R. von Koenigswald, R. E. F. Leakey.

1970 . . . , and R. J. G. Savage, Eds. Fossil Vertebrates of Africa. Vol. II. London: Academic Press.
Contributions by: C. S. Churcher, H. B. S. Cooke and S. C. Coryndon, A. W. Gentry, P. H. Greenwood and E. J. Todd, L. S. B. Leakey.

1970 . . . , R. J. G. Savage, and S. C. Coryndon, Eds. Fossil Vertebrates of Africa. Vol. III. London: Academic Press.

Contributions by: J. C. Rage, P. M. Butler and M. Greenwood, G. Petter, M. G. Leakey and R. F. Leakey, S. C. Coryndon and Y. Coppens.

B. Papers and Articles on Palaeontology

1925 Digging for Dinosaurs in Tanganyika. Illustrated London News 166:92–93. 1 pl., 12 figs.

1924 Expedition to Tendaguru to extract Dinosaurs, especially *Gigantosaurus.* Explanation of techniques for plastering fossil bone.

1931 . . . , Hopwood, A. T., and Reck, H. New Yields from the Oldoway Bone Beds, Tanganyika Territory. Nature 128:1075. ARCH.

Bed I—*Deinotherium, Hipparion, Elephas recki* and pre-Chellean tools. Bed II—Chellean tools. Bed III—Chelles-Acheulean transition. Bed IV—*Pelorovis, Hippopotamus gorgops.* Bed V—Kenya Upper Aurignacian tools. Oldoway Beds I–IV Lower to Middle Pleistocene.

1935 Does the Chalicothere—contemporary of the Okapi—still survive? Illustrated London News 187:730–750. 1 pl., 10 figs. ARCH.

Hippopotamus gorgops, Sivatherium, chalicotheres and tools from Bed II (BKII and SHKII) Olduvai.

1935 The largest elephant tusk discovered in England. How such relics are preserved. Illustrated London News 189:201. 5 figs.

Large tusk of *Elephas antiquus* from Swanscombe. Excavation and plastering of the specimen.

1942 Fossil Suidae from Oldoway. Journal of the East Africa Natural History Society 16:178–196. 3 pls., 2 figs.

Descriptions of *Mesochoerus olduvaiensis,* gen. et sp. nov., *Afrochoerus nicoli* gen. et sp. nov., *Phacochoerus complectidens,* sp. nov., *Potamochoerus majus* (Hopwood), *Notochoerus dietrichi* Hopwood, *Sus limnetes* Hopwood, and *Phacochoerus aethiopicus.* Fauna of Beds I, II and IV.

1943 Notes on *Simopithecus oswaldi* Andrews from the Type Site. Journal of the East Africa Natural History Society 17:39–44. 5 pls.

Description of male and female crania and juvenile mandible from Kanjera.

1943 New Fossil Suidae from Shungura, Omo. Journal of the East Africa Natural History Society 17:45–61. 7 pls.

Descriptions of: *Gerontochoerus scotti,* gen. et sp. nov., *Pronotochoerus jacksoni,* gen. et sp. nov., *Mesochoerus heseloni,* sp. nov. *Sus limnetes* also present.

1948 Palaeontological Notes—Miocene Fossils. Nature East Africa 4:8.

1400 specimens from Rusinga Miocene. 56 hominoids and large collection of fauna and flora.

1950 The Food of *Proconsul:* Fossilized Fruits and Seeds which are expected to Throw Light on the Life of the Prehistoric Primates of Kenya. Illustrated London News 217:334–335. 4 figs.

Recently discovered fossil fruits from Rusinga and Mfwangano Miocene.

1952 Lower Miocene Invertebrates from Kenya. Nature 169:624–625. 2 figs.

Report on fossil insects from Rusinga, illustrating fossilized caterpillar and pedipalpid. Material includes: Orthoptera, Coleoptera, Blattoidea, etc.

1953 Preserved in Three Dimensions. Fossil Insects of 25,000,000 B.C. Illustrated London News 222:445. 10 figs.

Soft-bodied invertebrates preserved from the Miocene of Mfwango Island, Kenya. Forms include: pedipalpid, ticks, caterpillars, grasshoppers, beetle, etc., similar to living forms. Possible clue to palaeoclimate.

1954 The Giant Animals of Prehistoric Tanganyika, and the Hunting Grounds of Chellean Man. New Discoveries in the Olduvai Gorge. Illustrated London News 224:1047–1051. 11 figs.

Giant animals from Bed II: *Pelorovis, Metridochoerus, Notochoerus, Eurygnathohippus, Sivatherium.* Reconstruction of scenes.

1955 . . . and Clark, W. Le Gros. British Kenya Miocene Expeditions. Interim Report. Nature 175:234.

Reports of expeditions in Kavirondo, Maralal, Moruorot, Karungu.

1958 . . . and Whitworth, T. Notes on the Genus *Simopithecus* with a Description of a New Species from Olduvai. Coryndon Memorial Museum Occasional Papers, no. 6. 14 pp., 10 pls., 4 tbls.

Simopithecus distinct from *Papio* and *Theropithecus.* Two species, *S. oswaldi,* containing *S. o. oswaldi;* from Kanjera; *S. o. mariae* from Olorgesaillie; *S. o. olduvaiensis* from Olduvai. *S. jonathani* sp. nov. from Olduvai Bed IV much larger and different morphology.

1958 More Giant Animals from Olduvai. Light on Chellean Man's Way of Life, and An Ivory Hand Axe of 1/2 Million Years Ago. Illustrated London News 233:41–43. 6 figs.

Finds from BK II including tusk of *Potamochoerus majus,* remains of *Pelorovis,* giant baboon. Hippo ivory hand axe from BK II. Reconstruction picture.

1958 Notes on an Aberrant Lower Third Molar of *Sus cristata.* South African Journal of Science 54 (6):134. 1 pl.

Tooth shows characters normally diagnostic of *Mesochoerus.* Emphasizes unreliability of individual molar cusp patterns taken in isolation as taxonomic characters in Suidae.

1959 A Preliminary Re-assessment of the Fossil Fauna from Broken Hill, Northern Rhodesia. *In:* Further Excavations at Broken Hill, Northern Rhodesia. J. D. Clark, Ed. Journal of the Royal Anthropological Institute 89:201–232.

Upper Pleistocene age with some extinct forms. Similar to Saldhana.

1962 Primates. *In:* W. W. Bishop, The Mammalian Fauna and Geomorphological Relations of the Napak Volcanics, Karamoja. Records of the Geological Survey of Uganda 1957–58, 6–9.

Description of *Mioeuoticus bishopi* gen. et sp. nov., a Miocene galagid.

1966 Africa and Pleistocene Overkill. Nature 212:1615–1616.

Comment on note to Nature by P. Martin. Out-of-date faunal lists used for conclusions. Desiccation at end of Bed IV times more likely to have killed off fauna than overkill by Palaeolithic Man.

H. PAL. 1967 Overkill at Olduvai Gorge. Nature 215:212–213.

Reply to letter in Nature by P. S. Martin refuting ability of Acheulean man to overkill.

1967 Notes on the Mammalian Fauna from the Miocene and Pleistocene of East Africa. *In:* Background to Evolution in Africa. W. W. Bishop and J. D. Clark, Eds. Chicago: University of Chicago Press, 7–29.

Faunal list of Lower and Upper Miocene faunas (Part 1) and Lower and Middle Pleistocene faunas (Part 2) from East Africa with authors, localities, etc.

1970 Additional Information on the Status of *Giraffa jumae* from East Africa. *In:* Fossil Vertebrates of Africa. Vol. II. L. S. B. Leakey and R. J. G. Savage, Eds. London: Academic Press, 325–330. 4 pls., 1 tbl.

Additional diagnostic characters and geological age of new specimen from Olduvai, Bed II.

IV. PAPERS AND ARTICLES ON GEOLOGY AND DATING

1931 . . . , Hopwood, A. T., and Reck, H. Age of the Oldoway Bone Beds, Tanganyika. Nature 128:724. 3 figs. ARCH. H. PAL.

Oldoway human skeleton Upper Kamasian in age with Upper Chellean and Acheulean tools.

1931 East African Lakes. Geographical Journal 77:497–514. 4 pls., 5 figs.

Climatic changes and correlation of East African pluvials.

1934 Changes in the Physical Geography of East Africa in Human Times. Geographical Journal 84:296–310. 2 pls., 5 figs.

Lake Kamasia in Early Pleistocene and subsequent changes in rifting and watersheds in Eastern Rift.

1950 The Lower Limit of the Pleistocene in Africa. Proceedings of the 18th International Geological Congress (Great Britain 1948). Part 9, 62–65. 1 fig.

African Pleistocene is sub-divided, and correlation of local climatic changes with cultural and physical events elsewhere are indicated. For each zone the characteristic units are noted.

1955 The Climatic Sequence of the Pleistocene in East Africa. *In* Congrès Panafricain de Préhistoire: Actes de la IIeme Session. Algiers. L. Balout, Ed. Paris: Arts et Metiers Graphiques. 293–294.

Recognition of four pluvials in Pleistocene of East Africa and two post-Pleistocene wet phases, with new term Kanjeran for Lower Kamasian.

1961 . . . , Evernden, J. F., and Curtis, G. H. Age of Bed I, Olduvai Gorge, Tanganyika. Nature 191:478–479. 1 tbl.

Many Olduvai deposits suitable for K/Ar dating. Dated localities MK I, FLK I, FLK NN I, etc. Dates fall between 1.6 and 1.9 m.y., average 1.75 for hominid sites. Top of Bed I dated 1.23 m.y.

1961 The Age of *Zinjanthropus.* New Scientist 12:501.

Correction of Gentner and Lippolt's dating for Bed I Olduvai.

1962 Age of Basalt Underlying Bed I Olduvai. Nature 194:610–612.

Answer to communication by von Koenigswald, Gentner and Lippolt. Samples used for dating not from *in situ* thus dates invalid.

1965 Fleischer, R. L., Price, P. B., Walker, R. M. and . . . . Fission Track Dating of Bed I, Olduvai Gorge. Science 148:72–74. 1 pl., 1 fig.

Fission track dating on pumice from Bed I = 2.03 ± 0.28 m.y. compared with 1.75 m.y. K/Ar date from *Zinjanthropus* horizon.

ARCH. 1965 Fleischer, R. L., Price, P. B. and Walker, R. M. and . . . Fission Track Dating of a Mesolithic Knife. Nature 205:1138. 2 figs.

Fission track dating effective for dates between twenty years and 10^9 years. Obsidian mesolithic knife from Gamble's Cave II reheated to high temperature at time of deposition, dated by fission track from this point in time. Age computed at 3,700 (± 900) years B.P.

1968 . . . , Protsch, R., and Berger, R. Age of Bed V, Olduvai Gorge, Tanzania. Science 162:559–560.

C 14, dates of 10,400 years B.P. from bones from Bed V.

H. PAL. 1969 Fossil Hominid Taxonomy. Science 163:1360.

Letter of reply to C. L. Brace criticizing correlation of hominid-bearing strata and dates at Olduvai.

1969 Age of Bed V, Olduvai Gorge, Tanzania. Science 166:532.

Letter explaining complexity of Bed V. 30,000 B.P. for caliche overlying Bed V.

V. ETHNOGRAPHY

A. Books and Monographs on Ethnography

1949 . . . and Leakey, M. D. Some string figures from North East Angola with Diagrams. Diamang Museu Do Dundo Publiçacoes Culturais Lisbon. 7–24. 22 figs.

Inchokwe natives have "serial" figures illustrating stories. Not common in Africa but generally world wide ideas.

1952 Mau Mau and the Kikuyu. London: Metheun. 114 pp. New York: John Day.

Kikuyu social organization before Europeans and effect their impact. Causes of discontent and possible remedies.

In Preparation. The Kikuyu—Three volumes for posthumous publication.

B. Papers and Articles on Ethnography

1926 A New Classification of the Bow and Arrow in Africa. Journal of the Royal Anthropological Institute 56:259–294. 3 pls., 5 maps.

Pan-African survey of bow and arrow forms.

1930 Some Notes on the Masai of Kenya Colony. Journal of the Royal Anthropological Institute 60: 185–209.

Social life and customs of the Masai peoples on the Mau Escarpment, Kenya.

1931 The Kikuyu Problem of the Initiation of Girls. Journal of the Royal Anthropological Institute 61:277–285.

Kikuyu of Fort Hall, Nyeri and Kyambu. Effects of missionary teaching and government aids.

1954 The Religious Element in Mau Mau. The Manchester Guardian Weekly (1, 7, 1954) 77.

1956 The Economics of Kikuyu Tribal Life. East African Economic Review 3:165–180.

1959 Die Bestimmung der Völkes Afrikas. Emil Schultes, Africa: Von Aequator zum Kap der Guten Hoffnung. Zurich. 8 pp.

Text accompanying photographs of peoples of Africa, particularly the Masai.

VI. NATURAL HISTORY

A. Books on Natural History

1954 Ylla and . . . . Animals in Africa. London: Harvill Press. 146 pp., 24 pls.

Color and black and white photographs of major large game animals and birds of East Africa with short text on each species illustrated.

1969 Animals of East Africa. Washington, D.C.: National Geographic Society, 199 pp., many pls.

Emphasizes the lesser known and not generally romanticized aspects of African faunas including fossil ancestors. Well balanced overall coverage regarding African faunas.

B. Papers and Articles on Natural History

1943 Notes on the Falconidae in the Coryndon Museum. Journal of the East Africa Natural History Society 17:103–122.

Birds from East Africa in the Museum collection.

1949 Taxonomy and Animal Ecology. Proceedings of the African Regional Science Conference. Johannesburg.

1949 Museum Services as Background to Research. Proceedings of the African Regional Science Conference. Johannesburg.

1969 Presumed Super-foetation in an *Erythrocebus patas* Monkey. Nature 223:754.

Evidence from caged monkey.

VII. MISCELLANEOUS

A. Books

1936 Kenya: Contrasts and Problems. London: Metheun. 189 pp. New Edition 1966: Cambridge, Massachusetts: Schenkman.

Natural and political history of Kenya, with insights into African traditions and the effect on these of colonialization. Suggestions for future administration. Belief in the People's of Kenya to lead Africa in co-operation and in successful running of country.

1954 Defeating Mau Mau. London: Metheun, 152 pp.

B. Papers and Articles

1931 Some Aspects of Black and White in Kenya. Bulletin of the John Ryland Library 15:2.

1941 Report on the Museum, 1st January to 30th June 1941. Journal of the East Africa Natural History Society 70:58.

1947 Report on the Pan-African Congress on Prehistory, January 1947, Nairobi, Man 47:86–87.

Rules and Terminology proposed for meetings every four years.

1950 'Scarface'—A Story of Africa 500,000 Years Ago. *In* Empire Youth Annual. Raymond Fawcett, Ed. London: P. R. Gawthorn. 88–94.

1954 Kenya—The Causes of Mau Mau. The Observer London (25, 7, 1954).

1954 What Should Be Done In Kenya. The Observer London (1, 8, 1954).

1959 First Lessons in Kikuyu. Nairobi: East African Literature Bureau.

General grammar with exercises in nine lessons.

1966 'Olduvai Gorge 1951—1961.' Geographic Journal London 132:178–180.

Comment on review by W. W. Bishop.

1967 Olduvai. Antiquity 41 (163):227–228.

Letter in answer to review by C. McBurney of 'Olduvai Gorge 1951–1961.'

1968 Violence and the Decline of Faith. *In* Alternatives to Violence. Larry Ng, Ed. New York:Time-Life Books, 86–88.

Answers to present day violence in western civilization may be found in applying problems to powers of reason and faces of conscience leading to renewal of basic religious faith.

1970 The Contribution of East Africa to the Development of Science and Technology. Proceedings of the East African Academy 8:11–15.

East Africa the cradle of Man and thus the fount of his cultures.

1971 The Development of Community Man. *In* Health and Disease in Africa. G. C. Gould, Ed. Proceedings of the 1970 East African Research Council. Sci. Con. Sess. 23. East African Literature Bureau Nairobi 363–366.

Evidence of disease in fossil man. Disease more likely once community living starts, and where infants dependent for long time. Evidence from archaeology and palaeontology for community living; learning to control fire; disease more prevalent with development of home, then village.

1971 . . . , and Ardrey, R. Aggression and Violence in Man—A Dialogue. Munger Africana Library Notes 9, 24 pp.

1971 History of the National Museum of Kenya. Kenya Past and Present 1:1–2.

Growth of Museum from 1911–1971.

THE AUTHORS

Dr. Peter Andrews's first interests were in forestry and ecology. Later he took up paleontological research and was encouraged by Louis Leakey to work on aspects of the Miocene fossil hominoids from East Africa. He was formerly attached to the Centre for Prehistory and Paleontology in Nairobi and is presently on the scientific staff of the British Museum (Natural History) in London.

Professor William Bishop has worked in East Africa for twenty years as a geologist, geomorphologist,and paleontologist. He has worked with the Uganda Geological Survey, and from 1962 to 1965 was Director of the Uganda Museum. He has been a Reader in Geology at Bedford College, London University, and is currently teaching geology at Queen Mary College, London University. In addition, Professor Bishop has continuously and vigorously conducted research into the fossil-bearing sediments of the East African rifts. Working closely with paleoanthropologists, he has contributed greatly to the field both in his research and as an editor of numerous symposium volumes.

Dr. C. K. Brain, a scientist and naturalist of unusual breadth, is Director of the Transvaal Museum in Pretoria. He has studied the cave deposits which yielded the Australopithecine fossils in South Africa, and has done research on the ecology and behavior of living monkeys. In recent years he has directed systematic paleontological and paleoenvironmental research at the Swartkrans fossil site and has pioneered studies designed to show how fossil assemblages are formed.

Professor J. Desmond Clark went out to Africa as a young man after graduating in Archaeology and Anthropology at Cambridge University in 1937. Since that time he has devoted his life to the study of African prehistory. He has done intensive work in more areas of the continent than any other investigator—Somalia, Zambia, Angola, the Sahara, Sudan, and Ethiopia. His monographs and books are the most authoritative writings on these regions. Professor Clark also developed one of Africa's finest archaeological museums, in Livingstone, Zambia. Since 1961 he has been teaching at the University of California, Berkeley.

Shirley Coryndon, an expert on the paleontology and evolution of the Hippopotamidae, began her work as a volunteer in the National Museum at Kenya. She later became Louis Leakey's research assistant—a job that she carried out with much skill during the crucial years that saw the discovery of *Zinjanthropus,* the founding of the Centre for Prehistory and Palaeontology, and the first excavations at Fort Ternan.

Professor Michael Day is a medical anatomist who began to study fossil hominids when he was invited by Louis Leakey to assist in the investigation of postcranial bones from Olduvai. He has pursued this line of research and has compiled the invaluable *Guide to Fossil Man.* He is now head of the Department of Anatomy at St. Thomas Hospital Medical School, London, and is doing research on the hominids from the Koobi Fora Formation.

Dr. Jane Goodall has, from a very early age, devoted her life to the study of animals. Louis Leakey helped her start on her studies of chimpanzees in the wild, and her work at the Gombe Stream Reserve in Tanzania has continued for some fifteen years. Her work is excellently summarized in her book *In the Shadow of Man,* as well as in numerous scientific papers and articles. Dr. Goodall is engaged in a broad range of studies of animal behavior and has written an account of carnivores in East Africa. She now divides her time between a teaching and research appointment at Stanford University and field work in Africa.

Professor Richard Hay became involved in the study of early man in 1962 when he accepted Louis Leakey's invitation to do research on the geology of Olduvai Gorge. As an expert on the chemical alteration of volcanic ashes, he was well suited for the work. He has achieved a reconstruction of paleoenvironments that will serve as a model for studies elsewhere. Because paleoanthropology and geology merge in such work, he often collaborates with Dr. Mary Leakey. Professor Hay teaches geology at the University of California, Berkeley.

Professor F. Clark Howell is a leading exponent of an interdisciplinary approach to the problem of exploring man's origins. His early research was concerned with Neanderthal fossils, but his subsequent research has involved as much archaeology, geology, and paleontology as physical anthropology. He has organized research at such important Middle Pleistocene sites as Isimila in Tanzania and Torralba-Ambrona in Spain. Since 1967 he has been one of the co-leaders of the International Omo Research project in southern Ethiopia. The book he edited with F. Bourlière, *African Ecology and Human Evolution,* can fairly be said to have started a new movement in paleoanthropology. Professor Howell teaches anthropology at the University of California, Berkeley.

Professor Glynn Isaac has strong interests in geology, zoology, and archaeology. In 1961 Louis Leakey appointed him Warden of Prehistoric Sites in Kenya, and he also served as Louis's deputy at the newly formed Centre for Prehistory and Paleontology. Since then he has worked at a succession of archaeological sites in East Africa. At present Professor Isaac is teaching archaeology at the University of California, Berkeley. Since 1970 he has been co-leader with Richard Leakey of the East Rudolf Research Project.

Dr. Mary Leakey is an archaeologist who excavated various important sites in Britain and in 1935 began excavating in prehistoric sites in Africa. At Olorgesailie in 1943 she pioneered the technique of excavating early Pleistocene occupation sites. In 1959 she found the *Zinjanthropus* cranium at Olduvai and has in the ensuing years uncovered dozens of occupation sites. Volume 3 of Cambridge University's series on Olduvai Gorge deals with her findings in Beds I and II. Dr. Leakey is at present continuing her work at Olduvai.

Richard Leakey, Director of the National Museum of Kenya, is leader of the research team working on the eastern side of Lake Turkana (formerly Rudolf). He has worked in the West Natron area and west of Lake Baringo, and from 1968 onwards he has demonstrated the extraordinary potential of the East Rudolf area. In addition, Dr. Leakey coordinates an interdisciplinary research project.

Elizabeth McCown is a physical anthropologist associated with the University of California, Berkeley. Her principal interest is comparative anatomy of primates. Since 1973 she has been managing editor of the Perspectives on Human Evolution series.

Professor Harry Merrick began his studies in African prehistory. His field work has been in the Nakuru area and Lukenya Hill in Kenya. He has also actively engaged in research on the very early archaeological sites of the Shungura Formation in the Omo Valley. He currently teaches anthropology at Yale University.

Professor Phillip Tobias was trained as a medical doctor and initially studied cytogenetics and then the physical anthropology and biology of living and recent peoples of Africa, including the Khoisan. However, from his student days onwards, he was involved with excavations at Sterkfontein and Makapansgat, two of the most important early hominid sites in Africa. He was invited by Louis and Mary Leakey to make detailed comparative studies of the Olduvai hominid fossils. Professor Tobias is the author of numerous papers and monographs, including a monographic treatment of the *Zinjanthropus* fossil, which appeared in 1967 as Volume 2 of the Cambridge University series on Olduvai Gorge. He currently directs new investigations at several of the southern African sites and teaches anatomy at Witwatersrand University.

Drs. John and Judith Van Couvering have contributed jointly and individually to the study of evolution in Africa. Louis Leakey recruited them to work on the Miocene sediments and fossils of Western Kenya. As a geologist, John Van Couvering has done much to clarify intercontinental correlations through the scrutiny of paleontological and geological data. He is currently doing research with the Museum of the University of Colorado. Judith Van Couvering is primarily a paleontologist. She is now working on broad ecological issues that affect our understanding of evolution in Africa and is also associated with the Museum of the University of Colorado.

Professor Alan Walker studied geology and primate evolution. For many years he taught medical anatomy in Uganda and became extensively involved in the study both of living primates and the fossils of their forebears. He is currently part of the research team responsible for the description and interpretation of the fossil hominids from the Koobi Fora Formation. He teaches in the Anthropology Department and Medical School of Harvard University.

INDEX